Remote Sensing Image Processing Algorithms for Detecting Air Turbulence Patterns

Maged Marghany
President, Innovative Information Sdn. Bhd.
Kuala Lumpur, Malaysia

T0392870

CRC Press
Taylor & Francis Group
Boca Raton London New York

CRC Press is an imprint of the
Taylor & Francis Group, an **informa** business

First edition published 2025
by CRC Press
2385 NW Executive Center Drive, Suite 320, Boca Raton FL 33431

and by CRC Press
4 Park Square, Milton Park, Abingdon, Oxon, OX14 4RN

CRC Press is an imprint of Taylor & Francis Group, LLC

Library of Congress Cataloging-in-Publication Data (applied for)

ISBN: 978-1-032-34458-4 (hbk)
ISBN: 978-1-032-34463-8 (pbk)
ISBN: 978-1-003-32227-6 (ebk)

DOI: 10.1201/9781003322276

Typeset in Times New Roman
by Prime Publishing Services

Dedicated to

My Mother Faridah

and

Richard Feynman and Michio Kaku taught me that a real distinguished professor contributes to knowledge genuinely, rather than relying on exploiting students and colleagues or resorting to false publications to boost their H-index. This underscores the value of integrity and authentic contributions in the academic realm.

Preface

This book was accomplished when I was ranked among the top 2% of scientists in a global record compiled four times by Stanford University. Currently, there are no books that have tackled the matter of air turbulence detection in remote sensing images. Most of the work done involves conventional optical remote sensing data with classical methods such as library spectral signatures, band ratios, and principle component analysis, without designating new methods and technology. Only a few pieces of literature have attempted to implement optical and microwave remote sensing images for air turbulence detections. This book aims to deliver new procedures for image processing for air turbulence detections using advanced remote sensing images and quantum image processing. Currently, there is a huge gap between research work in the field of air turbulence detection and advanced remote sensing technology. Most of the theories are not operated in terms of software modules. Most of the software packages in the field of remote sensing images cannot deal with advanced image processing techniques in air turbulence detections due to heavy mathematics work. In this view, this book will fill a gap between advanced remote sensing technology and air turbulence detection.

In this regard, this book presents a broad overview of quantum theory for understanding the automatic exploration of air turbulence detections in remote sensing data. The quantum theory can explain the electromagnetic wave ranges of various remote sensing satellite data from short wavelength to long wavelength. Principles of remote sensing, imaging, and automatic air turbulence detection are addressed based on quantum image processing, especially the Schrödinger equation. Finally, the four-dimensional exploration of air turbulence detections is also discussed based on quantum computing.

Chapter 1 delves into the intricacies of turbulence, addressing its definition and its substantial impact on everyday activities, such as its influence on airplane mobility. The distinction between laminar and turbulent flow is explored, emphasizing how the Reynolds number elucidates the transition from laminar to turbulent flow. Furthermore, the chapter delves into the principles of mathematical models, including the equations governing turbulence motion, the Navier-Stokes equations, the Spalart-Allmaras (SA) Model, k–ε Turbulence and k–ω Models, and the turbulence diffusion equation. These mathematical descriptions pose a challenge in capturing the complexity and nonlinearity inherent in turbulence phenomena.

Chapter 2, therefore, introduces the initial concept of the quantum mechanics of the air composition, ending by quantizing the gravity wave based on the wave-particle duality. The new theory of geostrophic wind flows is introduced based on the stress-energy tensor, named the Marghany quantum energy wind-driven theory.

Consequently, Chapter 3 debates the quantization of remote sensing technology, starting from the quantization of Maxwell's Equations by involving Feynman's speculations and wave-particle duality. This chapter also presents a novel definition of remote sensing addressed as "Marghany's Pioneering Concept in Remote Sensing." Thus, Chapter 4 unveils the influence of HAARP on the ionospheric layer, presenting a developed quantum computing algorithm based on SWAP gates to explore the mechanism of HAARP signals in ionospheric layer disturbances. The HAARP signal has been observed to generate turbulence in the plasma flow across the ionospheric layer.

Subsequently, Chapter 5 illustrates a novel method for producing extreme heatwaves using HAARP turbulence plasma, proven mathematically by the author. This novel theory is named the quantized Marghany heatwave mechanism. Additionally, Chapter 6 utilizes quantum spectral energy for modeling cloud types and cloud brightness temperature from MODIS data, leading to insightful conclusions. By employing quantum principles, one gains a refined understanding of the intricate relationships between cloud properties and their spectral signatures. This approach allows for enhanced classification of diverse cloud types based on their distinctive quantum manners. Furthermore, the application of quantum spectral energy provides a novel perspective on cloud brightness temperature patterns, shedding light on the underlying physics governing cloud patterns. Therefore, Chapter 7 unveiled an innovative approach for the automated detection of cloud streets using optical satellite data, exemplified by MODIS. The newly developed algorithm hinges on the concept of quantum spectral energy, gleaned from the fluctuations in brightness temperature spanning cloud street formations and their adjacent surroundings. The quantum edge detection algorithm, a pivotal component, exhibits a remarkable potential for the automated identification of cloud streets within optical remote sensing datasets, including MODIS and Meteosat data.

Chapter 8, is an innovative tutorial on the creation of an automatic detection algorithm for identifying von Kármán cloud vortex streets across multiple satellite images, leveraging the capabilities of quantum computing. The algorithm, named Vortex Distance Decomposition (VDD), was developed by the author and further optimized using the Quantum Approximate Optimization Algorithm (QAOA). In this regard, Chapter 9, then, has delved into the fundamentals of radar, with a particular focus on grasping the concept of radar resolution. The crux of understanding radar lies in comprehending the radar equation and the speculation surrounding Bragg scattering. By understanding the radar equation, which relates the power received by the radar to various parameters such as transmitted power, antenna gain, and target range, one can effectively analyze and interpret radar data. This knowledge serves as a foundational framework for a deeper understanding of the application of SAR data in studying air turbulence phenomena, notably gravity waves. Chapter 10, hence, introduces a groundbreaking method utilizing Quantum Particle Swarm Optimization (QSPSO) for the automatic detection of atmospheric gravity waves in multiSAR data across different wavelength and polarization bands. Notably, it reveals the first documented occurrences of atmospheric gravity waves in the coastal waters of Malaysia, Pattani (Thailand), and Kelantan, situated in the South China Sea.

In Chapter 11, an inventive method introduces a groundbreaking technique for the automated identification of tropical cyclonic vorticity flows in Synthetic Aperture Radar (SAR) data, leveraging platforms like RADARSAT-2 SAR ScanSAR wide mode in HH polarization data. The Quantum Multiobjective Evolutionary Algorithm (QMEA) is developed and evaluated using SAR imagery captured over typhoon Gani in the Pacific Ocean.

In the final Chapter 12, 4D images unveil the Reynolds number derived from 4D quantum hologram interferometry. The cyclonic turbulence spectra, recognized as evidence of quantum entanglement theory, serve as a key indicator of the complex and dynamic nature of cyclonic chaotic flows. The outcomes demonstrate the impact of 4D cyclonic chaotic flows in generating phenomena such as eddies, vorticity, and cyclonic turbulence boundary layer.

I wish to convey my appreciation to editorial project manager Vijay Primlani, who allowed the publishing of this book. Without his intense commitment, this book would not have become such a precious piece of novel knowledge.

Distinguished Prof. Dr. Maged Marghany

Contents

Preface v

1. Mathematical Modeling Principles for Turbulence Description **1**

1.1 What is Turbulence? 2
1.2 Can Turbulence Damage Aeroplanes? 2
1.3 What is the Origin of Air Turbulence? 3
1.4 Distinguishing Between Laminar and Turbulent Flows: An Exploration of Variances 4
1.5 What is the Nature of Turbulence? 7
1.6 Turbulent Boundary Layers 9
1.7 Causes of Atmospheric Turbulence 12
1.8 Air Turbulence Intensity Classes 17
1.9 Clear Air Turbulence 18
1.10 Turbulence and Thunderstorm 20
1.11 Wake Turbulence 21
1.12 Equation of Turbulence Motion 23
1.13 Turbulent Transport in the Mass Conservation Equation 25
1.14 Richardson's Four-Thirds Power Law 26
1.15 Kolmogorov Speculation 28
1.16 Navier-Stokes Equations 31
1.17 Spalart-Allmaras (SA) Model 34
1.18 The k–ε Turbulence and k–ω Models 36
1.19 What Serves to Make the Most Challenging Navier-Stokes Equations Extremely Complex? 37
1.20 Turbulent Diffusion 38

2. Quantum Mechanics of Atmospheric Turbulence **43**

2.1 What are the Compensates of Air? 43
2.2 Distribution of Air Chemical Composition through Atmospheric Layers 45
2.3 Molecules of Greenhouse Gases/Effect 45
2.4 Initiating of Air Quantum Turbulence 46
2.5 What is the Magic of Quantum Mechanical Pressure? 53
2.6 Marghany's Quantization of Earth's Rotation and its Impact on Wind Mobility Patterns 58
2.7 Quantum Geostrophic Turbulence 62
2.8 Quantum Decomposition of Energy Spectral Turbulence 67
2.9 Physical Characteristics of Vorticity 69
2.10 Can Quantize Vorticity? 73

2.11 Quantized Atmospheric Wave Turbulences 81
2.12 Wave-Particle Duality in Atmospheric Wave Turbulence Propagation 83

3. Quantum Mechanics in Remote Sensing: Theoretical Perspectives 91

3.1 Maxwell's Equations 91
3.2 Magnetic Fields Sources 94
3.3 Derivation of Electromagnetic Equations from Maxwell's Equations 99
3.4 Quantization of Maxwell's Equations 101
3.5 Feynman's Identification of Maxwell's Equations 103
3.6 Quantum Electromagnetic Radiation 106
3.7 Why can Quantum Mechanics Effectively Demonstrate Black Body Radiation? 110
3.8 The Photoelectric Effect 113
3.9 Photovoltaic Effect 117
3.10 De Broglie Spectulation of Electromagentic Spectra 118
3.11 Wave-Particle Duality 122
3.12 The Uncertainty Principle 124
3.13 Marghany's Pioneering Concept in Remote Sensing 125

4. Quantization of Haarp-Ionosphere Disturbance: Generating Turbulence in Ionospheric Plasma 128

4.1 What is Ionosphere? 128
4.2 Mechanism of Ionospheric Layer Ionization 129
4.3 Ionization-related Layers 129
4.4 Signal Attenuation in Ionospheric Layer 135
4.5 Ionospheric Oscillations 135
4.6 What Renders HAARP Particularly Incredible? 136
4.7 How does HAARP Operate? 137
4.8 How does HAARP Evoke the Ionosphere? 138
4.9 Quantization of HAARP-Ionosphere Turbulence Plasma Generation 141
4.10 How does Turbulence Plasma Wave Track from Space? 145
4.11 Quantum Algorithm for Retrieving Ionosphere Turbulent Plasma Wave Due to HAARP 146

5. Quantization of HAARP-Inducing Turbulence Plasma Extreme Heat Waves 157

5.1 Heat Wave: What is it? 157
5.2 Meteorology of Heatwaves 158
5.3 Mechanisms that Cause Heatwaves 159
5.4 Heat Wave Index 160
5.5 Heatwaves on Earth: Are Solar Storms Responsible? 161
5.6 Are the Ocean Heat Contents Blamed for the Earth's Heat Waves? 162
5.7 Quantized Marghany-Heatwaves 164
5.8 Satellite Data are Used for Heatwave Simulation 167
5.9 Quantum Simulation of Heatwave 169
5.10 Quantized Heatwave Owing to Ionospheric Turbulence Plasma 171
5.11 How Could HAARP Lead to Worldwide Heatwaves? 175

6. Utilizing Quantum Spectral Energy Signatures for Identifying Cloud Patterns 183

6.1 Types of Clouds 183

6.2 What is the Magical Association between Clouds and Turbulence? 186
6.3 Can Turbulence be Caused by Clouds? 187
6.4 What Sorts of Clouds Yield the Most Turbulence? 187
6.5 Have any Recent Developments in Cloud Types Involved Turbulence? 188
6.6 Satellite Observation of Clouds 197
6.7 Quantization of IR Emission from Clouds 199
6.8 Radiative Transfer Theory for Cloud Imaging Mechanism 202
6.9 Spectral Regions Relevant to Clouds 202
6.10 Entanglement of Cloud Spectral Absorption 203
6.11 How does Entanglement Form Spectral Discrimination of Clouds? 205
6.12 Simulation of Quantum Cloud Spectral Libraries 207
6.13 Cloud Quantum Spectral Energy Reflectance 210

7. Quantum Edge Detection Algorithm for Automated Cloud Street Identifications 217

7.1 What are Cloud Streets? 217
7.2 Mechanism of Cloud Street Formations 217
7.3 Characteristics of Cloud Streets 219
7.4 Mathematical Description of Cloud Streets Causing Turbulence 222
7.5 Observational Mechanisms Employed by Satellites to Monitor Cloud Streets 224
7.6 Quantized Spectral Signature of Cloud Streets 226
7.7 Quantum Edge Detection Algorithm for Cloud Streets 230
7.8 Automatic Detection Quantum Edge Algorithm of Cloud Street in MODIS Data 232
7.9 Why Quantum Edge Detection Algorithm Accurately Detect Cloud Streets? 237

8. Development of the Superposition Vortex Distance Decomposition Algorithm for Detecting von Kármán Vortex Cloud Streets in Multi-Satellite Imagery 241

8.1 What is the Magic of von Kármán Vortex Street 241
8.2 Mathematical Description of Kármán Vortex Streets 242
8.3 Mathematical Model for Dividing Streamline of Atmospheric von Kármán Vortex Streets 244
8.4 What is the Correlation between Froude Number and Temperature Fluctuations? 246
8.5 Formulation of Attributes: Vortex Streets 247
8.6 Quantum Vorticity States in Remote Sensing: A Novel Theoretical Framework 248
8.7 Entanglement Between Vortices and Electromagnetic Waves in Remote Sensing Data 249
8.8 Developing Quantum Distance Estimation Algorithm for Automatic Detection of von Kármán Vortex Street Features from Satellite Data 250
8.9 Quantum Amplitude Embedding for Identifying von Kármán Vortex Streets Features in Multi-Satellite Images 253
8.10 Tested Remote Sensing Satellite Data 255
8.11 What makes the Quantum Superposition VDD Algorithm a Valuable Tool for Automated von Kármán Cloud Vortex Streets Detection? 259

9. Exploring the Principles of Synthetic Aperture Radar 263

9.1 Principles of Microwave Bands 263
9.2 Radio Detecting and Ranging 264
9.3 What is meant by Synthetic Aperture Radar? 265
9.4 Radar Resolution 267
9.5 Radar Range Equation 272

9.6 Exploring Radar Backscattering 273
9.7 Mechanism of Surface Backscattering 275
9.8 What is the Backscatter Coefficient? 275
9.9 Principles of SAR Bragg Scattering 276
9.10 SAR Polarization 278
9.11 Speckles 278

10. Developing Quantum Soliton-Inspired Particle Swarm Optimization Algorithm for Automatic Detection of Atmospheric Gravity Waves in Multisar Satellite Data 281

10.1 Atmospheric Waves 281
10.2 Atmospheric Tide 281
10.3 Harmony in the Skies: Unveiling the Cosmic Symphony of Solar Atmospheric Tides 283
10.4 Mathematical Descriptions of Atmospheric Tide 284
10.5 Generation Atmospheric Waves 286
10.6 Deriving the Linearity of Gravity Waves 287
10.7 The Influence of Wind on the Propagation of Atmospheric Gravity Waves 289
10.8 Mechanisms of SAR Imaging of Atmospheric Gravity Waves 292
10.9 Quantized SAR Imaging Mechanism of Atmospheric Gravity Waves 293
10.10 Automatic Detection of Atmospheric Gravity Waves Using Quantum Particle Swarm Optimization Algorithm 295
10.11 Why QSPSO Cable for Automatic Detection of Atmospheric Gravity Waves in SAR data? 303

11. Quantum Multiobjective Algorithm for Automatic Detection of Tropical Cyclone in Synthetic Aperture Radar Satellite Data 307

11.1 What are the Distinctions between Cyclones, Hurricanes, and Typhoons? 307
11.2 What are Tropical Cyclones? 309
11.3 Formation Mechanism 309
11.4 What is Rapid intensification? 310
11.5 Tropical Cyclone Anatomy: Unveiling the Inner Workings of These Powerful Storms 311
11.6 Mathematical Description of Tropical Cyclones 314
11.7 How Does SAR Imagine Tropical Cyclone Pattern? 317
11.8 Quantum-Enhanced SAR Image Processing for Automated Tropical Cyclone Detection 319
11.9 Quantum Multiobjective Evolutionary Algorithm (QMEA) 321
11.10 Generation of Quantum Turbulent Flow Population Pattern 323
11.11 Pareto Optimal Solution for Quantum Non-dominated Sort and Elitism (QNSGA-II) 325
11.12 Automated Identification of Cyclonic Turbulent Patterns in Synthetic Aperture Radar (SAR) Imagery 327
11.13 The Significance of Pareto Optimization in the Automatic Detection of Cyclonic Clusters Using QNSGA-II 331

12. Four-Dimensional Quantum Hologram Radar Interferometry Radar for Tropical Cyclonic Tracking 336

12.1 What Characteristics Define a Space with Four Dimensions? 337

12.2 Topology of 4-D Reconstruction 339
12.3 Is the Existence of N-dimensional Space a Reality? 340
12.4 What is the Role of Calabi-Yau Manifolds in Hologram Construction? 342
12.5 Hologram Quantum Interferometry 343
12.6 Quantized Radar Hologram Interferometry 344
12.7 Marghany 4D Quantized Hologram Interferometry Algorithm 348
12.8 Marghany 4-D Quantized Phase Unwrapping Algorithm 349
12.9 4D Hologram Interferometry of Tropical Cyclone 351
12.10 4-D Goni Turbulence based on Quantum Hologram Hypercube Interferometry 354

Index 359

CHAPTER

1

Mathematical Modeling Principles for Turbulence Description

Settled in a comfortable aeroplane seat, watching a movie or playing some amazing games, you turn to receive a snack and some beverage served by the steward when, unexpectedly, the safety belt warning turns on and a voice over the intercom says,“Ladies and gentlemen, we are experiencing turbulence...”At that very moment, the aeroplane starts bobbing up and down, pushing the service cart to the back of the aisle. It is a scary moment (Figure 1.1). Turbulence can be scary; what is the science behind it?

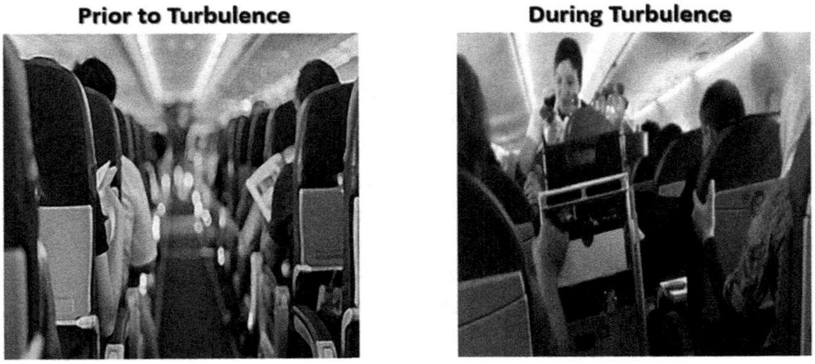

Figure 1.1: Aeroplane scary moment because of turbulence.

Learning the science behind turbulence makes it less scary. Turbulence is the unsteady dynamic movements in the atmospheric layers and in water. The changes in airflow around the plane cause turbulence. The changes in the motion of air (as around parts of an aeroplane in flight) relative to the surface of a body immersed in it create such scary turbulence (Figure 1.2).

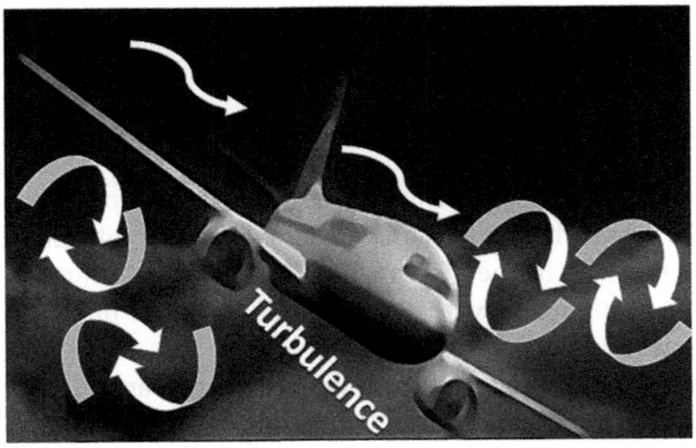

Figure 1.2: Scary turbulence around the fuselage.

1.1 What is Turbulence?

The straightforward explanation of turbulence is an abrupt, brutal shift in airflow that forms up and down currents triggered by fluctuating atmospheric mobility (Figure 1.3). In other words, turbulence can also be defined as that sudden jerking passengers occasionally feels in an aeroplane with the shaking of the aircraft (Figure 1.3). On May 20, 2024, Singapore Airlines flight SQ321 from London Heathrow to Singapore Changi encountered severe turbulence, leading to an emergency landing in Bangkok. Unfortunately, a 73-year-old passenger died, and several others were injured.

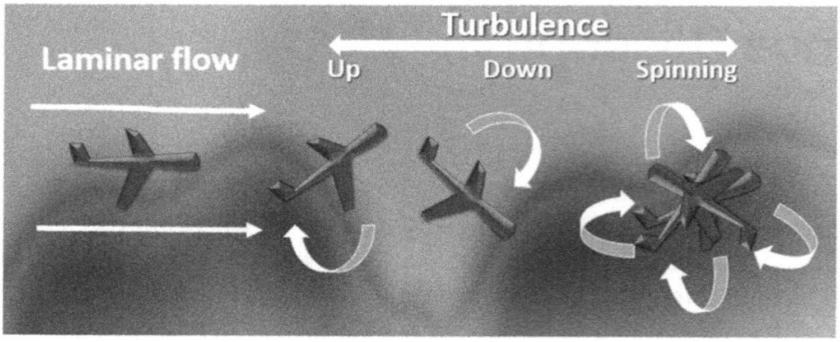

Figure 1.3: Abrupt change of airflow around the aeroplane.

From a geographical point of view, we can define turbulence as chaotic and capricious eddies of air, distressed from a quieter condition by countless forces. For instance, rough air occurs everywhere, from ground level to far above the spinning altitude across atmospheric layers. Therefore, atmospheric turbulence and small winds that fluctuate in momentum and path revealed scale and irregular air motions. In this view, turbulence is imperative since it mingles and agitates the atmosphere and instigates water vapor, smoke, and other substances, along with energy, to be perfectly distributed vertically and horizontally through the atmospheric boundary layers.

On the other hand, when the air current mobility that allows the aircraft to fly smoothly is interrupted, resulting in the aircraft pitching and shaking, this phenomenon is called turbulence. It is also acknowledged as an air pocket, which can trigger a sudden loss of altitude momentarily. Another definition is: turbulence is one of the foremost unpredictable of all the weather phenomena that are of significance to pilots. It is an irregular motion of the air resulting from eddies and vertical airflows. It can be insignificant as a couple of annoying bumps, or severe enough to momentarily throw an aeroplane out of control or to cause structural damage. Turbulence contains such natural phenomena as fronts, wind shear, thunderstorms, etc.

Atmospheric turbulence is characterized by irregular fluctuations stirring in atmospheric airflow. These fluctuations are changes in random and continuous patterns, which are superimposed on the mean dynamic movement of the air [1–3]. Therefore, turbulence is still a debated topic in the academic world. There are many dynamic causes for turbulence, therefore it is hard to define. Some of these obstruse definition can still raise countless significant questions.

1.2 Can Turbulence Damage Aeroplanes?

The primary concern among airline passengers often revolves around the question of whether turbulence can lead to the crash of flights. While turbulence may evoke anxiety among passengers, it is essential to understand that it does not pose a threat to the safety of the aircraft. Modern airplanes are meticulously designed to withstand various forms of turbulence, and pilots undergo thorough training to handle such situations effectively.

Despite the initial appearance of violent turbulence, aircraft are equipped to navigate through these conditions without compromising safety. Passengers are advised to follow safety protocols, such as fastening their seat belts, and remaining calm until the turbulence subsides. Although passengers may experience moments of discomfort due to the shaking of the plane, it's important to note that turbulence is a normal occurrence during flights.

The duration of turbulence can vary, depending on the nature of irregular airflow or air pockets. Irrespective of whether the turbulence lasts for a short or extended period, the aircraft and its systems are designed to handle such situations, ensuring the safety and well-being of everyone on board.

1.3 What is the Origin of Air Turbulence?

Now, the question is, what is the origin of turbulence? Turbulence originates from the unsteadiness of laminar flows (Figure 1.4). At adequately large values of the characteristic dimensionless number governing the flow, the flow undergoes a transition from laminar flow to turbulent flow. The *Reynolds number* R_e is the key dimensionless equation to identify the turbulence as a function of velocity U; length of scale L; and the kinematic viscosity v in which

$$R_e = UL[v]^{-1} \tag{1.1}$$

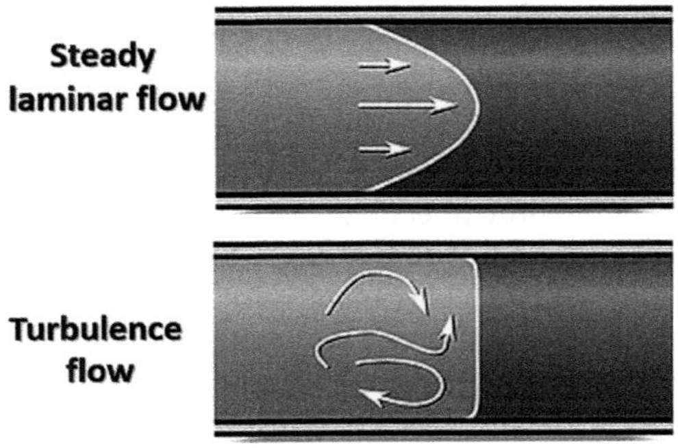

Figure 1.4: Laminar and turbulence flows.

Therefore, kinematic viscosity v signifies the ratio of dynamic (viscosity μ/density ρ) of the fluid. Equation 1.1 is the simple determination of turbulence when it exceeds a critical value. In this view, at Reynolds numbers between about 2000 and 4000, the flow is unstable because of the onset of turbulence. Scientists sometimes refer to these flows as transitional flows. Turbulent flow occurs if the Reynolds number is greater than 3500 (Table 1.1). For instance, implementing equation 1.1, $R_e = \frac{\text{Density}(\rho) \times \text{Diameter}(D) \times \text{Flow Speed}(U)}{\text{Viscosity}(v)}$ then a numerical calculation is showed as $R_e = \frac{1.25\,\text{kg/m}^3 \times 0.1\,\text{m} \times 35\,\text{m/s}}{1.83 \times 10^{-5}\,\text{Ns/m}^2} = 2.39 \times 10^5$. In this circumstance, turbulence occurs when the Reynolds number R_e is about 2300. Therefore, does a high Reynolds number mean turbulence?

Turbulence dominates the viscous regions in high Reynolds number flows. Instability in the shear layer creates turbulent fluctuations of flow-field properties. When is the flow laminar? If the Reynolds number is less than 2000, the flow is laminar which is also known as viscous flow. However, a streamlined flow occurs as well, as Reynold number is less than 1000 (or 2000) [1, 3, 5].

Table 1.1: Reynolds numbers for different internal flow regimes

Sort of Flow	Reynolds Number Range
Laminar	Up to $R_e = 2300$
Transition	$2300 < R_e < 4000$
Turbulent	$R_e < 4000$

The above perspective brings about the following significant question: what is the Reynolds number for airflow over airfoil transitions from laminar to turbulent? As long the Reynolds number R_e clarifies the flow patterns, then the airfoil can be presumed as free surface flow. Laminar flow occurs when $R_e < 2000$, and turbulent flow occurs when $R_e > 4000$, which reveals the internal flow. However, for external flow, laminar flow occurs up to $R_e = 5 \times 10^5$. The transitions region occurs when $R_e = 5 \times 10^5$ becomes the threshold value. We can view it this way, transition flow build-up and the Reynolds number ranges between 2100 and 4000. In this zone, the fluid can have either a laminar or turbulent flow. The sort of flow, therefore,would be dominated by factors,for instance, pipe roughness and flow consistency, which are not factored in when computing the Reynolds number. In this zone, the pressure ditches and thus fluctuates either linearly with velocity or with the square of the velocity of the fluid. We can say that in the transition region, the laminar boundary layer is in a state of unstable equilibrium turbulent flow. Thus, even the most infinitesimal vicissitudes in the surrounding form of the fluid can start the flow to become turbulent. That is why a small transition region occurs wherein the flow changes to turbulence. The region where laminar to turbulent transition is relevant is the boundary layer. This can lead to the question: why is the Reynolds number for the transition from laminar to turbulent flow in internal flows much smaller than that of external flows? The scientists use the thickness of the boundary layer as the length scale to form the Reynolds number relevant to transition, which fits all flow circumstances. For internal flows, the dimension used is perpendicular to the flow, for example, the diameter of the pipe. It is often a dimension in the flow direction, such as the chord of a wing, for external flows. We constrain the dynamics inside the boundary layer to existing within that length scale. The velocity gains from zero at the wall to the maximum value, which is impartial outside the boundary layer over that distance. In this circumstance, we use the Reynolds number for more than just boundary layers. Although a length scale is not specifically relevant to boundary layers but is explored [1, 4, 7, 8].

The following examples can address the accomplished frames of internal and external flow based on the Reynolds number. In the internal flow phase through a pipe, the boundary layer increases until its thickness is about half the diameter of the pipe. The boundary layer thickness is half the pipe diameter. In this circumstance, internal flow merges with the boundary layer on the opposite side. In this understanding, the internal diameter of a pipe is used to form the Reynolds number, which is relevant to boundary layer stability and laminar to the turbulent transition zone. The boundary layer cannot keep growing past this point, and the flow becomes fully developed.

In the case of external flow, the length scale is used to form the Reynolds number because the boundary layer thickness is much less than the Reynolds number. The diameter labels cylinders and spheres, while the chord (the distance from the leading edge to the trailing edge) recognizes wings. Therefore, the Reynolds numbers are numerically much larger than what would apply to boundary layer transition [3–5].

Therefore, the Reynolds number for transition is much higher for the external flow than for the internal flow. This is because it used a length scale that is much larger than the boundary layer thickness for the external flow. However, for the internal flow, the length scale is like the boundary layer thickness.

1.4 Distinguishing Between Laminar and Turbulent Flows: An Exploration of Variances

Consistent with the above perspective, laminar and turbulent flows are two commonly exploited terms in fluid/air dynamics. In this view, as well as the fluid moving through any pipe or tube, it either moves in a laminar fashion or a turbulent flow. We could describe laminar flow as the flow of fluid whenever each particle belonging to the fluid is a follower of a consistent course, routes which usually under no circumstances obstruct one another. One consequence of laminar movement would be that the speed belonging to the fluid is constant inside the fluid, whereas turbulent flow could be described as the uneven, unfrequented movement of fluid, which is a small whirlpool area. The speed of such a fluid is unquestionably not necessarily constant at every point. Table 1.2 summarizes the main distinguishing characteristics of laminar and turbulent flows [2, 5, 7].

Tables 1.1 and 1.2. reveal that in contrast to laminar flow, the fluid no longer travels in layers, and mixing across the tube is highly efficient. Flows at Reynolds numbers larger than 4000 are typically (but not necessarily) turbulent, while those at low Reynolds numbers below 2300 usually remain laminar. Once fluid flows in parallel layers, the laminar flow occurs in the circumstance of steady motion. In this scenario, there are not any crosscurrents perpendicular to the direction of flow, nor eddies or swirls of fluids. Therefore, the motion of the particles of the fluid is extremely orderly, with all particles getting hold of straight lines parallel to the pipe walls to form a streamlined flow. The question is: can mixing occur in the laminar flow? A slightly lateral commixture (mixing perpendicular to the flow direction) crops up by the encounter of diffusion between layers of the liquid. Diffusion commixture occurs slowly, but if the diameter of the pipe of the tube is small, then this diffusive mixing is often important.

Turbulence is characterized by chaotic property changes, which consist of a rapid variation of pressure and flow velocity in space and time. In contrast to laminar flow, the fluid no longer travels in layers, and mixing across the tube is highly efficient. Flows at Reynolds numbers larger than 4000 are typically (but not necessarily) turbulent, while those at low Reynolds numbers below 2300 usually remain laminar (Table 1.2) [5, 7, 9]. Flow between Reynolds numbers 2300 to 4000 and known as a transition (Table 1.1).

Table 1.2: Summarizing the main differences between laminar and turbulent flows

Types of flow/ Parameters	Laminar Flow	Turbulent Flow
Flow Fashion	The fluid flow in which the adjacent layers of the fluid do not mix and move parallel to each other is known as laminar flow.	The fluid flow in which the adjacent layers of the fluid cross each other and do not move parallel to each other is identified as turbulent flow.
Flow Layer Styles	In the laminar flow, the fluid layer moves in a straight line.	In turbulent flow, the fluid layers do not move in a straight line. They move randomly in a zigzag manner.
Reynolds Numbers	The fluid flow having a Reynolds number less than 2000 is called laminar flow.	The fluid flow having a Reynolds number greater than 4000 is called turbulent flow.
Shear Stress	Shear stress in laminar flow depends only on the viscosity of the fluid and is independent of the density.	The shear stress in turbulent flow depends upon its density.
Mobility	The laminar flow always occurs when the fluid flows with low velocity and in small diameter pipes.	The turbulent flow occurs when the velocity of the fluid is high, and it flows through larger diameter pipes.
Mixing	The fluid flow is orderly, i.e., there is no mixing of adjacent layers of the fluid, and they move parallel to each other and along the walls of the pipe.	The fluid does not flow in a definite order. There is a mixing of different layers and they do not move parallel to each other but cross each other.

In turbulent flow, the fluid particles do not obey smooth paths. In its place, there is a slice of local arbitrary flow diagonally along the path of the mean motion. These disorder motions frequently have the pattern of swirls (vortices) superimposed on the average fluid mobility [1–4]. Therefore, laminar and turbulent flows can exist in the same system network when operated at different flow rates. In this understanding, it demonstrates the conventional examples of both laminar and turbulent flows in cigarette smoke (Figure 1.5).

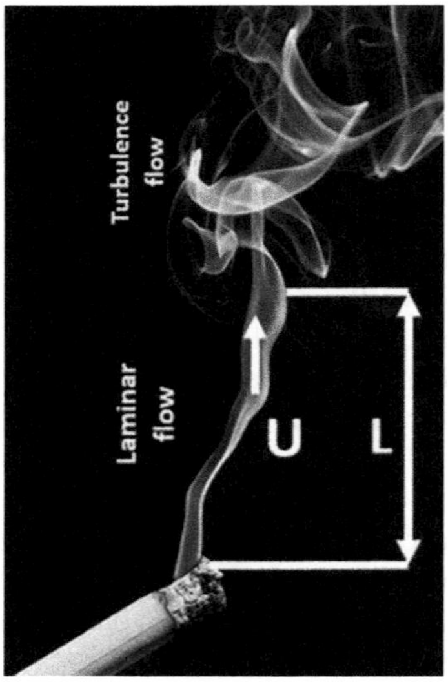

Figure 1.5: Laminar and turbulence are associated with cigarette smoke.

Let us assume that the velocity of smoke at the initial stage is about 30 cm/s, the length scale is approximately 3 cm, and the kinematic viscosity is 0.14 cm^2/s. Thanks to the Reynolds number, the calculated ratio between inertia and viscous forces, can decide either the laminar flow or the turbulence flow. Using equation 1.1, then $R_e = \frac{30 \times 3}{0.14} = 642$, designates transition flow from the laminar flow to the turbulence flow. In this scenario, hot smoke is less dense than the surrounding cooler air, which forms rapid smoke velocity [3, 5, 8, 10]. A net upward buoyant force acts on smoke correspondingly as on a helium balloon, which inductees a vertical acceleration. In this scene, smoke should shortly speed up.

The middle part of the plume which is the hottest is then accelerated. Concerning the edges, the velocity becomes closer to that of the surrounding air. This revolution of the smoke plume from place to place, known as shear, initiates turbulence. Whether the flow remains laminar or turns turbulent, it must be a function of a quantity of R_e (Figure 1.6). A typical example of flow classification with the Reynolds number is the flow past a two-dimensional cylinder. It is interesting to show that the fluctuation of Reynolds intervals creates different flow regimes. In the laminar steady flow regime, a limit emerges when the Reynolds number (R_e) is less than 5. When the Reynolds number falls within the range of 5 to 45, a pair of vortices becomes linked with the laminar steady flow. Consequently, the laminar vortex street associated with unsteady flow exists when $45 < R_e < 150$. And the transitional unsteady flow occurs. However, turbulent unsteady flow and turbulent vortex are developed and the Reynolds number intervals are $3 \times 10^5 < R_e < 3 \times 10^6$ and $R_e > 3 \times 10^6$, respectively (Figure 1.6).

Now the significant question is: what is the importance of turbulence for atmosphere heat transfer? In turbulent flow, the instabilities deliver a supplementary mechanism for heat transfer along with molecular diffusion. On the contrary, heat is transferred between dissimilar streams across molecular diffusion in laminar flow. The movement of molecules occurs from the higher concentration zone to the lower concentration region. In fact, in laminar flow, fluid particles flow in an orderly manner along certain path lines. Hence, the turbulent flow continuously has a higher heat transfer coefficient in contrast to the laminar flows. Therefore, in turbulence flow, the fluctuated liquid particles can form eddies (Figure 1.7), which rapidly transport mass, momentum, and energy across different regions of the flow.

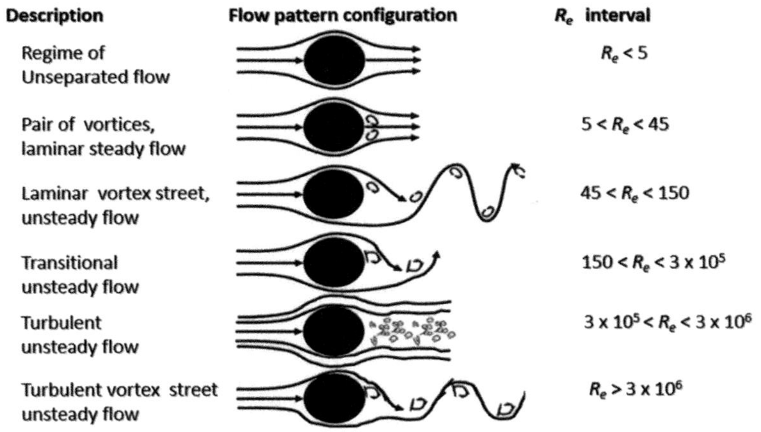

Figure 1.6: Different flow pattern regimes with different Reynolds number intervals.

Figure 1.7: Eddies generation due to fluctuated flow around the circle.

Compared to the eddies, molecular diffusion is an incredibly slow process and is regularly abandoned in computing the heat transfer rate in turbulent flows. Turbulence plays a great role in air heat transfer.

1.5 What is the Nature of Turbulence?

The example of a smoke plume from the cigarette (Figure 1.5) leads to a significant question: what is the nature of the turbulence? It is difficult to deliver specific identical answers owing to the complexity of the turbulence regime. However, it may be effective just to list some characteristics of turbulent flow to understand the nature of turbulence.

The most noticeable of the turbulence characteristics is irregularity. In this scenario, fluid or gas particles move randomly in space without a specific direction. Scientists handled turbulence problems statistically rather than deterministically. Turbulent flow is chaotic, however, not all chaotic flows are turbulent. In this sense, scientists consider the turbulence regime a continuum phenomenon, governed by the fluid equation of the motions [6, 9].

Turbulent flow occurs at low viscosity and at higher characteristic linear dimensions that align with higher chaotic movements. In this scenario, rapid mixing is associated with the highest rate of mass momentum and energy transport in the turbulence flow. It is well-known that this phenomenon is diffusivity.

Therefore, the diffusivity of the turbulence plays a tremendous role in avoiding boundary layer disjunction on airfoils at large angles of attack. Further, it surges the rate of heat transfers in entire varieties of machinery. It is also the foremost source of resistance to the flow in pipelines. Last, it swells the rate of momentum transfer between wind stress and ocean currents. Turbulent flow is differentiated by a strong three-dimensional vortex group mechanism known as vortex stretching (Figure 1.8). The question is: what is vortex stretching? In fluid dynamics, vortex stretching is the expansion of vortices in three-dimensional fluid flow, allied with consistent growth of the constituent of vorticity in the stretching direction—owing to the conservation of angular momentum. In this context, the affinity of turbulent flows to cause fine-scale movements from large-scale energy injection is regularly observed as a scale-wise surge of kinetic energy plunged by vorticity stretching.

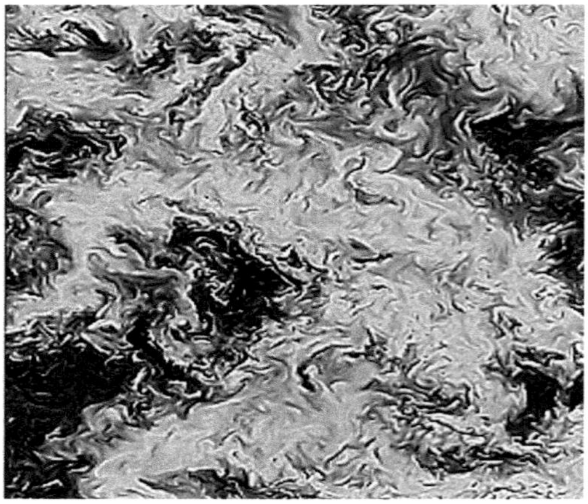

Figure 1.8: Vortex stretching regime.

However, the random vorticity fluctuations that identified turbulence could not sustain themselves if they develop in the two-dimensional flow where there is absence of vortex stretching. For instance, the cyclone is substantially a 2-D flow (Figure 1.9), which is not turbulence itself because of the absence of vortex stretching. Consistent with this perspective, the vortex must illustrate turbulence stretching [5–9]. Therefore, ocean waves are not considered turbulence flow since they are irrotational flows (Figure 1.10). In an irrotational flow, there can be circular paths for the fluid, but each fluid particle does not rotate. Needless to say, the dominated vortex stretching in turbulence flow creates rotational motion (Figure 1.10).

Figure 1.9: 2-D cyclone flow.

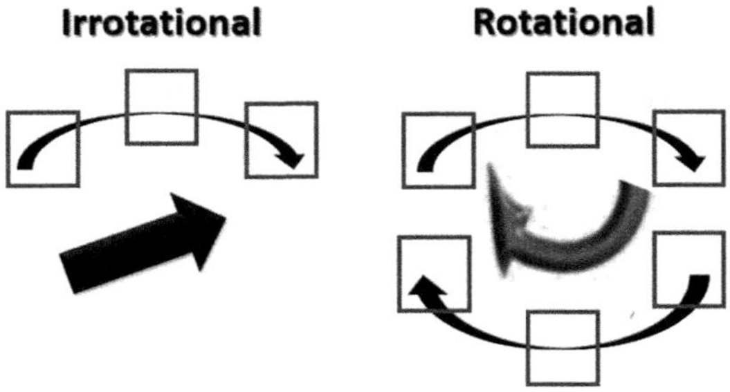

Figure 1.10: Irrotational and rotational flows.

The core discrepancy between turbulence and random waves is that the latter are nondissipative, while turbulence is essentially dissipative. The kinetic energy of the turbulent flow is converted into internal energy through viscous shear stress and hence is dissipative. The deformation performance of the viscous shear stress increases the internal energy of the fluid at the expense of the kinetic energy of the turbulence.

The origin of the turbulence as a function of the Reynolds number brings about a specific criteria of turbulence regime. As indicated previously in section 1.4, if the Reynolds number is greater than $R_e > 3500$, the flow is then turbulent. However, the major properties of turbulent regimes do not rely on the molecular characteristics of the fluid in which the turbulence develops. In this regard, every flow regime has specific exclusive characteristics that are amalgamated with its original and boundary conditions because of the nonlinearity of motion equations. In this understanding, there are no cosmopolitan solution to problems in the turbulent regime, and no well-defined solution to the Navier-Stocks equations can be had. As a result, every flow pattern is different owing to all turbulent flows having many distinguished characteristics in common. In other words, it cannot state the turbulence flow regime in specific laws because of the uniqueness of any description and classification of turbulent flows, such as boundary layers, boundary conditions, jets, and wake. It restricts turbulence mathematical theories to handle all sorts of turbulence flows. The determination of turbulence characteristics is a function of its derived environment [2, 5, 9].

1.6 Turbulent Boundary Layers

Let us assume that fluid flows over a stationary surface, e.g., the flat plate, the bed of a stream, or the pipe wall. Once the fluid is flowing, we assume it to do so in layers, as illustrated in Figure 1.11. The layer of the fluid to bear with the surface over which it is moving (the bottom-layer) has zero velocity, consistent with the 'no-slip condition'. Whereas the free surface of the flowing fluid (the upmost layer) has a changeable rate. Thus, every layer drags the layer slightly below it. If a horizontal plane is perpendicular to the plane of Figure 1.11, this dragging force is going to be shear on the plane. This results in every layer applying a shear force on the layer just under it. The shear stress is given by Newton's Law of viscosity:

$$\tau = \mu \frac{du}{dz} \tag{1.2}$$

Equation 1.2 reveals the shear stress τ directly proportional with fluid viscosity μ and its velocity gradient du regarding the distance change dz. Shear stress, therefore, is defined as a force per unit area, acting parallel to a microscopic surface element. Friction primarily instigates shear stress in fluids between fluid particles owing to fluid viscosity. When the fluid approaches, the surface ends up resting by the shear stress at the wall. The boundary layer is the region in

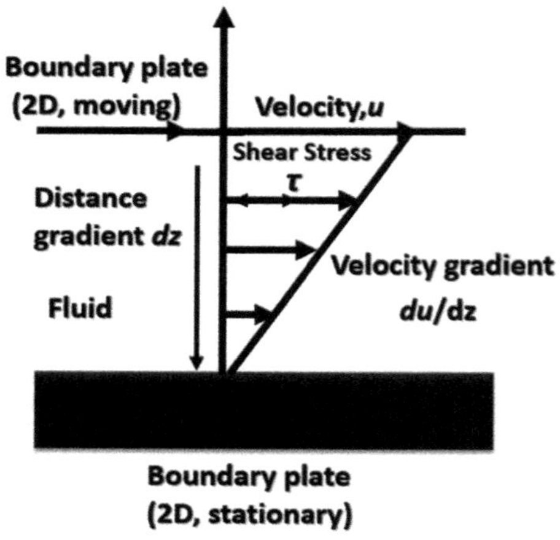

Figure 1.11: Simplification of boundary layer concept.

which flow adjusts from zero velocity at the wall to a maximum in the flow's mainstream. The concept of boundary layers is important in all viscous fluid dynamics and the theory of heat transfer [2, 6].

We can derive the velocity from equation 1.2 as:

$$\frac{\tau}{\mu} = \frac{du}{dz} \tag{1.3}$$

Consequently, u can be obtained at any height z above the bottom plate by integrating concerning z that is given by:

$$u_z = \int \frac{du}{dz} dz = \frac{\tau}{\mu} \int dz + c$$

$$= \frac{\tau}{\mu} \int dz + c = \frac{\tau}{\mu} z + c \qquad (1.4) \tag{1.4}$$

here c is the constant of integration and u is the velocity at $z = 0$ (i.e., where $u = 0$) such that:

$$u_z = \frac{\tau}{\mu} z \tag{1.5}$$

Equation 1.5 demonstrates that the velocity differs linearly from zero at the bottom ($z = 0$) to approximately maximum at the highest position of the plate. Therefore, as the applied force of shear stress τ upsurges so does the speed rate at each point directly above the lower plate. In this scenario, the viscosity upsurges the velocity at any point above the lower plate shrinks (Figure 1.11).

For all Newtonian fluids in laminar flow, the shear stress is proportional to the strain rate in the fluid where the viscosity is the constant of proportionality. However, for non-Newtonian fluids, this is no longer the case as, for these fluids, the viscosity is not constant. It imparted the shear stress onto the boundary because of this loss of velocity. Therefore,boundary layers can be laminar or turbulent, relying on the value of the Reynolds number, demonstrated early in section 1.4. The boundary layer of turbulence is above three layers: the overlap layer; buffer layer, and viscous sub-layer (Figure 1.12). The viscous sub-layer has the smallest velocity as compared to overlap and buffer layers [2, 7].

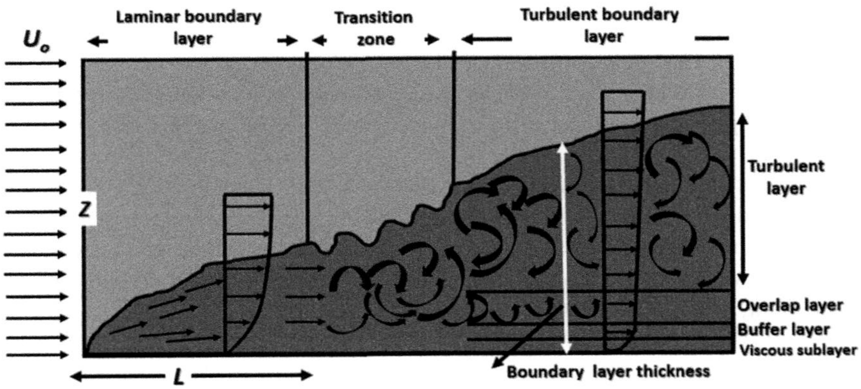

Figure 1.12: Boundary layer and boundary layer thickness of turbulent flow.

The boundary layer is a function of its thickness. We define the thickness of the boundary ∂ layer as the distance from the wall to where the free stream flows (Figure 1.12). Laminar boundary layers over a flat are governed by:

$$\partial \approx 5.0 \times \left(R_{e_L}\right)^{-0.5} \times L \tag{1.6}$$

Here R_{e_L} is the Reynolds number based on the length of the plate L. On the contrary, the boundary layer thickness of turbulent flow is given by:

$$\partial \approx 0.37 \times \left(R_{e_L}\right)^{-0.2} \times L \tag{1.7}$$

According to formula 1.7, the turbulent boundary layer thickness formula adopts that the flow is turbulent from the beginning of the boundary layer.

Let us consider water flow with a velocity of 0.1 ms^{-1}, which propagates through a flat plate length 1 m long at the temperature of 20°C with the kinematic viscosity equal to 1×10^{-6} m^2s^{-1}. Implementing equation 1.1 to estimate Reynolds number as shown by:

$$R_{e_{0.5L}} = \frac{0.1\ (\mathrm{ms}^{-1}) \times 0.5\ (\mathrm{m})}{1 \times 10^{-6}\ \left(\mathrm{m}^2\mathrm{s}^{-1}\right)} = 50000 \tag{1.8}$$

Calculation of formula 1.8 convinces the laminar circumstances, the boundary layer thickness is given by formula 1.7:

$$\partial \approx 5.0 \times 0.5 \times [50000]^{-0.5} = 0.011\ \mathrm{m} \tag{1.9}$$

Using the similar conditions of formula 1.8 with a change in the stream flow to 1 ms^{-1}. The question now is, at what specific distance L could be the transition from laminar to the boundary layer of turbulent in the circumstance of R_{e_L} ~500000?

$$L = 500000 \times 1 \times 10^{-6}\left[\mathrm{m}^2\mathrm{s}^{-1}\right] \times 1^{-1}\left[\mathrm{ms}^{-1}\right] = 0.5\ \mathrm{m} \tag{1.10}$$

The boundary layer is laminar for lower Reynolds numbers, and the stream-wise velocity consistently fluctuates as fluid flows away from the wall. For example, the Reynolds number surges, and the flow becomes unstable. Beyond doubt, the boundary layer is turbulent for higher Reynolds numbers, and unsteady swirling distinguished the stream-wise velocity flows interior of the boundary layer [3, 5, 9].

1.7 Causes of Atmospheric Turbulence

Various circumstances can create turbulence. These circumstances can range from the ground surface up to the highest atmospheric top layer, and there is no actual method to forecast turbulence or watch it on a map. What are the causes of turbulence? The keystone cause of the turbulence is wind shear.

Wind gradient, sometimes referred to as wind shear, is a variance in wind velocity and/or wind direction over a moderately abrupt distance through the atmosphere (Figure 1.13). In this view, we usually designate atmospheric wind shear as either vertical or horizontal wind shear. It is densely allied with vertical wind shear that originates turbulence with the vertical and horizontal transport of momentum, heat, and water vapor.

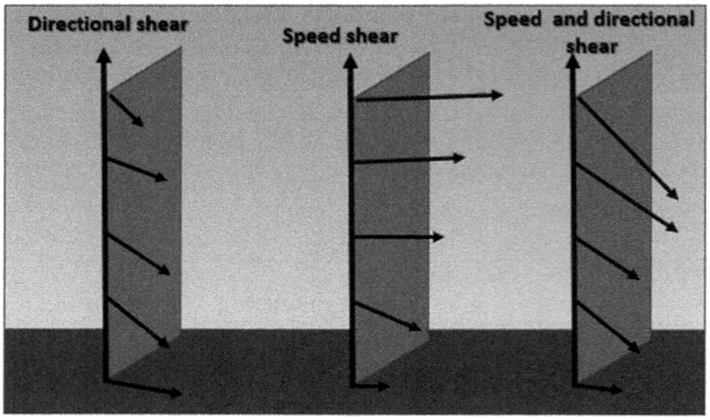

Figure 1.13: Vertical wind shear.

A narrow zone of abrupt velocity alteration is identified as a shear line. In this context, wind shear exists along troughs and lows, around jet streams (Figure 1.14), and in zones of temperature inversions. Turbulence allied with depressions and troughs is basically due to horizontal directional and velocity shear. In this sense, turbulence originates along troughs at any altitude, within lows at any altitude, and poleward of lows in the mid and upper latitudes. Turbulence is boosted by an "arching" (enlarged) jet stream about the troughs and ridges (Figure 1.14).

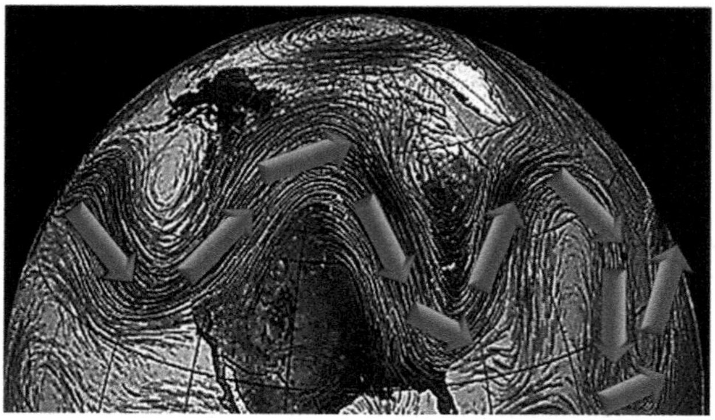

Figure 1.14: Turbulence around the jet stream.

Temperature inversions, therefore, are zones with vertical wind shear potential. The greatest shear, and thus the greatest turbulence, is found at the tops of the inversion layer. Strong stability prevents the mixing of the stable low layer with the warmer layer above. Turbulence, consequently, is allied with temperature inversions regularly arising because of radiational cooling, which is the nighttime cooling of the Earth's surface, generating a surface-established inversion (Figure 1.15). Consequently, it places turbulence where there are great horizontal variances in temperature between warm and cold air masses (Figure 1.16). Turbulence is a long strong isotach gradient zone (Figure 1.17). In this view, the bold line with the arrowhead suggests the jet stream axis or contour of an extreme wind velocity. On the right side of the axis (looking downstream), for instance, the isotach gradient is weaker than the gradient on the left [2–5]. This specifies that the horizontal wind shear is normally greater on the left side of the jet axis than on the right. This is because steadiness deliberations restrain the lateral shear on the right side of the jet flow (Figure 1.17).

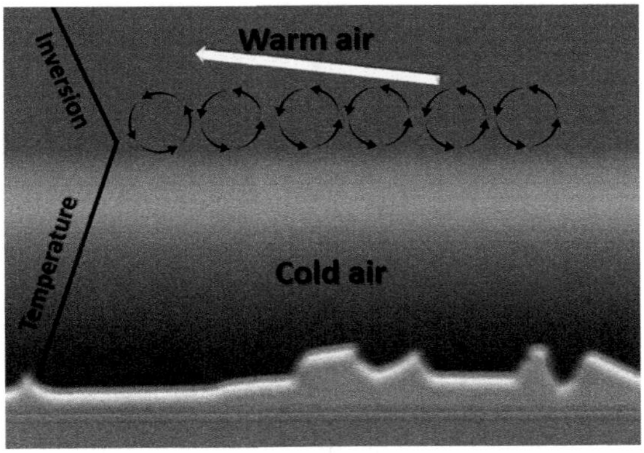

Figure 1.15: Temperature inversion.

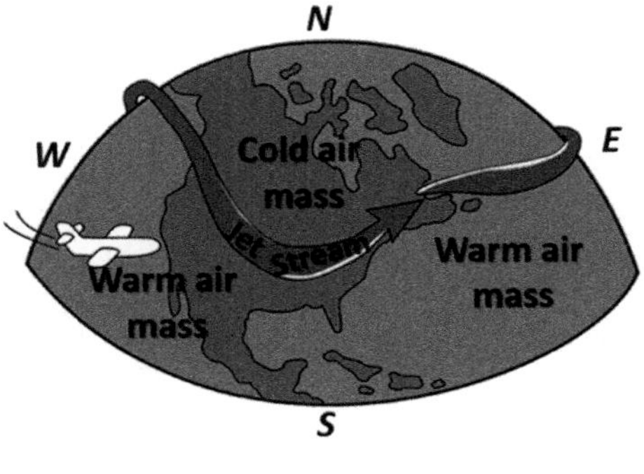

Figure 1.16: Temperature variances between warm and cold air masses.

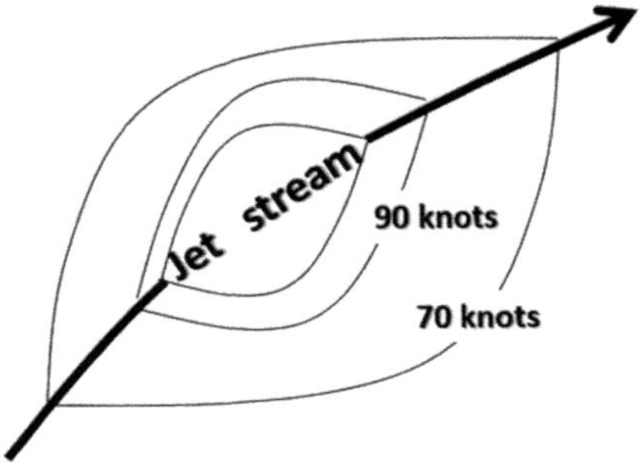

Figure 1.17: Example of isotach gradient contours.

As part of regular wind flow, a jet stream is a hub of extremely strong horizontal shear winds, which blows in a wavy pattern (Figures 1.14 to 1.16). Most often, turbulence is traced on the poleward side of the cyclonic jet stream (Figure 1.18). It regularly pinpointed turbulence on the equatorward side of the anticyclonic jet stream.

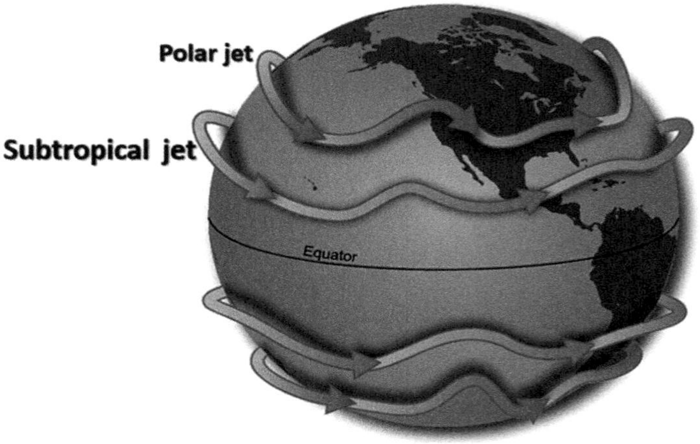

Figure 1.18: Polar and subtropical jets.

The boundary between different variations of cold air and warm air masses is known as the front turbulence zone (Figure 1.19). The elevating of the warm air by the inclined frontal surface and friction between the dual contrasting air masses generate turbulence in the frontal zone. This turbulence is further noticeable when the warm air is moist and unsteady and would be tremendously critical if thunderstorms grow. Turbulence, therefore, is regularly allied with cold fronts. However, turbulence can be found, to a lesser degree, in a warm front as well.

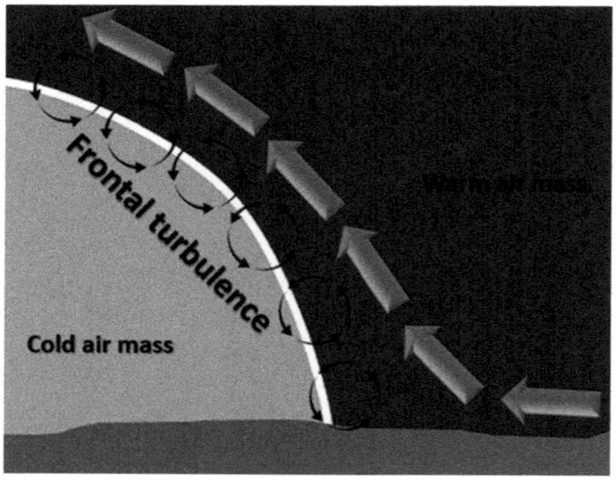

Figure 1.19: Frontal turbulence.

We expect turbulence, as the Sun heats the surface of the Earth irregularly. For instance, the grass-covered fields are heated slower than sandy, barren ground and rocky zones but are heated faster than water. The dynamic fluctuations of warm air rising, and cooler air descending can create isolated convective currents. These convective current motions are accountable for unsteady conditions and a plane hovering in and out of them. Therefore, convective currents create such thermal turbulence, which can pronounce a tremendous effect on the flight path of an aeroplane approaching a runway. In this understanding, it subjects the aircraft to thermal turbulence of fluctuating intensity established in motion over the runway closer to the path. These convective currents perhaps shift the plane from its standard glide route with the consequence that it would either undershoot or overshoot the runway [2, 5, 11].

Thermal turbulence mostly extends from the base to the top of the convection layer, with smooth circumstances originating overhead. For instance, if cumulus, towering cumulus, or cumulonimbus clouds exist, the turbulent layer outspreads from the surface to cloud tops. The thermal turbulence strength upsurges as the convective updraft strength growths. However, convective currents consequential in "dry thermals" have perhaps not been observed in cumuliform clouds. Favorable circumstances for dry convection, therefore, contain warm surface temperatures, uneven surface hearing, and steep surface-based lapse rates. In the circumstance of expected thermal turbulence, many pilots prefer to fly in the early morning or the evening when the thermal activity is not as severe.

Another cause of air turbulence is mountain wave events. When the shear winds flow perpendicular to the mountain ridgelines, very strong wavy shear wind is induced, known as mountain wave winds (Figure 1.20). In other words, mountain waves form when strong winds with at least 25 knots blow at the top of the mountain and perpendicular to a mountain range [3, 5, 9, 12].

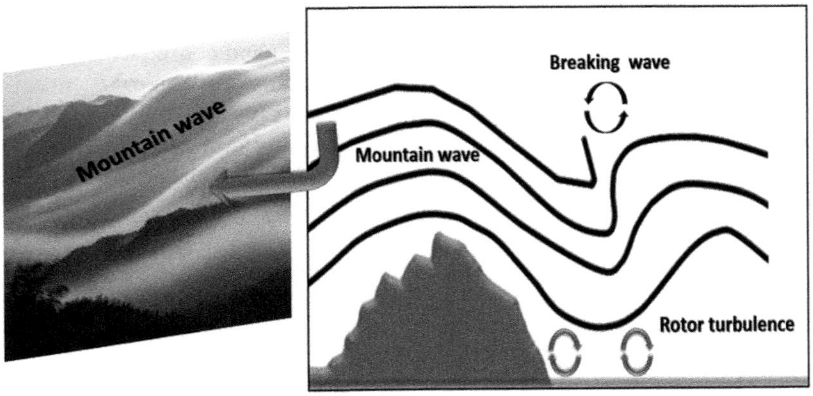

Figure 1.20: Mountain waves.

Mountain wave winds also arise on the downwind side or Lee of the mountain. Now, the question is: how do mountain waves work? Mountain waves are the consequence of curving air being forced to rise on the windward side of a mountain barrier, then as a consequence of convinced atmospheric circumstances, sinking on the leeward side. Therefore, in these breaking waves, the atmospheric flow turns turbulent. Consequently, the maximum turbulence generally befalls near the mid-level of the storm, between 12,000 and 20,000 feet. Subsequently, the mountain wave is of utmost severity in the excessive vertical cloud growth. In this scenario, clouds are found near the top of the rotor circulation, and under higher lenticular clouds (Figure 1.21) they contain strong turbulence and should be avoided by pilots. In this understanding, rotors, air spinning quickly comparable to a wheel (Figure 1.21) and can arise where wind velocities revolutionize in a wave or where friction reduces the wind near the terrain.

Nonetheless imperceptible at times, the beginning of mountain waves, breaking waves, and rotors can be revealed by concrete signs in the sky. Altocumulus standing lenticular clouds (ACSL) (Figure 1.21) can grow the highest of discrete waves,creating mountain waves under the impact of sufficient moisture, up and downwind of mountains [5, 9, 13].

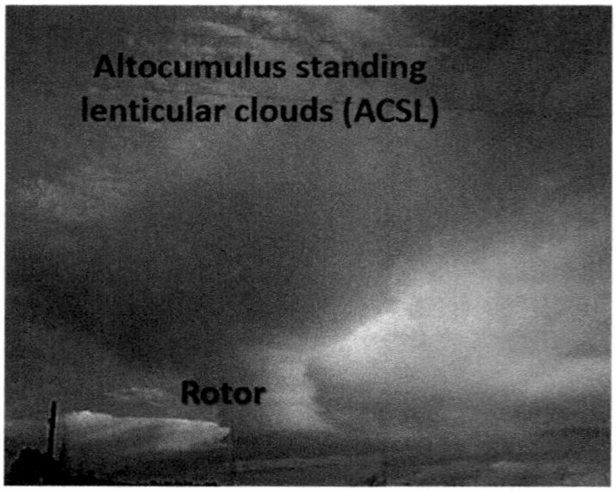

Figure 1.21: Turbulence associated with altocumulus standing lenticular clouds (ACSL) and rotors.

How far can mountain wave turbulence travel? Strong lift and clear air turbulence perhaps spread up to 100,000 feet. Vertically propagating waves with sufficient amplitude perhaps break and bear consequence in severe clear air turbulence between 20,000 and 40,000 feet [3, 5, 9, 11, 13].

Lastly, the main cause for turbulence is also well understood as mechanical (frictional) turbulence. In this regard, friction between the air and the ground (Figure 1.22), specifically irregular terrain and man-made obstacles, initiates eddies (Figure 1.23) and consequently turbulence in the lower levels. In other words, friction eddy effects on ground cause mechanical frictional turbulence. This sort of turbulence is mainly caused by wind shear blowing over tall building structures causing a series of eddy patterns (Figure 1.23). Consequently, irregular horizontal air movements can be transformed into a complicated pattern of eddies, which can be identified as small-scale, short-term, random, and frequent fluctuations in the wind shear flow velocity.

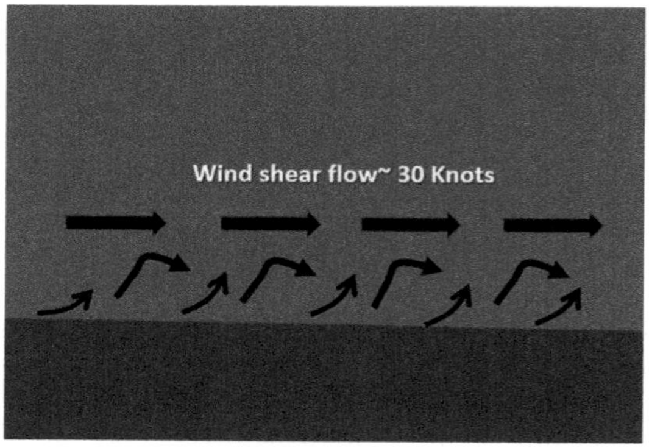

Figure 1.22: Mechanical (frictional) turbulence.

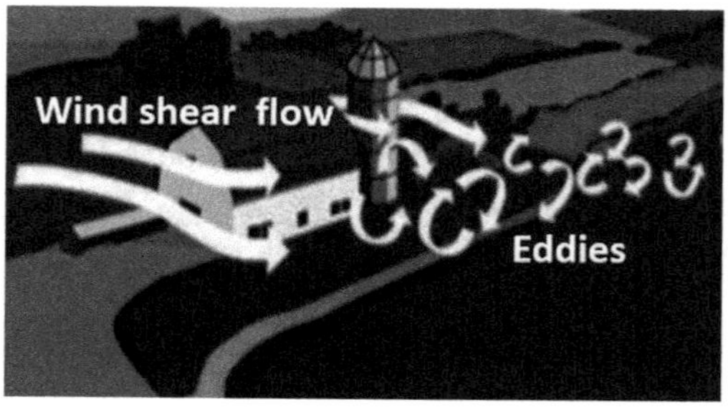

Figure 1.23: Turbulence eddies due to the tall building.

1.8 Air Turbulence Intensity Classes

Turbulence intensity is identified as the ratio of the standard deviation of changeable wind speed over time to the average wind velocity, and it signifies the intensity of wind speed instability. The wind speed instability shows the range of air turbulence intensity. In this understanding, the author consigns a modification of turbulence intensity as the ratio of the time fluctuation of shear stress to the mean estimation of shear stress. This can deliver a new definition of turbulence intensity as a function of the shear stress:

$$I_{\tau} = \frac{\sqrt{\sum_{i=1}^{N}\left(\tau_i(\vec{u}) - \sum_{i=1}^{N}\tau_i(\vec{u})N^{-1}\right)^2}}{\sum_{i=1}^{N}\tau_i(\vec{u})N^{-1}} \tag{1.11}$$

Equation 1.11 proves the new approach to estimating turbulence intensity based on the number of shear stress τ observed population (N) $i = 1, 2, 3......, N$. In equation 1.11, we assume the shear stress constitutive law as linear. If the I_{τ} value is 0.05%, the intensity is light. However, moderate turbulence has an intensity ranging between 0.1% to 0.3%. Conversely, severe turbulence has an intensity higher than 0.3% and less than 5%. Therefore, the fully turbulent flows have intensities ranging between 5% and 10%.

The Newtonian fluid of the dynamic viscosity is impartial to flow velocity while non-Newtonian flows are not valid, and one should subscribe to the reformation as expressed by:

$$\tau_i(\vec{u}) = \mu(\vec{u})\nabla\vec{u} \tag{1.11.1}$$

Equation 1.11.1 exposes that it can express the airflow viscosity as a function of fluid velocity, which states the shear stress and point as a function of wind speed. In this understanding, we can consider shear stress as a Newtonian motion in circumstances we can acknowledge it as a constant for the gradient motion $\nabla\vec{u}$. Let us consider a two-dimensional Cartesian coordinate x and y, with time (t) the turbulence intensity as a function of the wind shear velocity can be given by:

$$\begin{pmatrix} I(\tau_{xx}) & I(\tau_{xy}) \\ I(\tau_{yx}) & I(\tau_{yy}) \end{pmatrix} = \begin{pmatrix} x & 0 \\ 0 & -t \end{pmatrix} \cdot \begin{pmatrix} \frac{\partial I(\tau_{xx})}{\partial x} & \frac{\partial I(\tau_{xx})}{\partial y} \\ \frac{\partial I(\tau_{yy})}{\partial x} & \frac{\partial I(\tau_{yy})}{\partial y} \end{pmatrix} \tag{1.12}$$

Here equation 1.12 signifies the intensity of turbulence flow as the Newtonian flow. Therefore, equation 1.12 can be turned into an anisotropic flow with the viscosity tensor $\begin{pmatrix} \mu_{xx} & \mu_{xy} \\ \mu_{yx} & \mu_{yy} \end{pmatrix}$ as:

$$\begin{pmatrix} I(\tau_{xx}) & I(\tau_{xy}) \\ I(\tau_{yx}) & I(\tau_{yy}) \end{pmatrix} = \begin{pmatrix} \mu_{xx} & \mu_{xy} \\ \mu_{yx} & \mu_{yy} \end{pmatrix} \cdot \begin{pmatrix} \frac{\partial I(\tau_{xx})}{\partial x} & \frac{\partial I(\tau_{xx})}{\partial y} \\ \frac{\partial I(\tau_{yy})}{\partial x} & \frac{\partial I(\tau_{yy})}{\partial y} \end{pmatrix} \tag{1.13}$$

It can represent four categories of turbulence intensities (i) light; (ii) moderate; (iii) severe; and (iv) extreme. These air turbulence intensity classes are a function of air steady and the causing turbulence factors. For instance, the flight passengers perhaps experience slight strain against seatbelts, which is acknowledged as light turbulence. This occurs because the aeroplane changes its altitude slightly. Consequently, if the aeroplane changes altitude intensely, it experiences moderate turbulence. If the aeroplane changes attitude abruptly and throws out of control, and the passengers encounter influential tosses despite the seatbelt, the flight is under the effect of severe turbulence. If violent, disordered air movement rocked the fuselage, then it experiences extreme turbulence which can damage the fuselage [10, 13].

1.9 Clear Air Turbulence

The author has always experienced turbulence in a plane while flying across, what looks like a picture-perfect visible blue sky. This sort of turbulence occurs in clear and cloudless sky conditions and is known as clear air turbulence. Contrary to what we have shown in previous sections, why does clear air turbulence occur? It arises when two different air-moving masses meet at different speeds above 15,000 feet involving thunderstorms. It occurs when there is a cloud-free sky as well and 75% of the entire clear air turbulence occurs in clear air [8, 13, 16].

The most common example is the meeting of jet streams, which are dominated by fast motions of wind flows. These jet streams circulate the Earth in the atmospheric zone known as the tropopause (Figure 1.24). What is tropopause? It is a boundary between the troposphere and the stratosphere layers. The troposphere is the layer of the Earth's atmosphere closest to the surface, while the stratosphere is the layer of the atmosphere above the troposphere (Figure 1.25).

The troposphere encompasses upward to approximately 10 km (6.2 miles or about 33,000 feet) above sea level (Figure 1.25). The height of the top of the troposphere, therefore, differs with altitude (it is lowest over the poles and highest at the equator) and by season (it is lower in winter and higher in summer). It can be as high as 20 km (12 miles or 65,000 feet) near the equator, and as low as 7 km (4 miles or 23,000 feet) over the poles in winter [2, 9, 13, 17].

Air is warmest at the bottom of the troposphere near ground level. It gets colder as one rises through the troposphere. That is why the peaks of tall mountains can be snow-covered even in the summer. Air pressure and the density of the air also decrease with altitude. That is why the cabins of high-flying jet aircraft are pressurized [2, 5, 13].

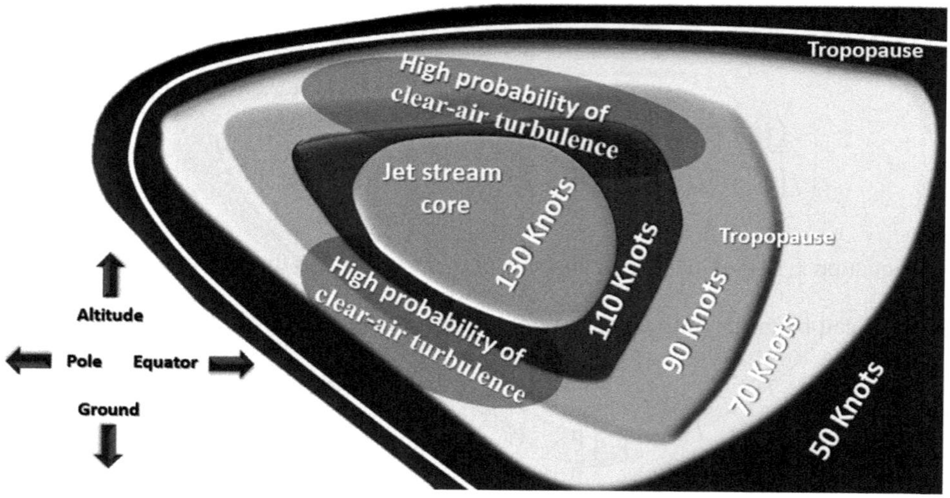

Figure 1.24: Clear air turbulence crosses the jet stream.

At the bottom of the troposphere, numerous meteorological features occur, for instance, precipitation, winds, storms, and clouds still, few of these features are observed in the stratosphere and mesosphere layers. In this regard, why the atmosphere of the troposphere is turbulent? The atmosphere of the troposphere is turbulent because of the convection currents ascending from variance in light air mass density and heavy air mass density between the equator and the poles.

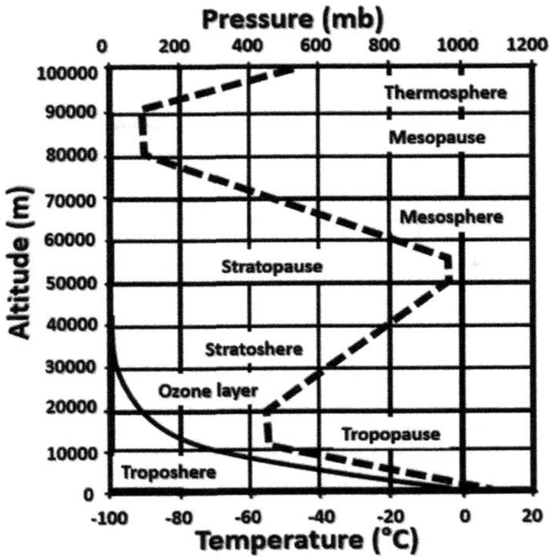

Figure 1.25: Atmospheric structural layers.

Now, the critical query is: why is there less turbulence in the stratosphere? The keystone of this answer is low temperature with light air density that allow aeroplanes to hover faster while sustaining the basic lift. Therefore, tropopause and the stratosphere are characterized by less turbulence because of the steady temperature rates. In this regard, the bottom of the stratosphere is about 10 km (6.2 miles or about 33,000 feet) above the ground at middle latitudes. The top of the stratosphere occurs at an altitude of 50 km (31 miles). The height of the bottom of the stratosphere varies with latitude and with the seasons. The lower boundary of the stratosphere can be as high as 20 km (12 miles or 65,000 feet) near the equator and as low as 7 km (4 miles or 23,000 feet) at the poles in winter. The lower boundary of the stratosphere is called the tropopause; the upper boundary is called the stratopause (Figure 1.25).

It is well understood that imbalanced heating of the surface and the atmosphere causes turbulent mixing and overturning in the troposphere layer. In other words, the adiabatic lapse rate increases with height because of the vertical reduction of the temperature, turbulence is created. In this understanding, the lower layer of warmer air rises due to low air mass density and mixes with cooler air aloft causing convective overturn flow [7, 15, 17].

It is needless to mention that strong vertical wind shear, atmospheric gravity internal wave events, propagation and breaking of the mountain and/or lee waves, strongly anticyclonic flows, and strong mesoscale and synoptic-scale convection are keystones that caused clear air turbulence. Yet this chapter does not answer: how does gravity wave cause turbulence? Gravity waves usually occur in the tropopause because of the fluctuation of warm and cool air masses. In other words, tropopause separates two different sorts of air masses. The top of the tropopause is characterized by warmer air masses while the bottom is dominated by cooler air masses. In this circumstance, the wind speed fluctuates faster from colder air with height and then decreases in air warmer mass above the tropopause. Consequently, the differences in air temperature and wind speed below tropopause and above it create such phenomena known as gravity waves (Figure 1.26) [3, 5, 13].

Countless kinds of waves and tides also dominated the stratosphere within the atmosphere. Several waves and tides carry energy from the troposphere rising into the stratosphere. Continuously, others convey energy from the stratosphere upward into the mesosphere. In this view, the waves and tides influence the wind flows within the stratosphere and may conjointly instigate regional heating of this layer of the atmosphere. Subsequently, a rare sort of electrical discharge, somewhat corresponding to lightning, ensues within the stratosphere [7, 14, 17]. These "blue jets" are created on top of thunderstorms and extend from the rock bottom of the stratosphere up to altitudes of 40 or 50 km (25 to 31 miles) (Figure 1.27).

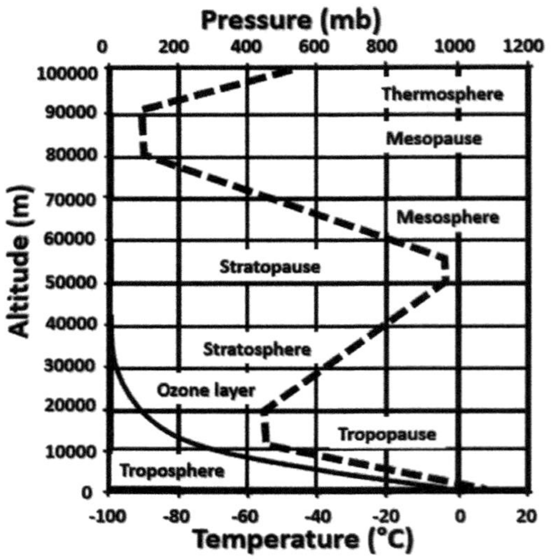

Figure 1.26: Atmospheric gravity waves.

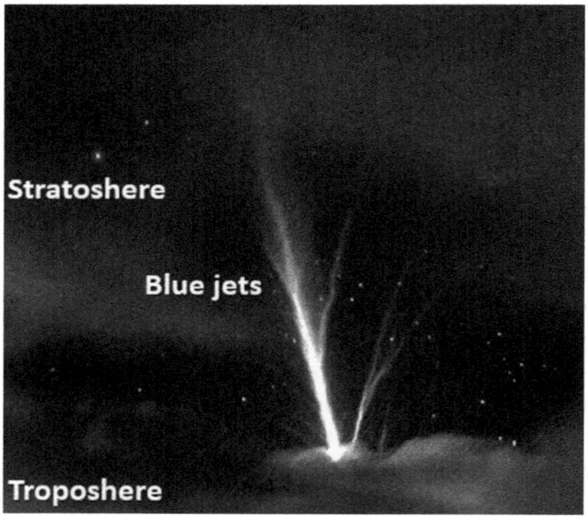

Figure 1.27: Occurrence of blue jets between troposphere to the above stratosphere.

1.10 Turbulence and Thunderstorm

Turbulence related to thunderstorms (Figure 1.28), is often extraordinarily risky, having the potential to trigger over-pitching of aeroplanes or loss of maneuver. Thunderstorms also are associated with heavy falls; heavy cloud covers and lighting (Figure 1.28). Thunderstorm vertical currents are also sturdy enough to spin an aeroplane up or down vertically as much as 2000 to 6000 feet. The greatest turbulence occurs within the locality of adjacent rising and descending currents of air. Gust loads are often severe and adequate to stall an aeroplane hovering at rough wind flow (maneuvering) velocity or to cripple it at design traveling velocity [5, 7, 14].

Extraordinary turbulence flow, therefore, typically arises close to the mid-level of the storm, between 12,000 and 20,000 feet, and is most severe in clouds of the greatest upward growth. It finds no severe turbulence within the cloud. They often expected it to be up to 20 miles from severe thunderstorms and can be greater downwind than into the wind. Severe turbulence and

Figure 1.28: Turbulence is associated with thunderstorms with heavy rain and lightning.

strong out-flowing winds can also exist below a thunderstorm. Microbursts (Figure 1.29) are often particularly risky owing to the severe wind shear related to them. A microburst is a restricted column of downdraft wind inside a thunderstorm and is frequently less than or equal to 2.5 miles in diameter. Microbursts can create extensive destruction on the ground, and in some occurrences, can be life-intimidating [5, 7, 9, 14].

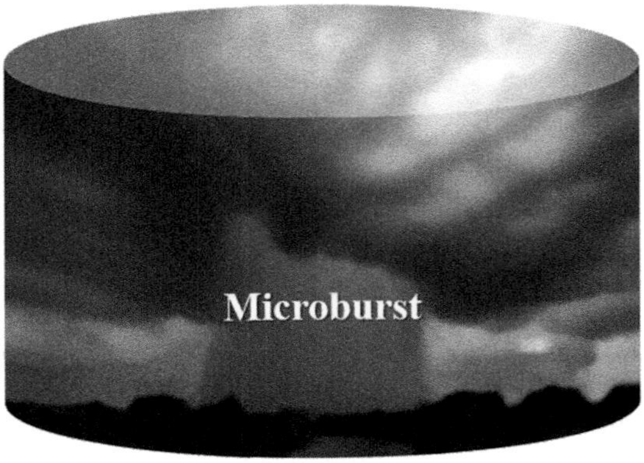

Figure 1.29: Example of microbursts.

Jet planes can securely fly over thunderstorms if their flight altitude is directly above the turbulent top clouds. Therefore, the most strong and turbulent storms are frequently the highest storms, so route flights continually request to fly around them [13–17].

1.11 Wake Turbulence

Another common turbulence feature is the wake turbulence. It is mainly generated by aeroplane lift and formatted in two counter-rotating vortices behind the aeroplane (Figure 1.30). Therefore, the question is, how can an aeroplane generate such a couple of wake turbulence as dual vortices?

Lift generates certain changeable pressure over the wing surface wherein, the highest pressure occurs under the wing, while the lowest pressure occurs at the top of the upper wing surface. As a result, pressure discrepancy causes the roll-up of the airflow behind the wing, stemming in swirling air masses sprawling downstream of the wingtips. Subsequently, the roll-up is accomplished, and the wake contains dual counter-spinning cylindrical vortices (Figure 1.31). This commences the wake vortex.

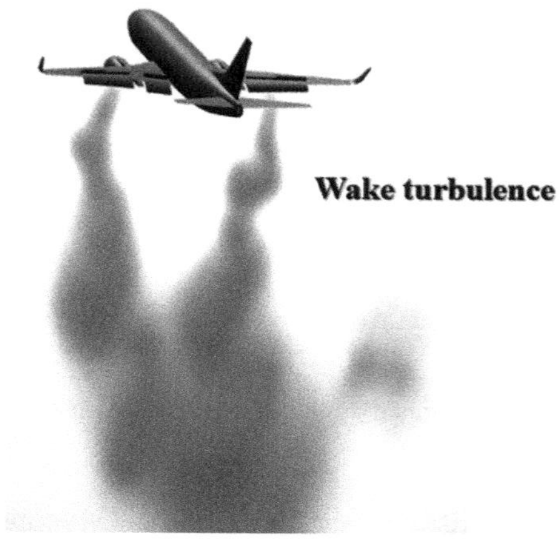

Figure 1.30: Wakes turbulence behind the aeroplane.

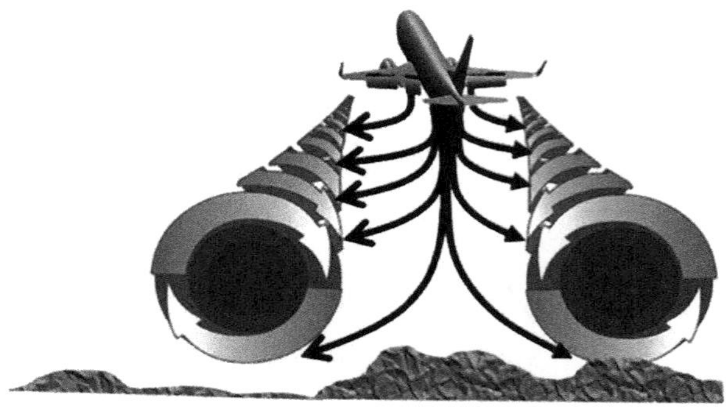

Figure 1.31: Simplification of dual wake vortex generation.

The aeroplane forms the wingtip vortices that gradually descend behind the aeroplane as it traverses. Subsequently, the vortices decay, resulting in the aeroplane touching down. The intensity of the vortex significantly increases when the generating aircraft is streamlined and aerodynamically efficient, as opposed to being bulky or poorly designed. Turbulence generated by a poorly designed aircraft accelerates the wake's dissipation. Consequently, the maximum velocity of the vortex reaches approximately 329.184 km/hour [13, 18, 20].

However, wake turbulence decay as soon as an aeroplane generates it. It is for a short duration but is extraordinarily turbulent. Wingtip vortices, on the other hand, are rather more stable and might sustain in the air for up to 3 minutes. It is thus not factual turbulence within the aerodynamic sense, as factual turbulence would be chaotic. Instead, it has the likeness of air turbulence when an aeroplane hovers through this region of disturbed air [5, 18, 20].

In this regard, as a wing generates lift, it creates wingtip vortices. The lower pressure, producing a vortex to the path from every wingtip, draws air from below the wing round the wingtip into the region on top of the wing. The strength of wingtip vortices is determined primarily by the load and wind speed of the aeroplane. Wingtip vortices, hence, form the primary and most dangerous element of wake turbulence [18–20].

Particularly, wake turbulence is risky within the region behind the aeroplane, within the take-off or landing phases of flight. Throughout take-off and landing, an aircraft operates at a high angle of attack. This flight angle maximizes the formation of robust vortices. Within the vicinity of an airport, there may be multiple planes, all functioning at gradual velocity and low altitude; this consigns an additional risk of wake turbulence with the abridged height from which to retrieve from any dilemmas [17–20].

According to the above perspective, the aeroplane is functioning perfectly at a level of over 600 m at the top of the terrain with less risk. It could be attributed to the fact that vortices sink at an altitude range from 90 to 150 meters per minute. Therefore, they stabilize approximately at altitude levels between 150 to 270 m below the aeroplane level.

1.12 Equation of Turbulence Motion

The equation of motion fully governed the turbulence characteristics and features that have been described in the previous section. To understand the dynamic of turbulence motion, let us begin by assuming that the velocity flow ($\vec{V}$) is the summation of the mean component ($\bar{\vec{v}}$) and a fluctuating component $\vec{v}'$, which is planned as:

$$\vec{V} = \bar{\vec{v}} + \vec{v}' \quad (1.14)$$

Equation 1.14 reveals that $\bar{\vec{v}}$ is perhaps a function only of space not of time average. It could be a seasonal mean over a finite period or of some mode of the set mean. However, $\vec{v}'$ deviates from that mean, which the average of the deviation average is accomplished by definition, zero; $\vec{v}' = 0$.

Consequently, if $\vec{u}, \vec{v}$, and $\vec{w}$ are the constituents of the rapid velocity vector along the rectangular coordinate axes *x*, *y* and *z*, correspondingly. In this sense, it perhaps articulated the components of the instantaneous velocity vector in standings of the deviations and time-mean as follows:

$$\vec{u} = \bar{\vec{u}} + \vec{u}', \qquad \vec{v} = \bar{\vec{v}} + \vec{v}', \qquad \vec{w} = \bar{\vec{w}} + \vec{w}' \quad (1.15)$$

We can characterize the turbulent mobility associated with the eddies using statistical notions since they are approximately random. In this perspective, the determined flow velocity is continuous, and we can calculate the mean by integrating the data. However, the measured velocity statistics are a series of discrete points, u_i. Let us assume a spectral average across the time interval t to $t + T$, where T is far longer than any turbulence time-scale but much less than the time-scale for mean-flow unsteadiness, such as atmospheric wave gravity. The scientific explanation of mean velocity, turbulent fluctuations, turbulent strength, and turbulent intensities can mathematically be written as:

$$\bar{u} = \int_{t}^{t+T} u(t)dt = N^{-1}\sum_{1}^{N} u_i \quad (1.16)$$

$$u'(t) = u(t) - \bar{u} \quad (1.17)$$

$$u_i^{'} = u_i - \bar{u} \quad (1.18)$$

$$\sqrt{u} = \sqrt{\overline{u'(t)^2}} - \sqrt{N^{-1}\sum_{i=1}^{N}(u_i^{'})^2} = 0 \quad (1.19)$$

$$\text{Turbulent intensity}: \sqrt{u}(\bar{u})^{-1} \quad (1.20)$$

The first term in equation 1.16 signifies a continuous record while the second term represents discrete, equi-spaced points. Turbulent fluctuation, therefore, can be addressed in two terms:

(i) continuous record (equation 1.17), and (ii) discrete points (equation 1.18), respectively. Equation 1.19 denotes a scenario of turbulent strength as a function of a continuous record, and equi-spaced points, respectively. In this view, equation 1.19 also demonstrates the standard deviation of the set of "random" velocity fluctuations u_i'. Subsequently, equation 1.20 behaves as a turbulent intensity index. A similar concept can determine the lateral and vertical air velocities, $v(t)$ and $w(t)$, respectively. In practical, larger $\sqrt{u}$ shows a higher spectrum of turbulence [5, 7, 13, 15, 19]. Both demos in Figure 1.32 have the same mean velocity, however, the one in Figure 1.32b has more turbulence.

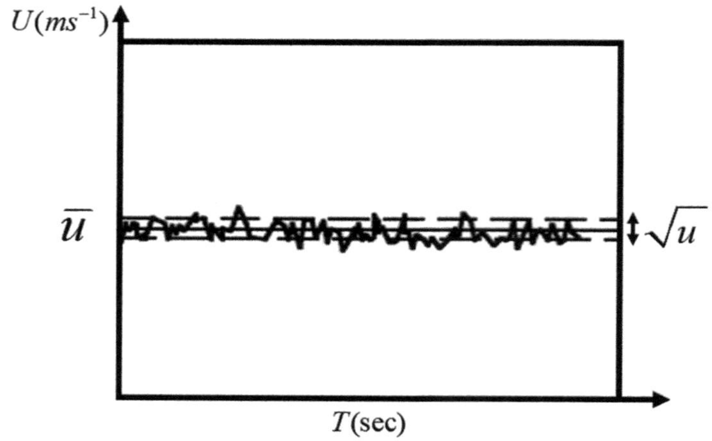

(a)

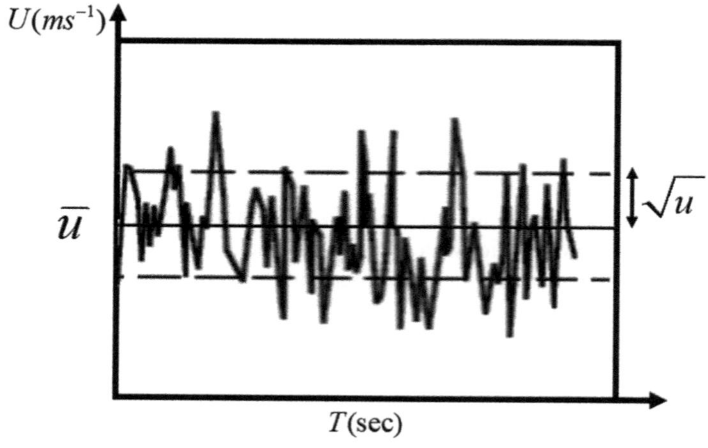

(b)

Figure 1.32: Turbulent mean velocity spectra (a) weak turbulent flow and (b) high turbulent flow.

Once airflow is turbulent, particles have higher transverse mobility, which speeds up the level of energy and momentum transmission among particles, boosting heat transfer and friction coefficient. In this view, let us assume that the air temperature and pressure have fluctuated in terms of temperature and pressure, i.e. $T(°\text{C}) = \overline{T}(°\text{C}) + T'(°\text{C})$ and $P = \overline{P} + P'$, respectively. Osborne Reynolds suggested this decomposition of a flow variable into a mean value and a turbulent fluctuation in 1895, and it is widely regarded as the beginning of the systematic mathematical analysis of turbulent flow as a sub-field of fluid dynamics, as detailed in section 1.3. The turbulent fluctuations are treated as stochastic variables, while the mean values are treated as predictable variables determined by dynamics laws. For a period, the heat flux and momentum transfer in the corresponding point to the flow (expressed by the shear stress τ) are computed by:

$$q = v_y' \rho c_p T'(°C) = -k_{turb} \frac{\partial \overline{T}(°C)}{\partial y}; \tag{1.21}$$

$$q = -\rho v_y' v_x' = -\mu_{turb} \frac{\partial \overline{v_x}}{\partial y}; \tag{1.22}$$

Both equations 1.21, and 1.22 demonstrate heat flux (q) which c_p is the heat capacity at constant pressure, ρ is the density of the airflow. Therefore, $-\mu_{turb}$ shows the coefficient of turbulent viscosity while $-k_{turb}$ is the turbulent thermal conductivity. Air's thermal conductivity is a measure of its ability to conduct heat, which occurs at a slower rate in low thermal conductivity air particles than in high thermal conductivity ones [2, 7, 13].

1.13 Turbulent Transport in the Mass Conservation Equation

Mass concentration fluctuations occur when turbulence occurs. We can group the thermal mass concentration into a temporal mean and chaotic fluctuations around that mean as shown in equation 1.15. The turbulence of mass concentration can be expressed as:

$$M_c(t) = \overline{M}_c + M'(t) \tag{1.23}$$

Therefore, mass conservation can be abridged in one-dimensional as:

$$\frac{\partial M_c}{\partial t} + \frac{\partial (uM_c)}{\partial x} - \frac{\partial}{\partial x} D_x \frac{\partial (M_c)}{\partial x} = 0 \tag{1.24}$$

Equation 1.24 validates the conversation of the mass concentration M_c transport under turbulence flow u in the x-direction with changeable time ∂t. Equation 1.24 can be changed using decomposition velocity as detailed in section 1.12. In this view, it can express the decomposition of velocity and concentration as:

$$\frac{\partial\left(\overline{M_c} + M_c'\right)}{\partial t} + \frac{\partial\left((\overline{u} + u')\left(\overline{M_c} + M_c'\right)\right)}{\partial x} - \frac{\partial}{\partial x} D_x \frac{\partial\left(\overline{M_c} + M_c'\right)}{\partial x} = 0 \tag{1.25}$$

here D_x is turbulent diffusivity, which is also known as a turbulent diffusion coefficient. In this regard, the D_x is defined as:

$$D_{t,x} \sim u' l_x \tag{1.26}$$

Equation 1.26 expresses the turbulent flux in the Fick law of diffusion, in which the flux is proportional to the mean concentration gradient, and is a counter gradient as:

$$\overline{u'M_C} \sim u'M_c' = -u'l_x \frac{\partial \overline{M_C}}{\partial x} \tag{1.27}$$

As a consequence, it can explicitly describe the turbulent flux as an additional diffusion term, including:

$$\overline{u'M_C'} = -D_{t,x} \frac{\partial \overline{M_C}}{\partial x}. \tag{1.28}$$

Simply, equation 1.28 exhibits that the turbulent flux is reliant on the turbulence strength (u') and the magnitude of the turbulence l_x. The friction velocity, as discussed previously, regulates the strength of turbulence, i.e., $u' \sim u_*$. In a turbulent flow, many length scales of turbulence occur, thus we must choose the length scale that is most consequential to the turbulent flux to explore equation 1.26. Because the formula of 1.26 states that effective diffusivity rises with flux scale, this will be the longest length scale in the system [3, 5, 21]. Thus, the dominating length-scale of turbulent transport would be determined by the domain's geometric limitations, which determine the domain's largest turbulent scale. Replacing the turbulent correlation in equation 1.25 with a steady correlation in equation 1.28. In this regard, it can describe the effect of turbulence on the transport equation by:

$$\frac{\partial \overline{M_C}}{\partial t} + \frac{\partial\left(u\overline{M_C}\right)}{\partial x} - \frac{\partial}{\partial x}\left(D_{t,x} + D_x\right)\frac{\partial \overline{M_C}}{\partial x} \tag{1.29}$$

As a result, equation 1.29 demonstrated that the influence of turbulence on the transport equation is perhaps simply approximated by increasing the coefficient of diffusion by a factor determined by the turbulence's strength and intensity. Turbulent diffusivity, $D_{t,x}$ is significantly higher than its molecular analogue, to where the latter can be neglected. The solutions developed for the transport equation can be implemented in turbulent flow, however, the molecular diffusivity has been substituted with its turbulent relative [7, 22]. Last of all, the identical argument can efficiently verify that the turbulent diffusivity in the vertical and lateral dimensions will scale as,

$$D_{t,y} \sim v'l_y \tag{1.30}$$

$$D_{t,z} \sim w'l_z \tag{1.31}$$

Needless to say, the turbulent diffusivity must be anisotropic $D_{t,x} \neq D_{t,y} \neq D_{t,z}$ since, in both length scale $l_x \neq l_y \neq l_z$ and intensity $u' \neq v' \neq w'$, turbulence is regularly anisotropic [21–24].

1.14 Richardson's Four-Thirds Power Law

Consistent with the above perspective, a turbulent diffusion coefficient is commonly used to characterize turbulent transport. In a phenomenological sense, this turbulent diffusion coefficient resembles molecular diffusivities, but it has no physical meaning because it is a consequence of flow patterns rather than a fluid property. The turbulent diffusivity idea presupposes a constitutive relationship between a turbulent flow and the gradient of a mean variable, akin to the flux-gradient link seen in molecular transport. This premise is essentially an approximation of the ideal scenario. Turbulent diffusivity remains the most straightforward method for analyzing turbulent flows qualitatively, and it has proposed various models to compute it. This coefficient, for instance, can be determined using Richardson's four-thirds power law and is governed by the random walk principle in vast atmospheric space or huge water bodies such as the ocean [5, 7, 13, 25].

In his pioneering work, Richardson introduced the concept of turbulent relative dispersion to explore the enormous changes in atmospheric turbulent diffusion observed at different spatial scales. Richardson, therefore, provided a diffusion equation for the probability density function of pair separation, $p(r,t)$. Such an equation can convert into the form if isotropy and assumed as:

$$\frac{\partial p(r,t)}{\partial t} = r^{-2}\frac{\partial}{\partial r}\left[r^2 D(r)\frac{\partial p(r,t)}{\partial r}\right] \tag{1.32}$$

Equation 1.32 accounts the large increase in measured values of turbulent diffusivity in the atmosphere for by the scale-dependent eddy diffusivity *D(r)*. In 1926, Richardson derived the renowned scaling law $D(r,t) \alpha r^{4/3}$ from experimental data. The well-known non-Gaussian distribution is derived from the statement of $D(r)$ as a function of r and the use of equation 1.32.

$$p(r,t) \alpha t^{-9/2} e^{\left(\frac{-Cr^{2/3}}{t}\right)} \tag{1.33}$$

Therefore, the mean square particle spacing grows because of equation 1.33, which is expressed mathematically by:

$$R(t) \equiv \langle r^2(t) \rangle = C_2 \varepsilon t^3 \tag{1.34}$$

For the pair dispersion, this is the well-known Richardson's t^3 law. The Richardson constant is C_2, and the mean energy dissipation is ε. Even though Richardson's law has been proposed for a long time, its value C_2 remains highly unknown. In the kinematic simulation of many models, C_2 values could range from approximately 10^{-2} to 10^{-1}, although an energy flux ε for kinematic models is difficult to quantify. Closure prediction, conversely, yields a value $C_2 \sim O(1)$ or even greater [25–27].

It is noteworthy to mention that the naïve technique, which should lead to the identification C_2 by examining the evolution of $R^2(t)$ relative to time t, is sensitive to specific pair physical separation, resulting in Richardson's constant predictions that are extremely suspicious. The fact that the data somehow does not fulfill the Richardson rule for generic beginning pattern, physical separation could be accounted for mainly because of the system's finite size effects (in space and time). Hence, the average will be skewed unless t is large enough that all particle pairs have "lost" their initial conditions. This explains why $R^2(t)$ flattens out over time at tiny intervals. The Richardson regime, therefore, is a transition from the starting circumstances. As the initial gap grows, the crossover's range expands. Unfortunately, because of the imperfect extension of the inertial range, we are unable to significantly increase the time t. Perhaps there could be a solution to overcome this obstacle by utilizing statistical techniques known as Lyapunov Exponents, which are based on statistics at a fixed scale. In this view, Finite-Scale the Lyapunov Exponent, or FSLE for brief, is an approach that relies on exit-time statistics at a fixed scale of trajectory separation that was previously described in the context of chaotic dynamical systems theory. Consequently, the matching mean square relative separation is anticipated to obey equation 1.34. In this understanding, a mathematical formula that recounts the FSLE to the Richardson's constant can be expressed as:

$$C_2 = 1.75 c^3 \varepsilon^{-1} \left(\frac{\rho^{2/3} - 1}{\rho^{2/3} \ln \rho} \right)^3 \tag{1.35}$$

here c^3 is a constant depending on the details of the numerical experiment. Because of finite-resolution limitations, the factor ρ cannot be arbitrarily close to 1; yet it must be minimally impacted enough to avoid contamination effects between different scales of motion. Therefore, equation 1.35 presents Richardson's constant C_2 to be approximately equal to 0.5 with a mainly estimated error of ±0.2 [7, 25, 27].

To end, it would also be fascinating to examine if the Richardson law governs the behavior of pair physical separation in buoyancy-dominated boundary layers. The influence of buoyancy in this scenario could change the formulation for the eddy diffusivity field, $D(r)$, resulting in an essentially new regime that has yet to be investigated.

Richardson's conception of turbulence was that a turbulent flow comprises different-sized "eddies." The sizes establish the eddies' distinctive length scale, which is also specified by flow velocity scales and time scales (yield time) that are reliant on the length scale. It divided the kinetic energy of the initial large eddy among the smaller eddies that arose from it, as the large eddies are unstable and eventually break up, resulting in the formation of smaller eddies. These smaller eddies examine the same process, producing smaller eddies that inherit the energy of the previous eddy, and so on. In this system, it transfers energy from huge scales of motion to smaller scales until it achieves a short enough length scale to be useful [2, 7, 25, 27].

1.15 Kolmogorov Speculation

Consistent with the previous discussion about turbulence, turbulent motions crop up on a variety of scales. From the macroscale, where energy is transported, to the microscale, where energy is dissipated by viscosity. Turbulent movements, according to Kolmogorov (Andrey Nikolaevich Kolmogorov, a Russian mathematician who made fundamental contributions to the mathematics of probability theory and turbulence), span a wide variety of scales. Consistent with the macroscale, where energy is generated, to the microscale, where viscosity degrades energy. For instance, consider a cumulus cloud (Figure 1.33). The cloud's macroscale can be measured in kilometers, and it can expand or endure for long periods. Within the cloud, eddies can form on millimeter scales. Microscales may be significantly smaller for smaller flows, such as those in pipes. The macroscale structures contain most of the kinetic energy of the turbulent flow. An inertial mechanism "cascades" energy from macroscale structures to microscale structures. The turbulent energy cascade is the name for this process [5, 7, 28].

Figure 1.33: Cumulus cloud assumed by Kolmogorov concept.

Kolmogorov hypothesized that small-scale turbulent movements are statistically isotropic for higher shear rates of Reynolds numbers, as shown in sections 1.5 and 1.6. In other words, there is no privileged spatial path that could be discriminated against the flow turbulent pattern exploiting Reynolds numbers. In this view, the vast scales of a flow are not isotropic because of the geometric parameters of the boundaries (L influences them). Kolmogorov proposed that in Richardson's energy cascade, this topological and orientation data is missing when the scale is shrunk, resulting in small-scale statistics that have a general strength: they are the equivalent of entirely turbulent motion when the Reynolds number is sufficiently high. In this understanding, we know the tiniest lengths in a turbulent flow as Kolmogorov microscales [5, 29]. These are tiny enough that molecular diffusion gains over viscous dissipation disperse energy, and it converts turbulent kinetic energy to heat. In this regard, the Kolmogorov microscales, or the smallest scales in turbulent flow, are:

$$\eta = \left(\frac{\nu^3}{\varepsilon} \right)^{1/4} \tag{1.36}$$

$$\tau_\eta = \left(\frac{\nu}{\varepsilon} \right)^{1/2} \tag{1.37}$$

$$u_\eta = (\nu\varepsilon)^{1/4} \tag{1.38}$$

The equations from 1.36 to 1.38 represent the Kolmogorov length scale, Kolmogorov timescale, and Kolmogorov velocity scale, respectively. The average rate of dissipation of turbulence kinetic energy (TKE) per unit mass is represented by $\varepsilon(m^2 s^{-3})$. Practically, determined root-mean-square (rms) velocity fluctuations $\langle \vec{u}' \rangle_{rms} = \frac{1}{\sqrt{3}}\sqrt{\left[u_1^{'} + u_2^{'} + u_3^{'}\right]^2}$ characterize the TKE rate of dissipation, which can be physically quantified or estimated using a turbulence model. Therefore, η, τ_η, and u_η describe the tiniest eddies in the flow, i.e., the rating scale for which it dispersed energy. Consequently, the Taylor dissipation law estimates the TKE dissipation rate as $\varepsilon \approx c_\varepsilon U^3 [L]^{-1}$, where L is the integral length scale, which is an archetypal length scale of the turbulent flow, U_L is the characteristic velocity of the greatest eddy and c_s is constantly under certain scenarios. In this context, a length scale used to characterize a turbulent fluid flow is the Taylor microscale, also known as the turbulence length scale. The Taylor microscale is a length scale wherein dynamic flow has a substantial influence on the dynamics of vortices in a flow. It commonly exploited this length scale to describe turbulent flow that has a Kolmogorov spectrum of velocity fluctuations. The viscosity does not affect length scales larger than the Taylor microscale in such a flow. The inertial range is the term used to describe the flow's longer length scales [2, 7, 29]. Turbulent motions below the Taylor microscale are subject to high viscous weights, and kinetic energy vanishes into thermal fluctuations.

The Taylor microscale is difficult to calculate since it requires the development of a particular flow correlation function, which would then expand in a Taylor series and the first non-zero term implemented to designate an osculating parabola. The Taylor microscale is proportional to $R_e^{-0.5}$, and the Kolmogorov microscale is proportional to $R_e^{-3/4}$, where R is the integral scale Reynolds number. The Reynolds number of turbulence determined using the Taylor microscale λ is casted as:

$$\begin{aligned} R_e(\lambda) &= \frac{1}{\sqrt{3}}\sqrt{\left[u_1^{'} + u_2^{'} + u_3^{'}\right]^2}\,\nu^{-1}\lambda \approx \frac{1}{\sqrt{3}}\sqrt{\left[u_1^{'} + u_2^{'} + u_3^{'}\right]^2}\,\nu^{-1}\sqrt{10\nu k\varepsilon^{-1}} \\ &= \nu^{-1}\sqrt{10\nu k\varepsilon^{-1}}\,\langle \vec{u}' \rangle_{rms} \end{aligned} \tag{1.39}$$

For the variable strain rate field, the Taylor microscale poses an excellent approximation of λ:

$$\lambda = \sqrt{\langle \vec{u} \rangle_{rms}^2}\left[\sqrt{\left(\frac{\partial \langle \vec{u} \rangle_{rms}}{\partial x}\right)^2}\right]^{-1} \tag{1.39.1}$$

Let us assume $L_0, \ell,$ and ℓ_0 that the initial (outer) (Figure 1.34), current, and inner turbulent eddy diameters, respectively, satisfy the constraint $L_0 > \ell > \ell_0$. In this view, Kolmogorov devised the renowned formula for the energy spectrum of isotropic and homogeneous turbulence using dimensional analysis and scaling constraints, which is mathematically described by:

$$E(k) = c_k \varepsilon^{2/3} k^{-5/3} \tag{1.40}$$

Equation 1.40 reveals that the inertial range $\ell_0^{-1} \le k \le L_0^{-1}$ contains the spectrum $E(k)$. Kolmogorov postulated that turbulent motions in the inertial subrange, where $L_0 > \ell > \ell_0$, are indeed homogeneous and isotropic. In this understanding, it can move energy from eddy to eddy without dissipation. In other words, the total energy transfused into the greatest system must be exactly equivalent to the quantity of energy dissipated as warm air. At vast scales, the atmosphere does not appear to have an inverse –5/3 spectrum, and there is no well-defined inverse energy cascade in the sense stated above. Although there is a –3 cascade in the atmosphere, it is unclear

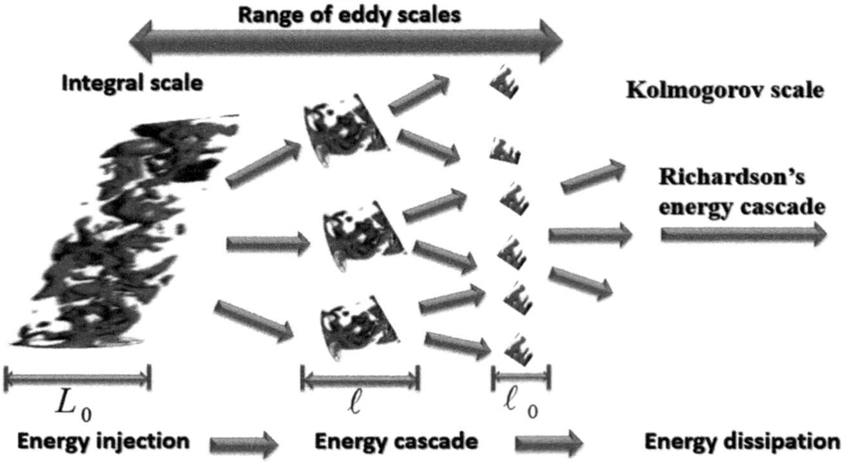

Figure 1.34: Richardson–Kolmogorov energy cascade.

if it causes this to a conventional forward enstrophy cascade [3, 5, 7, 30]. The enstrophy cascade rate, which is equal to the rate at which enstrophy is generated by stirring, is constant in the enstrophy inertial range. It seems to be hypothesized that the enstrophy rate is determined by:

$$E(k) = k^{-3}\eta^{2/3}k_\eta \tag{1.41}$$

Equation 1.41 exposes that energy provided at a certain rate is conveyed to massive scales, while it rerouted enstrophy to small scales, where it could be dispersed through viscosity. Therefore, k_η is a universal constant, corresponding to the Kolmogorov constant.

In this understanding, the viscosity of the airflow determines the size of the diminished turbulent flow. As viscosity falls, the Kolmogorov length scale diminishes. In this circumstance, the viscous intensities are smaller than the inertial forces in high Reynolds number flows. Smaller-scale motions, therefore, must be generated until viscosity effects become significant and energy is dissipated. In a turbulent flow, the ratio of greatest to the tiniest length scales is proportionate to the Reynolds number [3, 7, 27, 30]. The correlation between the Kolmogorov length scale and Reynolds number grows with the three-quarters power:

$$L\eta^{-1} \sim \frac{UL}{\nu} = R_e^{3/4} \tag{1.42}$$

Direct numerical simulations of turbulent flow are extremely difficult because of equation 1.40. For instance, consider a flow with a Reynolds number of 10^6. In this circumstance, the ratio of L/l must be proportional to $10^{18/4}$, which seems complicated to handle by computer simulation. In this scenario, it must generate a grid with at least 1014 grid points since a three-dimensional problem is considered. However, this is considerably beyond the capacity and capabilities of current computers.

Because the key kinematic variable is the average power dissipation rate, the Kolmogorov theory is a mean-field theory. Since the energy dissipation rate in fluid turbulence varies in space and time, we can conceive the microscales as spatially and temporally variable quantities. Mean-field values are commonly used since they describe the approximate value of the precise concentration in a specific flow.

1.16 Navier-Stokes Equations

The Navier-Stokes Equations in physics describe the motion of viscous fluid substances and are designated after Claude-Louis Navier and George Gabriel Stokes. Adapting Newton's second law to fluid movement, including the hypothesis that the stress in the fluid is the total of a diffusing viscous term (roughly equivalent to the differential of velocity) and a pressure variable explaining the flow of a viscous rise to these equilibrium equations. The key distinction between them and the conventional Euler formulas for viscid flow is that Navier-Stokes equations are not conservation equations in the Froude (without extrinsic force), but a vicious dissipation state, in the view that they are being situated into the quasi-linear homogeneous sense:

$$\vec{y}_t + \vec{A}(\vec{y})\vec{y}_x = 0 \tag{1.43}$$

The Navier-Stokes equations are a set of nonlinear partial differential equations for abstract vector fields of any dimension that are used in mathematics. They are a mathematical expression in physics and engineering using continuity dynamics to explain the movement of liquids or quasi-gases (where the phase velocity is short enough to be conceived as a continuity mean rather than a particle moving). The equations are a description of Newton's second law, with the forces approximated, such as in a viscous Newtonian fluid—as the total of forces from pressure, viscous stress, and an external body force in a viscous Newtonian fluid. Because the Clay Mathematics Institute sets issues in three dimensions, only one solution is possible for an incompressible and homogeneous fluid.

Turbulence is the chaotic behavior exhibited in many time-dependent fluid flows. We commonly assumed that it is related to the fluid's overall inertia: the culmination of time-dependent and convective acceleration; thus, flows with minor inertial effects are laminar (the Reynolds number quantifies how much inertia affects the flow). The Navier-Stokes equations are thought to fully explain turbulence, albeit this is not proven [2, 5, 31].

The exact simulation of the Navier-Stokes equations, therefore, for turbulent flow is incredibly challenging, and because of various mixing-length scales included in the turbulent movement, the steady solution requires such fine mesh resolution that the computation complexity for determination or direct numerical simulation would become markedly prohibitive. Using a laminar solver to address turbulent flow consequences in a time-unsteady suspension that strives to resolve successfully.

According to the mean flow, it addressed turbulent flow in section 1.12 specifically using equation 1.15, by replacing $V = \bar{V} + V'$ the Navier-Stokes equations and averaging, the mean-flow equations for turbulent flow are determined. The Reynolds-averaged Navier-Stokes (RANS) equations result from this procedure:

$$\nabla \cdot \vec{V} = 0 \quad \text{i.e. } \nabla \cdot \bar{V} = 0 \text{ and } \nabla \cdot V' = 0 \tag{1.44}$$

Equation 1.44 reveals the continuity of movement that addresses the mean-flow. Therefore, the momentum flow pattern can be mathematically expressed as:

$$\rho \frac{D\vec{V}}{Dt} + \rho \frac{\partial}{\partial x_j}\left(\overline{u_i' u_j'}\right) = -\rho g \hat{k} - \nabla \bar{p} + \mu \nabla^2 \bar{V}$$

or

$$\rho \frac{D\vec{V}}{Dt} = -\rho g \hat{k} - \nabla \bar{p} + \nabla \cdot \tau_{ij} \tag{1.45}$$

where

$$\tau_{ij} = \mu \left[\frac{\partial u_i}{\partial x_j} + \frac{\partial u_j}{\partial x_i} \right] - \rho \overline{u_i' u_j'} \tag{1.45.1}$$

Equation 1.45 reveals turbulence flow owing to the existence of Reynolds stress terms $\overline{u_i'u_j'}$ where the differential operators ∇ and ∇^2 are the gradient and Laplace operators, respectively. In this circumstance, the most imperative inspiration of turbulence on the mean-flow is a growth in the fluid stress owing to what it termed as deceptive stresses. The equations (1.44) and (1.45) fulfill mass, momentum, and energy conservation. These formulas are Newton's laws of motion in a continuous form. Turbulence has the same effect as viscosity because it transports momentum from one site to another. The problem is uncertain because is unknown $\overline{u_i'u_j'}$: Closure is the major hurdle in turbulent flow analysis! The time dependence of $\overline{u_i'u_j'}$ commonly comprises averages of a series of nonlinear derivative terms. The universal form of $\overline{u_i'u_j'}$ is:

$$\frac{\partial\overline{u_i'u_j'}}{\partial t}+\bar{u}_k\frac{\partial\overline{u_i'u_j'}}{\partial x_k}=\overline{u_i'u_k'}\frac{\partial\overline{u_i'}}{\partial x_k}+\overline{\frac{p'}{\rho}\left(\frac{\partial u_i'}{\partial x_j}+\frac{\partial u_j'}{\partial x_i}\right)}-$$
$$\frac{\partial}{\partial x_k}\left(\overline{u_i'u_j'u'_k}+\frac{\overline{p'u_i'}}{\rho}\partial_{jk}+\frac{\overline{p'u_j'}}{\rho}\partial_{ik}-\nu\frac{\partial\overline{u_i'u_j'}}{\partial x_k}\right)-$$
$$2\nu\overline{\frac{\partial u_i'}{\partial x_k}\frac{\partial u_j'}{\partial x_k}} \tag{1.46}$$

Since equation 1.46 has the potential to be highly sophisticated, it also permits some flexibility in application, allowing particular terms in the strain and/or stress tensors to be neglected. In this scenario, an additional turbulence model is required to describe these nonlinear terms in the stress tensor. Therefore, most of the turbulence models utilized in the RANS equations are developed from actual data rather than fundamental theories. Boussinesq proposed the notion of eddy viscosity, which should minimize the Reynolds stress tensor to a simpler form, to solve the RANS equations immediately after their invention. Theoretically, eddy viscosity signifies turbulent energy transmission in a turbulent flow through moving convection. This links the Reynolds stress term to the turbulent kinetic energy, which is described below:

$$k=-\rho\mu\,\frac{\partial\bar{u}_i}{\partial x} \tag{1.47}$$

$$-\overline{v_i'v_j'}=\nu_t\left(\frac{\partial\overline{v_i}}{\partial x_j}+\frac{\partial\overline{v_j}}{\partial x_i}\right)-\frac{2}{3}\left(0.5\overline{v_i'v_j'}\right)\partial_{ij} \tag{1.48}$$

Equation 1.48 exposes that the flow velocity squared term is the proportionality constant for the Kronecker delta term: $k=\left(0.5\overline{v_i'v_j'}\right)$. Therefore, the turbulent eddy viscosity ν_t can also be described as kinematic viscosity since it is the constant for the flow gradient terms, which might be a simple constant for viscous turbulent flow. These variables determine the turbulent kinetic energy and anisotropic turbulence when multiplied by the density in the Reynolds stress equation. This procedure may describe mixing layers, jets, turbulent boundary layers, channel flows, and many other systems with the free shear flow [31–34].

Equations (1.44) and (1.45) are deterministic in their classical version, and they can be susceptible to boundary conditions in equation 1.49, which is associated with the so-entitled initial and boundary value problem.

$$\vec{u}(0,\vec{x})=\vec{u}_0\left(\vec{x}\right),\vec{u}\left(\vec{x},t\right)\big|_{\partial G}=0,t>0,\vec{x}\in G \tag{1.49}$$

This view illustrates that G is the relevant domain of flow. The Navier-Stokes equations are invariant under a potential movement in time and space, and so-termed "scaling invariant spaces" define a series of natural Navier-Stokes equations that monotonically mimic the behavioral patterns of a diverse range of turbulent occurrences. A simple assumption would determine the associations between normal/shear stresses and the rate of deformation (velocity field variation) in a viscous flow. The stresses are proportional to the rate of deformation (Newtonian fluid). The dynamic viscosity of the fluid is the proportional constant for the relationship ν. The renowned Navier-Stokes equations are descended as consequences of:

$$\rho\left(\frac{\partial u}{\partial t}+u\frac{\partial u}{\partial x}+v\frac{\partial u}{\partial y}+w\frac{\partial u}{\partial z}\right)=-\frac{\partial P}{\partial x}+\rho g_x+\nu\left(\frac{\partial^2 u}{\partial x^2}+\frac{\partial^2 u}{\partial y^2}+\frac{\partial^2 u}{\partial z^2}\right) \quad (1.50)$$

$$\rho\left(\frac{\partial v}{\partial t}+u\frac{\partial v}{\partial x}+v\frac{\partial v}{\partial y}+w\frac{\partial v}{\partial z}\right)=-\frac{\partial P}{\partial y}+\rho g_y+\nu\left(\frac{\partial^2 v}{\partial x^2}+\frac{\partial^2 v}{\partial y^2}+\frac{\partial^2 v}{\partial z^2}\right) \quad (1.51)$$

$$\rho\left(\frac{\partial w}{\partial t}+u\frac{\partial w}{\partial x}+v\frac{\partial w}{\partial y}+w\frac{\partial w}{\partial z}\right)=-\frac{\partial P}{\partial z}+\rho g_z+\nu\left(\frac{\partial^2 w}{\partial x^2}+\frac{\partial^2 w}{\partial y^2}+\frac{\partial^2 w}{\partial z^2}\right) \quad (1.52)$$

The Navier-Stokes equations (1.50 to 1.52) represent the component of turbulent shear flow u, v, and w, respectively in Cartesian three-dimensional (3-D) through x, y, and z. 3-D Navier-Stokes equations show that the potential gradient can be exploited to quantify the mass forces, for instance, gravity acceleration g which is defined as:

$$F_G=-\nabla\Pi\cdot\rho V=-\nabla(gz)\cdot\rho V \quad (1.53)$$

Here V is the volume, F_G which is the potential gradient as a function of the gravitational potential $\Pi = gz$. Surface forces act on contact surfaces and can be decomposed on normal and shear components [3, 7, 31, 34]. The momentum description of RANS in three-dimensional 3-D can be mathematically expressed as:

$$\rho\left(\frac{\partial u_x}{\partial \tau}+u_x\frac{\partial u_x}{\partial x}+v_y\frac{\partial u_x}{\partial y}+w_z\frac{\partial u_x}{\partial z}\right)=-\frac{\partial P}{\partial x}+\frac{\partial}{\partial x}\left[(\nu+\nu_t)\frac{\partial u_y}{\partial x}\right]+\frac{\partial}{\partial y}\left[(\nu+\nu_t)\frac{\partial v_y}{\partial y}\right]+ \frac{\partial}{\partial z}\left[(\nu+\nu_t)\frac{\partial v_y}{\partial z}\right]+\overline{S_x} \quad (1.54)$$

$$\rho\left(\frac{\partial v_y}{\partial \tau}+u_x\frac{\partial v_y}{\partial x}+v_y\frac{\partial v_y}{\partial y}+w_z\frac{\partial v_y}{\partial z}\right)=-\frac{\partial P}{\partial y}+\frac{\partial}{\partial x}\left[(\nu+\nu_t)\frac{\partial V_y}{\partial x}\right]+\frac{\partial}{\partial y}\left[(\nu+\nu_t)\frac{\partial v_y}{\partial y}\right] +\frac{\partial}{\partial z}\left[(\nu+\nu_t)\frac{\partial v_y}{\partial z}\right]+\overline{S_y} \quad (1.55)$$

$$\rho\left(\frac{\partial w_z}{\partial \tau}+u_x\frac{\partial w_z}{\partial x}+v_y\frac{\partial w_z}{\partial y}+w_z\frac{\partial w_z}{\partial z}\right)=-\frac{\partial P}{\partial z}+\frac{\partial}{\partial x}\left[(\nu+\nu_t)\frac{\partial w_z}{\partial x}\right]+\frac{\partial}{\partial y}\left[(\nu+\nu_t)\frac{\partial w_z}{\partial y}\right] +\frac{\partial}{\partial z}[(\nu+\nu_t)\frac{\partial w_z}{\partial z}]+\overline{S_z} \quad (1.56)$$

In equations, 1.54 to 1.56, ν and $\nu_t = 0.03874\ \rho\nu d$ are laminar and turbulent viscosity, respectively while ρ, ν, and d are air density, air velocity and distance to the closet surface. Furthermore, $\overline{S}$ is the mean of the strain rate tensor [32–34].

1.17 Spalart-Allmaras (SA) Model

As discussed in the earlier stages, excessive flow velocity, vast length scales, surface roughness, and low viscosity are all components of a transition from laminar to turbulent flow. However, what happens to the fluid once turbulence begins? In this view, Navier-Stokes equations and RANS cannot, nevertheless, be fulfilled in boundary conditions for each system or flow scenario. Therefore, the SA simulation method is a one-equation solution that consumes a shorter time to solve than the exhaustive Navier-Stokes and RANS equations [33, 35, 37]. This alternative approach also would simplify various Reynolds stress equation terms into concise formulations, as discussed hereunder:

$$\frac{D\tilde{\nu}}{Dt} = G_\nu \left\{ \frac{\partial}{\partial x_j}\left[(\mu + \rho\tilde{\nu}) \frac{\partial \tilde{\nu}}{\partial x_j} \right] + C_{b2}\rho \left(\frac{\partial \tilde{\nu}}{\partial x_j} \right)^2 \right\} - Y_\nu + S_{\tilde{\nu}} \tag{1.57}$$

This solution aims to predict the turbulent eddy viscosity, which is determined by measuring:

$$\mu_t = \rho\tilde{\nu}\left[\frac{(\tilde{\nu}/\nu)^3}{(\tilde{\nu}/\nu)^3 + C_{v1}^3} \right] \tag{1.57.1}$$

$$S_{\tilde{\nu}} = \sqrt{2\left(0.5\left(\frac{\partial u_i}{\partial x_j} - \frac{\partial u_j}{\partial x_i}\right)\right)^2} + \tilde{\nu}\left[0.41d^2\right]^{-1}\left[1 - \frac{\tilde{\nu}\nu^{-1}}{\tilde{\nu}\nu^{-1}\left[\frac{(\tilde{\nu}/\nu)^3}{(\tilde{\nu}/\nu)^3 + C_{v1}^3}\right]} \right] \tag{1.57.2}$$

The model solves for the undamped turbulent kinematic viscosity, $\tilde{\nu}$. This module of 1.57 contains the typical SA model excluding the trip term. The default values for modeling parameters are: $C_{b2} = 0.622$, and $C_{v1} = 7.1$. In these equations, the molecular kinematic viscosity is $\nu = \mu/\rho$, the fluid density is ρ, and μ is the fluid viscosity. The other parameters in the model are given in terms of the distance from the field point to the nearest wall d, and the vorticity magnitude $\sqrt{2\left(0.5\left(\frac{\partial u_i}{\partial x_j} - \frac{\partial u_j}{\partial x_i}\right)\right)^2}$.

Therefore, for enhanced wall treatment (EWT), each wall-adjacent cell centroid Y_ν should be located within the viscous sub-layer, i.e. $Y_\nu \approx 1$. Consequently, EWT can inevitably acclimatize cells sited in the log-law layer. Prior to computer grid design, determine the size of wall-adjacent cells as follows:

$$Y_\nu^+ = \frac{Y_\nu u_\tau}{\nu} \Rightarrow Y_\nu = \frac{Y_\nu^+ \nu}{u_\tau} \tag{1.57.3}$$

The circumstance physics and near-wall mesh are such that Y_ν^+ is expected to fluctuate dramatically throughout a sizeable portion of the wall section. Make the mesh coarse or fine enough to prevent inserting wall-adjacent cells in the buffer layer i.e. $5 < Y_\nu^+ < 30$.

$$U_\tau = \sqrt{\frac{\tau_w}{\rho}} = U_e\sqrt{\frac{\overline{C}_f}{2}} \tag{1.57.3.1}$$

Empirical correlations can be exploited to identify the skin friction coefficient $\overline{C}_f$ for flat plate and duct; respectively as:

$$\frac{\bar{C}_f}{2} \approx \frac{0.037}{\text{Re}_L^{1/5}} \tag{1.57.3.1.1}$$

$$\frac{\bar{C}_f}{2} \approx \frac{0.039}{\text{Re}_{D_h}^{1/4}} \tag{1.57.3.1.2}$$

If the distinctive Reynolds number is minimal or if near-wall parameters must be resolved, then EWT is required to be implemented. Consequently, the boundary conditions of the Spalart-Allmaras (SA) simulation method are:

$$\nu_{t,wall} = 0 \qquad \tilde{\nu}_{farfield} = 3\nu_\infty : to : 5\nu_\infty \tag{1.58}$$

It is worth noting that these boundary conditions on the SA turbulence field variable approximate turbulent kinematic viscosities of:

$$\nu_{t,wall} = 0 \qquad \tilde{\nu}_{t,farfield} = 0.210438\nu_\infty : to : 1.3\nu_\infty \tag{1.59}$$

It should be recognized that in the freestream, this model contains source terms (generation and degradation) that are non-zero, even when the vorticity is zero. The source terms, on the other hand, are almost small: equivalent to d^{-1}. It is imperative to note that calculating minimum distance (d) by searching along grid lines or locating the nearest wall gridpoint (or cell centre) is inaccurate and does not represent computing the real minimum distance to the nearest wall (Figure 1.35). The previous approaches can generate grid-dependent variations in findings. Erroneously computed minimum distance functions could yield false results in circumstances when the grid lines are not completely parallel to the body surface, or when the nearest body does not reside in the grid-scale domain [33–36].

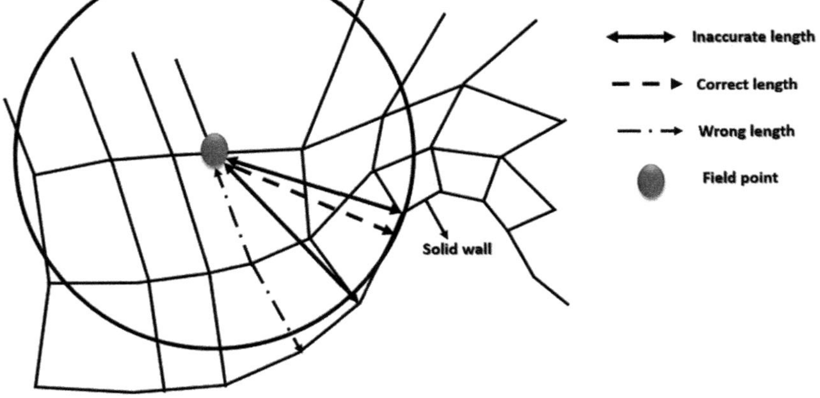

Figure 1.35: Minimum distance to the nearest wall.

To detect transitions, Spalart and Allmaras propose employing an equivalent "turbulence index" i_t at walls:

$$i_t = \left[0.41\sqrt{\nu\left(\sqrt{2\left(0.5\left(\frac{\partial u_i}{\partial x_j} - \frac{\partial u_j}{\partial x_i}\right)\right)^2}\right)}\right]^{-1} \frac{\partial \tilde{\nu}}{\partial n} \tag{1.60}$$

In the domain of normal wall-direction (n), the index would be practically zero for a laminar zone and equal to one for a turbulent region. The SA turbulence model has the following advantages:

(i) Dimensionality reduction, as discussed above, reduces issue complexity and overall simulation duration.
(ii) There is no need to calculate a length scale using the local thickness of the shear layer along with a turbulent flow.
(iii) Applicable to boundary layer flows with flexible pressure gradients.

Despite its advantages, it has numerous constraints, including the inability to capture the decay of turbulent flows and the doubtful application beyond wall-bounded systems. Nonetheless, research into modifications of Spalart-Allmaras turbulence models seems to be continuing, and their applications in large-scale systems with turbulent flow are being explored [34–37].

1.18 The k–ε Turbulence and k–ω Models

The most common model to retrieve the characteristics of the mean turbulent flow pattern is also the k–ε turbulencemodel. Consequently, this model pertains to the RANS series of turbulence models, which reveals all turbulence consequences and involves two equations. In other words, the k–ε turbulencemodel is acquainted with two supplementary transport models and two reliant variable quantities: the turbulent kinetic energy, k, and the turbulent dissipation rate, ε. The turbulent viscosity is exhibited as:

$$\mu_t = 0.09\rho\frac{k^2}{\omega} \tag{1.61}$$

The transport equation for k can be expressed as:

$$\rho\frac{\partial k}{\partial t} + \rho\vec{u}.\nabla k = \nabla.\left(\left(\mu + \frac{\mu_t}{\sigma_k}\right)\nabla k\right) + \beta_k - \rho\varepsilon \tag{1.62}$$

The production term β_k is given by:

$$\beta_k = \mu_t\left(\nabla\vec{u} : (\nabla\vec{u} + \nabla\vec{u})'\right) - 0.67\left(\nabla.\vec{u}\right)^2) - 0.67\rho k\nabla.\vec{u} \tag{1.62.1}$$

Therefore, the transport equation for ε says:

$$\rho\frac{\partial\varepsilon}{\partial t} + \rho\vec{u}.\nabla\varepsilon = \nabla.\left(\left(\mu + \frac{\mu_t}{1.3}\right)\nabla\varepsilon\right) + 1.44\frac{\varepsilon}{k}\beta_k - 1.92\rho\frac{\varepsilon^2}{k} \tag{1.63}$$

However, it executes weakly in complex flows with a high-pressure gradient, separation, and a robust streamlined curvature. In other words, often these two-equation models, including k–ε models, overestimate turbulent stresses in wake (velocity-defect) areas, resulting in a poor model performance for boundary layers with deleterious pressure gradients and detached flows. Therefore, all the k–ε models, as well as the Reynolds stress model, provide conventional and non-equilibrium wall functions as alternatives [38–40].

The k–omega (k–ω) turbulence model is a common two-equations turbulence model in computational fluid dynamics that is used as an approximation for the Reynolds-averaged Navier-Stokes equations (RANS equations). The model aims to anticipate turbulence using two partial differential equations for two variables, k, and ω, where k is the TKE and (ω) is the specific rate of dissipation (of the TKE k into internal thermal energy). In this view, the dissipation rate ω can be given by:

$$\omega \approx \frac{\varepsilon}{k} \propto \frac{1}{\tau} \tag{1.64}$$

$$\rho \frac{\partial k}{\partial t} = \tau_{ij} \frac{\partial \overline{u}_i}{\partial x_j} - \rho \beta^* k \omega + \frac{\partial}{\partial x_j}\left[\left(\mu + \frac{\mu_t}{\sigma_k}\right)\frac{\partial k}{\partial x_j}\right] \tag{1.65}$$

$$\rho \frac{\partial \omega}{\partial t} = \alpha \frac{\omega}{k} \tau_{ij} \frac{\partial \overline{u}_i}{\partial x_j} - \rho \beta \omega^2 + \frac{\partial}{\partial x_j}\left[\left(\mu + \frac{\mu_t}{\sigma_\omega}\right)\frac{\partial \omega}{\partial x_j}\right] \tag{1.66}$$

Equation 1.65 demonstrates TKE, while equation 1.66 reveals a specific dissipation rate (Wilcox, D.C. (1988)). In this regard, the closure coefficients, and auxiliary relations in equations 1.65 and 1.66 are as follows: $\beta = \frac{3}{40}, \beta^* = \frac{9}{100}, \sigma = 0.5, \sigma^* = 0.5$, and $\varepsilon = \beta^* \omega k$, respectively. Based on Wilcox [43], a two-transport-equation model is used to solve for k and ω the dissipation rate (ε/k). This is the standard k–ω model. Outstanding performance for wall-bounded and low Reynolds number flows. It demonstrates the possibility of transition prediction. Transitional, free shear and compressible flows are all taken into consideration. Consequently, k–ε models have achieved prominence following circumstances: (i) they can be integrated into the wall without the need for damping functions, and (ii) they are accurate and resilient for a wide variety of boundary layer flows with pressure gradient [38–43].

1.19 What Serves to Make the Most Challenging Navier-Stokes Equations Extremely Complex?

Mathematicians and physicists conclude that comprehending the answers to the Navier-Stokes equations can provide clarification for and anticipate both the airflow and the turbulence. Even though these formulations have been written in the nineteenth century, our comprehension of them continues to be a challenge. The goal is to make significant attempts to develop computational approaches that might reveal the insights hidden in the Navier-Stokes equations. The crucial question is: how are these equations, which interpret everyday phenomena including a river flowing across a tube, extremely complicated to grasp mathematically than Einstein's mathematical formalism, which incorporates mind-boggling monsters such as black holes? Turbulence, which we experience in everyday phenomena is the explanation. All of us have encountered it, whether piloting through turbulent air at 10 km height or observing a vortex swirl in the drainage channels. Nevertheless, an acquaintance has not engendered understanding: one of the least relevant aspects of the physical universe is turbulence [1, 3, 27, 30, 40].

Although there is a simple solution for Navier-Stokes equations, precisely understanding how a running river behaves remains a mystery. In other words, "the nature of a running river, for instance, puzzles scientists. The core equations that describe river flows and how to flow explain the unexpected in general, but bridging from the equations that instruct river flow how and where to run to any discussion of how fluids physically run is exceedingly puzzling."

Consequently, how else can mathematicians and physicists demonstrate that options exist? Begin by considering what may serve as solutions to cease to exist. The Navier-Stokes equations have been exploited to compute fluctuations in quantities such as the rate of flow as a function of velocity and pressure (equations 1.54 to 1.56). In this view, mathematicians are concerned about the following scenario: implementation of formulas, then after a constrained significant period, particle in the fluid is noticed traveling endlessly fast. That would be an issue since the change of an infinite number cannot be determined owing to computing procedures that cannot divide by zero. Mathematicians call such scenarios "effusion," since, in a freakout scenario, the Navier-Stokes equations disintegrate and thus have no direct indication [5, 37, 38, 43].

Claiming how no debacle develops and that options have always been possible is analogous to demonstrating that the velocity profile of every particle within the stream remains confined under a specific finite number. Consequently, the kinetic energy in the stream, for instance, is the most crucial of these measures.

The intricacy of the Navier-Stokes equations is, in some circumstances, a precise approximation of the complexity of the turbulent flows it is designed to explain. However, turbulence is intended to depict the fluctuations of kinetic energy from the biggest to tiniest scales. From a mathematical point of view, there is no existing solution and knowledge in the tiniest scales of eddy flow. However, mathematicians would like to be able to evaluate if the Navier-Stokes equations can be traced completely through, to observe precisely how a flow swaps moment by moment, and even trace the origin of turbulence.

Since the Navier-Stokes equation is nonlinear, it is particularly difficult to tackle. It would (and can) be considerably easier to solve if the inertial terms were not there (either owing to the geometry or if the inertial terms are meaningless).

1.20 Turbulent Diffusion

The transmission of mass, heat, or momentum across a system caused by random and chaotic time-variant motions is known as turbulent diffusion. It arises when turbulent fluid systems approach significant consequences because of shear flow induced by partial pressure gradients, density gradients, and high velocities. It proceeds significantly faster than molecular diffusion and is hence particularly critical for mixing and transport difficulties in systems dealing with combustion, pollutants, dissolved oxygen, and industrial solutions. In these domains, turbulent diffusion is a useful technique for effectively diminishing the quantities of a species in fluid or surroundings, which is required for vigorous stirring throughout processing or quick pollution or contamination elimination for security.

Nevertheless, owing to the impossibility of quantifying both an instantaneous and expected fluid velocity concurrently, it has been exceedingly difficult to construct a solid and completely functioning model that can be applied to the diffusion of a species in all turbulent systems. This occurs in turbulent flow because of various properties, for instance, unpredictability, quick diffusivity, high amounts of fluctuating vorticity, and kinetic energy dissipation.

Atmospheric dispersion, often known as diffusion, is the exploration of how pollutants mix in the atmosphere. Many elements are considered in this modeling process, including the level of the atmosphere(s) being mixed, the stability of the environment, and the kind of pollutant and source being mixed. Both the Eulerian and Lagrangian models (described below) have been used to predict air diffusion, and they are critical for understanding how pollutants react and mix in diverse settings.

The Eulerian technique of turbulent diffusion relies on an infinite volume in a definite time and place, where physical characteristics such as mass, momentum, and temperature are quantified. The model is valuable since Eulerian statistics are always quantifiable and have a variety of applications in chemical processes. Likewise,molecular models should also adhere to the similar approaches as the continuity equation (in which an element's or species' advection is compensated by its diffusion, formation through interaction, and transfer from other source materials or sites) and the Navier-Stokes equations:

$$\frac{\partial c_i}{\partial t}+\frac{\partial}{\partial x_i}\left(u_j c_i\right)=D_i\frac{\partial^2 c_i}{\partial x_j \partial x_i}+R_i(c_1,c_2,......,c_N,T)+\gamma_i \tag{1.67}$$

here c_i reveals the concentration of the species per unit volume of fluid velocity and is not considered as a mixing ratio (kg/kg) in flow pattern circumstances in direction x_j. Therefore, D_j, R_i and γ_i demonstrate the molecular diffusion constant; rate of created reaction; and rate of c_i origination by the input source; respectively. Only the advection components on the left side of equation 1.67 persist if an inert species (no reaction) is considered with no sources and expect molecular diffusion to be minimal. In this understanding, the solution to this model appears simple at first sight but the random component of the movement is excluded as well as the velocity gradient $u_j = \bar{u} + u'_j$, which is usually attributed to turbulent instability. In consequence, the Eulerian model's concentration solution would include a random component $c_j = \bar{c} + c'_j$. This delivers a closure issue with infinite elements and formulas, making it virtually impossible to model for a definite under the provided constraint of assumptions [3–8]. Fortunately, a closure approximation exists in embracing the concept of eddy diffusivity K_{jj} and relevant statistical predictions for the stochastic concentration and movement constituents of turbulent mixing:

$$\langle u'_j c' \rangle = -K_{jj} \frac{\partial (c)}{\partial x_j} \tag{1.68}$$

Swapping equation 1.68 into the first continuity equation while disregarding reactions, sources, and molecular diffusion yields the succeeding differential equation in eddy diffusion only by using the turbulent diffusion approximation:

$$\frac{\partial c_i}{\partial t} + \bar{u}_j \frac{\partial (c)}{\partial x_j} - \frac{\partial}{\partial x_j}\left(K_{jj} \frac{\partial (c)}{\partial x_j} \right) = 0 \tag{1.69}$$

Therefore, equation 1.69 reveals that the eddy diffusivity, unlike the molecular diffusion constant *D*, is a matrix description that might fluctuate in space and thus cannot be evaluated beyond the exterior derivation [3, 5, 13, 15].

Let us assume that the particle sits at a location $X'(x_1, x_2, x_3)$ at time t_0 in which the particle's mobility at a specific time t is demonstrated $\psi(x_1, x_2, x_3)\, dx_1 dx_2\, dx_3 = \psi(x,t)dx$. In this sense, the Lagrangian model of turbulent diffusion employs a rotating conceptual framework to track the particles' trajectories and dislocations as they drift, as well as the statistics of each particle individually. Consequently, the probability density function (pdf) *P* of particle acceleration displacements in space and time can be defined as:

$$\psi(x,t) = \int_{-\infty}^{\infty}\int_{-\infty}^{\infty}\int_{-\infty}^{\infty} P(x,t|x',t')\psi(x',t')\,dx' \tag{1.70}$$

The particle concentration at a given location x and time t could perhaps be determined by adding the probabilities of the number of particles detected, as shown below:

$$\langle c(x,t) \rangle = \sum_{i=1}^{m} \psi_i(x,t) \tag{1.71}$$

Returning to the pdf integral at equation 1.70, and equation 1.71 is then solved as:

$$\langle c(x,t) \rangle = \int_{-\infty}^{\infty}\int_{-\infty}^{\infty}\int_{-\infty}^{\infty} P(x,t|x_0,t_0)\langle c(x_0,t_0)\rangle\, dx_0 + \int_{-\infty}^{\infty}\int_{-\infty}^{\infty}\int_{-\infty}^{\infty}\int_{t_0}^{t} P(x,t|x',t')\gamma(x',t')\,dt dx' \tag{1.72}$$

Thus, in the statistics of their motion, equation 1.72 is exploited to evaluate the site and speed of particles relative to their neighbors and environments, and it approximates the random concentrations and velocities allied with turbulent diffusion. The resulting equations for both the Eulerian and Lagrangian models for assessing the statistics of species in turbulent flow culminate in striking similar formulas for computing the average concentration at a place from a continuous source. Both solutions yield a Gaussian Plume and are almost similar if the variances in the x, y, and z directions are attributed to the eddy diffusivity:

$$\langle c(x,y,z) \rangle = W\left[2\pi\sigma_y\sigma_z\right]^{-1} e^{\left[-\left(y^2\left[\sigma_y^2\right]^{-1} + z^2\left[\sigma_z^2\right]^{-1}\right)\right]} \tag{1.73}$$

where

$$\sigma_y^2 = 2K_{yy}x[\bar{u}]^{-1}\, \sigma_z^2 = 2K_{zz}x[\bar{u}]^{-1} \quad (1.73.1)$$

In direction *i*, equation 1.73 illustrates that wind speed, *W* is species emission rate with the variance of fluctuation σ_i. The variances and diffusivities of turbulent diffusion are measured and utilized to calculate a reliable estimation of concentrations at a specific point from a source under various external factors such as directional flow speed (wind) and environmental conditions. This model is extremely valuable in atmospheric sciences, particularly when addressing pollutants in air pollution emitted by sources, for example, combustion stacks, rivers, or chains of vehicles on a road [12–15].

To date, investigation in turbulence research has been sparse, owing to the complexity of implementing mathematical equations for turbulent flow and diffusion. Previously, laboratory studies employed data from the steady influx in streams or fluids with a high Reynolds number streaming through pipes, but these approaches are complicated to procure precise data accurately. This is because these approaches employ ideal flow, which cannot imitate the turbulent flow conditions required for constructing turbulent diffusion simulations. Scientists have been able to simulate turbulent flow to an accurate perception of turbulent diffusion in the atmosphere and fluids owing to developments in computer-aided modeling and programming. These algorithms may derive spatial correlations and variations, besides acquiring concentration and velocity data. As technology and computer capabilities advance, these algorithms will improve substantially and will more than likely be at the core of future work on simulating turbulent diffusion.

This chapter delves into the intricacies of turbulence, addressing its definition and its substantial impact on everyday activities, such as its influence on airplane mobility. The distinction between laminar and turbulent flow is explored, emphasizing how the Reynolds number elucidates the transition from laminar to turbulent flow. Furthermore, the chapter delves into the principles of mathematical models, including the equations governing turbulence motion, the Navier-Stokes equations, the Spalart-Allmaras (SA) Model, k–ε Turbulence and k–ω Models, and the turbulence diffusion equation. These mathematical descriptions pose a challenge in capturing the complexity and nonlinearity inherent in turbulence phenomena. For instance, the Navier-Stokes equations, being nonlinear, present a significant challenge in solving and their complexity would be notably reduced without the presence of inertial terms, either due to geometry or if these terms are inconsequential.

Exploring the intricacies of turbulence nonlinearity may contribute to a better understanding of the more complex realm of quantum mechanics in atmospheric turbulence, a topic that will be addressed in the upcoming chapter.

References

[1] Sreenivasan, K.R. (1999). Fluid turbulence. *Reviews of Modern Physics*, 71(2), S383.
[2] Bradshaw, P. (2013). *An Introduction to Turbulence and Its Measurement: Thermodynamics and Fluid Mechanics Series*. Elsevier.
[3] Lesieur, M. (2008). Introduction to turbulence in fluid mechanics. *In:* Turbulence in Fluids. Fourth Revised and Enlarged Edition, 1–23.
[4] Davidson, P.A. (2015). *Turbulence: An Introduction for Scientists and Engineers*. Oxford University Press.
[5] Vinnichenko, N. (2013). *Turbulence in the Free Atmosphere*. Springer Science & Business Media.
[6] Atlas, D. (1969). Clear air turbulence detection methods: A review. *In:* Clear Air Turbulence and Its Detection: Proceedings of a Symposium on Clear Air Turbulence and Its Detection, Organized and Sponsored by the Flight Sciences Laboratories, Boeing Scientific Research Laboratories, Office of the Vice President—Research and Development. The Boeing Company, Seattle, Washington, August 14–16, 1968 (pp. 381–401). Springer US.

[7] Wyngaard, J.C. (2010). *Turbulence in the Atmosphere.* Cambridge University Press.

[8] Nieuwstadt, F.T. and Van Dop, H. (Eds.). (2012). Atmospheric Turbulence and Air Pollution Modelling: A Course Held in The Hague, 21–25 September, 1981 (Vol. 1). Springer Science & Business Media.

[9] Reiter, E.R. and Burns, A. (1965). Atmospheric structure and clear-air turbulence. Doctoral dissertation, Colorado State University. Libraries.

[10] Dutton, J.A. and Panofsky, H.A. (1970). Clear air turbulence: A mystery may be unfolding: High altitude turbulence poses serious problems for aviation and atmospheric science. *Science*, 167(3920), 937–944.

[11] Comings, E.W., Clapp, J.T. and Taylor, J.F. (1948). Air turbulence and transfer processes. *Industrial & Engineering Chemistry*, 40(6), 1076–1082.

[12] Rieutord, M., Dubrulle, B. and Lévêque, E. (2006). An introduction to turbulence in fluids, and modelling aspects. *European Astronomical Society Publications Series*, 21, 7–42.

[13] Libby, P.A. (1996). *An Introduction to Turbulence*. CRC Press.

[14] Coantic, M.F. (1978). An introduction to turbulence in geophysics and air-sea interactions. Presented as Lecture Series at California University.

[15] Bailly, C. and Comte-Bellot, G. (2015). *Turbulence*. Springer.

[16] Davies, J.T. (2012). *Turbulence Phenomena: An Introduction to the Eddy Transfer of Momentum, Mass, and Heat, Particularly at Interfaces*. Elsevier.

[17] Reiter, E.R. and Hayman, R.W. (1962). *On the Nature of Clear-Air Turbulence (CAT)* (p. 0047). Colorado State University Research Foundation.

[18] Patnaik, B.S.V. and Wei, G.W. (2002). Controlling wake turbulence. *Physical Review Letters*, 88(5), 054502.

[19] Lee, S., Churchfield, M., Moriarty, P., Jonkman, J. and Michalakes, J. (2012, January). Atmospheric and wake turbulence impacts on wind turbine fatigue loadings. *In:* 50th AIAA Aerospace Sciences Meeting including the New Horizons Forum and Aerospace Exposition (p. 540).

[20] Olsen, J. (Ed.). (2012). Aircraft Wake Turbulence and Its Detection: Proceedings of a Symposium on Aircraft Wake Turbulence Held in Seattle, Washington, September 1–3, 1970. Sponsored Jointly by the Flight Sciences Laboratory, Boeing Scientific Research Laboratories and the Air Force Office of Scientific Research. Springer Science & Business Media.

[21] Reeks, M.W. (1991). On a kinetic equation for the transport of particles in turbulent flows. *Physics of Fluids A: Fluid Dynamics*, 3(3), 446–456.

[22] Sturgess, G. and Mcmanus, K. (1984, January). Calculations of turbulent mass transport in a bluff-body diffusion-flame combustor. *In:* 22nd Aerospace Sciences Meeting (p. 372).

[23] Besnard, D., Harlow, F.H., Rauenzahn, R.M. and Zemach, C. (1992). *Turbulence Transport Equations for Variable-density Turbulence and Their Relationship to Two-field models* (No. LA-12303-MS). Los Alamos National Lab. (LANL), Los Alamos, NM (United States).

[24] Alexander, L.G., Baron, T. and Comings, E.W. (1953). Transport of momentum, mass, and heat in turbulent jets. University of Illinois. Engineering Experiment Station. Bulletin no. 413.

[25] Nakao, H. and Imamura, T. (1992). Mechanism in leading to Richardson's four-thirds law. *Journal of the Physical Society of Japan*, 61(8), 2772–2778.

[26] Nakao, H. (1993). Time dependence of Eulerian velocity correlation for the duration of Richardson's Law. *Journal of the Physical Society of Japan*, 62(1), 6–9.

[27] Gioia, G., Lacorata, G., Filho, E.M., Mazzino, A. and Rizza, U. (2004). Richardson's law in large-eddy simulations of boundary-layer flows. *Boundary-layer Meteorology*, 113, 187–199.

[28] Tsugé, S. (2003). The Kolmogorov turbulence theory in the light of six-dimensional Navier-Stokes' equation. arXiv preprint nlin/0303013.

[29] Chen, W. (2006). A speculative study of 2/3-order fractional Laplacian modeling of turbulence: Some thoughts and conjectures. *Chaos: An Interdisciplinary Journal of Nonlinear Science*, 16(2).

[30] Basu, A. and Bhattacharjee, J.K. (2019). Kolmogorov or Bolgiano-Obukhov scaling: Universal energy spectra in stably stratified turbulent fluids. *Physical Review E*, 100(3), 033117.

[31] Constantin, P. and Foias, C. (1988). *Navier-Stokes Equations*. University of Chicago Press.

[32] Temam, R. (2001). *Navier-Stokes Equations: Theory and Numerical Analysis* (Vol. 343). American Mathematical Society.

[33] Doering, C.R. and Gibbon, J.D. (1995). *Applied Analysis of the Navier-Stokes Equations* (No. 12). Cambridge University Press.

[34] Chorin, A.J. (1968). Numerical solution of the Navier-Stokes equations. *Mathematics of Computation*, 22(104), 745–762.

[35] Javaherchi, T. (2010). *Review of Spalart-Allmaras Turbulence Model and Its Modifications*. University of Washington.

[36] Crivellini, A., D'Alessandro, V. and Bassi, F. (2013). A Spalart-Allmaras turbulence model implementation in a discontinuous Galerkin solver for incompressible flows. *Journal of Computational Physics*, 241, 388–415.

[37] Allmaras, S.R. and Johnson, F.T. (2012, July). Modifications and clarifications for the implementation of the Spalart-Allmaras turbulence model. *In:* Seventh International Conference on Computational Fluid Dynamics (ICCFD7) (Vol. 1902).

[38] Umlauf, L., Burchard, H. and Hutter, K. (2003). Extending the k–ω turbulence model towards oceanic applications. *Ocean Modelling*, 5(3), 195–218.

[39] Wright, N.G. and Easom, G.J. (2003). Non-linear k–ε turbulence model results for flow over a building at full-scale. *Applied Mathematical Modelling*, 27(12), 1013–1033.

[40] Shaheed, R., Mohammadian, A. and Kheirkhah Gildeh, H. (2019). A comparison of standard k–ε and realizable k–ε turbulence models in curved and confluent channels. *Environmental Fluid Mechanics*, 19, 543–568.

[41] Menter, F.R. (1992). Improved two-equation k-omega turbulence models for aerodynamic flows (No. A-92183). NASA Technical Reports Server (NTRS)

[42] Bredberg, J., Peng, S.H. and Davidson, L. (2002). An improved k−ω turbulence model applied to recirculating flows. *International Journal of Heat and Fluid Flow*, 23(6), 731–743.

[43] Wilcox, D.C. (1988), Re-assessment of the scale-determining equation for advanced turbulence models. *AIAA Journal*, 26(11), 1299–1310.

CHAPTER

2

Quantum Mechanics of Atmospheric Turbulence

It is well agreed that the turbulence of classical fluids is a common occurrence that may be witnessed in the flow of a stream or river, as Leonardo da Vinci did in his renowned drawings. In the previous chapter, it was revealed that turbulence is an unresolved issue in mathematics rather than physics. The argument is that mathematicians are stumped as to whether the Navier-Stokes equation consistently enables solutions that are well-behaved at fine adequate lengths and time scales.

Consequently, can quantum address the nature of air turbulence? To comprehend the phenomenon of air/atmosphere turbulence from the standpoint of quantum mechanics, it is imperative to primarily scrutinize the nature of both mechanics. In this aspect, the atmosphere is a composite of various atomic gases, such as oxygen, nitrogen, and others, which form the foundation of quantum mechanics. In other words, Quantum mechanics is a fundamental theory of physics that describes the physical aspects of nature at the atomic and subatomic particle scales. It is the cornerstone of all quantum physics, including quantum chemistry, quantum field theory, quantum technology, and quantum information science. Nevertheless, theorists have gradually demonstrated in recent years how it relates to all aspects of life, individually and collectively.

2.1 What are the Compensates of Air?

Only five gases make up approximately all of the Earth's atmosphere: nitrogen, oxygen, water vapor, argon, and carbon dioxide. Numerous other molecules are included as well. The nature of the air differs globally from one region to another, as well as whether it is day or night [1]. The primary component of air is nitrogen gas and, nitrogen, oxygen, water vapor, argon, and carbon dioxide constitute approximately 99% of the composition of air (Figure 2.1). Trace gases contain

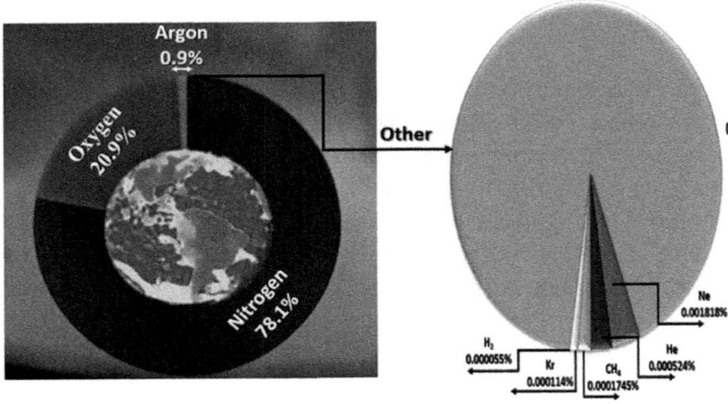

Figure 2.1: Natural air compositions.

neon, methane, helium, krypton, hydrogen, xenon, ozone, and a large array of other atoms and molecules (Table 2.1). In this understanding, quantum mechanics is the only keystone explanation of the dynamic mobilities of atoms and molecules rather than classical physics.

The three primary components of the atmosphere are nitrogen, oxygen, and argon. Water content fluctuates but averages roughly 0.25% of the atmosphere. Trace gases include carbon dioxide and all other elements and molecules. Carbon dioxide, methane, nitrous oxide, and ozone are examples of trace gases. Other noble gases, except for argon, are trace elements. Neon, helium, krypton, and xenon are examples of them [2]. Chlorine and its compounds, fluorine and its compounds, elemental mercury vapor, sulphur dioxide, and hydrogen sulphide are all industrial pollutants. Spores, pollen, volcanic ash, and salt from sea spray are other components of the atmosphere.

Table 2.1: Air chemical compositions

Gasses	Symbols	Percentage (%)
Nitrogen	N_2	78.084%
Oxygen	O_2	20.9476%
Argon	Ar	0.934%
Carbon Dioxide	CO_2	0.04%
Neon	Ne	0.001818%
Methane	CH_4	0.0002%
Helium	He	0.000524%
Krypton	Kr	0.000114%
Hydrogen	H_2	0.00005%
Xenon	Xe	0.0000087%
Ozone	O_3	0.000007%
Nitrogen Dioxide	NO_2	0.000002%
Iodine	I_2	0.000001%
Carbon Monoxide	CO	Trace
Ammonia	NH_3	Trace

Even though water vapor (H_2O) is not listed in Table 2.1, air can comprise up to 5% water vapor, normally varying from 1–3%. In this sense, water vapor is the third most prevalent gas in the 1–5%, which modifies the additional percentages correspondingly. The water content changes with air temperature. Dry air, for instance, has a higher density than damp air. However, humid air that includes real water droplets can be denser than humid air that merely contains water vapor. In other words, water vapor would be almost absent in arid or extremely cold areas. Water vapor contributes to a significant amount of atmospheric gases in warm and tropical regions [1–3].

The 5% of water vapor in the air is substantially influenced by temperature. Consequently, there is relatively little water vapor when it is cold. However, water is more prevalent than argon in hot, humid areas.

The fluctuations of air thermodynamics are the keystone of atmospheric turbulence and shall be addressed later. Atmospheric turbulence is effectively created by a uniform composition of the homosphere. On the contrary, the heterosphere is dominated by an irregular chemical composition, which fluctuates largely corresponding to altitude [1, 4].

Moreover, the homosphere consists of the deeper layers of the atmosphere, including the troposphere, stratosphere, mesosphere, and lower thermosphere. At approximately 100 kilometers (62 miles), the turbopause is the boundary of space and approximately the edge of the homosphere. In this sense, the exosphere and thermosphere are located above this layer, and make up the heterosphere [2, 4]. Oxygen and nitrogen are heavier elements and are only found closer to the bottom region of the heterosphere up. The upper heterosphere is almost entirely made up of hydrogen.

Ozone (O_3), is spread erratically across the Earth's atmosphere. The ozone layer is a part of the stratosphere ranging in altitude from 15 to 35 kilometers (9.3 to 21.7 miles). Its thickness, nevertheless, fluctuates regionally and periodically. With a concentration of 2 to 8 parts per million, the ozone layer holds approximately 90% of the O_3 in the atmosphere. Although the ozone concentration is substantially higher than in the troposphere, O_3 is considered only a trace gas in the ozone layer [1-3].

Carbon dioxide, methane, and nitrous oxide concentrations have been increasing on average. Ozone is concentrated in the stratosphere and surrounding cities. In addition to the elements in Table 2.1 and the previously mentioned krypton, xenon, nitrogen dioxide, and iodine, there are trace quantities of ammonia, carbon monoxide, and numerous additional gases [3–5].

Now the significant question is: why is it crucial to comprehend gas profusion? Weather forecasting and prediction require such information. In this understanding, the quantity of water vapor in the air is very significant in weather forecasting. The composition of the gas allows us to better understand the consequences of natural and man-made compounds emitted into the atmosphere. Since the composition of the atmosphere is essential to climate forecasting, changes in gases. Perhaps, deliver large-scale forecasting of climate change. Lastly, molecules and gases in the atmosphere are the cornerstone of air/atmosphere turbulence mobility.

2.2 Distribution of Air Chemical Composition through Atmospheric Layers

The amount of air chemical compositions is numerous and varied across the atmospheric layers. The chemical composition of the troposphere is known as the composition of air. The troposphere is the atmosphere's lowest layer, spanning from the surface to approximately 12 km (39000 feet). The troposphere accounts for almost 80% of the mass of the Earth's atmosphere. This layer contains nearly all the water vapor in the atmosphere.

Consequently, the tropopause is the boundary between the stratosphere and the troposphere, which extends vertically up to 50 to 55 kilometers (164000 to 180000 ft) with extremely little water in it. The majority of the atmospheric ozone O_3 is found in the stratosphere. On the contrary, little water and O_3 are found through the height of 80 to 85 km (260000 to 280000 ft), which is referred to as the mesosphere layer. However, water is completely absent across altitudes of 500 to 1000 kilometers, which extends from the mesopause to the thermopause and refers to the thermosphere. It hosts the ionosphere, which includes gas molecules that could be ionized [2–6].

The exosphere is the Earth's outer layer of the atmosphere, which fuses with the solar wind approximately 10,000 km (33,000,000 ft). The elements in this layer contain hydrogen, helium, nitrogen, oxygen, and carbon dioxide. Somewhere towards the layer's bottom, the components take on molecular forms. The density of "air" in the exosphere is too tiny for it to react as a gas whereas plasma is formed owing to the solar wind ionizes atoms.

2.3 Molecules of Greenhouse Gases/Effect

Greenhouse particles are the numerous gases that prevent departing long-wave infrared from freely escaping Earth's atmosphere. The Greenhouse Effect is the mechanical fact in the atmosphere that dictates how much heat can be retained well within the atmosphere. Consequently, the primary natural greenhouse gases are H_2O (water), CO_2 (carbon dioxide), CH_4 (methane), and nitrous oxide (N_2O) (Figure 2.2). There are others used in industry, for example, Chlorofluorocarbons (CFC's). In this regard, natural elements in the atmosphere contain H_2O, CO_2, CH_4, and N_2O. These are the gases that retain humans warm enough to live somewhat securely on Earth. This is referred to as the "natural" greenhouse effect.

If greenhouse gases or greenhouse impacts do not exist, the Earth would be a frozen planet devoid of sustaining life. In this circumstance, there would be only ice and stone with no plants,

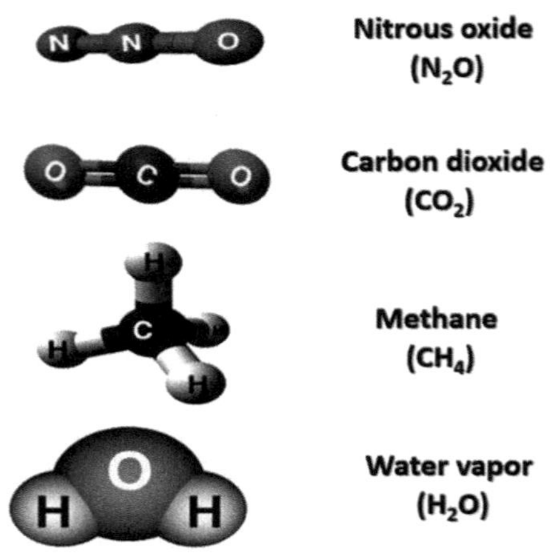

Figure 2.2: Greenhouse gas molecules.

forests, or animals. In other words, the temperature of the Earth would be approximately zero degrees F (–18°C) instead of its present 57°F (14°C) without a natural greenhouse effect. Thus, the debate is not whether we have a greenhouse effect, but whether human actions are causing it to be exacerbated by the emission of carbon dioxide through fossil fuel burning and deforestation [3, 5, 7].

However, the main challenge today is that humans swell the profusion of greenhouse gases in the atmosphere by burning fossil fuels and other pollution industrial activities. In this circumstance, the greenhouse effect is unnaturally boosted.

The greenhouse effect is the practice by which greenhouse gases interact with incident radiation of heat energy departing the planet's surface. In other words, GHGs can obstruct heat evaporation, preventing some overheating quickly and effectively. When heat energy strikes a greenhouse gas molecule, it can be remitted in any manner. Some energy is radiated back down, while others are radiated at an angle to interact with other molecules. Heat energy eventually escapes into space. Based on these criteria, the amount of greenhouse gases in the atmosphere along with other factors determines the radiative balance and/or temperature at which relative thermal equilibrium occurs for a planet.

2.4 Initiating of Air Quantum Turbulence

According to the aforementioned viewpoint, atmospheric strata include a fluctuation of distinct gas molecules. This is the cornerstone for incorporating quantum mechanics into the study of air/atmosphere turbulence dynamics. The question now is: is quantum gas? In a quantum gas, entirely of the molecules' properties are constrained to explicit values, or quantized, corresponding stages on a ladder or notes on a musical scale. In this understanding, chilling the gas to the lowest temperatures grants extreme control over its molecules. Consequently, quantum states of atoms and molecules are the orientation of quantum mechanics as it operates in spectroscopy, atomic and molecular electronic structures, and molecular characteristics. In other words, the main properties of atomic and molecular atmospheric gases can be demonstrated by the quantum mechanics of the atoms that construct those gases.

Consistent with this conception, CH_4 is constituted of one carbon atom (symbol C, atomic number 6) that is coupled to (connected to) four hydrogen atoms (symbol H, atomic number 1) (Figure 2.2). When methane is exposed to water, it does not form ions. In reality, it is not soluble in water. It does not come apart until heated to extremely high temperatures, like in a flame. The quantum mechanical model of the atom exploits sophisticated forms of orbitals, which are hardly alluded to as electron clouds, and electron-occupied volumes of space [5–7].

In a nutshell, turbulence with an uneven energy pattern is caused by particle movement. If the atmosphere is made up of atoms and molecules of various gases when these particles acquire a lot of kinetic energy, they act as turbulent flows. The kinetic energy gained causes electrons to vibrate through their orbits. Turbulence flow is caused by the aggregation of distinct vibrations of various gas electrons in the atmosphere. Without a doubt, the Sun's energy is the primary source of this energy [8].

Therefore, the physical characteristics of electrons and their orbitals can be addressed through (i) principal quantum number, (ii) angular momentum quantum number, (iii) magnetic quantum number, and (iv) spin quantum number (Table 2.2). The critical question is: what is the significance of quantum numbers to begin to understand quantum turbulence? Quantum numbers are the string of numbers exploited to quantify the location and energy of an electron in an atom.

Table 2.2: Quantum numbers that describe the electrons

Quantum Number	Symbol	Values
Principal quantum number	n	1,2,3,4,5,...........
Angular momentum quantum number	ℓ	0,1,2,3,......, (n–1)
Magnetic quantum number	m_1	$-\ell, ..., -1, 0, 1, ..., \ell$
Spin quantum number	m_s	$+\frac{1}{2}, -\frac{1}{2}$

Quantum numbers represent the values of a quantum system's preserved quantities. Electronic quantum numbers (quantum numbers explaining electrons) are a set of numerical values that offer solutions to the Schrödinger wave equation for hydrogen atoms. In this regard, the Schrödinger wave equation is a mathematical model that outlines the energy and orientation of an electron in space and time while accounting for the electron's matter wave nature within an atom [6–8].

It is based on three considerations: (i) the classical plane wave equation, (ii) Broglie's matter-wave hypothesis, and (iii) energy conservation. Therefore, the time-dependent Schrödinger equation is written as:

$$i\hbar \frac{d}{dt}\left|\Psi(t)\right\rangle = \hat{H}\left|\Psi(t)\right\rangle \tag{2.1}$$

Or

The position basis time-dependent Schrödinger equation is expressed as:

$$i\hbar \frac{\partial \Psi}{\partial t} = -\hbar^2 \left[2m\right]^{-1} \frac{\partial^2 \Psi}{\partial x^2} + P_o(x)\Psi(x,t) \equiv \hat{H}\Psi(x,t) \tag{2.2}$$

The Hamiltonian operator $\hat{H}$ is included in both equations 2.1 and 2.2. Therefore, i is an imaginary unit; P_o is potential energy; Ψ is the time-dependent wavefunction; $\hbar$ is the modified form of Planck's constant, i.e. $\hbar = \frac{h}{2\pi}$. In metrology it is exploited, simultaneously with other constants, to state the kilogram, an SI unit. The SI units are well-defined in just such a sense that, when the Planck constant is articulated in SI units, it has the precise value $h = 6.62607015 \times 10^{-34}$ J·Hz^{-1}. Thus, the exact value of $\hbar = 1.054571817... \times 10^{-34}\ \mathrm{J \cdot s} = 6.582119569... \times 10^{-16}\ \mathrm{eV \cdot s}$. In this perspective, the angular momentum of an electron coupled to an atomic nucleus, for instance, is quantized which can only be a multiple of $\hbar$ [5, 7, 9].

In compressed form, the time-independent Schrödinger equation is:

$$\hat{H}\Psi = E\Psi \tag{2.3}$$

Equation 2.3 reveals that the energy eigenvalue is proportional directly to the Hamiltonian operator $\hat{H}$ that presents the energy operator. In this regard, turbulence can be described using the Hamiltonian operator $\hat{H}$ as will be discussed later. Mathematically, an operator is a rule that transforms one attribute into another. For instance, if 'A' represents the potential energy of air gases that can turn property $P_0(x)$ into kinetic energy E, it is an operator $E(y)$. $P_0(x)= E(y)$. In this circumstance, the Hamiltonian operator is the total particle potential and kinetic energies measured across three coordinates and time, i.e., $\hat{H} = E + P_0$. This concept can be mathematically expressed in the Schrödinger equation as:

$$\hat{H} = -\hbar^2 [2m]^{-1} (\nabla)^2 + P_O(r,t) \tag{2.4}$$

For instance, Figure 2.3 demonstrates the accurate location of an electron. In this view, an electron can be in regions **A** and **C** since both regions have extreme amplitude Ψ and hence the maximum probability density of Electrons $|\Psi|^2$. In other words, wave function Amplitude Ψ = Ψ (*r*, *t*); where, '*r*' is the position of the particle in terms of *x*, *y*, and *z* directions [7, 9, 10].

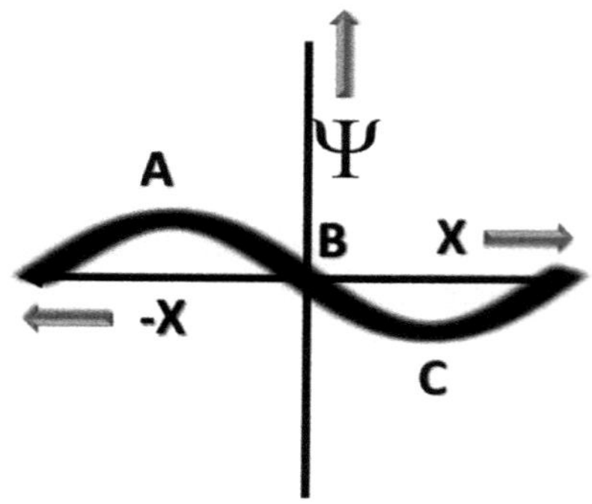

Figure 2.3: Wave function amplitude for determining particle location.

According to the above perspective, 'matter waves' are described using the wave function. Matter waves are very tiny particles in motion with a wave character—particle and wave dual nature. A wave function of the matter-wave is any variable attribute that makes up the matter waves. Amplitude, a wave attribute, is measured by tracking the particle's movement using its Cartesian coordinates concerning time. A wave function is the amplitude of a wave. The wave sort and amplitudes are determined by location and time. In other words, the wave functions are the three-dimensional probability amplitude waves that designate the provinces of space anywhere the electron is expected to be retrieved. In this understanding, air turbulence is a system state in which the probability density is given by $|\Psi|^2$ and is time-variant. The electron in a gas atom is a matter-wave with quantized angular momentum, energy, and so on. The electrons in their orbit move in such a way that the probability density varies exclusively concerning the radius and angles. In this view, the vibration energy of the electron creates irregular movement that is time-independent and resembles a stationary wave between two fixed endpoints. The wave function notion of matter waves is used to an atom's electrons to identify its changeable characteristics which are based on understanding air turbulence flow [7–10,12].

The fundamental question is now: how do quantum numbers show the particle's uneven energy distribution? Schrödinger's solution to the gas atom, for instance, hydrogen, delivers the energy levels of any gas atom. Consequently, every state of a gas atom can be well addressed by the wave functions (Figure 2.3). Subsequently, Schrödinger's solution to the gas atom issue delivers energy levels reliable with the analytically attained Rydberg principle. Thus, the energy levels are formulated as:

$$E(n) = -R_H\left[n^{-2}\right] \tag{2.5}$$

Equation 2.5, therefore, delivers the energy level as a function of the principal quantum number n, which is greater than or equal to 1 (Table 2.2). In this scenario, the Rydberg formula is formed as the variance in energy between slightly dual-energy echelons. In the Schrödinger solution, yet R_H is not a realistic constraint. Hence R_H can be determined by primary constants:

$$R_H = -\Delta E \times n^2 \tag{2.6}$$

$$R_H = -\mu e^4\left[8\varepsilon_0^2 h^2\right]^{-1} \tag{2.6.1}$$

Equation 2.6 says ε_o is a constant equal to 8.54×10^{-12} C^2/J m, with the units Coulombs squared per Joule–meter which is known as the permittivity of vacuum. Moreover, e is the charge on the electron where μ is the abridged mass of the proton m_p and the electron m_e, respectively as:

$$\mu = m_p m_e\left[m_p + m_e\right]^{-1} \tag{2.6.2}$$

Equation 2.6 demonstrates that when an electron is infused with energy, i.e. excited state, the electron moves from one primary shell to a higher shell, causing the value of n to rise. Similarly, as electrons lose energy, they return to lower shells, lowering the value of n. In this sense, absorption denotes the increase in value of n for an electron, accentuating the photons or energy absorbed by the electron. Likewise, the reduction in the value of n for an electron is known as emission, and here is where the electrons discharge their energy. This is an initial principle in understanding remote sensing mechanisms as discussed further in the following sections [11, 14].

The azimuthal quantum number, also known as the (angular momentum quantum number or orbital quantum number), characterizes the subshell, and yields the magnitude of the orbital angular momentum through the following mathematical relationship:

$$L^2 = \hbar^2\left(\ell^2 + \ell\right) \tag{2.7}$$

Equation 2.7 reveals that in spectroscopy and chemistry, ℓ = 0 is known as the *s* orbital, ℓ = 1, the p orbital, ℓ = 2, the *d* orbital, and ℓ = 3, the *f* orbital (Figure 2.4).The azimuthal quantum number, therefore, also determines the shape of the orbital.

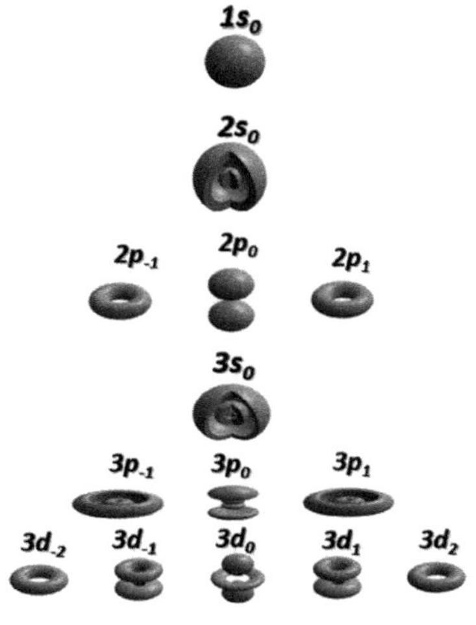

Figure 2.4: Angular momentum quantum orbits.

Figure 2.4 demonstrates that 14 atomic single-electron orbitals for the lowest three n quantum numbers are compiled. The three blocks represent the quantum numbers n=1, n=2, and n=3. In ascending sequence, each row displays one ℓ-quantum number ℓ=0 (s), ℓ=1 (p), and ℓ=2 (d) with m quantum numbers. The hue reflects the complex-valued phase of the wave function. In contrast to the real-valued orbitals in chemistry, which are superpositions of many m-eigen states, the probability densities illustrated here are symmetric around the z-axis. The solid bodies encircle the volume where the continuous probability density surpasses a well-chosen threshold. The orbital formulae are described in detail in the Hydrogen atom. To show the inner structure, the 2s and 3s orbitals are ripped open (Figure 2.4).

The electron cloud shape is determined by the angular quantum number, ℓ. In this sense, any number of n has several ℓ values ranging from 0 to (n-1). There are names for each value ℓ in chemistry. The initial value, ℓ = 0, is known as an$s$ orbital, which is spherical. The second, ℓ = 1, is recognized as a p orbital, which is predictably polar, forming a teardrop petal shape with the tip facing the nucleus. Therefore, a d orbital is a ℓ = 2 orbitals. These orbitals have more 'petals' resembling a cloverleaf than the p orbital form. Ring forms can also be seen at the base of the petals. The orbital after ℓ = 3 is known as an f orbital. These are orbitals. These orbitals are similar to d orbitals, but with extra 'petals.' Higher values ℓ have names that are listed alphabetically [9–12].

The allowable subshells under different combinations of 'n' and 'ℓ' are mentioned above. It may be recognized that the '2d' orbital cannot exist as the value of 'ℓ' is always smaller than that of 'n' (Figure 2.5).

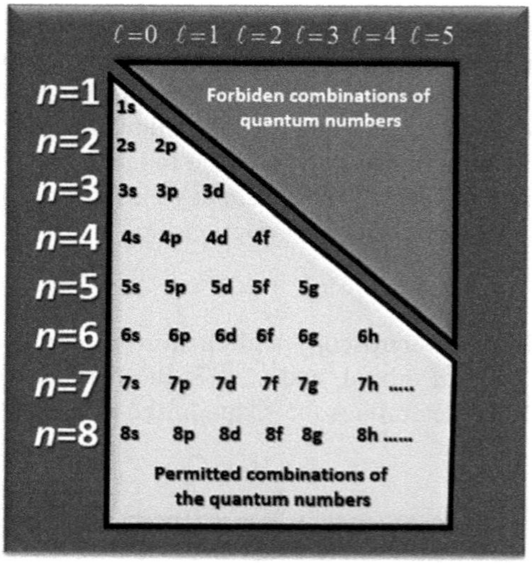

Figure 2.5: Permitted and forbidden combinations of the quantum numbers.

Consequently, the magnetic quantum number describes the precise orbital (or "cloud") within that subshell and produces the orbital angular momentum projection along a specified axis:

$$L_z = m_\ell \hbar \tag{2.8}$$

Equation 2.8 reveals that m_ℓ ranges from $-\ell$ to ℓ, with integer intervals. Since the s subshell (ℓ = 0) includes just one orbital, the m_ℓ of an electron in an s orbital has always been 0. The p subshell (ℓ = 1) involves 3 orbitals (featured as three "dumbbell-shaped" clouds in various systems), hence the m_ℓ electron in a p orbital will be –1, 0, or 1. The d subshell (ℓ = 2) has five orbitals, each with a m_ℓ value of –2, –1, 0, 1, and 2. In this view, the lowest four n quantum numbers are represented by a series of 16 atomic single-electron orbitals. Figure 2.6 depicts just the scenarios with a single radial node. In ascending sequence, each row displays one l-quantum number ℓ

=0 (*s*), ℓ=1 (*p*), ℓ=2 (*d*), and ℓ=3 (*f*) with *m* quantum numbers. The hue represents the wave function's complex-valued phase. Unlike real-valued orbitals, which are superpositions of many *m*-eigen states in chemistry, the probability densities illustrated here are symmetric around the *z*-axis. The solid bodies surround the volume where the continuous probability density exceeds a certain threshold [12–14].

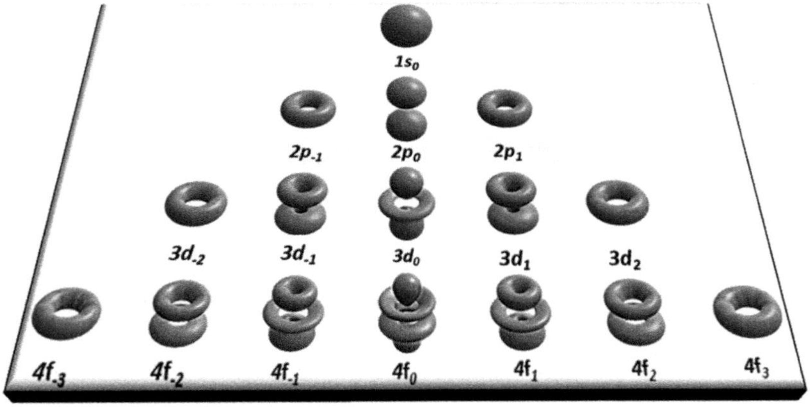

Figure 2.6: The magnetic quantum number.

Needless to say, the magnetic quantum number is determined by the azimuthal (or orbital angular momentum) quantum number. For a specific amount ℓ, the value m_ℓ falls between –l and +l. As a result, it is substantially reliant on the value of *n*. The actual population of orbitals in a specified subshell is dictated by the orbital's 'ℓ' value. It is calculated using the formula (2 ℓ+ 1). The '3*d*' subshell (*n*=3, ℓ=2), for instance, comprises 5 orbitals (2*2 + 1). Every orbital can hold a pair of electrons. As a consequence, the 3*d* subshell can accommodate a maximum of 10 electrons [13, 15].

Since these massive atoms have more than one electron, the fourth quantum number, *s*, arises to overcome the disability of the Schrödinger Equation to be exactly solved in this case. We will be able to grasp several of the features of atoms and how they create molecules by using some simple laws. The spin quantum number specifies the electron's inherent spin angular momentum within every orbital and yields the spin angular momentum *S* projection along the particular axis [13–15]:

$$S_z = m_s\hbar. \tag{2.9}$$

Generally, m_s values vary significantly from –*s* to *s*, where *s* is the spin quantum number aligned with the particle's inherent spin angular momentum:

$$m_s = -s, -s+1, -s=2, \ldots\ldots, s-2, s-1, s. \tag{2.9.1}$$

Since an electron has a spin number of $s = \frac{1}{2}$, m_s will also be $\frac{1}{2}$, corresponding to "spin up" and "spin down" states (Figure 2.7). Since the Pauli exclusion principle requires that each electron in each specific orbital have a distinct quantum number, an orbital can never contain more than two electrons. In other words, the Pauli exclusion principle declares that an electron in an atom can be designated by an exclusive combination of quantum numbers. This signifies that in an atom with more than one electron, no dual electrons can have the equivalent combination of 4 quantum numbers [12, 14, 16].

Consistent with the spin-statistics proposition, particles with integer spin inhabit symmetric quantum states, whereas particles with half-integer spin dominate antisymmetric states; additionally,

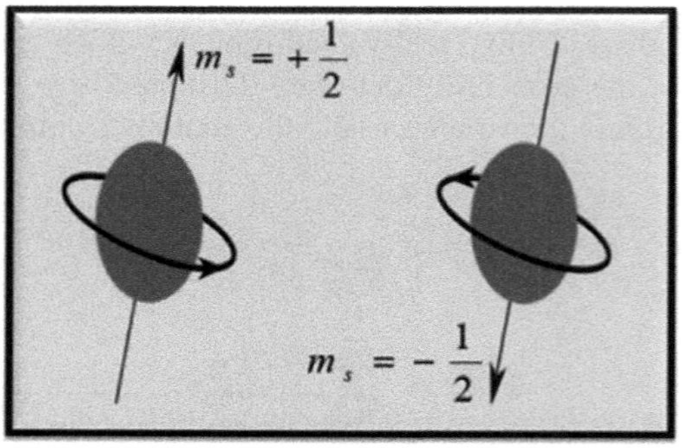

Figure 2.7: Spin quantum numbers.

only integer or half-integer values of spin are allowed under the rules of quantum mechanics. The Pauli principle is derived by applying a rotation operator in imaginary time to particles with half-integer spin in relativistic quantum field theory.

What triggers electrons to occupy orbitals? This is significant for establishing the sign of m_s. Electrons primarily inhabit orbitals singly, then pair together (Figure 2.8). Separately box depicts a single orbital, and every orbital can have only an electron pair. There are three '*p*' orbitals, and the filling of the '*p*' orbitals is depicted on the left side. There are five '*d*' orbitals, and the diagram on the right shows how the '*d*' orbitals fill as the number of electrons rises [13–16].

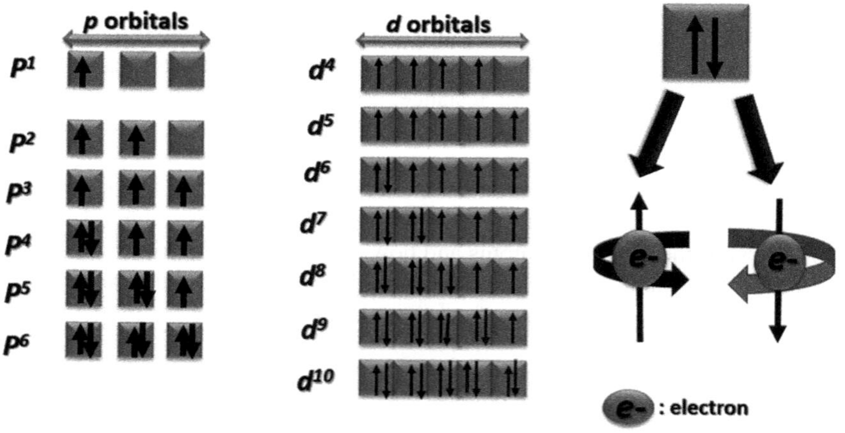

Figure 2.8: Detailed explanations of spin quantum numbers through several orbits.

It is worth notable that the opposite orientation of dual electrons in one orbit. In other words, one electron spins up, while the other spins down. In this understanding, spinning up occurs as the final electron enters vertically and $m_s = +\frac{1}{2}$. On the contrary, $m_s = -\frac{1}{2}$ as long the final electron spins down.

Therefore, the spin of an electron enables it to function as such a little magnet, and the spin regulates an atom's magnetic property. Electrons are extremely tiny particles with limited mobility. This movement, nevertheless, creates a weak magnetic field, which is known as diamagnetic atoms. Since entirely electrons are paired in every orbital the total spin would be zero, and they oppose magnetic fields [14, 16].

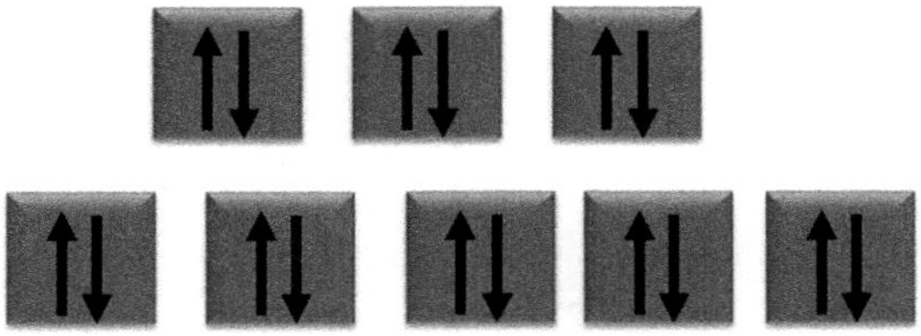

Figure 2.9: Paired electrons diamagnetic.

Let's assume an atom with unpaired electrons within the orbitals. In their orbitals, these unpaired electrons are alone. Accordingly, these atoms with unpaired electrons are paramagnetic. Unpaired electrons in paramagnetic atoms are attracted to a magnetic field. Since the electron in the orbital has a net spin, the spins do not cancel out. As a consequence, the entire atom would have a net spin [12–16].

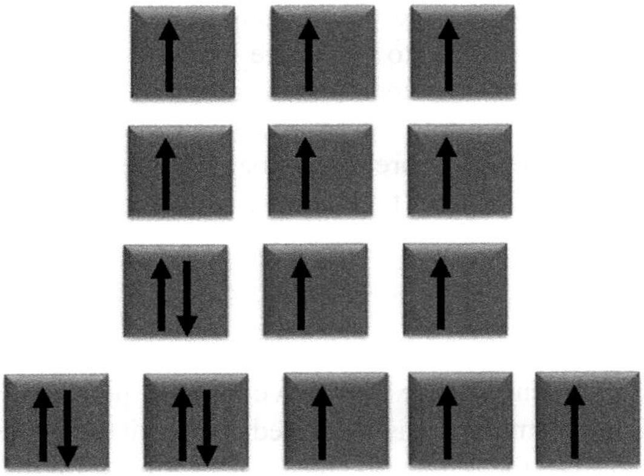

Figure 2.10: Unpaired electrons are diamagnetic.

2.5 What is the Magic of Quantum Mechanical Pressure?

The core question is, what would atmospheric pressure mean? The option is identified as a potential definition of the common expression of pressure. Consistent with this origin, pressure (P) is the pressure acting perpendicular to an object's surface per square meter about which that force is released (Figure 2.11). Thus a scalar quantity is termed pressure, which unifies the vector area component to the dynamic pressure operating on it. In this understanding, the pressure is the scalar proportionality regular that associates the dual typical vectors mathematically as:

$$d\vec{F}_n + Pn\frac{\partial F}{\partial P} = 0 \text{ or } d\vec{F}_n = -Pn\frac{\partial F}{\partial P} \tag{2.10}$$

The negative sign in equation 2.10 refers to the fact that the force is applied to the surface component, whilst the normal vector is directed outward. Equation 2.10 has relevance because the total force exerted by the object on any surface in contact with the object is the surface integral over the surface of the right-hand side of the preceding equation 2.10. It is improper (albeit common) to state, "the pressure is directed in a certain direction." As a kind of scalar, pressure has no direction. The prior association to the quantity provides a direction for the force, but not for the pressure.

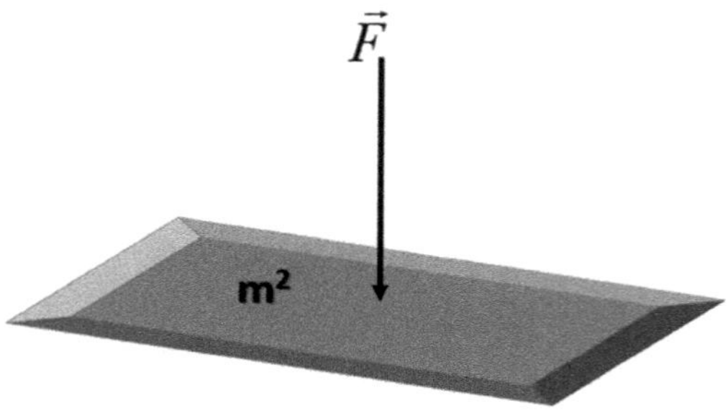

Figure 2.11: Concept of pressure as force per unit area.

When the orientation is changed of the surface component, the resultant force changes direction, but somehow the pressure persists unrevealing. Throughout every location, pressure is distributed to solid boundaries or arbitrary portions of fluid, for instance, equal to these boundaries or portions. It is a primary thermodynamic quantity that is conjugated to volume. In this sense, vapor pressure is the pressure of a vapor in thermodynamic equilibrium with its condensed phases in the locked system. Entirely liquids and solids incline to evaporate into gaseous forms, and all gases lean-to condense back into liquids or solids. The temperature at which the vapor pressure matches the ambient air pressure is known as the atmospheric pressure boiling point of a liquid and is also known as the normal boiling point. Therefore, molecules in an ideal gas have no volume and do not interact. Pressure changes linearly with temperature and quantity, and inversely with volume, according to the ideal gas law:

$$P = \frac{nRT}{V_a}, \tag{2.11}$$

here T, and V_a are the absolute temperature and air volume, respectively. Therefore, the amount of substance n in a specific given molecule is described in chemistry as the number or quantity of discrete atomic-scale particles per the Avogadro constant NA. Contextually, the particles or entities might be molecules, atoms, ions, electrons, or anything else. The Avogadro constant NA, therefore, has been estimated to be 6.022140761023 mol^{-1}. The mole (mol) is a unit of substantial quantity in the International System of Units, defined (since 2019) by holding the Avogadro constant. The amount of material is also referred to as the chemical amount. To prevent misunderstandings, the composition of the particles should be indicated in any determination of substance quantity: accordingly, a sample of 1 mol of molecules of oxygen (O_2) has a mass of about 32 grams, and yet a sample of 1 mol of atoms of oxygen (O) has a mass of approximately 16 grams.

The symbol R represents the molar gas constant which is also known as the gas constant, universal gas constant, or ideal gas constant. As a corollary, the SI value of the molar gas constant R is 8.31446261815324 $J.K^{-1}\ mol^{-1}$.

In a static air composition stage, the air as a whole does not seem to be fluctuating. On the other hand, individual air gas molecules have random mobility, which is not observable since an enormously vast number of random motions of molecules in every direction in the atmosphere. Consistent with pressure being a scalar quantity rather than a vector quantity. It has magnitude but no sense of direction related to that as well. At a spot within a gas, pressure force occurs in all directions. The pressure resultant force is perpendicular to the surface of a gas. Mathematically, the stress tensor σ_t is a related quantity that correlates the vector force $\vec{F}$ to the vector area meter square (m^2) via a linear relationship: $\vec{F} = \sigma_t.\vec{m}^2$. As said by general relativity theory, pressure increases the impact of a gravitational field and so contributes to the mass-energy consequence of gravity. In this understanding, the stress-energy tensor, also known as the stress-energy-momentum tensor or the energy-momentum tensor, is a physical quantity that characterizes the density and flux of energy and momentum in space-time by generalizing Newtonian physics' stress tensor. In this sense, this density and flux of energy and momentum are the sources of air turbulence. The pressure stress is an energy-like quantity that may be calculated by taking the volume integral of the microscopic locally fluctuating stress field. In this view, momentum flux density spins to create high pressure and low pressure. The fluctuations in pressure assist air particles to flow from high momentum density to a lower one. In other words, the momentum flux density operator has the properties of a stress tensor operator. In this regard, the energy density of the system is thus possible to obtain the Hamiltonian density from the stress-energy tensor.

$$\frac{\partial H}{\partial t} + \nabla.\left(\frac{\partial \ell}{\partial \nabla \phi_\alpha}\dot{\phi}_\alpha\right) = 0 \tag{2.12}$$

Equation 2,12 says $\frac{\partial \ell}{\partial \nabla \phi_\alpha}\dot{\phi}_\alpha$ the energy flux density of the atmospheric system, which ℓ is Lagrangian density, which is a function of the set of fields ϕ_α. In this regard, the stress-energy tensor can be mathematically given by a diagonal matrix:

$$\left(E^{\alpha\beta}\right)_{\alpha,\beta=0,1,2,3} = \begin{pmatrix} \rho & 0 & 0 & 0 \\ 0 & P & 0 & 0 \\ 0 & 0 & P & 0 \\ 0 & 0 & 0 & P \end{pmatrix} \tag{2.13}$$

Equation 2.13 demonstrates that the stress-energy tensor $E^{\alpha\beta}$ is defined as the tensor of order two, that delivers the flux of the αth constituent of the momentum vector diagonally to an air surface. Accordingly, $E^{\alpha\beta}$ induces such as the energy-momentum of the air molecules cause particle mobility dynamic movements as a function of air pressure as can be seen in equation 2.13.

Consistent with the above perspective, the mobility of quantum mechanics of air turbulence is initiated by pressure. In this regard, the weight of the air molecules overhead creates air pressure, which induces the stress-energy tensor. Nonetheless, tiny air molecules have weight, and the massive quantities of air molecules that construct Earth's atmosphere's layers together have plenty of weight, which bears down on whatever lies below. In other words, the force per unit area exerted on the Earth's surface by the weight of the air above the surface is defined as atmospheric or air pressure. The force exerted by an air mass is caused by the molecules that comprise it, as well as their size, speed, and quantity in the air. The previous section has demonstrated the principle of quantum mechanics in understanding the mechanism of air particle compositions.

From the point of view of classical physics, ascending and descending air creates areas of high and low pressure. As the air heats, it rises, resulting in low pressure at the surface. As the air cools, it falls, resulting in high pressure near the surface. Moreover, the atmosphere of the Earth imposes pressure on the surface. Pressure is expressed in hector Pascals (hPa), commonly known as millibars. The standard pressure at sea level is 1013 hPa, however, there are extensive zones of either high or low pressure. Since these regions are mostly related to one another, what constitutes a high would differ accordingly. The keystone question is: would there be a quantum mechanical foundation for air pressure?

Wind pressure is calculated using the equation $P = 0.00256 \times W^2$, where W is the wind speed in miles per hour (mph). Wind pressure is measured in pounds per square foot (psf). For instance, if the wind speed is 70 miles per hour, the wind pressure is $0.00256 \times 702 = 12.5$ pounds per square foot. In this understanding, the dynamic pressure P_d is given by:

$$P_d = \frac{1}{2}\rho W^2 \tag{2.14}$$

here ρ is air density that presents the ratio of air mass (m) per air volume V_a as ($\frac{m}{V_a}$). Therefore, dynamic pressure is the ratio of kinetic energy per volume $\frac{E}{V_a}$ which creates the stress-energy tensor across air composite molecules. In other words, wind generation velocity from the point of view of the quantum theory is discrete packets of energy, quanta, and the atmospheric system, corresponding to an oscillator, would only have a discrete set of energy levels, i.e. states of different energy, which described in sequences of wave functions. An energy state is, obviously, a form of energy and, hence, Boltzmann's Law concerns. More specifically, if the various energy levels are denoted, i.e. the energies of the various molecular states, by $\left[\Psi|E_1\rangle, \Psi|E_2\rangle, \ldots\ldots., \Psi|E_n\rangle\right]$ and if Boltzmann's Law applies, then the probability of finding a molecule in the particular state $\Psi|E_1\rangle$ would be proportional e^{-kT}.

In these constrain, the wave function of the stress-energy tensor can be described in the time-independent Schrödinger equation is:

$$\Psi = \psi e^{-\left(2\pi i \frac{E^{\alpha\beta}}{\hbar}\right)t} \tag{2.15}$$

$$\Psi = \psi e^{-\left(2\pi i \frac{E^{\alpha\beta}}{\hbar}\right)t} + \psi e^{-\left(2\pi i \Pi\left(E^{\alpha\beta}\right)\right)t} \tag{2.16}$$

Equation 2.15 demonstrates only the initial state of the creation of turbulent flow due to the stress-energy tensor across air different particles. Air particles spin up and down causing fluctuation of $\Psi|E_n\rangle$. In this regard, the wave function of an initial stage of air turbulent flow may be expressed as a function of momentum and time. The momentum flux would be a function of the stress-energy tensor of air particles or air molecules. In other words, the momentum flux can be identified as the force-density operator on air particles. The kinetic impact of the stress is grasped by deliberating the quantum mechanical momentum flux density $\Pi\left(E^{\alpha\beta}\right)$. Consequently, the quantum mechanical momentum flux density of the air might take the form:

$$\Pi\left(E^{\alpha\beta}\right) = -\frac{\left(E^{\alpha\beta}\right)_{\alpha,\beta=0,1,2,3}}{\hbar} \lim_{\partial E^{\alpha\beta} \to E^{\alpha\beta}} \left(\nabla - \nabla'\right) \otimes \left(\nabla - \nabla'\right) \Psi^*\left(E^{\alpha\beta}\right) \Psi\left(\partial\left(E^{\alpha\beta}\right)\right) \tag{2.17}$$

The momentum flux can be determined by the wavefunction of the initial stress-energy tensor $\Psi^*\left(E^{\alpha\beta}\right)$ and the wavefunction of the dynamical change in the stress-energy tensor $\Psi\left(\partial\left(E^{\alpha\beta}\right)\right)$. In other words, the wave function can be determined by the kinetic energy of air particle movements due to changes in pressure dynamics and its momentum flux density in space and time. The quantum state of dynamic air pressure is obtained from wave function. The wind initiates turbulence and begins with dynamic changes of pressure which add stress fluctuations as a function of heat flux. In this view, the quantum definition of wind generation can be addressed as discrete heat flux generates vibration and spinning of air molecules. If the air molecules lose more discrete heat flux, their densities increase causing the great quantum state of a stress-energy tensor which induces a high quantum state of wind stress. In this circumstance, the air particles move from the high quantum state of dynamic pressure toward the lowest quantum state of low heat fluxes causing wind flows. This explanation delivers a novel definition of the mechanism of wind flow, which perhaps is named as "quantized Marghany wind generation mechanism".

In the kinetic theory of gases, we have defined a term called root-mean-square velocity, as the net movement of a velocity vector is zero. The kinetic energy of a particle is reliant on the temperature and molecular mass and velocity of an atom, so the temperature will increase the diffusion rate will increase, and molecular mass will increase keeping the temperature constant the diffusion rate will decrease, rate of diffusion would increase as the root-mean-square velocity increase.

In this regard, quantum diffusion speculation can be implemented to have a resounding grasp of the quantization Marghany wind generating mechanism. Let us assume there is a random variation in air molecule density in the quantum state $\Psi(|\Pi\rangle)$ across the lattice Λ, i.e. $\Psi(|\Pi\rangle) \in \Lambda$. In this view, the wave function of momentum flux density involves discrete sequences of wave functions $[\Psi|\Pi_1\rangle, \Psi|\Pi_2\rangle, \ldots\ldots, \Psi|\Pi_n\rangle]$, which randomly fluctuate from higher-pressure order to lower-pressure order under the circumstance $\sum_{i=1}^{n} \Psi(|E_{i+1}^{\alpha\beta}\rangle) < \Psi(||E_i^{\alpha\beta}\rangle\rangle) \in \Lambda$ (Figure 2.12).

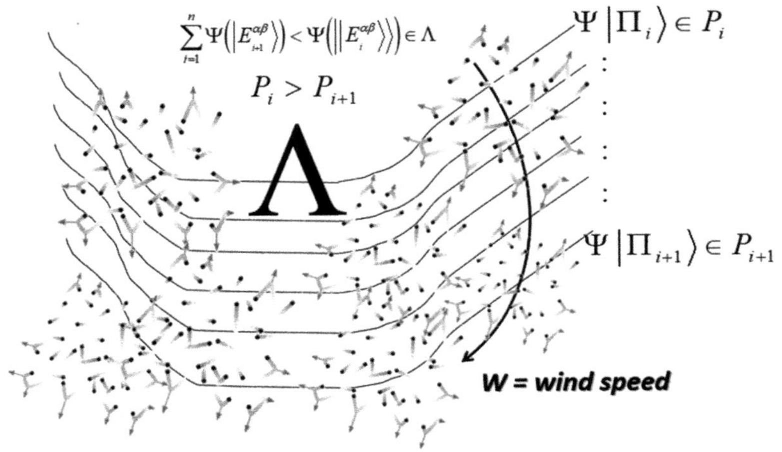

Figure 2.12: Quantization of Marghany wind generation mechanism due to momentum flux density.

In this understanding, the quantum dynamic-wind mobility can be identified as the quantum diffusion of air high-density particles as a great stress-energy tensor owing to high-dynamic pressure and wind stress zones to the lowest ones. This can be expressed mathematically in Hamiltonians as:

$$(H\psi)(E^{\alpha\beta}) := (\Delta\psi)e^{(E^{\alpha\beta})t} + \psi(E^{\alpha\beta})e^{(E^{\alpha\beta})t} \tag{2.18}$$

here $\nu(E^{\alpha\beta})$ is the standard normal distribution of stress-energy tensor from one region to another region i.e., $\nu(E^{\alpha\beta})$: $E^{\alpha\beta} \in P_d + \tau_w \in \Lambda$. In other words, Hamiltonians describe the probability density function of quantum stress-energy tensor exponential random growth from microscopic scale to large scale as time-dependent. In this understanding, the fluctuations in stress-energy tensor as a function of dynamic pressure lead to wind dynamic flow: $\partial P_d \rightarrow E^{\alpha\beta} \rightarrow W$. Mathematically, wind velocity can be estimated as a function of dynamic pressure using equation 2.14:

$$W = \sqrt{\frac{2Q}{\rho}} \tag{2.19}$$

In equation 2.19, Q is dynamic pressure in pascals (i.e., kg/m·s^2), ρ air mass density (e.g. in kg/m^3, in SI units), and W is the wind speed in ms^{-1}. Equation 2.19 reveals wind speed is just considered as velocity pressure. In this sense, wind speed perhaps spreads from high pressure to the lowest one which is recognized as the air particles' diffusion rate. As a consequence, the rate of diffusion of a gas is inversely proportional to both the time and the square root of molecular mass. It is also inversely proportional to the square root of density. But it is directly proportional to pressure.

$$R_G = k\sqrt{\frac{2Q}{\rho}} \tag{2.20}$$

Substitute equation 2.19 into 2.20:

$$R_G = k\langle W^2\rangle^{0.5} = k\langle W\rangle \tag{2.21}$$

Equation 2.21 delivers the rate of diffusion R_G is proportional directly to wind velocity W and k is a constant that is contingent on temperature but is impartial to pressure. Therefore, kinematic pressure can be expressed as $P_k = P\rho^{-1} = m^2s^{-2}$, where P_a is the pressure and ρ constant mass density. In this view, kinematic pressure is exploited similarly to kinematic viscosity ν to retrieve the Navier–Stokes equation deprived of evident presentation of the density ρ, which is casted as:

$$\frac{\partial W}{\partial t} + (W\nabla)W = -\nabla P_k + \nu\nabla^2 W. \tag{2.22}$$

Therefore, the stress-energy tensor, in general relativity, is that as the sole function of delineating the energy density of the corpse that triggers the curvature of wind pattern fluctuations in space-time (Figure 2.12).

2.6 Marghany's Quantization of Earth's Rotation and its Impact on Wind Mobility Patterns

In consistence with the above perspective, is there a force that deflects the winding path into the curved path? In addition to the air particles getting accommodated by the curvature path, the flow turns into turbulence. This can be explained from the point of view of the Earth's rotation and curvature.

In Einstein's field equations, the stress-energy tensor defines how matter affects spacetime geometry and vice versa. Matter and spacetime are not the same things. Spacetime can be bent locally without the presence of matter. Otherwise, the presence of matter alters the spacetime curvature. The stress-energy tensor illustrates how they interact with one another. It may be thought of as a form of interface between matter and spacetime, defining physical quantities that are transferred between them.

In general relativity, the stress-energy tensor consists mainly of its matrix under pressure and energy density, which confirms the curvature of the spacetime owing to the dynamic pressure of air molecules. Consequently, there is impartiality between the stress-energy tensor viscosity and velocity gradient. In this context, the air particles would travel in space-time as waves on the surface of the water. Therefore, there must be energy dissipation and velocity variation. Therefore, time-space or wind flow is directly related to the stress-energy tensor, since it is the former that distresses the curvature of space-time, and hence the viscosity [15, 17].

The curvature of spacetime influences the motion of massive air particles under high stress-energy tensor within it; in turn, as massive air masses move in spacetime, the curvature swaps, and the geometry of spacetime is in continuous evolution. Gravity then delivers a depiction of the dynamic interface between air masses and spacetime. In this circumstance, the wind flows in straight lines in the sense that the trajectory movements are as short as possible. Wind trajectory movements sense in which such a space is curved, which is the total of interior angles of a triangle is dissimilar commencing (Figure 2.13). Consequently, in such a case, it is larger than π. Needless to say, gravity is the curvature of space-time.

Figure 2.13: The curvature of wind flow in space-time.

Fortunately for us, Earth's gravity is substantial and sufficiently preserves its atmosphere in a locked state. Mars, in contrast, would be less than 50% of Earth as well as around one-tenth of the Earth's mass. With less mass, there is less gravitational force. Mars's atmosphere is just around one-hundredth the density of the Earth. Incidentally, most of Mars is generally CO_2.

In other words, gravity affects gas molecules. This is why the atmosphere does not just drift away into space. The gist is that in the lower atmosphere, the time between collisions is very short, and the distances are quite small. The atmospheric pressure is proportional to gravity. On aggregate, raising the gravity would lead to twice the pressure. This is essentially relevant when the atmosphere is relatively thin in comparison to the radius, and gravitational acceleration would be almost equal throughout the entire atmosphere. Generally, the air pressure would be zero if gravity did not exist. The question therefore is, how does gravity influence airflow? A density gradient is created in the air as gravity hugs the layer of air to the Earth's surface. Gravity pulls on the air near the earth, which is compressed by the air rising in the sky. Therefore, air near the ground is denser and has a higher pressure than air at higher elevations [16–18].

Is there another force curve in the airflow path than gravity? The rotation of the Earth is crucial to the Coriolis effect. At the equator, the Earth revolves faster than at the poles. Because the Earth is broader towards the equator, the equatorial areas reach speeds of about 1,600 kilometers per hour. The earth revolves at a pace of 0.00008 kilometers per hour near the poles.

Since the Earth spins, its surface is an accelerating reference frame (we call this a non-inertial frame). When an item is in a non-inertial frame, false forces must be delivered. We call this fake for the Coriolis force when an object travels relative to the axis of rotation in a rotating frame [10–12].

Consequently, the Coriolis force is perpendicular to the axis of the object. The Coriolis force operates north-south as the Earth spins on its axis from west to east. In this view, the Coriolis force is zero at the Equator. The Coriolis effect causes wind (and water) patterns that travel east toward the equator and west toward the poles near the earth's surface. These prevailing wind patterns are important for transporting clouds over the world and, as a result, causing weather patterns in various places.

Generally, the Coriolis effect depicts convection, circulation, and deflection of air. In the Northern Hemisphere, the Coriolis effect causes air to rotate counterclockwise around large-scale low-pressure systems and clockwise around large-scale high-pressure systems. The flow direction is reversed in the Southern Hemisphere (Figure 2.14). In other words, the amount of deflection caused by the air is proportional to both its speed and latitude. As a result, slowly blowing winds will be deflected just somewhat, but stronger winds will be deflected substantially. Similarly, winds flying closer to the poles will be deflected more than winds flowing closer to the equator. At the equator, the Coriolis force is zero. Can quantum mechanics explain these assertions precisely?

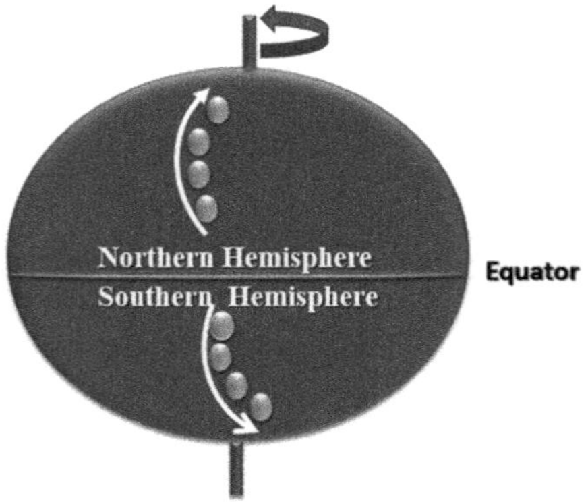

Figure 2.14: Deflection of air particles in the northern hemisphere and the southern hemisphere due to the Coriolis effect.

The physical effect of the stress-energy tensor would be demonstrated by involving both torque term and Coriolis terms. In classical physics, the Coriolis effect rises due to the earth spinning around its axis. This would not be a complete approach to understanding the occurrence of the Coriolis effect. Torque is a property of the stress-energy term, and the Coriolis forces are generated as secondary properties of the torquing of matter-energy in spacetime. Consequently, the ensuing Coriolis effects are caused by torquing on spacetime, and hence spacetime geometry is modified. In this topology, Coriolis forces emerge from the rotational effects of torque. In this understanding, the Coriolis force can generate torque on the spinning air molecules. To this end, let us assume a horizontal hoop of mass m and radius r which is spinning with angular velocity about its vertical axis at co-latitude theta (θ) [15–18]. In this sense, the Coriolis frequency, also known as the Coriolis parameter or Coriolis coefficient, equals twice the Earth's rotation rate multiplied by the sine of the latitude. Assume the air mass is moving with v such that the centripetal and Coriolis forces acting on it are balanced. Then there is:

$$\frac{v^2}{r} = 2(\Omega \sin\theta)v \tag{2.23}$$

The magnitude of the spin rate of the Earth is given by:

$$v = r\omega \tag{2.23.1}$$

here r is the radius of curvature of the path of an object that moves on the Earth's surface.

In this regard, the Coriolis parameter can be obtained by:

$$f = \omega = 2\Omega \sin\theta \tag{2.24}$$

Thus Ω is the angular velocity or frequency essential to maintain a body in a set spiral of latitude or zonal boundary. If the Coriolis parameter is great, the influence of the earth's rotation on the body is large since a greater angular frequency is required to maintain equilibrium with the Coriolis forces. Let us assume that a parcel of air traveling north latitude 33.7° N at 6 ms^{-1} would have a lateral acceleration of 4.48×10^{-4} ms^{-2} due to the Coriolis force [21].

Conversely, if the Coriolis parameter is minimal, the consequence of the earth's rotation is minor since the Coriolis force eliminates only a small fraction of the centripetal force on the body. Therefore, the magnitude of f has a significant impact on the relevant dynamics that contribute to the body's motion [18, 20, 22].

Torque can be estimated as a function of Coriolis force as:

$$T_r = r.f = r.2\Omega \sin\theta\left[\frac{\pi}{2} - \gamma\right] \tag{2.25}$$

here γ is a phase angle. For an accelerated observer of air particle velocity

$$v \equiv -a - f + 2(a \bullet \vec{v})\vec{v} \tag{2.26}$$

Equation 2.26 reveals the relativistic correction to an inertial frame $2(a \bullet v)v$ for geostrophic wind [26–28]. In classical geostrophic wind, the velocity is given by

$$v_g = \frac{\partial P}{\partial x}[2\Omega\rho\sin\theta]^{-1} \tag{2.27}$$

Equation 2.27 can be modified by implementing the inertial acceleration as:

$$v_g = \frac{\partial P}{\partial x}[2\Omega\rho\sin\theta]^{-1} + 2(a \bullet \vec{v})\vec{v} \tag{2.28}$$

here a is the four accelerations, i.e., $a_1 \times a_2 \times a_3 \times a_4$. In this view, the Coriolis force is a force that acts on geostrophic wind motion within a frame of reference that rotates concerning an inertial frame. In other words, in a reference frame with clockwise rotation, the force acts to the left of the motion of the object. In other words, $\vec{v} = W$ and $a \approx \frac{W^2}{r}$ are the velocity and acceleration of the point of interest concerning the inertial frame; $\vec{v}$ is the speed of the wind in question, while a is its angular momentum, which is given by Newton's theory of relativity. Consequently, geostrophic wind would be created along an inertial frame owing to the rotation of the Earth. The appearance of inertia would seem to create a force in a direction orthogonal to wind motion, the Coriolis effect from the perspective of a rotating frame of reference [12, 22, 24].

In accordance with the torque speculation of the Earth's rotation, the newly developed wind speed can be termed as geostrophic-torque velocity v_{gT_r}, which can be mathematically expressed as:

$$v_{gT_r} = \frac{\partial P}{\partial x}\left[r.2\Omega\sin\theta\left[\frac{\pi}{2} - \gamma\right]\right]^{-1} \tag{2.29}$$

In this understanding, wind velocity can be considered as the sum of geostrophic velocity, geostrophic-torque, and ageostrophic-torque velocities. In consideration of this speculation, the real wind velocity comes out to be:

$$W_s = v_g + v_{gT_r} + v_{ageo-torque} \tag{2.30}$$

Thus, the total stress-energy $E^{\alpha\beta}$ includes torque and Coriolis force is given by:

$$E^{\alpha\beta} = \kappa\left[v_{gT_r} \times 2\Omega\sin\theta\right]\left[\Psi\left(E^{\alpha\beta}\right) + \left(r.2\Omega\sin\theta\left[\frac{\pi}{2} - \gamma\right]\right)\left[\frac{\sqrt{\hbar\left(\Omega^2 r\cos\theta\right)}}{v_{gT_r} \times 2\Omega\sin\theta}\right]\right] \tag{2.31}$$

The fluctuation of atmospheric pressure across the variety of isobars can create a total of stress-energy which is affected by both torque and Coriolis force, as seen above. In equation (2.31) κ is the coupling constant equal 8π and $E^{\alpha\beta}$ would be illustrated in the scalar curvature path owing to the Coriolis force effect. In this understanding, the stress-energy of geostrophic wind perhaps creates twist displacement and hence shear stress and shear-a strain which accommodates highly viscous air gases [17, 19, 22]. Needless to say, when torque and Coriolis effects are included in Einstein's field equations, the spacetime manifold corresponds closely with the Coriolis structure of atmospheric dynamics as will be addressed in the following chapters.

The wind flow can create various levels of energy through different altitudes. The simplification form of wind-induced energy can be delivered based on the well-known Einstein equation of energy $E = mc^2$. In this understanding, the wind-derived energy in different altitudes can have a new expression derived by the author:

$$E_W = \frac{\partial P_a}{\partial x}\left[\frac{\partial W}{\partial t}\right]^{-1}\sqrt{E^{\alpha\beta}\left[\frac{\partial P}{\partial x}[a]^{-1}\right]^{-1}} \quad (2.32)$$

Equation 2.32 demonstrates that the new formula for stress-energy and wind acceleration $a = \frac{\partial W}{\partial t}$ sustain energy wind-driven theory introduced by the author and quoted as the "Marghany quantum energy wind-driven theory". In this speculation, the curvature energy preserved by the wind would not be circulated in a simple back-and-forth pattern between high – the stress-energy tensor zone in the poles to low – the stress-energy tensor zones in the equator. In other words, the high and low stress-energy tensor zones contribute to the curvature of the spacetime of the Earth's atmosphere in addition to its spinning to create continuous deflection in air circulation.

2.7 Quantum Geostrophic Turbulence

In classical physics, the transition to turbulence occurs in a succession of phases, commencing with the obliquely symmetric (no waves present) and ending with the chaotic flow. The succession, therefore, comprises particularly of amplitude hesitancy of periodic flow, the structural hesitancy of semi-periodic flow, and a conversion zone anywhere where the features steadily switch to chaotic behavior. Consequently, the rotation rate rises and the spectra in the transition zone are characterized by a progressive fusion of the excitation energy with the spectral peaks representing the flow fields flow.

The nature and applicability of the instability that occurs in a spinning differentially heated annulus of the atmosphere have sparked interest in its behavior. The interior flow density and pressure surfaces are aligned under optimum lateral heating and rotation circumstances so that potential energy in the density field can be converted to kinetic energy. Consequently, the flow instability that arises is known as baroclinic instability, which is the primary mechanism of large-scale midlatitude atmospheric circulation. However, one of the more difficult concepts to comprehend in geophysical fluid mechanics is baroclinic instability. However, it is also one of the most essential notions in this discipline since it is the primary driver of large-scale atmospheric circulations and key ocean circulations. This mechanism is responsible for the atmosphere's redistribution of heat from low to high latitudes, the maintenance of the synoptic weather system, and the formation of some oceanic eddies [17, 19–23].

The atmosphere is not two-dimensional, but it is 'quasi-two-dimensional' due to the influences of layering and spinning, and it resembles the quasi-geostrophic equations. Let us thus broaden our two-dimensional theory to incorporate stratification consequences. In this regard, a characterization of the atmosphere, as well as its state, is requested without the use of a scale variable like the Reynolds number. Therefore, the momentum diffusivity to thermal diffusivity ratio is required to characterize baroclinic instability in a quasi-geostrophic atmosphere. The Prandtl number (Pr) might be used in this scenario. In practice, the Prandtl number (Pr) is described as the ratio of momentum diffusivity to thermal diffusivity and is stated as follows [25]:

$$\mathrm{Pr} = \frac{\nu}{\alpha_{thermal}} = \frac{c_p\mu}{k_c} \quad (2.33)$$

In equation 2.33, ν; $\alpha_{thermal}$; c_p; and k_c are kinematic viscosity; thermal diffusivity; specific heat; and thermal conductivity respectively. The Prandtl numbers (Pr) in the temperature range between –100 °C and +500 °C may be determined using the following formula for air at a pressure of 1 bar [28].

$$\mathrm{Pr}_{air} = \frac{10^9}{1.1\bullet\vartheta^3 - 1200\bullet\vartheta^2 + 322000\bullet\vartheta + 1.393.10^9} \quad (2.34)$$

The temperature ϑ should be expressed in degrees Celsius. The variation values are limited to 0.1%. Small Prandtl numbers, Pr << 1, indicate that thermal diffusivity is dominant. With large levels of Pr >> 1, momentum diffusivity dominates the behavior. For instance, the reported value for liquid mercury implies that heat conduction is more important than convection, implying that thermal diffusivity is dominating. However, in engine oil, convection is more effective than pure conduction at transferring energy from a region, therefore momentum diffusivity is dominant [26].

The turbulent Prandtl number (Pr_t) is a non-dimensional quantity that refers to the ratio of momentum to heat transfer eddy diffusivity, which has an average value of 0.85 based on investigational data. However, it ranges from 0.7 to 0.9 relying on the fluid's Prandtl number Pr. Therefore,it is essential for tackling the turbulent boundary layer flow heat transfer problem. The Reynolds analogy is the simplest model for Pr_t, yielding a turbulent Prandtl number of 1. Thus, the initiation of eddy diffusivity and, subsequently, the turbulent Prandtl number serves to define a simplistic causal relation between excessive shear stress and heat flux in a turbulent flow. If the momentum and thermal eddy diffusivities are both 0, the turbulent flow computations are equivalent to laminar equations [25–27]. The eddy diffusivities for momentum transfer ε_M and heat transfer ε_H are specifically defined as:

$$-\overline{u'v'} = \varepsilon_M \frac{\partial \overline{u}}{\partial y} \tag{2.35}$$

And

$$-\overline{v'\vartheta'} = \varepsilon_H \frac{\partial \overline{\vartheta}}{\partial y} \tag{2.36}$$

In accordance with equations 2.35 and 2.36, the apparent turbulent shear stress and the apparent turbulent heat flux are $-\overline{u'v'}$ and $-\overline{v'\vartheta'}$, respectively [25]. In this circumstance, the turbulent Prandtl number Pr_t is fairly well understood as:

$$\mathrm{Pr}_t = \varepsilon_M \left[\varepsilon_H\right]^{-1} \tag{2.37}$$

The velocity and temperature profiles are comparable when the Prandtl number and turbulent Prandtl number are both equal to unity as demonstrated in the Reynolds analogy. This greatly simplifies the response to the heat transfer problem. If somehow the Prandtl and turbulent Prandtl numbers are not identical, calculating the turbulent Prandtl number enables one to compute the momentum and thermal equations [27]. Consequently, the turbulent momentum boundary layer equation can be given by:

$$\overline{u}\frac{\partial \overline{u}}{\partial x} + \overline{v}\frac{\partial \overline{u}}{\partial y} = -\left[\rho\right]^{-1}\frac{d\overline{P}}{dx} + \frac{\partial}{\partial y}\left[\left(\nu + \varepsilon_M\right)\frac{\partial \overline{u}}{\partial y}\right] \tag{2.38}$$

and the turbulent thermal boundary layer equation is casted as:

$$\overline{u}\frac{\partial \overline{\vartheta}}{\partial x} + \overline{v}\frac{\partial \overline{\vartheta}}{\partial y} = \frac{\partial}{\partial y}\left[\left(\alpha_{thermal} + \varepsilon_H\right)\frac{\partial \overline{\vartheta}}{\partial y}\right] \tag{2.39}$$

Consequently, using the turbulent Prandtl number in equation 2.37 in both equations 2.38 and 2.39 one can obtain:

$$\overline{u}\frac{\partial \overline{u}}{\partial x} + \overline{v}\frac{\partial \overline{u}}{\partial y} = -\left[\rho\right]^{-1}\frac{d\overline{P}}{dx} + \frac{\partial}{\partial y}\left[\left(\nu + \frac{\varepsilon_M}{\mathrm{Pr}_t}\right)\frac{\partial \overline{u}}{\partial y}\right] \tag{2.40}$$

$$\overline{u}\frac{\partial \overline{\vartheta}}{\partial x} + \overline{v}\frac{\partial \overline{\vartheta}}{\partial y} = \frac{\partial}{\partial y}\left[\left(\alpha_{thermal} + \frac{\varepsilon_H}{\mathrm{Pr}_t}\right)\frac{\partial \overline{\vartheta}}{\partial y}\right] \tag{2.41}$$

Geostrophic turbulence is similar to two-dimensional turbulence in that energy is transported to enormous scales. The transfer occurs in both the vertical and horizontal planes and the flow barotropizesas it transfers to larger scales. In this scenario, let us assume that q is the inverse of geostrophic turbulence energy spectra wave number and $u = \frac{\partial \psi}{\partial y}$ and $v = \frac{\partial \psi}{\partial x}$, which represent the stream function components in x and y directions; respectively. Therefore, consider a quasi-geostrophic atmosphere with the following unforced, inviscid governing equation:

$$q = \nabla^2 \psi + [\mathrm{Pr}_t]^2 \frac{\partial^2 \psi}{\partial z^2} \tag{2.42}$$

where

$$\frac{Dq}{Dt} = 0, \tag{2.43}$$

Therefore, the two-dimensional material derivative is given by:

$$\frac{D}{Dt} = \frac{\partial}{\partial t} + \vec{u}.\nabla \tag{2.44}$$

Consequently, the vertical atmospheric boundary conditions are as follows:

$$\frac{D}{Dt}\left(\frac{\partial \psi}{\partial z}\right) = 0, \quad \text{at } z = 0, H_a \tag{2.45}$$

Equations 2.42 to 2.45 are equivalent to the equations of motion for absolute two-dimensional flow. Particularly, with either periodic lateral boundary conditions, or conditions of no-normal flow, two quadratic invariants of the motion have occurred. Multiplying equation 2.42 by $-\psi$ and q and then integrating throughout the domain of the motion and volume V, the energy $\widehat{E}$ and the enstrophy $\widehat{Z}$ can be obtained by:

$$\widehat{E} = \int (\nabla_3 \psi)^2 dV, \qquad \widehat{Z} = \int (\nabla_3^2 \psi)^2 dV \tag{2.46}$$

$$\nabla_3 = \vec{i}\frac{\partial}{\partial x} + \vec{j}\frac{\partial}{\partial y} + \vec{k}\frac{\partial}{\partial\left[z[\mathrm{Pr}_t]^{-1}\right]} \tag{2.46.1}$$

The energy and enstrophy invariants are demonstrated in Equation 2.46. Enstrophy, in practice, is a measure of the kinetic energy of a geostrophic flow caused by turbulence. The enstrophy invariant occurs in response to the vortex, the stretching factor, which is so crucial in three-dimensional turbulence, dissipating completely in two dimensions. Because vorticity is preserved on parcels, it is obvious that the integral of any vorticity function, when integrated across the area, is zero [27–30]. The vortex phenomenon will be discussed further in the following sections.

Equation 2.46 denotes any dynamical behavior observed in two-dimensional equations that are simply determined by energy/enstrophy constraints would have an analogous in quasi-geostrophic flow. Particularly, transfer of energy to large scales and enstrophy to small scales would occur in quasi-geostrophic flow, with matching spectra $k^{-\frac{5}{3}}$ and k^{-3} precisely as these transfers are performed through a discrete spectra energy quanta. In other words, at wavenumbers greater than the wavenumber corresponding to the instability scale, the energy per unit horizontal wavenumber $\widehat{E}(k)$ decays as rapidly as k^{-3}. This anticipated kinetic energy spectrum is broadly compatible with observations at the synoptic scale. On the contrary, the transition to the smoother $k^{-\frac{5}{3}}$ mesoscale spectrum has been construed as the key signature of small-scale geostrophic flows triggered by convective occurrences. Therefore, these synoptic-to-mesoscale transition hypotheses rely on turbulent dynamics and extensive interactions between synoptic and mesoscale flows [28–30].

In the quasi-geostrophic scenario, nevertheless, the three-dimensional wavenumber is significant, with the perpendicular direction scaled by the turbulent Prandtl ratio. Accordingly, the energy cascade to broader horizontal scales is typically convoyed by a cascade to larger vertical scales—a flow barotropization. Consequently, in two layers ψ_1 and ψ_2 the total energy and enstrophy $\hat{Z}_1$ and $\hat{Z}_2$ of quasigeostrophic flow can be expressed as:

$$\hat{E}(k_d) = 0.5\int_{\Lambda}\left[(\nabla\psi_1)^2 + (\nabla\psi_2)^2 + 0.5k_d^2(\psi_1 - \psi_2)^2\right]d\Lambda, \tag{2.47}$$

The kinetic energy is represented by the first two components in the energy formula 2.47, and the available potential energy is proportional to the temperature discrepancy. Thus, the baroclinic radius of deformation is inversely proportional to the wavenumber .

$$\hat{Z}_1 = \int_{\Lambda} q_1^2 d\Lambda, \qquad \hat{Z}_2 = \int_{\Lambda} q_2^2 d\Lambda. \tag{2.48}$$

Expression 2.48, therefore, demonstrates two layers of enstrophy. Yet, the analogy in two-dimensional and quasi-geostrophic cascades would not be extended much further, since potential vorticity in the latter is advected solely through horizontal flow. Therefore, the kinematics of quasi-geostrophic turbulence would not be isotropic in three dimensions k_d. Let us consider $\vartheta_b = \psi_1 - \psi_2$ is the two layers' baroclinic streamfunction while $\psi = \psi_1 + \psi_2$ is the barotropic stream function. The potential vorticities for every layer could, therefore, be expressed as follows:

$$p_{v_1} = \nabla^2\psi + (\nabla^2 - k_d^2)\vartheta_b, \qquad p_{v_2} = \nabla^2\psi - (\nabla^2 - k_d^2)\vartheta_b \tag{2.49}$$

In potential vorticities of two-layer expressions, the mode of barotropic stream function ϑ_b is accompanied by $-k_d^2$ everywhere the Laplacian operator ∇ accomplishes on it. In this circumstance, the expression of the operative horizontal wavenumber of ϑ_b is shifted as $k^2 \to k^2 + k_d^2$. In this perspective, the kinematics can be recast as evolution equations for ϑ_b and ψ as follows:

$$\frac{\partial}{\partial t}\nabla^2\psi + J(\psi, \nabla^2\psi) + J\left(\vartheta_b, (\nabla^2 - k_d^2)\vartheta_b\right) = 0, \tag{2.50}$$

$$\frac{\partial}{\partial t}(\nabla^2 - k_d^2)\vartheta_b + J(\vartheta_b, \nabla^2\psi) + J\left(\psi, (\nabla^2 - k_d^2)\vartheta_b\right) = 0. \tag{2.51}$$

$$J = \int\left(k^{-1} - \frac{\int k^{-1}\hat{Z}(k^{-1})dk^{-1}}{\int \hat{Z}(k^{-1})dk^{-1}}\right)^2 \hat{Z}(k^{-1})dk^{-1}. \tag{2.52}$$

here $\hat{Z}(k^{-1})dk^{-1}$ presents the enstrophy. Consequently, expressions 2.50 and 2.51 demonstrate nonlinear interactions between ϑ_b and ψ in geostrophic turbulence flow. In this view, ϑ_b is a baroclinic mode with an effective vertical wavenumber of one. On the other hand, ψ is the barotropic mode with a 'vertical wavenumber', $k^{z'}$, of zero, It is explicit from expression 2.51 that geostrophic turbulence is similarly comprised of a total of interacting triads, however, there are potentially two distinct forms: (i)$(\psi, \psi) \to \psi$, and (ii)$(\vartheta_b, \vartheta_b) \to \psi$ or $(\psi, \vartheta_b) \to \vartheta_b$. In these circumstances, the initial form is a barotropic triad, which essentially unites the barotropic mode. A baroclinic triad, therefore, is represented by the other two states. In the circumstances of the triad interactions, both ψ and ϑ_b are not equivalent. In this scenario, three vertical wavenumbers would count to zero other than ψ be affected by a vertical wavenumber of zero while ϑ_b having a vertical wavenumber of plus or minus one [12, 15, 20, 28, 30, 31].

Geostrophic turbulence is evidently a consequence of the quantum equation of motion. In common belief, the equation of motion is thought to be a direct solution to the ground state function [32]. In other words, the equation of motion begins with a ground state wave function that has been optimized at a particular level of description. To present both ϑ_b and ψ in the quantum equation of motion, the general form of Marghany Operator $\widehat{\mathfrak{M}}^{\dagger}_{\vartheta_b} = |\vartheta_b\rangle\langle\psi|$ is introduced in which $|\vartheta_b\rangle$ presents the motion generation due to air molecules fluctuations in the baroclinic state and $\langle\psi|$ is the quantum state of barotropic mode. Under the circumstance of $|\vartheta_b\rangle > \langle\psi|$ the energy of the quantum geostrophic turbulence transfers from $|\vartheta_b\rangle \rightarrow \langle\psi|$. In this scenario $\widehat{\mathfrak{M}}^{\dagger}_{\vartheta_b} > 1$ and initiates the turbulence flow. On the other hand, geostrophic turbulence does not occur if $|\vartheta_b\rangle = 0$ i.e. $\widehat{\mathfrak{M}}^{\dagger}_{\vartheta_b} = 0$. In this understanding, the transfer of quantum energy from $|\vartheta_b\rangle \rightarrow \langle\psi|$ can be expressed mathematically by the Hamiltonian and the general operator form of Marghany $\widehat{\mathfrak{M}}^{\dagger}_{\vartheta_b}$ as:

$$\left(\hat{\mathrm{H}}, \widehat{\mathfrak{M}}^{\dagger}_{\vartheta_b}\right)|\vartheta_b\rangle = \hat{\mathrm{H}}, \widehat{\mathfrak{M}}^{\dagger}_{\vartheta_b}|\vartheta_b\rangle - \widehat{\mathfrak{M}}^{\dagger}_{\vartheta_b}, \hat{\mathrm{H}}, \langle\psi| = E_{\vartheta_b} \widehat{\mathfrak{M}}^{\dagger}_{\vartheta_b}|\vartheta_b\rangle. \tag{2.53}$$

The energy of the quantum geostrophic turbulence between two quantum states of $|\vartheta_b\rangle$ and $\langle\psi|$ is formulated as:

$$E_{\vartheta_b} = \frac{|\vartheta_b\rangle\left(\hat{\mathrm{H}}, \widehat{\mathfrak{M}}^{\dagger}_{\vartheta_b}\right)\langle\psi|}{|\vartheta_b\rangle\left(\widehat{\mathfrak{M}}^{\dagger}_{\vartheta_b}\right)\langle\psi|} \tag{2.54}$$

Equation 2.54 summarizes that turbulence energy leads to geostrophic turbulence in a two-layer quantum system. Because of efficient energy of the baroclinic mode E_{ϑ_b} is greater than that of the barotropic mode E_{ψ}, energy will be transferred to the barotropic mode, which can be addressed as a novel process named "quantum bororopization". Because this quantum mechanical is essentially instability of E_{ϑ_b}, the change from quantum baroclinic state $E_{\vartheta_b}|\vartheta_b\rangle$ to weak barotropic quantum state energy inclines to arise at the scale of the radius of deformation $E_{\vartheta_b}|\vartheta_b\rangle$. The quantized Marghany quasi-geostrophic turbulence concept can be addressed based on the dynamical quantum energy changes from a higher level at the quantum baroclinic mode $E_{\vartheta_b}|\vartheta_b\rangle$ to a lower level at the quantum barotropic state $E_{\psi}|\psi\rangle$ (Figure 2.15). At large horizontal scales, a quantum source of baroclinic state energy could be the differential electron excitation of heating between the pole and the equator in the atmosphere [33]. In this circumstance, quantum baroclinic energy state instability generates a non-domestic transfer of quantum energy to the strain scale, triggering both quantum baroclinic and barotropic energy modes. Therefore, in the quantum barotropic phase, energy is subsequently transmitted back to large scales, where it is eventually dissipated by quantum energy friction between $E_{\vartheta_b}|\vartheta_b\rangle$ and $E_{\psi}|\psi\rangle$.

Simultaneously, quantum enstrophy $E_{\widehat{Z}}\langle\widehat{Z}| \approx E_{\psi}\langle\psi|$ collapses to lower and smaller scales within every quantum energy layer, until the quantum spectral energy is wide enough so that non-geostrophic influences grow important and quantum enstrophy is dispersed and dissipated by three-dimensional processes.

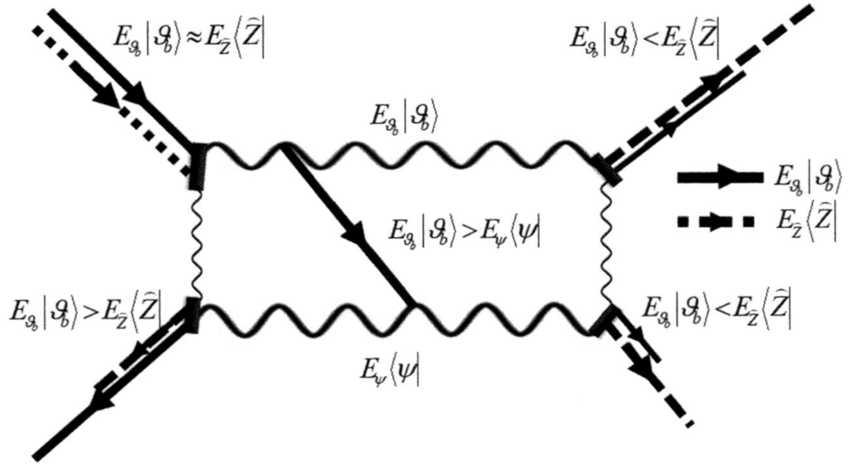

Figure 2.15: Quantized Marghany quasi-geostrophic turbulence speculation.

According to the above perspective, the Reynolds number can be modified based on quantized Marghany quasi-geostrophic turbulent energy wavelength as follows:

$$R_e = \frac{\partial P_a}{\partial x} \times \left[\eta W_s E_\vartheta \left| \lambda_\vartheta \right\rangle \right]^{-1} \tag{2.55}$$

Equation 2.55 reveals the clear transformation of the quantum energy of quasi-geostrophic flow into turbulence flow due to fluctuation of air pressure ∂P across the isobar distance changes ∂x and inverse changes in air viscosity η; actual wind speed W_s; and spectra wavelength of quantized Marghany quasi-geostrophic turbulent λ_ϑ in a quantum baroclinic state $E_{\vartheta_b}\left|\vartheta_b\right\rangle$ (Figure 2.15). In this understanding of strong vorticity, the viscosity of the center of vorticity is smaller than that around it owing to the influence of the curvature of the space-time of the Earth's rotation as discussed in section 2.5. Needless to say, stress-energy tensors are involved principally in its matrix in pressure and energy density, which causes fluctuation pressure on air composite molecules as the source of the curvature of wind speed trajectory movement across the Earth's space-time [32–35]. In the next section, the vorticity turbulent generation owing to the fluctuation of quasi-geostrophic would be addressed as confirmation of the curvature of the Earth's surface as space-time curvature.

2.8 Quantum Decomposition of Energy Spectral Turbulence

Now the query is: what atmospheric dynamic system may quasi-geostrophic turbulence generate? From the point of view of quantum mechanics, energy would be transferred to another quantum energy state. The quantum energy state of quasi-geostrophic turbulence would be quantized vertically into discrete energy with the equilibrium of inertia, buoyancy, pressure gradient, and Coriolis forces. The decomposition of these quantum spectra energy is well-known as large-scale inertia–gravity waves. On the other hand, the curvature of Earth's surface and Earth's rotation as explained previously assist in generating a quasi-2-D flow in which the quantum energy is trapped on a large scale. On the contrary, there is no quantum energy transfer to small scales as shown in Figure 2.15. Consequently, large-scale-nonlinear turbulence occurrences cause a wider scale of quantum energy transferred [33–35].

A sequence of weakly nonlinear inertia-gravity waves dominates the mesoscale energy. Therefore, the dispersion and polarization relations of inertia-gravity waves would relate the mesoscale spectra of horizontal wind variations as a function of horizontal wavenumber, vertical wavenumber, and frequency. The dominance of inertia-gravity waves at the mesoscale, on the other hand, appeared incongruous under the weak quantum energy state of enstrophy circumstances. In this understanding, inertia-gravity waves interact extremely weakly with the geostrophic flow because they have tiny amplitudes in the troposphere and lower stratosphere [25–28].

Even though there is swelling recognition of rare instances of inertia–gravity waves directly influencing sensitive weather patterns, strong interactions between inertia–gravity waves and the geostrophic flow are naturally restricted to the middle and upper atmosphere, where the wave amplitude becomes sufficient to permit the breaking of inertia–gravity waves and the concurrent drag force on the geostrophic flow. In this view, let us assume the dissociated superposition of a geostrophic flow and inertia–gravity waves. Therefore, the quantum energy state of inertia–gravity wave composites of the kernel of the quantum kinetic and quantum potential energies, are based on the quantized Marghany quasi-geostrophic turbulence speculation. In other words, the quantum of the inertia–gravity wave energy spectrum is equivalent to the total energy spectrum in the discrete mesoscale range.

According to the above perspective, geostrophic flows contain just a rotating component because they are horizontally nondivergent, whereas inertia–gravity waves have both a rotational and a divergent component as follows:

$$u = -\psi_y + \phi_x \tag{2.56}$$

$$v = \psi_x + \phi_y \tag{2.57}$$

In terms of the velocity field, the functions ψ and ϕ are uniquely regulated with particularly periodic boundary conditions. The superposition of the energy eigenstates of geostrophic turbulence energy $|E_g\rangle$ and inertia-gravity waves $|E_w\rangle$ can be presented in the modification form of quantized Marghany quasi-geostrophic turbulence speculation as casted by:

$$\left(\hat{\mathrm{H}}, \widehat{\mathfrak{M}}^{\dagger}_{g_b}\right)\left|E_g\right\rangle = \hat{\mathrm{H}}, \widehat{\mathfrak{M}}^{\dagger}_{E}\left|E\right\rangle - \widehat{\mathfrak{M}}^{\dagger}_{w}, \hat{\mathrm{H}}, \left|E_w\right\rangle. \tag{2.58}$$

The quantized Marghany decomposition of superposition (Figure 2.16) of the energy eigenstates of geostrophic turbulence energy $|E_g\rangle$ and inertia-gravity waves $|E_w\rangle$ in equation 2.58 reveals that inertia-gravity waves dominate the mesoscale range, although they do not deliver considerable energy at synoptic scales. At synoptic scales, the residual quantized Marghany decomposition of superposition of the energy eigenstates is considered as the subjugated entire spectrum $\Psi|E\rangle$ minus the inertia-gravity wave $\Psi|E_w\rangle$ [28, 33, 35].

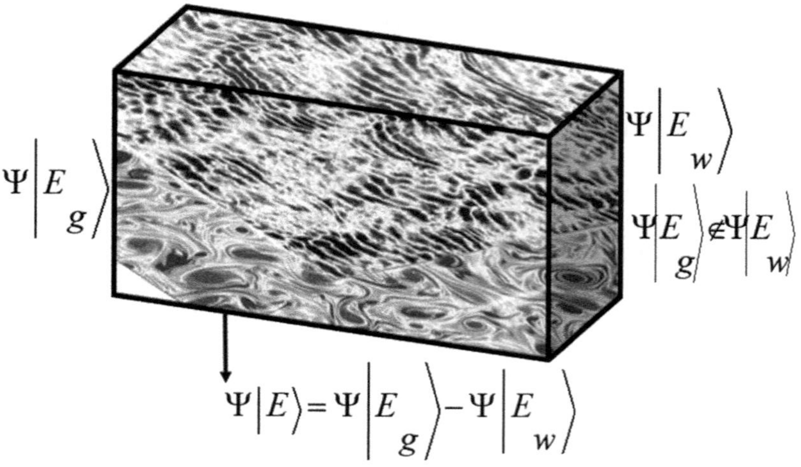

Figure 2.16: Superposition of quantum energy states of quasi-geostrophic turbulence and inertia-gravity wave.

Because the spectrum in the synoptic range is exclusively rotating and hence horizontally nondivergent, this component would be confidently assigned to geostrophic flows. The geostrophic component of total energy continues to diminish significantly at the transition scale—the transformation appears to be due to inertia–gravity waves becoming dominant in the mesoscale range. Therefore, the superposition of energy eigenstates for both $\Psi\left|E_g\right\rangle$ and $\Psi\left|E_w\right\rangle$ can be given by:

$$\Psi\left|E\right\rangle = \frac{1}{\sqrt{2}}\Psi\left|E_g\right\rangle - \frac{1}{\sqrt{2}}\Psi\left|E_w\right\rangle \tag{2.59}$$

Equations 2.58 and 2.59 prove the quantum constitution delivered by quantized Marghany quasi-geostrophic turbulence. Quantized Marghany quasi-geostrophic turbulence in a higher quantum energy state of $E_{\vartheta_b}\left|\vartheta_b\right\rangle$ is considered a superposition state with other quantum energy states in the atmospheric system such as $E_{\hat{Z}}\left\langle\hat{Z}\right|$ and $E_\psi\left\langle\psi\right|$. In other words, the atmosphere, identical to the ocean, is a fluid that spins rapidly and is stratified. The utmost energy large-scale dissimilarities are caused by $E_{\vartheta_b}\left|\vartheta_b\right\rangle$ baroclinic instabilities in both fluids, even though somewhat rapid disturbance triggers small-scale inertia–gravity wave energy $\Psi\left|E_w\right\rangle$. Both fluids display a shift from geostrophic dynamics at large sizes to inertia–gravity wave dynamics at small scales [12, 24, 28, 30, 35].

2.9 Physical Characteristics of Vorticity

Before such a debate on quantum vorticity, the question of what vorticity arise? A vortex (plural vortices/vortexes) is a region in a fluid where the flow swirls about an axis of rotation that might be straight or curved. Vortices arise in agitated fluids and may be seen in smoke rings (Figure 2.17), whirlpools in a boat's wake (Figure 2.18), and winds enveloping a tropical cyclone, tornado, or dust monster. The significant question now is: Is vorticity a major component of turbulent flow? A cornerstone of the turbulent flow component is indeed vorticity. Turbulent flow is dominated by vortices. Vortices are characterized by the distribution of velocity, vorticity (the curl of the flow velocity), and the theory of circulation. The fluid flow velocity in most vortices is generally formed closer to its axis and diminishes in inverse proportion to the outer radius.

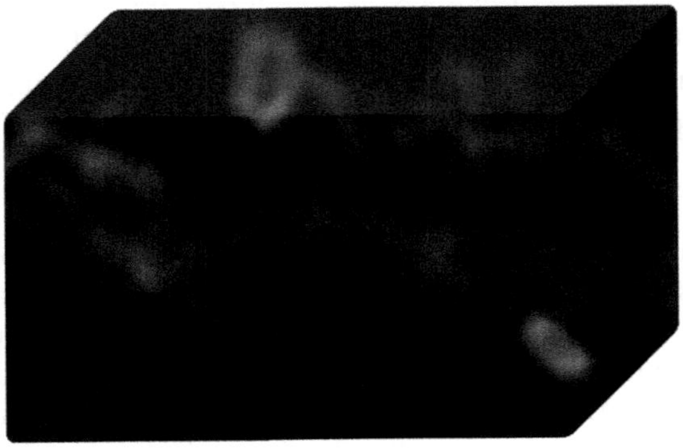

Figure 2.17: Vorticity in the form of smoke rings.

In the circumstance of the nonexistence of exterior forces, the fluid tends to flow into irrotational vortices, conceivably superimposed to larger-scale motions, counting larger-scale vortices due to viscous friction. In this view, once created, vortices can transfer, stretch, twist, and interact in complex circumstances. An energetic vortex conveys with it various angular and linear momentum, energy, and mass [36].

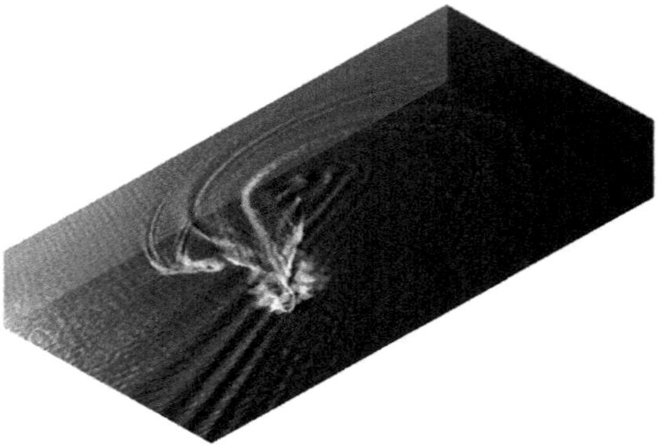

Figure 2.18: Vorticity in the forms of whirlpools in a boat's wake.

Figure 2.19: Vorticity in the form of a dust monster.

Therefore, the vorticity vector's direction is demarcated as the axis of rotation of this imaginary ball (consistent with the right-hand rule), and its length is twice the angular velocity of the ball. The vorticity $\vec{\omega}$ is described mathematically as the curl (or rotational) of the fluid's velocity field, which is commonly signified and specified by the vector analysis formula times, where nabla ∇ is the nabla operator and $\vec{u}$ is the local flow velocity.

$$\vec{\omega} \equiv \nabla \times \vec{u} \tag{2.60}$$

Technically speaking, del is not a unique operator, but instead a convenient mathematical symbol for these kinds of three operators that simplifies the formulation and memorizing of numerous equations. The del symbol (or nabla) could well be understood as a vector of partial derivative operators, and its three alternative meanings—gradient, divergence, and curl—can formally be seen as the product of the "del operator" with the field with a scalar, a dot product, and a cross product, respectively. These formal products are sometimes not compatible with other operators or products. These three product operators can be casted as: (i) *Curl*: *curl* $\vec{\omega} \equiv \nabla \times \vec{u}$, (ii) Gradient: grad $f = \nabla f$; and (iii) Divergence: div $\vec{\omega} = \nabla . \vec{u}$ [36–38].

If there is shear, the vorticity may be nonzero even though all particles are traveling down straight and parallel path lines (that is, if the flow speed varies across streamlines). For instance, in laminar flow within a pipe with a constant cross-section, all particles flow parallel to the axis, but rapidly near the axis and are practically steady in line with the walls [36]. The vorticity would be zero on the axis and maximal towards the walls, in which shear would be greatest. The vorticity equation of fluid dynamics depicts the growth of the vorticity of a fluid particle as it travels with the flow; that is, the fluid's local rotation (in terms of vector calculus, this is the curl of the flow velocity) [38]. The governing formula of vorticity is:

$$\begin{aligned}\frac{D\omega}{Dt} &= \frac{\partial\omega}{\partial t} + (\vec{u}.\nabla)\omega \\ &= (\omega.\nabla)\vec{u} - \omega(\nabla.\vec{u}) + \frac{1}{\rho^2}\nabla\rho\times\nabla P_a + \nabla\left(\frac{\nabla.\tau}{\rho}\right) + \nabla\times\left(\frac{\vec{B}}{\rho}\right)\end{aligned} \tag{2.61}$$

The left-hand side term in equation (2.61) $\frac{D\omega}{Dt}$ is the material derivative of the vorticity vector. $\frac{D\omega}{Dt}$, therefore, represents the rate at which the vorticity of a moving fluid particle changes. This variation might even be ascribed to unsteadiness in the flow $\frac{\partial\omega}{\partial t}$, or the motion of the fluid particle as it travels from one location to another $(\vec{u}.\nabla)\omega$, i.e., the convection term. Consequently, the term $\nabla \times B$ indicates variations caused by external body forces. These are forces, such as gravity, that are distributed throughout the three-dimensional space of the fluid. In contrast to forces, for instance, drag on a wall and surface tension around a meniscus that only operates on a surface and a line, respectively. Subsequently, on the right side of equation 2.61, the term $(\omega.\nabla)\vec{u}$ depicts the extending or bowing of vorticity caused by flow velocity gradients. It is worth noting that $(\omega.\nabla)\vec{u}$ is a vector quantity since $(\omega.\nabla)$ is a scalar differential operator, whereas $\nabla\vec{u}$ is a nine-element tensor variable. In this view, the baroclinic term is defined as $\frac{1}{\rho^2}\nabla\rho\times\nabla P_a$. It compensates for vorticity variations caused by the junction of density and pressure surfaces. The term $\nabla\left(\frac{\nabla.\tau}{\rho}\right)$, on the other hand, compensates for the diffusion of vorticity owing to viscous processes. In this context, the expression $\omega(\nabla.\vec{u})$ refers to the extension of vorticity generated by flow compressibility [36–40]. It derives from the Navier-Stokes equation for continuity, which is as follows:

$$\begin{aligned}&\frac{\partial\rho}{\partial t} + \nabla.(\rho\vec{u}) = 0 \\ &\Leftrightarrow \nabla.\vec{u} = -\frac{1}{\rho}\frac{\partial\rho}{\partial t} = \frac{1}{v}\frac{dv}{dt}\end{aligned} \tag{2.62}$$

where $v = \frac{1}{\rho}$ is the fluid element's specific volume $\nabla.\vec{u}$ may be seen as a measure of flow compressibility. A negative sign is sometimes added to the expression. Accordingly, the vorticity formula simplifies for an inviscid, barotropic fluid with conservative body forces as given by:

$$\frac{d}{dt}\left(\frac{\omega}{\rho}\right) = \left(\frac{\omega}{\rho}\right).\nabla\vec{u} \tag{2.63}$$

Similarly, the vorticity mathematical formula, if such fluid is incompressible and inviscid, with conservative body forces is expressed as:

$$\frac{d\omega}{dt} = \frac{d\omega}{dt} + (\vec{u}.\nabla)\omega = (\omega.\nabla)\vec{u} \tag{2.64}$$

The relative vorticity, as defined above, is the vorticity caused by the air velocity field relative to the Earth. Because the relative vorticity vector is often a scalar rotation quantity perpendicular to the ground, this air velocity field is frequently treated as a two-dimensional flow parallel to the ground. When the wind turns counterclockwise as seen from the earth's surface, vorticity is positive. Positive vorticity is referred to as cyclonic rotation in the northern hemisphere, while negative vorticity is referred to as anticyclonic rotation in the southern hemisphere [32, 34, 36, 40].

Since absolute vorticity is determined from air velocity relative to an inertial frame, it incorporates a factor, owing to Earth's rotation, known as the Coriolis parameter (section 2.5). The potential vorticity is calculated by dividing the absolute vorticity by the vertical separation between levels of constant (potential) temperature (or entropy). The absolute vorticity of an air mass changes when it is stretched (or compressed) vertically, but the potential vorticity is conserved in adiabatic flow. Since adiabatic flow prevails in the atmosphere, potential vorticity can be used as an approximate tracer of air masses in the atmosphere over a few days, especially when observed on levels of constant entropy [37–40].

Therefore, the atmosphere is assumed to be nearly barotropic through the barotropic vorticity equation. The height does not govern the direction and speed of the geostrophic wind. In this understanding, the geostrophic wind would not have a vertical wind shear. Consequently, in this sort of atmosphere, the centres of warm and cold temperature anomalies are dominated by high and low-pressure zones. For instance, the subtropical ridge and the Bermuda-Azores high have warm-core highs while shallow Arctic highs and tropical cyclones have cold-core lows and strengthening winds with height, with the reverse true for cold-core highs and warm-core lows; respectively. The barotropic vorticity equation is an alternative version of the vorticity equation for an inviscid, divergence-free flow (solenoidal velocity field) and is expressed as:

$$\frac{D(\zeta + f)}{Dt} = 0 \tag{2.65}$$

here ζ is relativity vorticity, which is demarcated as the vertical constituent of the curl of the fluid velocity and the Coriolis parameter. Consequently, equation 2.65 may be expressed as the following in terms of relative vorticity:

$$\frac{D(\zeta)}{Dt} = -v\frac{\partial f}{\partial y}, \tag{2.66}$$

The relativity vorticity is characterized by the fluctuation of the Coriolis parameter $\frac{\partial f}{\partial y}$ in the north-south direction with distance y, and the component of velocity v in this direction. Subsequently, the mechanism of vorticity transfer from the environment to an air parcel in convective movement is known as helicity. In this occurrence, the definition of helicity H_c is abridged to involve simply the horizontal constituent of wind $\vec{W}_h$ and horizontal vorticity ζ_h:

$$H_c = \int \vec{W}_h . \zeta_h d\vec{Z} = \int \vec{W}_h . \nabla \times \vec{W}_h \, d\vec{Z} \tag{2.67}$$

Consistent with formula 2.67, under the circumstances of $\vec{W}_h$ and $\nabla \times \vec{W}_h$ are perpendicular to each other $H_c = 0$ as long as $\vec{W}_h$ does not change with altitude fluctuation $d\vec{Z}$. This scenario renders their scaler product nil. On the other hand, H_c tends to be positive as long as $\vec{W}_h$ turns into a clockwise pattern with altitude changes $d\vec{Z}$. However, counterclockwise $\vec{W}_h$ causes negative H_c. In meteorology, H_c, therefore, is exploited as energy per unit of mass $\left(\text{m}^2\text{s}^{-2}\right)$. As a consequence, H_c is apprehended as a measure of energy transfer by the direction of wind shear fluctuation $d\vec{Z}$, encompassing heat and energy exchange [36, 37, 39]. This understanding is used to forecast the likelihood of tornadic development in a thundercloud. Vertical integration would be restricted below cloud tops (usually 3 km or 10,000 feet) in this instance, and horizontal wind would be determined relative to the storm by eliminating its movement:

$$H_c(S_R) = \int(\vec{W}_h - \vec{V}_C).\nabla \times \vec{W}_h d\vec{Z} \tag{2.68}$$

Storm Relative Helicity $H_c\ (S_R)$ used cloud motion $\vec{V}_C$ to determine tornadic development. In this regard, supercells are possible with weak tornadoes, according to the Fujita scale, if $H_c\ (S_R)$ range from 150 to 299. Therefore, strong tornadoes accompany supercell development in cases of $H_c\ (S_R)$ ranges from 300-499. Consequently, if $H_c\ (S_R)$ is larger than 450, violent tornadoes occur. Nevertheless, helicity is not the crucial constituent of severe thunderstorms, and these values would also be interpreted cautiously [41]. As a corollary, the Energy Helicity Index $E(H_c\ (S_R))_I$ was developed. $E(H_c\ (S_R))_I$ is calculated by multiplying $H_c\ (S_R)$ by E_{CAP} (Convective Available Potential Energy) and then dividing by a specified threshold as:

$$E(H_c(S_R))_I = \frac{H_c(S_R) \times E_{CAP}}{160000}. \tag{2.69}$$

Formula 2.69 demonstrates the occurrence of possible tornados if $E(H_c\ (S_R))_I$ is equivalent to one. Moderate to strong tornados occur under the circumstances of $E(H_c\ (S_R))_I$ ranges from 1 to 2. Consequently, as long as $E(H_c\ (S_R))_I$ is larger than 2, destructive tornados develop. In this sense, E_{CAP} resides inside the free convective layer (FCL), a potentially unsteady layer of the troposphere, where such a rising air parcel is warmer than the surrounding atmosphere. E_{CAP} is expressed in terms of joules per kilogram of air $\left(\frac{\mathrm{J}}{\mathrm{kg}}\right)$ but may also be expressed as $\mathrm{m}^2\mathrm{s}^2$. All values greater than zero $\left(\frac{\mathrm{J}}{\mathrm{kg}}\right)$ imply increased instability and the potential of thunderstorms and hail [37–41]. Generic E_{CAP} is computed by vertically integrating a parcel's local buoyancy from the level of free convection z_f to the equilibrium level z_n:

$$E_{CAP} = \int_{z_f}^{z_n} g\left(\frac{T_{v,parcel} - T_{v,env}}{T_{v,env}}\right) dz \tag{2.69.1}$$

The integral of formula 2.69.1 expresses the work accomplished by the buoyant force excluding the work executed in contradiction to gravity g, henceforth it determines the additional energy, which turns into kinetic energy. In this context, $T_{v,parcel}$ and $T_{v,env}$ are the virtual temperature of the specific parcel and the virtual temperature of the environment, respectively. Both temperature units in formula 2.69.1 would be in the Kelvin scale [39–43].

2.10 Can Quantize Vorticity?

Now the cornerstone question is: what is quantum vorticity? Let us first address the topological solitons as the keystone for vortex formation and development. Certainly, topological solitons arise because of symmetrical splitting in atmospheric parameters such as pressure, temperature, density, geostrophic turbulence energy, etc., which cause phase transitions through the atmospheric layers. For instance, the change of the phase transition of geostrophic turbulence energy can form a vortex flow across atmospheric layers. In other words, topological solitons are distinguished by a relevant topological index of pressure dynamic fluctuations Q, the maintenance of which is a solution to some structural features of the atmospheric physical space and the theory's field space. Therefore, the existence of the so-called topological energy bound, which asserts that the energy

E_w of any field configuration is constrained from below by the topological level, ensures the stability of topological solitons. Generally, the bound has the following formula form:

$$E_w \geq C|Q| \tag{2.70}$$

C is a numerical constant that is independent of the volume of the base space. Consequently, this bound is applicable to both infinite and finite volume base spaces. In this scenario, the volume-dependent bound encodes some information about the soliton's resistance to external pressure. Indeed, it demonstrates how energy increases when soliton is forced to inhabit a confined volume region. This inevitably leads to a significant quantity describing a soliton, and its compressibility, which is defined as [44]:

$$\kappa = -[V]^{-1}\left(\frac{\partial V}{\partial P_a}\right)_{Q,T} \tag{2.71}$$

here where V is the volume of the soliton, P_a is its pressure and T the temperature. When given non-zero pressure, solitons diminish their volumes. Sustaining the solitons in the decreased space demands more energy. Nevertheless, this energy is indeed restricted, even though it grows rapidly as V reduces (or, equivalently, as p increases). Consequently, all recognized solitons are compressible and can be squeezed to smaller sizes with a finite quantity of energy. Quantum vortices, therefore, are often a sort of topological soliton seen in superfluids and superconductors. Therefore, the quantization of vorticity is a direct result of the presence of a super mobility air pattern such as hurricanes and tornadoes. Indeed, quantum vortices characterize superfluid circulation, and their excitations are thought to be a consequence of superfluid phase transitions. In this perspective, a quantum vortex is thought of as a cavity with superflow swirling around the vortex axis (Figure 2.20); the inside of the vortex can comprise energized particles. Furthermore, when the water moves away from equilibrium, quantum vorticity increases line length by counter-current flow [39–44].

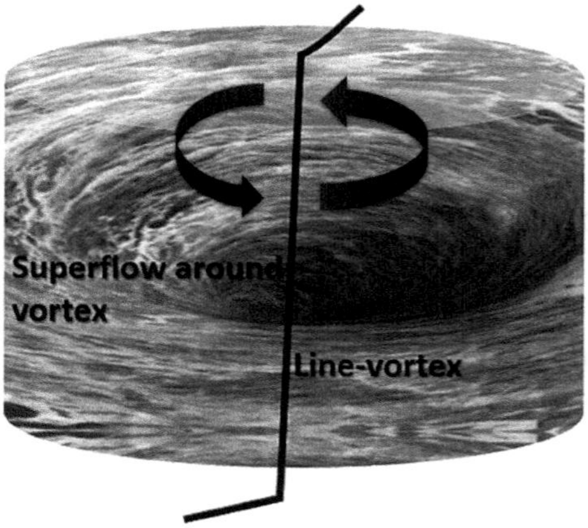

Figure 2.20: Simplification of superflow around the line vortex.

In this understanding, tornados and hurricanes (Figure 2.21), for instance, in the atmosphere do not transpire as we recognize them, nonetheless, differences in air flow are large-scale assemblies of quantized vortices. There are manifold potentials for the fine-scale assembly of these vortices, as a function of temperature and pressure. In this scenario, the super hurricane mobilities still conform to air mass-dependent dynamics but absent an upsetting circulation on account of homogenous temperature.

Figure 2.21: Tornados and hurricanes swirl as vortices.

A hurricane has the special property of exhibiting phase, which is specified by the wave function and its velocity. Therefore, the phase gradient (in the parabolic mass approximation) is proportional to the velocity variations. If the region encompassed is only allied, the circulation around any closed loop in the vorticity is zero. The superfluid is considered irrotational flow; however, if the bounded region consists of a smaller region with no vortex, such as a rod passing through a vortex, then the circulation (Figure 2.22) is mathematically described as:

$$\Gamma = \oint_C \vec{V}.\, d\vec{l} = \frac{\hbar}{m}\oint_C \nabla\phi_v.\, d\vec{l} = \frac{\hbar}{m}\Delta^{tot}\phi_v, \tag{2.72}$$

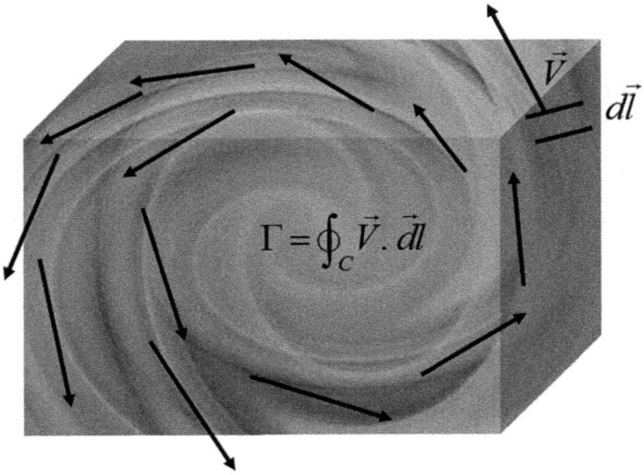

Figure 2.22: Vortex and circulation.

In this understanding, a quantum vortex is a quantized flux circulation of a physical property in physics. The entire phase difference of the air mass m oscillations encircling the vortex $\Delta^{tot}\phi$ is calculated as follows:

$$\Delta^{tot}\phi = 2\pi n \tag{2.73}$$

Because n is an integer, i.e., 0, 1, 3,...., equation 2.73 states that the wave-function must return to its original value after an integer number of rotations around the vortex. Consequently, the vortex circulation can be quantized as follows:

$$\Gamma = \oint_C \vec{V}.\, d\vec{l} \equiv \left[\frac{\partial \vec{V}}{\partial t}\right]\frac{2\pi\hbar}{Q} n. \tag{2.74}$$

The term $\left[\frac{\partial \vec{V}}{\partial t}\right][Q]^{-1}$ presents fluctuation in air mass as a function of the topological pressure index Q which is based on the Marghany quantum energy wind-driven theory. In this understanding, vortex circulation is proportional inversely to the pressure topology index and directly to the velocity of air vorticity [43–45].

Furthermore, the existence of $\hbar$ indicates that quantized vorticity is a result of quantum physics. In other terms, a quantized vortex has such quantized circulation. Only quantized vortices can sustain a superflow's rotating motion [40–45]. Furthermore, the ratio $\frac{\hbar}{Q}$ is fairly macroscopic, and the number of vortex lines varies with it. Therefore, a quantized vortex is a stable topological soliton of a Bose-Einstein condensate (BEC) with a wide wave-function [47]:

$$\psi(\vec{r},t) = |\psi(\vec{r},t)|\, e^{i\phi(\vec{r},t)} \tag{2.75}$$

Because the macroscopic wave function $\psi(\vec{r},t)$ has a single value for the space coordinate $\vec{r}$, equation 2.75 confirms that the closed-loop circulation in equation 2.74 is quantized by $\frac{\hbar}{Q}$. A quantized vortex is seen as an inviscid superflow vortex in this paradigm. As a result of its sparse core, the quantized vortex cannot be annihilated by viscous vorticity diffusion. Vortices, on the other hand, are not precisely defined as classical viscous fluids. They are, indeed, chaotic, arising and vanishing continuously. As a result, the circulation is not maintained and is not distinct for each vortex. In this view, the quantized vortex is described as a topological perturbation whose potential velocity is provided by:

$$\vec{v}_s = \frac{d\vec{V}}{dt}\left(\frac{\hbar}{Q}\right)\nabla\phi \tag{2.76}$$

As a corollary, the vorticity $\nabla \times v_s$ could well be characterized by the superflow velocity, which vanishes all over in a single zone of the level component of the modulation and demodulation, i.e., $\psi = \sqrt{\rho}e^{i\phi}$ as well as all spinning flow is only approved by quantized vortices. Density ρ at the core diminishes as long as ϕ revolves around the core by 2π. The specific wave function of this revolution might also be characterized in this sense as:

$$\psi(r_1,.....,r_N) = \Pi_{i=1}^{N}\, \phi(r_i) \tag{2.77}$$

here N is the overall number of particles which is totaled as:

$$N = \int dr\, |\psi|^2 \tag{2.77.1}$$

The kinetic energy, potential energy, and interaction energy would propel this wave function. In this aspect, the mean-field energy delivers the interaction energy as pursues:

$$\overline{E} = \frac{4\pi\hbar^2 l}{2Q} \times \frac{d\vec{V}}{d\vec{r}} \tag{2.78}$$

Equation 2.78 reveals the increment in the length of vorticity topography and velocity rotating of vorticity $\vec{V}(\vec{r})$ which can be described by the wave function $\psi(\vec{r})$. In this sense, wave function would include the interaction energy $\bar{E}$, and quantized vortices $\frac{\hbar^2}{2Q}$. In this regard, the Marghany quasi-geostrophic turbulence (section 2.7) can be modulated into vorticity formation based on the Schrödinger equation and described mathematically as:

$$\left(-\frac{\hbar^2}{2Q}\nabla^2\phi+\vec{V}(r)+E_{\vartheta_b}\left|\psi|\vartheta_b\rangle\right|^2\right)=n\bar{E}(|\psi\rangle), \tag{2.79}$$

Equation 2.79 is a time-independent Gross-Pitaevskii equation (GPE) that may be implemented with a continuity equation of density ρ and superflow velocity $\vec{v}_s$ as follows:

$$\frac{\partial\rho}{\partial t}+\nabla.(\rho\vec{v}_s)=0, \tag{2.80}$$

$$\frac{\partial}{\partial t}+\nabla v^2=-\frac{\nabla}{Q\rho}(0.5g\rho^2)+\frac{\nabla}{Q}\left(\frac{\nabla\ \sqrt{\rho}}{\sqrt{}}\right). \tag{2.81}$$

On the right side of equation 2.81, the first term signifies an operative pressure i.e., $p=\int_{z_0}^{z}dz\rho(z)g(z)$, which is the modification of the hydrostatic pressure through the vortex atmospheric latitude z from the surface z_0, which begins owing to the nonlinearity of the GP equation. As a matter of fact, the second concept can be defined as quantum pressure, which has no equivalent in ordinary fluid mechanics and is substantial at tiny scales similar to healing length, as well as at vortex cores, in which ρ dramatically swings in the scale of ξ [40, 45], as determined by:

$$\xi\approx\sqrt{\frac{\hbar^2}{2Q\bar{E}n}} \tag{2.82}$$

As a result, the irrotationality criterion for an atmospheric configuration with a vortex along the z-axis could well be extrapolated to:

$$\vec{\nabla}\times\vec{v}_s=z\frac{n\hbar}{Q}\partial^2(\rho), \tag{2.83}$$

here n is an integer that denotes the circulation frequency of the singularity in the wave function that generates angular momentum through vortex kinetic spinning. Consistent with this perspective, the wavefunction would include a phase parameter $e^{i\theta}$, where θ is the azimuthal angle. The azimuthal velocity could perhaps be obtained by utilizing:

$$v_\theta=\frac{\hbar}{Q}\rho^{-1}\frac{\partial}{\partial\theta}\phi \tag{2.84}$$

As stated by Equation 2.84, the velocity field of a regular spinning flow operates like a single axis, with the azimuthal velocity rising linearly as one moves away from the symmetry axis (Figure 2.23). A superflow has an unusual velocity pattern in which the azimuthal velocity diverges as one moves away from the rotation axis. To prevent the system's kinetic energy from diverging, the density of the superflow would have to be zero.

Consistent with the above perspective, the dynamics and configurations of the vortex cores in a nonlinear quantum vortex flow may indeed be specified in terms of potential vortex-vortex couple bonding (Figure 2.24] [41]. The realistic intervortex potential is likely to disrupt quantum phase transitions while also creating a diversity of few-vortex structures and several vortex formations (Figure 2.25] [42].

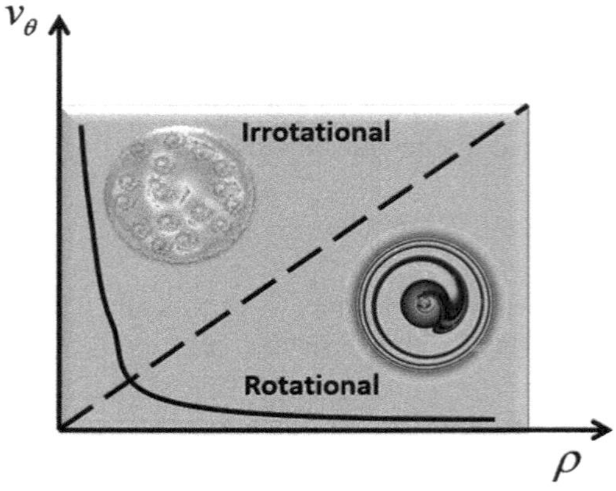

Figure 2.23: The azimuthal velocity of a vortex

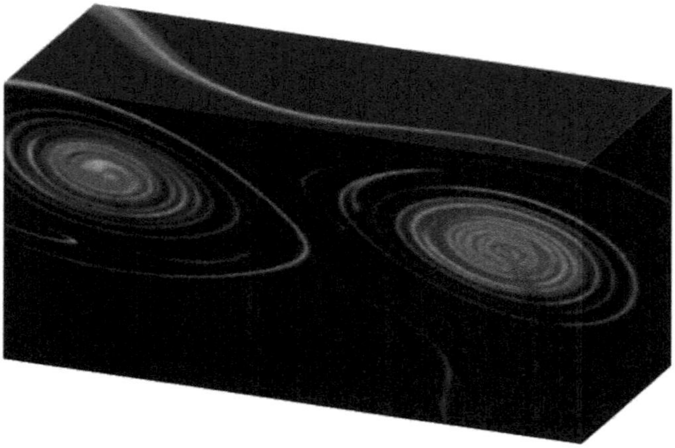

Figure 2.24: Two vortex motion.

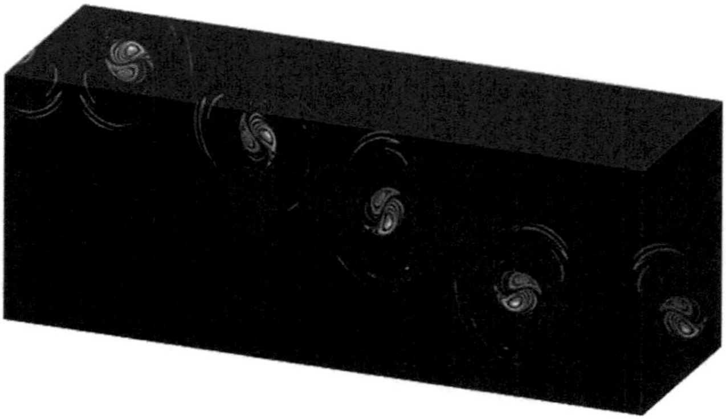

Figure 2.25: Diversity of few-vortex.

Can the quantized vortex describe turbulence as a quantum phenomenon? In this sense, Figure 2.26 delivers an overview of quantum turbulence flow caused by a virtual vortex tangle comprising quantized vortices. Quantum turbulence, in this paradigm, is an entanglement of these quantized vortices, rendering it a perfect form of turbulence. In this aspect, it is straightforward to designate classic turbulence as where several potential eddy swaps rapidly make the issue too complex to forecast what would occur. Turbulence in conventional fluids is commonly illustrated simply by introducing virtual vortex filaments, behind which there is a specific circulation of the fluid, to establish a thoughtful of what is ensuing in the fluid. Therefore, in quantum turbulence, these vortex lines are real—they perhaps indeed be experimental and have an identical convincing circulation—and they also offer the aggregate of the physics of quantum turbulence flows. Since it incorporates vortices, quantum turbulence is considered hydrodynamic. In this view, this turbulence pattern—a tangle of quantized vortex lines—can be agitated at minor length scales or vortex rings (Figure 2.27) or greater length scales as illustrated inside the formation of altocumulus clouds (Figure 2.28).

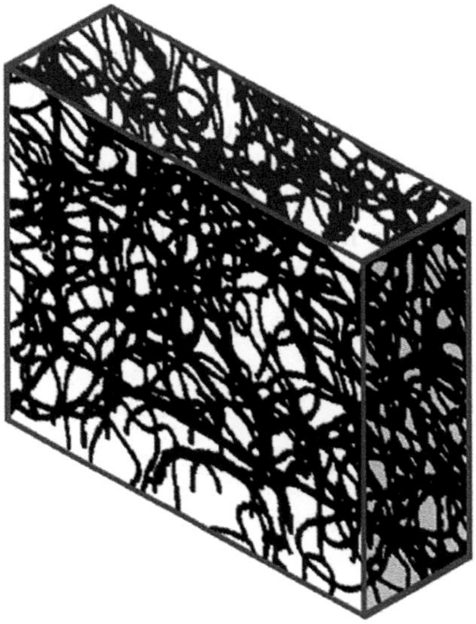

Figure 2.26: Vortex tangle creating quantum turbulence flow.

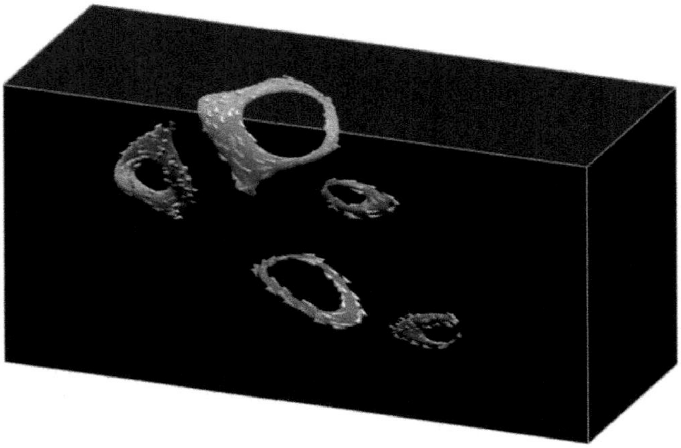

Figure 2.27: Vortex ring.

Figure 2.28: Large-scale vortexes formed due to altocumulus clouds.

Furthermore, an abrupt rotation of air particles driven by external forces may result in a vortex pattern. The polar vortex (Figure 2.29), for example, is driven by the Coriolis force effect. Thus, according to the Marghany quantum energy wind-driven theory, the excess stress-energy tensor creates significant low pressure and low air temperature, leading to the formation of a large pocket of freezing coherent air mass. Aside from the Earth's rotation, the stress-energy tensor causes a curvature of cold coherent air trajectory movements with two vortex centres, one over Baffin Island in Canada and the other over northeast Siberia.

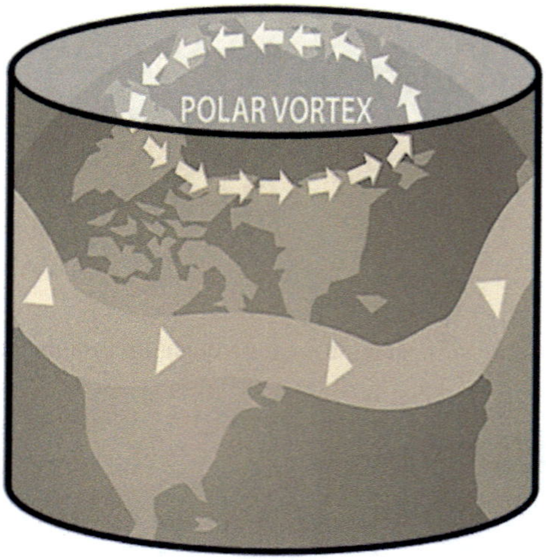

Figure 2.29: The polar vortex trajectory movement.

Along with the above proposition, the Arctic vortex spins counterclockwise due to an increment in the Earth's time-space curvature around the North Pole with wind speeds of 80 mph, which is stronger than the jet stream's standard 70 mph winds. In other words, quantum air particle diffusions with the fluctuation of the quantum thermodynamics under the circumstances of the strongest stress-energy tensor that induces torque and Coriolis forces are the primary causes of the polar vortex. In the fall, the stress-energy tensor causes the strongest circumpolar wind flows, causing the polar vortex to spin up farther into the stratosphere and potential vorticity values to rise, forming a cohesive air mass: the polar vortex. As winter approaches, winds near the poles weaken and the air in the vortex core cools [45, 47, 49].

The airflow slows, and the vortex ceases to expand. Heat and wind circulation return as late winter and early spring approach, forcing the vortex to shrink. Large chunks of vortex air are pushed out into narrow sections into lower latitudes during the last warming, or late winter. Strong potential vorticity gradients persist at the stratosphere's lowest level, and the bulk of air molecules remain confined until December in the Southern Hemisphere and April in the Northern Hemisphere, long after the mid-stratosphere vortex has broken up. Generally, the streamlines form circular paths around the vortex core [47–49]. In this scenario, quantized vortices can arise only along infinitely long (potentially curved) lines and around closed curves. The streamlines create concentric tubes in planes perpendicular to the axis of a thread vortex and concentric tori around circular "smoke ring" vortices. The jet streamlines around all over the polar vortex core create approximately circular loops in quantum mechanics, the wave function phase experiences a 2π (or integer multiple of 2π) winding, and the circulation integral and angular momentum are always quantized. A polar vortex may be produced and collapsed in time-independent investigations, and probability density can be briefly confined in the region of each loop vortex [49].

2.11 Quantized Atmospheric Wave Turbulences

Wave turbulence (WT) arises in configurations of significantly interacting nonlinear waves and therefore can culminate in energy streams that span length and frequency scales similar to vortex turbulence. Generally, energy flows through a nondissipative inertial range until it approaches a tiny sufficient dimension when viscosity takes over and terminates the cascade by dispersing the energy as heat. Wave turbulence in quantum fluids is of special interest, partly because revealing investigations might well be carried out on a laboratory scale, and partly because WT amongst Kelvin waves on quantized vortices is expected to play a significant role in the later phases of (vortex) quantum turbulence decay [50–52].

First and foremost, WT is significantly different from "mild turbulence." The former refers to a true physical event in a nonequilibrium statistical system whose primary construction blocks are random colliding waves. The latter refers to an intended target in which all interconnecting waves are mild and have random phases, permitting it to be specified by a wave kinetic equation. Consequently, in practical uses, WT may or may not be weak. WT systems frequently feature both random weak waves and strong coherent patterns, with these two constitutions interacting and transferring energy throughout the WT's entire lifecycle.

Weak turbulence, on the other hand, provides the conceptual foundation for WT and permits one to comprehend numerous (but not all) quantum sciences inspired by real systems of random waves. Weak turbulence is based on two basic assumptions: that the waves are weakly nonlinear and have random phases. It is also expected that the system is physically limitless and quantitatively identical. The minor turbulence formulation yields wave kinetic equations that govern the expansion of the wave spectrum. The kinetic equation might be three-wave, four-wave, or higher-order depending on the system. The kinetic equations often have strongly nonequilibrium steady-state solutions similar to the Kolmogorov cascades in conventional hydrodynamic turbulence, the so-called Kolmogorov-Zakharov (KZ) spectra, in addition to the regular thermodynamic Rayleigh-Jeans spectra, which symbolize a minimizing specific instance of general Bose-Einstein distribution [47, 49, 51].

In the previous section 2.9, the Gross–Pitaevskii nonlinear Schrödinger equation was exploited to present the quantum turbulence vortex flow [45]. Therefore, GP can exploit to describe the quantized wave turbulence flow as:

$$i\frac{\partial\psi}{\partial t}\left(\vec{V},t\right)+\nabla^2\psi\left(E_d,t\right)=\psi\left(\vec{V},t\right)\left|\psi\left(E,t\right)\right|^2 \tag{2.85}$$

Formula 2.85 delivers ψ as a complex function which is well-known as the condensate wave function. Therefore, wave energy dispersion is depicted by the first term $\nabla^2\psi\left(E_d,t\right)$. On the contrary, the strong correlation between waves or particles is demonstrated by $\psi(\vec{V},t)|\psi(E,t)|^2$. In this sense, the number of particles N fluctuations can be casted as:

$$N = \int |\psi|^2 d\vec{V} \tag{2.86}$$

Equation 2.86 reveals the dynamic number of particles N motions and dynamic fluctuations $d\vec{V}$. In this view, the entire energy for dynamic fluctuations $d\vec{V}$ mathematically is formulated as:

$$E = \int |\psi|^2 d\vec{V} + \frac{1}{2}\int |\psi|^4 d\vec{V}. \tag{2.87}$$

However, the decay of WT commences from the high-frequency close of the spectrum, whereas the utmost of the energy endures restricted at low frequencies. In this circumstance, the scaling index of the turbulent spectrum would rely on the spectral content of the driving force such as the energy stress-tensor in the curvature time-space as a function of pressure dynamic. Actually, there is a coherent WT spectrum of disordered waves, along with a power-law Kolmogorov-like energy cascade toward higher frequencies. Nevertheless, behind the appropriate circumstances, an instability opposing subharmonic creation might emerge, directing to an inverse cascade. It entails an energy transfer toward lower frequencies [45, 48, 50]. The crucial nature of the commencement of the inverse cascade, when atmospheric heater power is raised, might be tied to the necessity to combat dissipation. In other words, nonlinear wave interactions trigger a substantial transfer of energy between frequency ranges throughout decay. Consequently, the entire spectrum decays concurrently, although the top end decays rapidly owing to the stronger viscous influences at high frequencies.

Long-range vortex-vortex interactions provide the slowest time frames, which can be on the order of 1 s, whereas wave oscillations along quantized vortices with periods less than 10^{-9} s produce the most rapid kinetic wave turbulence [45, 50]. These waves are transverse, circularly polarized displacements that are regenerated by vortex tension created by a quantized vortex's kinetic energy per unit length. The approximate dispersion relation for such Kelvin waves along a rectilinear vortex is provided by:

$$\omega = \frac{\hbar\left[\frac{\partial W}{\partial t}\left[\frac{\partial P_a}{\partial x}\right]^{-1}\right]k^2}{4\pi}\left[\ln\left[ka_0\right]^{-1} + 1\right] \tag{2.88}$$

Formula 2.88 addresses that a quantized vortex tangle explicitly incorporates the interaction of a wide range of spatial and temporal scales as necessary for the atmospheric system to be referred to as turbulent. Here a_0 is a vortex cutoff parameter and is k spectra wave number. In this understanding, Kelvin waves propagating on quantized vortex lines have been widely explored as basic movements responsible for cascading energy to much lower scales than the mean intervortex separation scale. Evidently, energy is conveyed to greater vortices by the reconnection of discrete vortices. On the vortex lines, reconnection processes generate polychromatic spiral Kelvin waves (Figure 2.30). Kelvin waves interact nonlinearly, creating greater wave numbers until they lose energy to the atmospheric boundary layer [50–53].

Generally, if reconnection became conceivable, larger quantized vortex loops would generate small ones, and so on, until the loops became tiny enough to decay due to friction with the regular fluid or interactions with the boundaries. Reconnection is expected to eventually generate sinusoidal field oscillation (Figure 2.30) and create Kelvin waves on underdamped vortex lines that interact nonlinearly. Higher-frequency oscillations are formed by the nonlinear coupling between waves until they acquire a high sufficient frequency to transmit phonons into the surrounding fluid, which are subsequently subsumed by the boundary layers of the atmosphere [53–57].

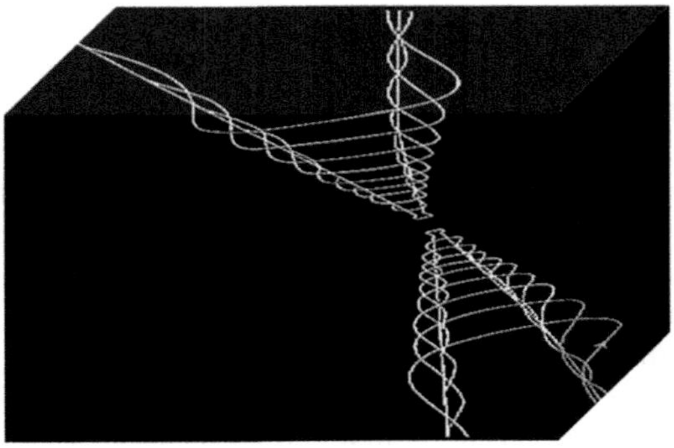

Figure 2.30: Kelvin wave formations after vortex line reconnections.

Internal, or baroclinic, Kelvin waves arise within the stably stratified ocean and atmosphere. Internal Kelvin waves are frequently seen in layers with substantial density gradients; the density gradient functions as an interface that permits internal gravity waves to occur.

2.12 Wave-Particle Duality in Atmospheric Wave Turbulence Propagation

What is a gravity atmosphere wave? The word “gravity” in the phrase gravity wave makes the notion sound more sophisticated than it is. That has nothing to do with presenting a clear relationship to gravity. Most airflow is influenced by gravity. When the phrase gravity is eliminated, just the term wave is left. Air may move in one of two directions: straight or waves. In this view, waves can be vertical or horizontal. We notice horizontal waves when we look at a 500-millibar chart with troughs and ridges (waves above or to a lesser extent horizontal plane). In this regard, let us consider what is a gravitational wave in the atmosphere. To initiate a gravity wave, a trigger device would generate vertical acceleration of air. Mountains and thunderstorm updrafts are two excellent examples of trigger mechanisms that cause gravity waves. To make a gravity wave, the steady air is always driven to rise (Figure 2.31). Why? Because rising air in unsteady air layers continues to ascend and does not produce a wave pattern. When air is driven to ascend in stable air, it has a natural propensity to sink back down over time. In other words, it is generally as if the parcel forced to rise is colder than the surrounding environment. Consequently, the momentum of the air delivered mostly by the trigger mechanism would cause the parcel to continue rising, and

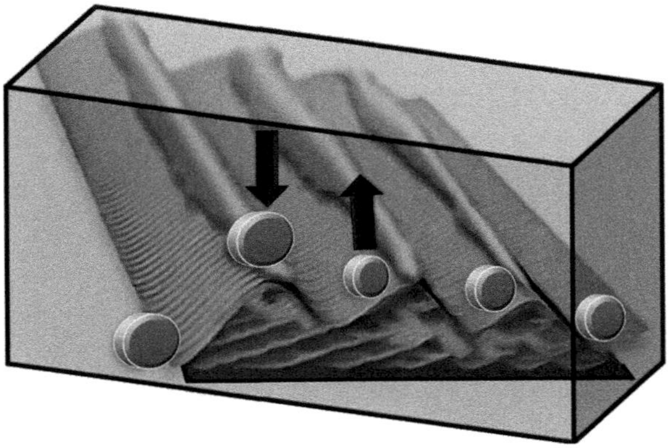

Figure 2.31: Concept of gravity wave.

the steadiness of the atmosphere would enable the parcel to descend even after raises, which is the initial stage in the formation of a wave [58–60].

It is essential to consider the principle of momentum. An air parcel that is rising or falling would “greatly exceed” its equilibrium state. The parcel of air in a gravity wave would strive to remain in a zone of the atmosphere where no forces are driving it to rise or sink. Once a force shifts the parcel out of its natural state of balance, this would seek to restore its stability. Nevertheless, because of momentum, this should exceed and undershoot that natural position every moment it rises or descends. The strength of the gravity wave decreases as one moves away from the point at which the trigger mechanism allows the parcel to ascend. The packet of air becomes closer to the ground as it travels further away [57, 59, 61].

The upward-flowing zone of a gravity wave is optimal for cloud formations (Figure 2.32), whereas the descending region is excellent for blue skies. Consequently, viewers can discern rows of clouds with clear zones in between (Figure 2.33). A gravity wave is just a wave that travels across the atmosphere’s stable layer. Gravity waves would be formed as thunderstorm updrafts strive to breach the tropopause. The tropopause is an extraordinarily stable zone of air. The combination of this stable air with the upward velocity of a thunderstorm updraft (trigger mechanism) can generate gravity waves to develop well within clouds as they strain to surge into the tropopause.

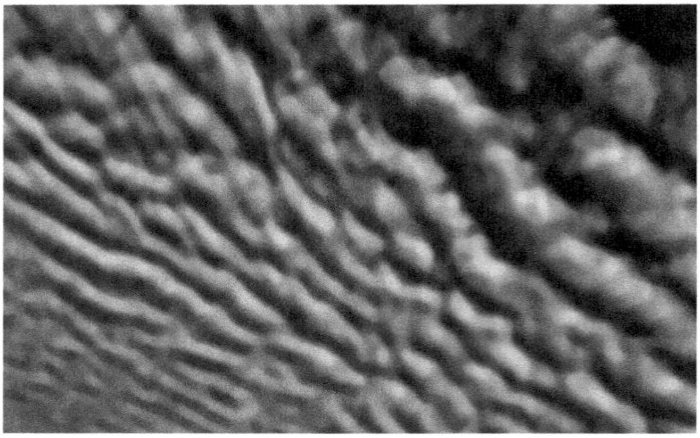

Figure 2.32: Gravity waves are optimal for cloud formations.

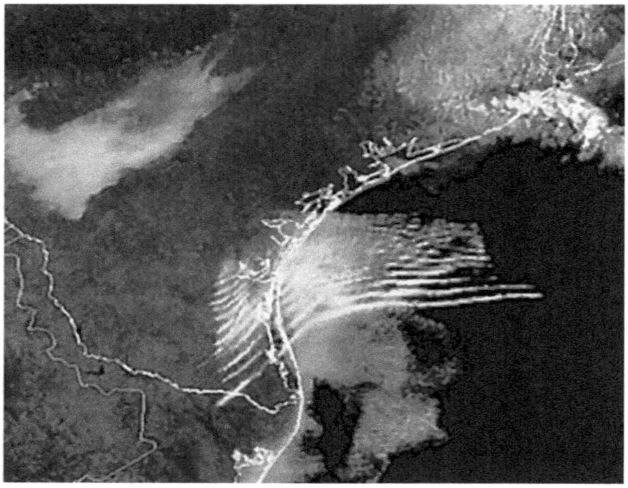

Figure 2.33: Rows of clouds as an indicator for gravity waves in a satellite image.

Now the crucial question is, "are gravity waves likely to express wave-particle duality, such as in the double-slit experiment?" In section 2.1, it was confirmed that air forms from different gas molecules. These gas molecules are spun due to external thermal energy that is received from the Sun and then form fluctuation layers of quantum pressure dynamics. In this perspective, the particles are represented as gas molecules (Figure 2.34). Furthermore, these particles swing randomly in space-time curvature and initiate waves under gravity's action. Why would it quantize the atmospheric gravity wave? It is considered a particle as it would collide and bounce off other molecular particles (Figure 2.35), cling together,swap energy, become bonded, etc. It is recognized as a wave (Figure 2.37) as it would diffract and interfere with itself [45, 59, 60].

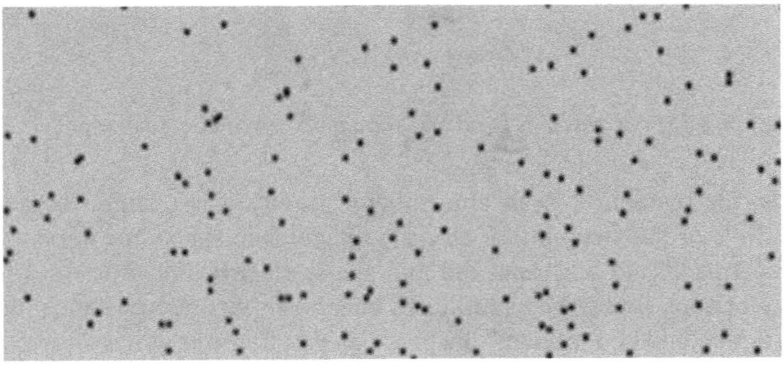

Figure 2.34: Random walk of atmospheric molecule gas particles.

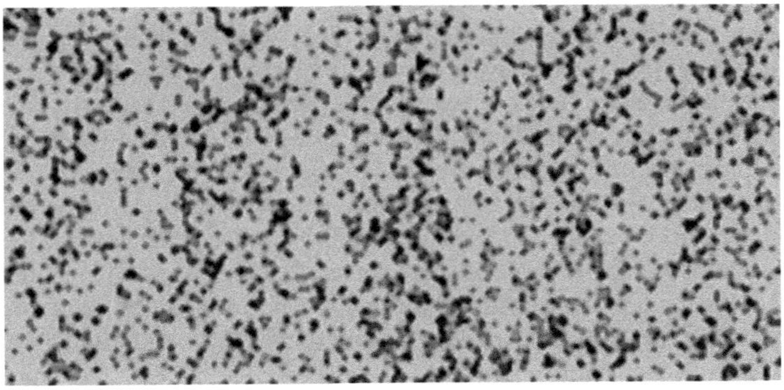

Figure 2.35: Particles collide and bounce off other molecular particles.

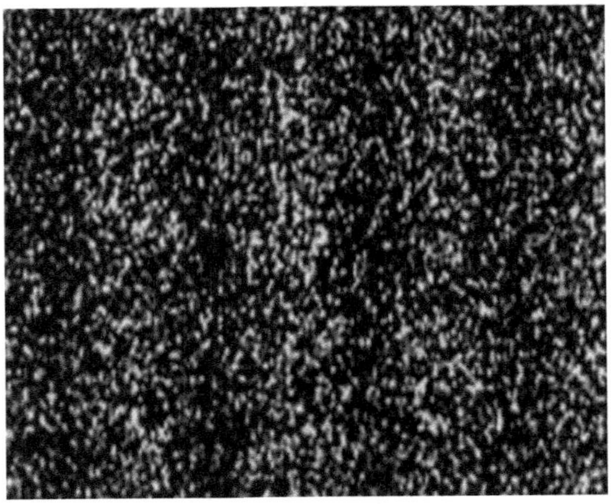

Figure 2.36: Air particles cling together and become bonded.

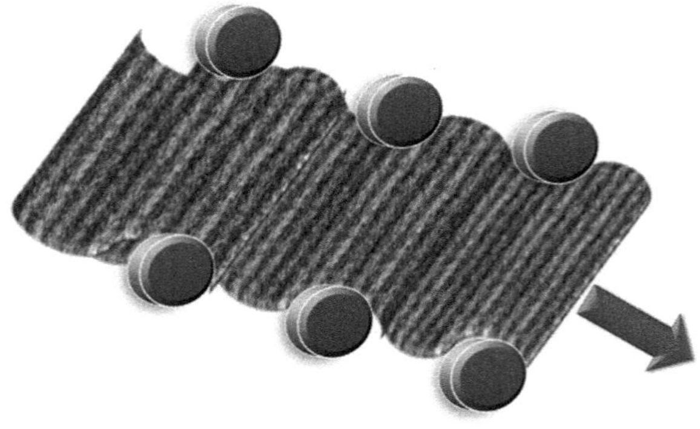

Figure 2.37: Air particles bond and form a wave-particle duality pattern.

In other words, the dynamics of an atmospheric gravity wave cannot be addressed without considering it a particle or the swinging of air gas molecules in spacetime curvature. It is believed that air gases are quantized into electrons and that single electrons or molecules may operate like particles, ionizing electrons if they have sufficient energy. In this circumstance, those air different particle characteristics should be gravitons, the particles that transmit gravity's force and are entirely predicted to develop a reliable grasp of gravity being such an intrinsically quantum force in nature [4, 19, 20, 60]. Identical to general relativity, this wave involves the inspiral phase, merger phase, and ringdown phase. In other words, we could confidently assume that it will continue to behave in the wave-like manner predicted by General Relativity. They vary from other waves in several ways: they are not scalar waves like water waves, nor are they vector waves like light, with in-phase, oscillating electric and magnetic fields. Tensor waves, on the other hand, force space to shrink and rarify in perpendicular directions when the wave passes through that range. In this regard, as the wave-particle dissipates, the atmospheric gravity waves convey the momentum of particles into the background of airflow [60–64]. Consequently, this wave-particle forces the airflow in the stratosphere. In this scenario, a sudden increase in stratospheric warming is caused by wave-particle momentum deposition by planetary-scale Rossby waves. In other words, the spinning of particles (Figure 2.38) across the momentum transform of gravity waves is demonstrated as a quasi-biennial oscillation (QBO). In the tropical stratosphere, the spinning of the equatorial air particles between easterlies and westerlies is well-known as a quasi-biennial oscillation (QBO) (Figure 2.39) [63–65]. These spinning air particles cause such oscillation with a mean period of 28 to 29 months. At the top of the lower stratosphere, stress energy induces spinning in air particles which permits them to propagate as waves downwards at approximately 1 km per month. However, in tropical tropopause stress energy is created by wave particles dissipated. As a result, downward stress energy creates lesser irregular kinetic oscillations in westerlies than in easterlies. The kinetic wave-particle energy of the easterly phase, on the other hand, is roughly twice as powerful as that of the westerly phase. Easterlies prevail at the top of the vertical QBO domain, while at the bottom westerlies are more likely to appear [4, 19, 45, 65].

In an irregular atmosphere state, the predicted value of air particle velocity u is determined by the gravity wave amplitude, which is a slowly changing envelope for ζ and is, therefore, Rayleigh distributed with parameter η_ζ. The mathematical formulation of the probability density of the atmosphere $P(a)$ condition in this circumstance is as follows:

$$P(a) = \frac{A}{\eta_\zeta^2} e^{\left(-\frac{A^2}{\eta_\zeta^2}\right)};\ A \geq 0, \tag{2.89}$$

here A is the gravity wave amplitude with the particle velocity that can be formulated as:

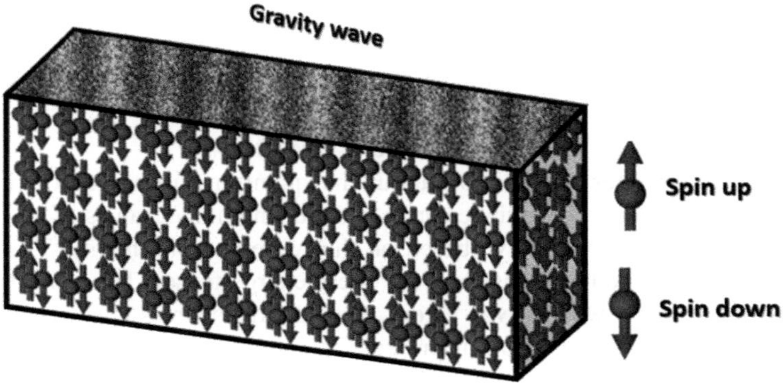

Figure 2.38: Wave particles spin up and down forming wave gravity.

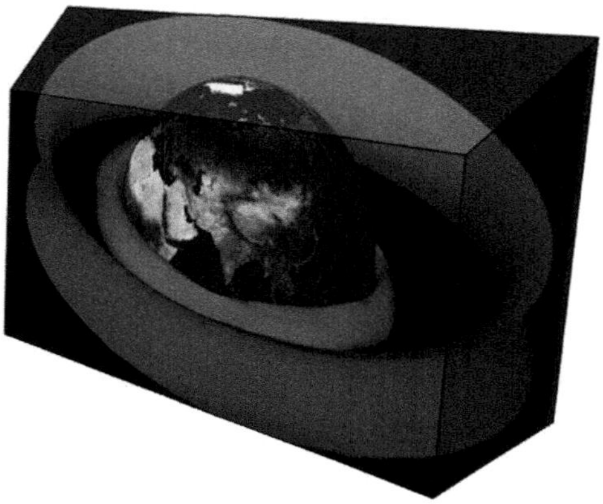

Figure 2.39: Concept of quasi-biennial oscillation (QBO).

$$\Psi | E(a) \rangle = \int_{0}^{\infty} dE^{\alpha\beta} \left| \Psi(x,t) \right|^{2} E\left[u|A \right] dA \tag{2.90}$$

For a given elevation, z, the decision of whether the point $E\left[u|A\right]$ of air particle is descended or not depends on the amplitude a. Hence equation 2.90 must be split into two integrals, one for $(-z) > A$ which can be evaluated analytically [20, 45]. However, for the points that are always descended, the mean horizontal velocity can be determined from:

$$\bar{u} = -A\omega k e^{2kz} \tag{2.91}$$

Because the gravity wave spectrum is narrow-banded, the components in equation 2.90 reflect virtually sinusoidal waves that develop periodically instead of a superposition of constituents that arise continuously. Consistent with this approach, the mean value of horizontal particle velocity in the propagation zone is:

$$\overline{u(0,z)} = \frac{A\omega}{\pi}\left[e^{kz}\cos\theta_0 - Ake^{2kz}\left(\frac{\pi}{2} - \theta_0\right)\right]. \tag{2.92}$$

Equation 2.92 demonstrates the view of the gravity wave symmetry over a full cycle with the phase angle $\theta_0 = (\omega t)_0$ [19, 46]. In other words, the full cycle of the gravity wave can be tackled as the value of the integral in the phase angle range $\left(-\frac{\pi}{2} < \theta < \frac{\pi}{2}\right)$. In this regard, wave period is $T = \frac{2\pi}{\omega}$ and wavelength is $\lambda = \frac{2\pi}{k}$. The particles spin and convey momentum in closed circles, hence the mean velocity when following a particle is zero [20, 45].

This chapter demonstrates speculation about the quantum turbulence of the atmosphere. Therefore, the chapter introduces the initial concept of the quantum mechanics of the air composition and ends by quantizing the gravity wave based on the wave-particle duality. The new theory of geostrophic wind flows is introduced based on the stress-energy tensor. The novel theory is named by Marghany quantum energy wind-driven theory. The quantized stress-energy tensor causes such discrete wind flows due to fluctuation of air pressure dynamics.

This discrete wind flow is titled "quantized Marghany quasi-geostrophic turbulence speculation". In this chapter offers a new concept, as the quasi-geostrophic are not only affected by the Coriolis force but also by torque force as demonstrated in the general relativity theory of Einstein.

Finally, this chapter discussed quantum turbulence as a study case of the quantum vortex, which would decompose into wave turbulence such as the Kelvin wave and the gravity wave. In this manner, it addressed gravity waves from the point of view of wave-particle duality. The next chapter will introduce the novelty of quantum remote sensing entanglement with quantum turbulence.

References

[1] Barry, R.G. and Chorley, R.J. (2009). *Atmosphere, Weather and Climate*. Routledge.

[2] Lide, D.R. (1997). éd. CRC Handbook of Chemistry and Physics, 78 e édition.

[3] Doshi, E.M. (2016). A review of climate changes due to the global warming. Doctoral dissertation, Sudan University of Science and Technology.

[4] Wallace, J.M. and Hobbs, P.V. (2006). *Atmospheric Science: An Introductory Survey* (Vol. 92). Elsevier.

[5] Lide, D.R. and Frederikse, F.P.R. (1996). CRC Handbook of Chemistry and Physics – 1996–1997. New York, 4.

[6] Haramein, N. and Rauscher, E.A. (2007). Spinors, twistors, quaternions and the spacetime torus topology. *Int. J. Comput. Anticip. Systems*, 1–18.

[7] Evans, M.W. (1992). Generally covariant dynamics. Preprint on www.aias.us and www.atomicprecision.com.

[8] Kopeikin, S. and Vlasov, I. (2004). Parametrized post-Newtonian theory of reference frames, multipolar expansions and equations of motion in the N-body problem. *Physics Reports*, 400(4–6), 209–318.

[9] Hasse, L. and Wagner, V. (1971). On the relationship between geostrophic and surface wind at sea. *Monthly Weather Review*, 99(4), 255–260.

[10] Roth, R., Hofmann, M. and Wode, C. (1999). Geostrophic wind, gradient wind, thermal wind and the vertical wind profile – A sample analysis within a planetary boundary layer over arctic sea-ice. *Boundary-layer Meteorology*, 92(2), 327–339.

[11] Rhines, P.B. (1979). Geostrophic turbulence. *Annual Review of Fluid Mechanics*, 11(1), 401–441.

[12] Haramein, N. and Rauscher, E.A. (2005). The origin of spin: A consideration of torque and Coriolis forces in Einstein's field equations and grand unification theory. *Beyond the Standard Model: Searching for Unity in Physics*, 153–168.

[13] Levine, I.N., Busch, D.H. and Shull, H. (2009). *Quantum Chemistry* (Vol. 6). Upper Saddle River, NJ: Pearson Prentice Hall.

[14] Lowe, J.P. and Peterson, K. (2011). *Quantum Chemistry*. Elsevier.

[15] Kauzmann, W. (2013). *Quantum Chemistry: An Introduction*. Elsevier.

[16] Helgaker, T., Klopper, W. and Tew, D.P. (2008). Quantitative quantum chemistry. *Molecular Physics*, 106(16–18), 2107–2143.

[17] Germann, T.C. and Miller, W.H. (1997). Quantum mechanical pressure-dependent reaction and recombination rates for O^+ OH→ H^+ O_2, HO_2. *The Journal of Physical Chemistry A*, 101(36), 6358–6367.

[18] Francisco, E., Recio, J.M., Blanco, M.A., Pendás, A.M. and Costales, A. (1998). Quantum-mechanical study of thermodynamic and bonding properties of MgF_2. *The Journal of Physical Chemistry A*, 102(9), 1595–1601.
[19] Sato, M. and Fujii, T. (2001). Quantum mechanical representation of acoustic streaming and acoustic radiation pressure. *Physical Review E*, 64(2), 026311.
[20] Charlo, D. and Clary, D.C. (2004). Quantum-mechanical calculations on pressure and temperature dependence of three-body recombination reactions: Application to ozone formation rates. *The Journal of Chemical Physics*, 120(6), 2700–2707.
[21] Meleshko, S.V. (2020). Complete group classification of the two-dimensional shallow water equations with constant Coriolis parameter in Lagrangian coordinates. *Communications in Nonlinear Science and Numerical Simulation*, 89, 105293.
[22] Burmasheva, N.Y.V. and Prosviryakov, E.Y. (2020). A class of exact solutions for two-dimensional equations of geophysical hydrodynamics with two Coriolis parameters. Russian Mathematics, 65, 8–16.
[23] Howard, L.N. and Drazin, P.G. (1963). *On instability of Parallel Flow of Inviscid Fluid in a Rotating System with Variable Coriolis Parameter*. Massachusetts Inst. of Tech. Cambridge.
[24] Deng, L., Li, T., Bi, M., Liu, J. and Peng, M. (2018). Dependence of tropical cyclone development on Coriolis parameter: A theoretical model. *Dynamics of Atmospheres and Oceans*, 81, 51–62.
[25] Kays, W.M. (1994). Turbulent Prandtl number. Where are we? *ASME Journal of Heat Transfer*, 116(2), 284–295.
[26] Verzicco, R. and Camussi, R. (1999). Prandtl number effects in convective turbulence. *Journal of Fluid Mechanics*, 383, 55–73.
[27] Kerr, R.M. and Herring, J.R. (2000). Prandtl number dependence of Nusselt number in direct numerical simulations. *Journal of Fluid Mechanics*, 419, 325–344.
[28] Wells, N.C. (2011). *The Atmosphere and Ocean: A Physical Introduction*. John Wiley & Sons.
[29] Lutgens, F.K., Tarbuck, E.J. andTusa, D. (1995). *The Atmosphere* (Vol. 462). Englewood Cliffs, NJ, USA: Prentice-Hall.
[30] Barry, R.G. and Chorley, R.J. (2009). *Atmosphere, Weather and Climate*. Routledge.
[31] Andrews, D.G., Holton, J.R. andLeovy, C.B. (1987). *Middle Atmosphere Dynamics* (No. 40). Academic Press.
[32] Hines, C.O. (1974). *The Upper Atmosphere in Motion: A Selection of Papers with annotation*. American Geophysical Union.
[33] Beretta, G.P., Gyftopoulos, E.P. and Park, J.L. (1985). Quantum thermodynamics. A new equation of motion for a general quantum system. *Il NuovoCimento B (1971–1996)*, 87(1), 77–97.
[34] Yarman, T. (2004). The general equation of motion via the special theory of relativity and quantum mechanics. *Ann. Fond. Louis de Broglie*, 29(3), 459–491.
[35] Beretta, G.P. (2005). On the general equation of motion of quantum thermodynamics and the distinction between quantal and nonquantal uncertainties. arXiv preprint quant-ph/0509116.
[36] Sokolovski, D. (1997). Beyond the Schrödinger equation: Quantum motion with traversal time control. *Physical Review Letters*, 79(25), 4946.
[37] Wu, J.Z. and Wu, J.M. (1996). Vorticity dynamics on boundaries. *Advances in Applied Mechanics*, 32, 119–275.
[38] Zheng, Y.Y., Wang, Y.N., Lu, Q.H. and Yuan, B. (2014). Simulation study of strengthening heat transfer of heat exchanger with notched turbolator. *In:* Applied Mechanics and Materials (Vol. 448, pp. 1278–1283). Trans Tech Publications Ltd.
[39] Chorin, A.J. (2013). *Vorticity and Turbulence* (Vol. 103). Springer Science & Business Media.
[40] Taylor, R.G. (1977). *Physical Fluid Dynamics*. Van Nostrand-Reinhold, New York.
[41] Markowski, P.M., Straka, J.M., Rasmussen, E.N. and Blanchard, D.O. (1998). Variability of storm-relative helicity during VORTEX. *Monthly Weather Review*, 126(11), 2959–2971.
[42] Kumjian, M.R. and Ryzhkov, A.V. (2009). Storm-relative helicity revealed from polarimetric radar measurements. *Journal of the Atmospheric Sciences*, 66(3), 667–685.
[43] Coffer, B.E., Parker, M.D., Thompson, R.L., Smith, B.T. and Jewell, R.E. (2019). Using near-ground storm relative helicity in supercell tornado forecasting. *Weather and Forecasting*, 34(5), 1417–1435.
[44] Adam, C., Naya, C., Oles, K., Romanczukiewicz, T., Sanchez-Guillen, J. and Wereszczynski, A. (2020). Incompressible topological solitons. *Physical Review D*, 102(10), 105007.
[45] Marghany, M. (2021). *Nonlinear Ocean Dynamics: Synthetic Aperture Radar*. Elsevier.
[46] Marghany, M. (2019). *Synthetic Aperture Radar Imaging Mechanism for Oil Spills*. Gulf Professional Publishing.

[47] Haddad, L.H. and Carr, L.D. (2015). The nonlinear Dirac equation in Bose–Einstein condensates: Vortex solutions and spectra in a weak harmonic trap. *New Journal of Physics*, 17(11), 113011.
[48] Haddad, L. H. (2012). The nonlinear Dirac equation in Bose-Einstein condensates (2012-Mines Theses & Dissertations). Colorado School of Mines. Reprint from https://people.mines.edu/lcarr/wp-content/uploads/sites/23/2018/11/haddad_thesis_2012.pdf.
[49 Borgh, M.O. and Ruostekoski, J. (2013). Topological interface physics of defects and textures in spinor Bose-Einstein condensates. *Physical Review A*, 87(3), 033617.
[50] Newell, A.C. and Rumpf, B. (2011). Wave turbulence. *Annual Review of Fluid Mechanics*, 43, 59–78.
[51] Zakharov, V., Dias, F. and Pushkarev, A. (2004). One-dimensional wave turbulence. *Physics Reports*, 398(1), 1–65.
[52] Mordant, N. (2008). Are there waves in elastic wave turbulence? *Physical Review Letters*, 100(23), 234505.
[53] L'vov, V.S. (2012). *Wave Turbulence Under Parametric Excitation: Applications to Magnets*. Springer Science & Business Media.
[54] Bellet, F., Godeferd, F.S., Scott, J.F. and Cambon, C. (2006). Wave turbulence in rapidly rotating flows. *Journal of Fluid Mechanics*, 562, 83–121.
[55] Kozik, E. and Svistunov, B. (2004). Kelvin-wave cascade and decay of superfluid turbulence. *Physical Review Letters*, 92(3), 035301.
[56] Forbes, J.M. (2000). Wave coupling between the lower and upper atmosphere: Case study of an ultra-fast Kelvin Wave. *Journal of Atmospheric and Solar-Terrestrial Physics*, 62(17–18), 1603–1621.
[57] Straub, K.H. and Kiladis, G.N. (2002). Observations of a convectively coupled Kelvin wave in the eastern Pacific ITCZ. *Journal of the Atmospheric Sciences*, 59(1), 30–53.
[58] Fritts, D.C. and Alexander, M.J. (2003). Gravity wave dynamics and effects in the middle atmosphere. *Reviews of Geophysics*, 41(1).
[59] Benney, D.J. (1962). Non-linear gravity wave interactions. *Journal of Fluid Mechanics*, 14(4), 577–584.
[60] Vincent, R.A. (1984). Gravity-wave motions in the mesosphere. *Journal of Atmospheric and Terrestrial Physics*, 46(2), 119–128.
[61] Kelley, M.C., Larsen, M.F., LaHoz, C. and McClure, J.P. (1981). Gravity wave initiation of equatorial spread F: A case study. *Journal of Geophysical Research: Space Physics*, 86(A11), 9087–9100.
[62] Holton, J.R. (1983). The influence of gravity wave breaking on the general circulation of the middle atmosphere. *Journal of Atmospheric Sciences*, 40(10), 2497–2507.
[63] Baldwin, M.P., Gray, L.J., Dunkerton, T.J., Hamilton, K., Haynes, P.H., Randel, W.J., ... and Takahashi, M. (2001). The quasi-biennial oscillation. *Reviews of Geophysics*, 39(2), 179–229.
[64] Lindzen, R.S. and Holton, J.R. (1968). A theory of the quasi-biennial oscillation. *Journal of Atmospheric Sciences*, 25(6), 1095–1107.
[65] Richter, J.H., Anstey, J.A., Butchart, N., Kawatani, Y., Meehl, G.A., Osprey, S. and Simpson, I.R. (2020). Progress in simulating the quasi-biennial oscillation in CMIP models. *Journal of Geophysical Research: Atmospheres*, 125(8), e2019JD032362.

CHAPTER

3

Quantum Mechanics in Remote Sensing: Theoretical Perspectives

In reality, the issue of turbulence also constitutes difficulties in terms of the analytical processing of observed data and the removal of pertinent data, like turbulent forms, patterns, or characteristics from the dynamical ambient. Experimental data of the vortex energy spectrum and/or its oscillations, in conjunction with ancillary distinctions with the well-known Kolmogorov's "five-thirds" law, constitute the most prevalent choice available to most scientists and hardware developers.

In the previous chapter, therefore, we demonstrated the nature of air turbulence from the view of quantum mechanics. In point of fact air composites of a mixture of gas atoms, which can be considered tiny particles. Therefore, remote sensing relies on the potential of photons' interactions with any particles that exist in universes. In other words, photons are particles acting as waves that can easily interact with other particles in the universe to deliver remote-sensing images and signals. Overall, it seems that diverse disciplines of science and engineering are trying to comprehend and investigate just particle interactions throughout the entire universe.

Nonetheless, the air turbulence numerical conventional approach, which is unquestionably accurate quantitatively does not offer the necessitous facts on the structural features and parametric dynamics of the turbulence being explored. Practically and figuratively speaking, the entire vision is needed for turbulence-accurate automatic detection and forecasting. The physics of electromagnetic (EM) wave propagation is bridging the gap between air turbulence imaging and observation technologies. In other words, how do we understand the EM wave propagation in turbulence media?

3.1 Maxwell's Equations

A foundational element of remote sensing technology is Maxwell's equations. In reality, all theories on remote sensing are founded on the physics of Maxwell's equations. In light of this, the chapter continues by providing a scientific justification for the mathematical and physical ideas underlying Maxwell's equations. In actuality, Maxwell's equations provide all explanations of electromagnetic phenomena. To later comprehend how remote sensing sensors operate, this section is devoted to thoroughly comprehending the theory of Maxwell's equation.

One of the most superior achievements in modern physics was made by Scottish physicist James Clerk Maxwell in the early 19th century when he cunningly contrived to encapsulate the entire field of electromagnetism in merely four incompatible formulas. The foundation of remote sensing sensor technology is Maxwell's four equations. In actuality, all the theories on remote sensing are founded on the physics of Maxwell's equations. They demonstrate how electromagnetic waves, the same phenomenon, are explained in various manners by electric and magnetic fields [1]. The incredibly thorough Maxwell's equations are next explained in this section.

3.1.1 Electromagnetic Waves

Electrons and magnets are subject to electric and magnetic forces. Magnetic fields are created by electric fields that are charging, and vice versa. Maxwell explained how both originate from a typical event, an electromagnetic wave that retains both electric and magnetic properties. A fluctuating electric energy field and a magnetic field that differ evenly but remain perpendicular to one another amply compensate electromagnetic waves [1-3].

Maxwell derived the speed of electromagnetic waves propagating through a vacuum and observed it to be practically identical to the velocity of light. Light, on the other hand, was positively confirmed as a transmitting electromagnetic distraction by the completed work of Hans Christian Ørsted and Faraday. Demonstratively, consistent with this precise perception, light waves, and all electromagnetic radiation propagate at a constant velocity of 300 million meters per second in a vacuum. As a practical consequence, this proper motion is predetermined by the absolute electric and magnetic properties of vacant space [1, 3].

3.1.2 Fields

Maxwell exploited Faraday's creative exploration of experimental measurements to describe electric and magnetic fields. From the practical standpoint of fundamental physics, fields are the creative process by which forces are merely perpetuated across considerable space. Gravity, for instance, acts across even virtually unlimited swaths of space, yielding a force of gravity. Identically, electric and magnetic fields can disturb charged particles from a considerable distance. It was convincingly demonstrated that the magnetic field exchanges the iron dust into looped contours stretching from the irresistible magnet's north to south poles (Figure 3.1).

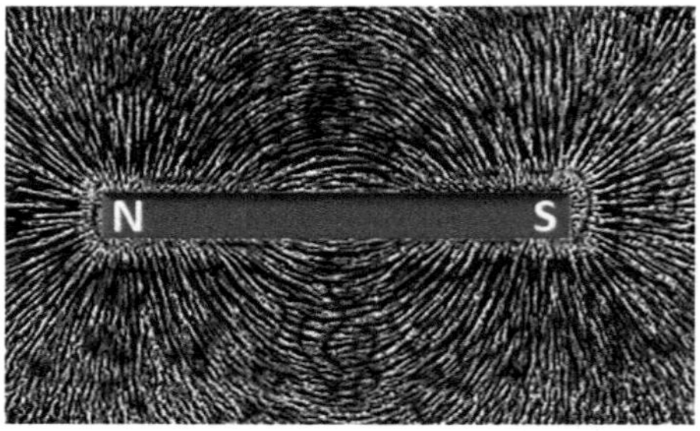

Figure 3.1: Magnetic field impact causes an iron dust loop.

Magnetic fields, as exposed in Figure 3.1, are expanses where an object reveals a magnetic impact. The fields cause distress to neighboring objects along a mechanism known as magnetic field lines. A magnetic object can draw attention to itself or displace other magnetic objects. The amount of gravity is proportional to an object's mass, whereas magnetic power is proportional to the materials from which the object is made. Faraday had represented these 'field lines' and established simple rules in this regard. In addition, he denoted analogous field lines for electrically charged forms. Maxwell, on the other hand, compiled these various concepts into Maxwell equations [3, 5].

3.1.3 Four Mathematical Maxwell Equations

Within only four essential equations, Maxwell was able to exhaustively express the myriad electromagnetic phenomena that exist precisely. In accordance with this prevailing theory, Maxwell's independent equations are expressed as follows:

$$\oint \vec{E}.d\vec{A} = \frac{Q}{\varepsilon_0} \tag{3.1}$$

$$\oint \vec{B}.d\vec{A} = 0 \tag{3.2}$$

$$\oint \vec{E}.ds = -\frac{d\Phi_B}{dt} \tag{3.3}$$

$$\oint B.ds = \mu_0 i + c^{-2} \frac{\partial}{\partial t} \int \vec{E}.d\vec{A} \tag{3.4}$$

Equation 3.1 conveys Gauss' Law, which stipulates that the electric flux Φ ($\Phi = EA\cos\theta$) (Figure 3.2) out of any closed surface $\vec{A}$ (Figure 3.3) is proportional to the entire charge Q enclosed within the surface and divided by permittivity (ε_0) (Figure 3.4). Gauss' Law can be employed to compute the electric fields ($\vec{E}$) surrounding charged objects using its integral form. While the electric field's area integral provides a measurement of the net charge contained, the electric field's divergence provides a measurement of the density of sources. Additionally, it affects charge conservation [4].

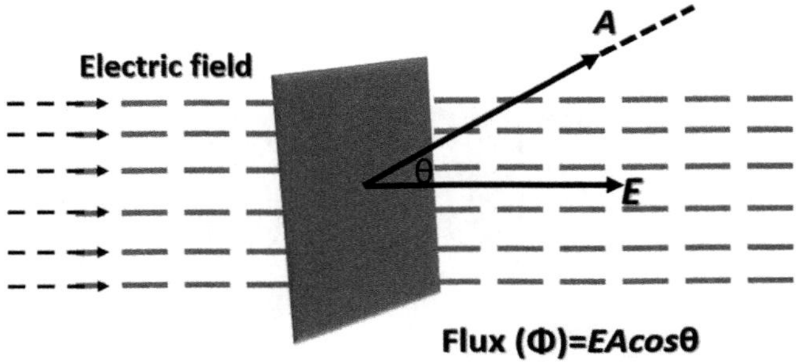

Figure 3.2: The electric flux through the constant surface.

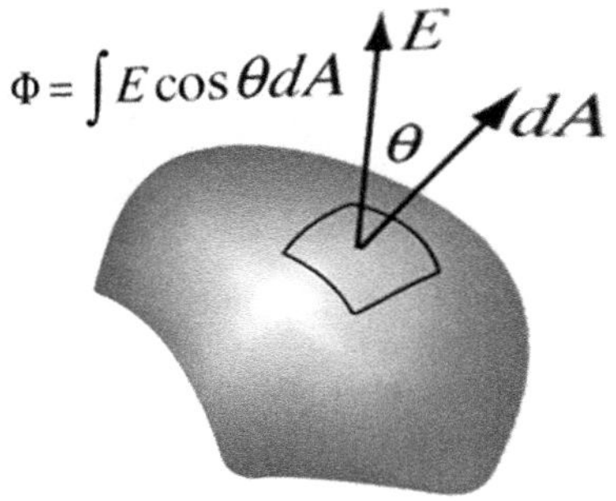

Figure 3.3: The electric flux across an unstable surface.

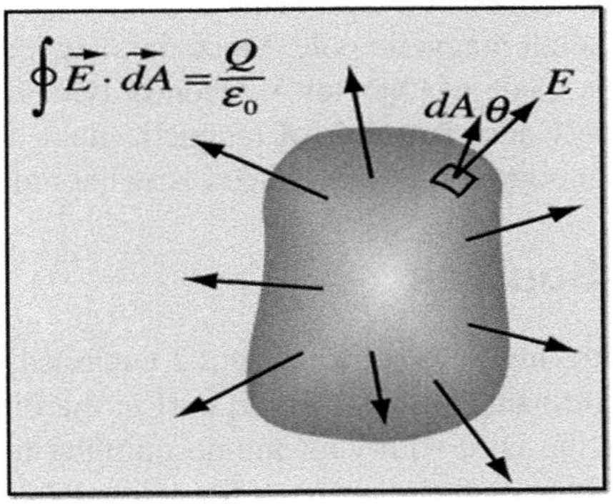

Figure 3.4: Gauss' Law integral form.

Therefore, equation 3.2 reveals the Gauss' Law for magnetism in which any closed surface has a zero net magnetic flux ($\vec{B}$) out. This essentially amounts to a claim on the magnetic field's origins. Any closed surface that has a magnetic dipole will have a magnetic flux directed inward toward the south pole that is equal to the flux directed outward from the north pole. For dipole sources, the net flow will always be zero. A non-zero area integral would result if the source were a magnetic monopole. The form of Gauss' Law for magnetic fields is a declaration that there are no magnetic monopoles since the divergence of a vector field is proportional to the density of a point source [1-4].

Consequently, Faraday's Law of Induction is expressed in equation 3.3. In this context, the rate of change of the magnetic flux (Φ_B) through the area (s) contained by the loop is equal to the negative of the line integral (Figure 3.5) of the electric field around a closed loop. Electric generators are built on Faraday's rule because the generated voltage or emf in the loop is equal to this line integral. Transformers and inductors are also built on their foundation.

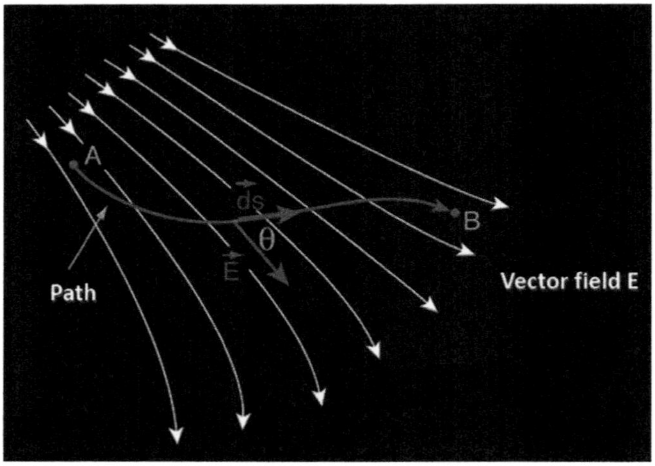

Figure 3.5: Line integral.

Lastly, equation 3.4 demonstrates Ampere's Law. In this sense, the line integral of the magnetic field encircling a closed loop in a static electric field is inversely correlated with the electric current passing through the loop. This is effective for computing the magnetic field for simple configurations. It is unquestionable that electric currents, which can be microscopic currents associated with electrons in atomic orbits or macroscopic currents in wires, generate magnetic fields. The Lorentz force law expresses the magnetic field B in terms of the force acting on a moving charge. There are numerous practical uses for magnetic field and charge interaction. Dipolar in nature, magnetic field sources have a north and south magnetic pole. Since the magnetic component of the Lorentz force law F(magnetic) = QvB is made of (Newton x second)/ (Coulomb x meter), the SI unit for magnetic field is the Tesla. The Gauss is a unit of magnetic field magnitude (1 Tesla = 10,000 Gauss) [1, 3, 5]. A significant important question indeed is: what are magnetic field sources?

3.2 Magnetic Fields Sources

Let us consider a long wire carrying an electric current I encircled by magnetic field lines that create concentric rings. The right hand's fingers would curl in the direction of the magnetic field B if you wrapped them around the wire with your thumb pointing in the direction of the current (Figure 3.6). The magnetic field is perpendicular to the wire. In this regard,Ampere's law can be used to calculate the magnetic field of an endlessly long straight wire. The magnetic field is expressed as follows:

$$B = \frac{\mu_0 I}{2\pi r} \tag{3.5}$$

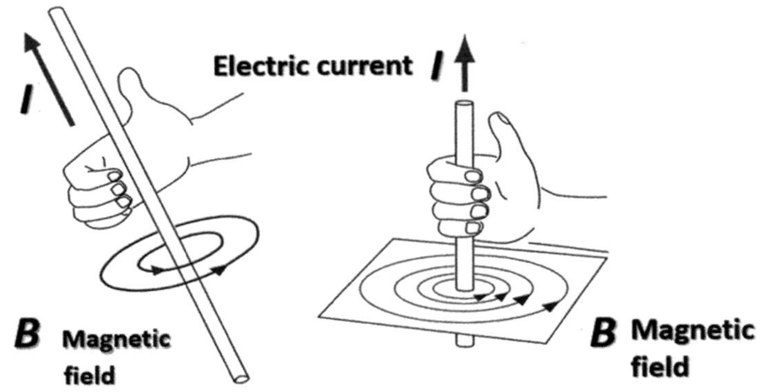

Figure 3.6: The magnetic field of current.

here r is radial distance, and μ_0 i.e. ($\mu_0 = 4\pi \times 10^{-7}\ N[A^{-2}]$) is the magnetic permeability for free space, which includes the force unit N for Newton and unit A is the Ampere, the unit of electric current. In other words, as said by Ampers' Law, the magnetic field is always perpendicular to the path of a circular path that is centered on the wire (Figure 3.7).

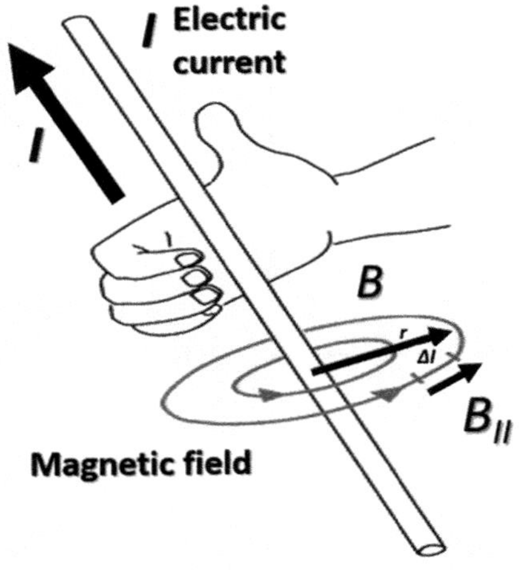

Figure 3.7: Ampers' Law, the magnetic field of current.

Therefore, all segments of the loop contribute a magnetic field in the same direction within the loop, as can be shown by looking at the direction of the magnetic field generated by a current-carrying length of wire. In a circular electric current loop, the magnetic field is more concentrated in the center than on the outside (Figure 3.8) [2, 4, 6]. The field is even more concentrated in what is known as a solenoid when numerous loops are interconnected.

Therefore, the Biot-Savart law is the formula used to determine the magnetic field generated by a current. Two researchers who studied the interaction between a straight, current-carrying wire and a permanent magnet are recognized by the law's honorific name. By using this law, we may determine the strength and direction of the magnetic field created by a current in a wire. In this regard, Biot-Savart law is mathematically expressed by:

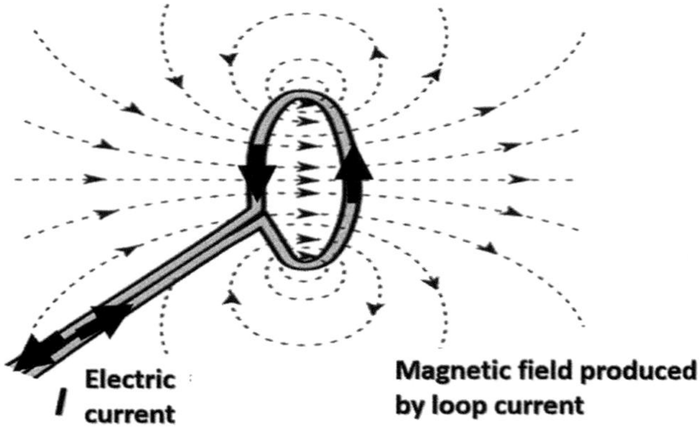

Figure 3.8: The magnetic field is generated by the loop current.

$$d\vec{B} = \frac{\mu_0 I d\vec{L} \times \hat{r}}{4\pi R^2} = \frac{\mu_0 I d\vec{L} \sin\theta}{4\pi R^2} \tag{3.6}$$

where $d\vec{L}$ is an element of a current-carrying wire; R^2 is the distance from $d\vec{L}$ to point P; and $\hat{r}$ is the unit vector from $d\vec{L}$ to point P (Figure 3.9). Therefore, the direction of $d\vec{B}$ is determined by implementing the right-hand rule to the vector product $d\vec{L} \times \hat{r}$ to produce the right side of equation 3.6, where θ is the angle between $d\vec{L}$ and $\hat{r}$.

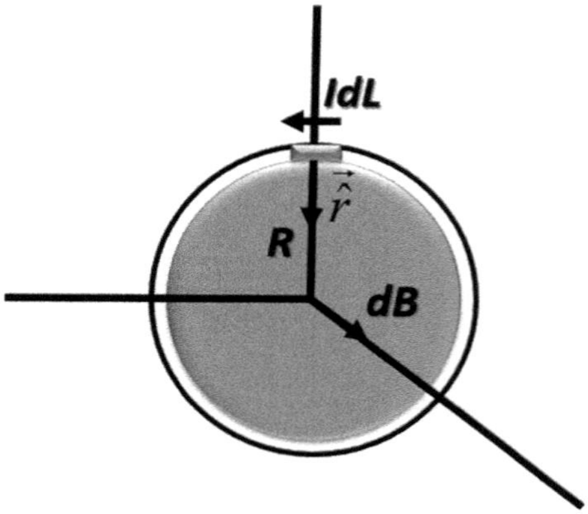

Figure 3.9: A current element produces a magnetic field given by the Biot-Savart law.

In the circumstance of the right angle for all points along the path and the distance to the field point is constant, the integral becomes:

$$B = \frac{\mu_0 I}{4\pi R^2} \oint dL = \frac{\mu_0 I}{4\pi R^2} 2\pi R = \frac{\mu_0 I}{2\pi R} \tag{3.7}$$

In this specific scenario, the symmetry endures such that the field contributions of the entire current elements around the circumference add directly to the centre. In this view, the line integral of the length merely remains the circumference of the circle.

Let us assume the element field rotates around as you progress around the loop, and by the symmetry offers a net zero field for the loop. At this point, the field endures in the *z*-direction, along the centerline of the loop (Figure 3.10). Therefore, integrating the *z*-component is required to apply the Biot-Savart law on the centerline of a current loop, which is casted as:

$$dB_z = \frac{\mu_0 IdL}{4\pi}\frac{R}{\left(z^2+R^2\right)^{\frac{3}{2}}} \quad (3.8)$$

All of the components in this element are constant due to the symmetry, except for the distance element *dL*, which, when integrated, only yields the circle's circumference. Thereafter, the magnetic field:

$$B_z = \frac{\mu_0}{4\pi}\frac{2\pi R^2 I}{\left(z^2+R^2\right)^{\frac{3}{2}}} \quad (3.9)$$

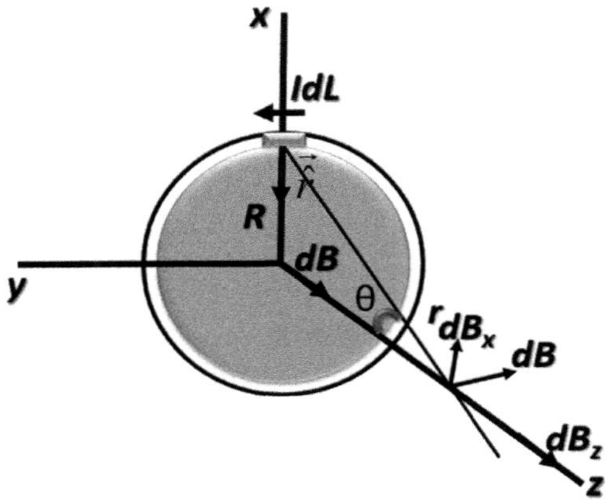

Figure 3.10: Field on current loop's axis.

To create a virtually uniform magnetic field akin to a bar magnet, a long, straight coil of wire can be utilized. Solenoids, which are these coils, have a plethora of practical uses. By including an iron core, the field can be significantly reinforced. These cores are typical of electromagnets. A rectangular evaluation of Ampere's Law with a side parallel to the solenoid field of length *L* results in a contribution BL inside the coil (Figure 3.11). The field makes a very small contribution because it is nearly perpendicular to the path's sides. The length inside the coil makes the largest contribution if the end is removed from the coil so far that the field is minimal [2, 4, 6]. This admittedly great demonstration of Ampere's Law provides:

$$BL = \mu_0 nI \quad (3.10)$$

where *n* is the number of turns in one unit of length inside of the solenoid. This is consistent with what we previously discovered for *B* on the solenoid's central axis. However, since segment 1's position, in this case, is random, needless to say, this equation accurately predicts the magnetic field throughout the entire infinite solenoid.

A bar magnet's magnetic field produces closed lines. Conventionally, it is assumed that the magnet's field flows from its north pole inward to its south pole. Ferromagnetic materials can be used to create permanent magnets. In ferromagnetic materials, the microscopic ordering of electron spins results in the creation of domains, which are areas of magnetic alignment. The magnetic field is strongest inside the magnetic substance, as seen by the magnetic field lines. Near the poles, there

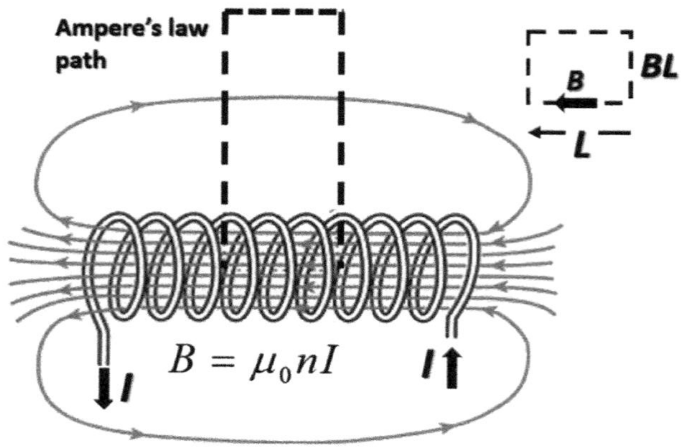

Figure 3.11: Solenoid field from Ampere's Law.

are the strongest external magnetic fields. A magnetic north pole will pull a magnet's south pole toward it while repelling another magnet's north pole. Iron core solenoids are the typical kind of electromagnets. The internal magnetic domains of the iron align with the smaller driving magnetic field generated by the current in the solenoid due to the iron core's ferromagnetic characteristics [2-5]. The result is a tens to even thousands-fold increase in the magnetic field. According to the solenoid field relationship:

$$B = k\mu_0 nI \tag{3.11}$$

The iron's relative permeability, or k, illustrates the iron core's amplifying effect.

Lastly, the magnetic field of the Earth resembles that of a bar magnet that has been tilted away from its spin axis by 11 degrees. The issue with that image is that iron melts at a temperature of about 770 °C. Since the Earth's core is warmer than that, it is not magnetic. Then how did the magnetic field of the Earth form (Figure 3.12)? Electric currents are surrounded by magnetic fields, leading us to hypothesize that the Earth's molten metallic core is the source of the magnetic field. An earth-like field is produced by a current loop.

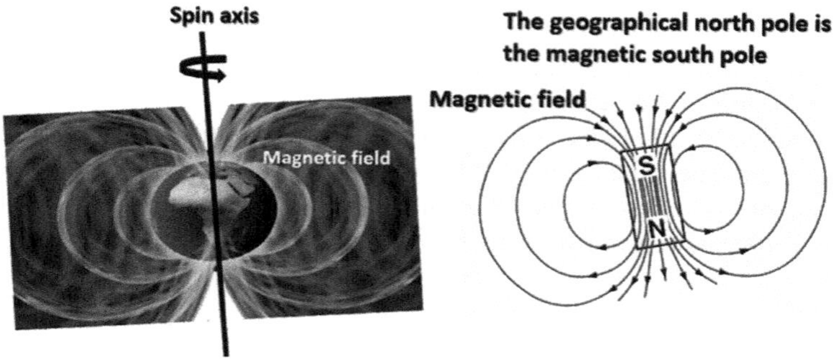

Figure 3.12: Earth magnetic field.

In the northern hemisphere, the magnetic field dips toward the Earth and is approximately half a Gauss in magnitude. On Earth's surface, the magnitude varies between 0.3 and 0.6 Gauss. The dynamo effect of circulating electric current is believed to be the source of the Earth's magnetic field, but its direction is not constant. Different orientations of permanent magnetization are present in rock samples from similar places and different ages. There is proof of 171 magnetic field reversals in the last 71 million years. The circumstances for the aurora occurrences at the poles are created by the interaction of the terrestrial magnetic field with solar wind particles [1, 5, 7]. A compass needle's north pole is also a magnetic north pole. It is drawn to the magnetic South Pole, which is located at the geographic North Pole (Figure 3.12) (opposite magnetic poles attract).

3.3 Derivation of Electromagnetic Equations from Maxwell's Equations

The wave equation for electromagnetic waves is contained in Maxwell's equations. Computing the wave equation in one approach:

First, consider Faraday's law's curl:

$$\nabla \times (\nabla \times E) = -\frac{\partial(\nabla \times B)}{\partial t} \tag{3.12}$$

Second, replace Ampere's law with a charge-free region:

$$\nabla \times (\nabla \times E) = -\frac{\partial^2 E}{\partial t^2}\left[c^{-2}\right] \tag{3.13}$$

Therefore, the three-dimensional wave equation within the vector pattern can be casted as:

$$\frac{\partial^2 \vec{E}_x}{\partial y^2} + \frac{\partial^2 \vec{E}_x}{\partial z^2} = -\frac{\partial^2 \vec{E}_x}{\partial t^2}\left[c^{-2}\right] \tag{3.14}$$

Therefore, equation 3.14 is the vector pattern of the three-dimensional wave equation. It is difficult to discern in this form. It appears more identifiable once restricted to a plane wave with only an x-direction field. In this understanding, the propagation is perpendicular to the x-axis and can be in any direction in the y-z plane because the electric field only has an x-direction, depending on the derivatives' values. The two-dimensional wave equation is represented by equation 3.14 in its general form. Consequently, the magnetic field wave in a plane perpendicular to the electric field takes on the same waveform (Figure 3.13). The magnetic field and the electric field are both perpendicular to the x-axis of motion [1-5]. The speed of light or other electromagnetic waves is denoted by the letter "c." Maxwell's equations lead to the wave equation for electromagnetic waves. The shape of the electric field's plane wave solution is

$$E = E\sin(kx - \omega t) \tag{3.15}$$

likewise, the magnetic field:

$$B = B\sin(kx - \omega t) \tag{3.16}$$

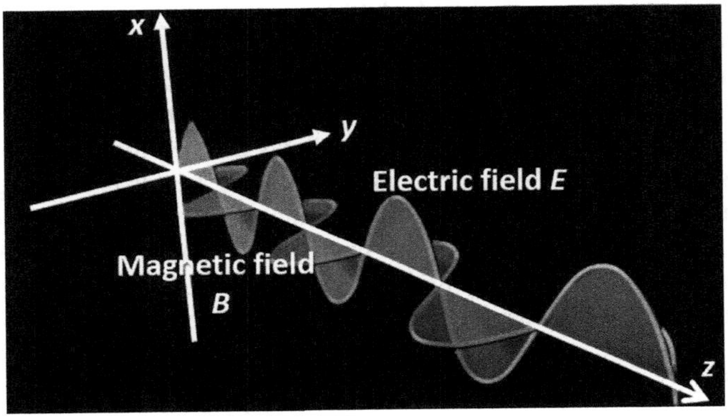

Figure 3.13: Electromagnetic wave propagation.

These solutions are required to be integrated for them to be consistent with Maxwell's equations:

$$\frac{E}{B} = c \tag{3.17}$$

Following the above perspective, in the orientation where the vector product $E \times B$ is in the direction of wave propagation, the magnetic field B is perpendicular to the electric field E. Energy, therefore, is carried through space by electromagnetic waves. Both the magnetic field B and the electric field E have an associated energy density S. The vector represents the rate of energy transmission per unit area:

$$\vec{S} = \vec{E} \times \vec{B}[\mu_0]^{-1} \tag{3.18}$$

Equation 3.18 is also known as the Poynting vector (Figure 3.14). Since the magnetic field is perpendicular to the electric field in this expression, the magnitude can be represented as follows:

$$S = [\mu_0]^{-1} EB \tag{3.19}$$

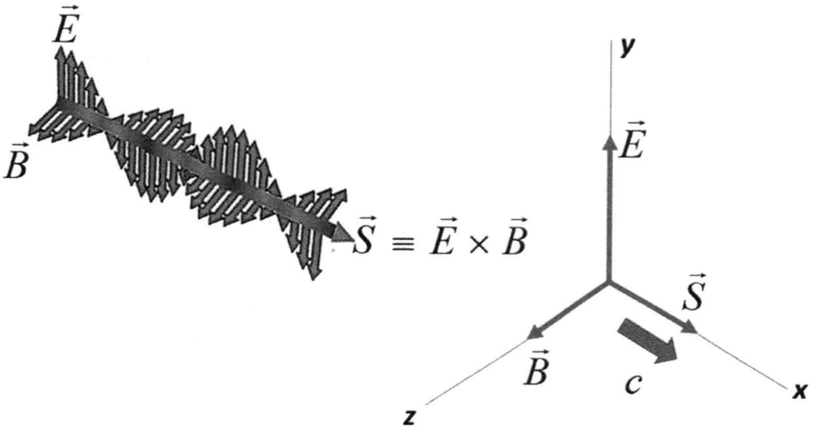

Figure 3.14: Concept of Poynting vector.

The wave's propagation direction and the rate of energy transport S are both perpendicular to one other (Figure 3.14). $B = [C]^{-1} E$ is a requirement of the wave equation for a plane wave to express the average wave intensity as follows:

$$S = [c\mu_0]^{-1} E^2 \overline{\sin^2(kx - \omega t)} = [c\mu_0]^{-1} 0.5E^2 \tag{3.20}$$

This takes advantage of the fact that a sinusoidal function's average square over all possible periods is merely half. Generally, both magnetic and electric fields are produced when there are changes in the electric field. Electromagnetic waves propagate as a result of this interaction between the induced electric and magnetic fields. Free space allows for the propagation of electromagnetic waves. The wave is carrying $E = hv$ energy, and the Poynting vector is pointing in that direction or energy-momentum. In this understanding, it is well known that electromagnetic waves transport energy. Poynting's theorem, which is the work-energy theorem in electrodynamics, provides a concise explanation of the functionality of the Poynting vector [4, 7]. The Poynting Theorem states that the Lorentz force works on a charge distribution at the following rate:

$$\frac{dW}{dt} = -0.5\frac{\partial}{\partial t}\int_V \left(\mu_0^{-1}B^2 + \varepsilon_0 E^2\right)dV - \mu_0^{-1}\int_S \left(\vec{E} \times \vec{B}\right).d\vec{S} \tag{3.21}$$

Equation 3.21 reveals that the energy (power) an electromagnetic wave imparts to a charge is equivalent to the decline in the energy contained in the fields throughout a volume *V* (first integral) less the energy emitted through the surface *S* enclosing the volume *V* (second integral). If $\frac{dW}{dt} = 0$, so there is no charge existing. In this scenario, the energy emitted out through a boundary containing the volume involves the energy decreasing in the field across a volume, which is considered. This is the principle of energy conservation law.

In this view, the second integral, or the surface integral of the Poynting vector, depicts the rate at which energy departs the surface. The rate at which energy travels through a surface per unit area is represented by the Poynting vector *S*. Thus, the Poynting lines reveal the electromagnetic energy that is transmitted through a surface at the expense of the energy that is stored in the fields. As it states that whatever flows out should be at the expense of what is left inside, it can also provide us with a continuity equation. In this view, the charged particles do not cooperate with the Poynting vector to move through the lines that have been considered. Take a look at Poynting's theorem. The energy stored in the fields will diminish as a result of the work required to transport the charge as well as the energy flux radiated out. Where does the field's diminished energy end? The residue is emitted through the surface ever since a portion of it serves a significant function. The Lorentz force acting on the charge causes the charged particles to travel in that direction. Lines of force are not present in the energy flow density lines. It has nothing to do with how the charge is moving. Energy always radiates whether there is a charge present or not, and this equates to a reduction in the energy held in the fields [7-10].

3.4 Quantization of Maxwell's Equations

Taking this into account, electromagnetic phenomena have been known to be explained by Maxwell's equations. Maxwell demonstrates how light is a manifestation of an electromagnetic field. The Maxwell theory has been proven to be effective in explaining several phenomena that also occur in solid materials. Maxwell demonstrated the electromagnetic nature of light. According to quantum mechanics, light is made up of photons with no mass. The electromagnetic theory's gauge invariance forbids it. The quantum electrodynamic theory covers the entire field of photons. Let's assume for this explanation that the photon when it undergoes a tremendous transformation at any point $\vec{r}$, and time t is characterized by a complex wave function, $\Psi(\vec{r},t)$, which satisfies the eigenvalue equation [11-13]. Consequently, the photon wave equation can be mathematically expressed as:

$$\Psi(\vec{r},t) = \hat{p}^{-1} mc\Psi(\vec{r},t) \tag{3.22}$$

here m and c are the mass and speed of the light respectively, while $\hat{p}$ is the photon potential energy. The electric and magnetic characteristics of the medium influence the speed of light there, and the formula for the speed of light in a vacuum is:

$$c = [\varepsilon_0 \mu_0]^{-05} = 299,792,458 \pm 1.2 \text{ ms}^{-1} \tag{3.22.1}$$

According to sub-formual 3.22.1, ε_0 and μ_0 are electric permittivity and magnetic permeability, respectively. In this view, comparisons to earlier standards for the length of the metre account for the majority of the uncertainties in sub-equation 3.22.1. The length of the metre has therefore been redefined to be consistent with the aforementioned number, and the speed of light 299, 792, 458 ms^{-1} has been accepted as a standard. Therefore, equation 3.22 can be casted as:

$$(i\Lambda, \Psi(\vec{r},t)) = \left[i\frac{E}{c}, \vec{p} \right]^{-1} mc(i\Lambda, \Psi(\vec{r},t)) \tag{3.23}$$

where Λ is a scalar function that requires to be defined and comprises a magnetic field density dimension. In this regard, the complex wave function $\Psi(\vec{r},t)$ can be defined as:

$$\Psi(\vec{r},t) = c^{-1}\vec{E} - i\vec{B} \tag{3.24}$$

The point of view of quantum mechanics $\vec{p}$ is given by:

$$\vec{p} = -i\hbar\vec{\nabla} \tag{3.25}$$

and

$$E = -\hbar\frac{\partial}{\partial t} \tag{3.26}$$

Utilizing equations 3.24 to 3.26, equation 3.23 can be formulated into the following new formulas of Maxwell's equations:

$$\vec{\nabla}\times\vec{B} = c^{-2}\frac{\partial\vec{E}}{\partial t} + \left[m\hbar^{-1}\right]\vec{E} - \vec{\nabla}\Lambda, \tag{3.27}$$

$$\vec{\nabla}\times\vec{E} = -\frac{\partial\vec{B}}{\partial t} + \left[mc^2\hbar^{-1}\right]\vec{B}, \tag{3.28}$$

$$\vec{\nabla}.\vec{B} = 0, \tag{3.29}$$

$$\vec{\nabla}.\vec{E} = \frac{\partial\Lambda}{\partial t} + \left[mc^2\hbar^{-1}\right]\Lambda. \tag{3.30}$$

The conceivable potential differences are generated using equation 3.27 if Λ and m equal to zero in which $V = mc^2\hbar^{-1}\phi_B$. In this understanding, over increasing distance, the magnetic field diminishes exponentially. The second term in equation 3.27's right-hand side, therefore, specifies the powerful photon's magnetic current density. In this sense, the voltage V is in charge of the one that causes the quantum Hall effect [11, 15, 17]. In this situation, $V = mc^2BA\times\ ^{1}$, whereas in a two-dimensional Hall effect, $V = IBA[Q]^{-1}$. As seen, by comparison, the photon's rest-mass energy equals:

$$m = i\hbar\left[c^2Q\right]^{-1}, \quad m = J\hbar\left[n_s e\right]^{-1} \tag{3.31}$$

Equation 3.31 demonstrates that J is the surface current density ($J \sim 10^{12}$ A/m^2), n_s is the surface number density ($n_s \sim 10^{15}$ m^{-2}), and Q is the total charge encircled by the sample in the magnetic field. It seems that when a static magnetic field is supplied to the sample, the photon's mass is incredibly small. Additionally, it exhibits the magnetic current flowing through the sample.

The work of Vlaenderen and Waser is generalized in equations (3.27–3.30) to include massive photons [9]. The equations (3.27) through (3.30) have a quantum signature since $\hbar = [2\pi]^{-1}h = [2\pi]^{-1}\times 6.62607015\times 10^{-34}$ J·Hz^{-1} is present. They, therefore, conclude that both equations govern the massive photon. In this regard, equations (3.27–3.30) are referred to as the quantized Maxwell's equations for the photon since they embody the Planck constant $h = 6.62607015\times 10^{-34}$ J·Hz^{-1}. The fact that the photon's wave function satisfies the aforementioned matter wave and Maxwell's equations illustrates the photon's dual existence as a particle and a wave [12-15]. Therefore, equations (3.27) through (3.31) are comparable to the initial Maxwell's equations, where $\left[mc^2\hbar^{-1}\right]$ is set to a general constant.

As well as the fact that Hubble's constant is given by $H = \left[0.5mc^2\hbar^{-1}\right]$, it is fascinating to perceive that the evolution of the electromagnetic field in an expanding universe during inflation provides field equations that are similar to equations (3.27–3.30), but with $\Lambda = 0$. In this circumstance, the magnetic field in equation 3.27 can be given by:

$$\oint \vec{B}.dl = Q\left[m\varepsilon_0^{-1}\hbar^{-1}\right] \tag{3.32}$$

According to equation 3.32, a stationary electric charge also generates a magnetic field. Owing to the spins of the static charges (electrons), ferromagnetic materials contain a magnetic field [13, 15]. Therefore, when a large photon is present, $J_s = mc^2e\hbar^{-1}n_s$, which does not equal zero, gives the current density aligned to this magnetic field.

As a result, the magnetic field is allowed to degrade exponentially over time. The photon's mass affects the decaying factor. Consequently, some remnant magnetic fields would still exist today. Because of this, finding the Hubble's constant, $m = 2H\hbar c^{-2}$, can be used to determine the photon mass [11, 16, 18]. In the point of fact, the H is $\sim 10^{35}s^{-1}$ leads to inflation in which the photon rest-mass energy is equivalent to 10^{11} GeV.

3.5 Feynman's Identification of Maxwell's Equations

Richard Feynman thought that light is composed of particles. It is crucial to understand that light exhibits both wave- and particle-like behavior. Feynman conjectured that both the photon and the electron migrate from one point and time to another. In other words, photons and electrons are in a dynamic system that changes over time. In addition, an electron emits or absorbs a photon at a certain location and time. A Feynman diagram illustrates these dynamic processes (Figure 3.15). On the other hand, the Feynman diagram shows a straight line for electron ejection and a wavy line for photon propagation [15-18].

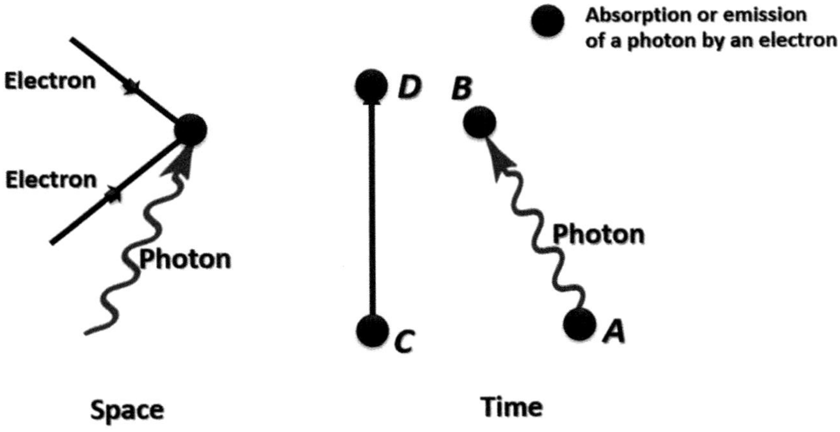

Figure 3.15: Simplification of Feynman diagram.

Feynman was also in charge of a different time-bending method of investigating quantum phenomena. On the other hand, Maxwell's equations for electromagnetism included two explanations that represented two distinct types of electromagnetic waves: retarded waves, which are the ones we can understand, and advanced waves, which travel backward in time from the object to the antenna, for instance. The advanced wave explanation had only been disregarded because it had been proven to have no connection to reality. Feynman did order that the advanced waves do exist. This eliminates a possible issue that could arise when an electron emits a photon—the recoil caused by the conservation of momentum in the electron.

According to this theory, each photon would have the energy of the absorbed one due to their simultaneous oppositional motion through time and space. The outcome is identical to that produced by a single conventional photon. Maxwell's equations, however, make sense because of it. In theory, advanced waves are capable of transporting a photon backward in time, but for it to work, we would need to find an area of space devoid of electromagnetic spectrum absorbers [16, 19]. Due to its large wavelength, the microwave spectrum is the only one that, when used in an electromagnetic beam, does not absorb in the atmosphere.

The actual issue is that by switching between classical notation and quantum notation, Feynman slinks into physical postulations that make it very difficult to distinguishable to verify Maxwell's equations. Let's consider that there is a particle at the point x^{μ} with the momentum p^{μ} that relies upon the function $F(x^{\mu})$. The scientific explanation of how forces act on particles may then be expressed mathematically as:

$$\frac{dp^{\mu}}{d\tau} = F_1^{\mu}(x^{\mu}) + F_2^{\mu\nu}(x^{\mu})p\nu + ... \tag{3.33}$$

The field F is represented by the tensor F_i. But because F_i is so broad, it is asked that further physical presumptions be revealed. The Lagrangian, whose velocity is quadratic, is assumed to be $\ell(x^{\mu}, \dot{x}^{\mu}, t)$. Equation 3.33's mathematical differentiation suggests that the force can only have linear momentum. This viewpoint formulates this linear momentum as follows:

$$\frac{dp^{\mu}}{d\tau} = F_1^{\mu} + F_2^{\mu\nu} p\nu. \tag{3.34}$$

Thus equation 3.34 is ascertained from conventional Hamiltonian mechanics as is abided by:

$$\left[x_i, v_i\right] = i\frac{\hbar}{m}\partial_{ij}. \tag{3.35}$$

Equation 3.35 is modified to incorporate quantum mechanics by following the Dirac recommendation to replace Poisson brackets with commutators. The resulting equations $\left[x_i, p_j\right] = i\hbar\partial_{ij}$, where x_i and p_j are often canonically conjugate, are known as canonical commutation relations. The momentum can be expressed using a Lagrangian formulation and is derived from:

$$p \equiv \frac{\partial \ell}{\partial \dot{x}} \tag{3.36}$$

Equation 3.36 explains that since the Lagrangian seems to have a quadratic velocity, the force has a maximum linear velocity. Equation 3.36 can be converted into the Lorentz force law under the circumstances of $F_1 = 0$ and $F_2 = F$ by:

$$\frac{dp^{\mu}}{d\tau} = F^{\mu\nu} p\nu. \tag{3.37}$$

Because there is no knowledge about the field's dynamics, deriving Maxwell's equations is a hassle. To bridge this gap, the Hamiltonian's simplicity is taken into account. As a corollary, the mathematical representation of quadratic momentum is described in the following:

$$\mathrm{H} = p^2(2m)^{-1} + \vec{A}_1.\vec{p} + A_2 \tag{3.38}$$

The Hamiltonian equation for momentum when $\vec{A}_1$ and A_2 are gathered into a four-vector A^{μ} is expressed as follows:

$$\frac{dp^{\mu}}{d\tau} = (dA)^{\mu\nu} p\nu \tag{3.39}$$

In equation 3.39, d is the external derivative. That seems to be, the Hamiltonian's simplicity forces the field F to be defined in terms of potential, $F = dA$. Since $dF = 0$, and $d^2 = 0$, which contains two of Maxwell's equations, Gauss's law of magnetism and Faraday's law. The crucial argument is that Feynman's derivation is diverse, but not entirely obscured. It is not, in particular, collaborating classical and quantum physics; the quantum equations that Feynman proposes correspond to conventional experiments worked from the Hamiltonian equation. In essence, Feynman adopted the Dirac quantization technique [15, 17, 19].

Since the space of potential theories is extremely confined, it is not astonishing that electromagnetism appears virtually 'for free.' We can obtain Maxwell's equations in the more general framework of quantum field theory by assuming locality, parity symmetry, Lorentz invariance, and the existence of a long-range force mediated by a spin 1 particle. Considering that the only classical physics we can experience are quantum fields with a meaningful chronological restriction, this has implications for classical physics. In practice, no one has ever quantified an electric field, only how it interacts with particles.

When applied to massless spin-one particles, such as a single electron, Dirac's particle method precisely predates Maxwell's equations. In other words, the Maxwell field is regarded as the quantum wave function for a certain photon at point $\vec{r}$ and time t. In this approach, the well-known Dirac theory of quantum is embodied in the quantization of a single photon's non-operator Maxwell field. The scientific explanation for Maxwell's equations can be stated mathematically as:

$$\frac{\partial}{\partial t}\vec{E}(\vec{r},t) = c\vec{\nabla}x\vec{B}(\vec{r},t) \tag{3.40}$$

$$\frac{\partial}{\partial t}\vec{B}(\vec{r},t) = c\vec{\nabla}x\vec{E}(\vec{r},t) \tag{3.41}$$

$$\vec{\nabla}.\vec{E}(\vec{r},t) = 0 \tag{3.42}$$

$$\vec{\nabla}.\vec{B}(\vec{r},t) = 0 \tag{3.43}$$

Then, in terms of the wave function $\Psi(\vec{r},t)$, Maxwell's equations can be expressed as follows:

$$\frac{i}{c}\frac{\partial}{\partial t}\Psi(\vec{r},t) = \vec{\nabla}x\Psi(\vec{r},t) \tag{3.44}$$

Therefore, $\Psi(\vec{r},t)$ can be written as:

$$\Psi(\vec{r},t) = \sum\vec{E} + i\vec{B} \tag{3.45}$$

Consequently,

$$i\frac{\partial}{\partial t}\vec{E} = \left(\vec{\nabla}x\vec{B}\right)c \tag{3.46}$$

$$i\frac{\partial}{\partial t}\vec{B} = \left(-\vec{\nabla}x\vec{E}\right)c \tag{3.47}$$

As addressed by Maxwell's equations, a photon is a primordial emission of the quantized electromagnetic field as will be debated in the following sections. In this perspective, the complex Maxwell equations can be expressed mathematically as:

$$i\frac{\partial}{\partial t}\vec{\Psi}_T(\vec{r},t) = c\vec{\nabla}\times\vec{\Psi}_T(\vec{r},t). \tag{3.48}$$

It should go without saying that the quantum wave function $\vec{\Psi}_T(\vec{r},t)$ of a photon is described by the classical Maxwell equations. In other words, the conventional Maxwell field is a single photon's quantum wave function $\vec{\Psi}_T(\vec{r},t)$, which turns into a three-dimensional vector owing to the photon's spin-one nature. Likewise, it is thought that the Maxwell field and wave function $\vec{\Psi}_T(\vec{r},t)$ from the Schrödinger equation are comparable as will be discussed in the following section. The probability amplitudes for distinct potential quantum occurrences are transformed in this approach, where the photon energy is generated within a given size. This means that a realistic description of the electron's location is impossible [18, 20].

3.6 Quantum Electromagnetic Radiation

Why is it vitally significant to comprehend quantum theory to understand remote sensing and electromagnetic waves? According to the above perspective, answering the question is great challenging. Thoughtful machinations and capabilities of remote sensing technologies are significantly assisted by quantum theories. Utilizing both optical and microwave remote sensing methods, this is accomplished. This section's major goal is to make clear how Maxwell's equations are investigated using quantum mechanics in the previous section (section 3.5) and can be used to quantize remote sensing principal theories. Because of this, comprehension of the electromagnetic (EM) spectrum and electromagnetic radiation, such as light, radar, and radio waves, is entirely reliant on quantum mechanics.

Max Planck's theories on energy quantization, the photoelectric effect, particle-wave duality, contemporary atomic theory, the uncertainty principle, the wave equation, the wave function, and the exclusion principle are the foundational hypotheses for the quantization of remote sensing imaging mechanisms.

Plank pioneered the notion that heated bodies emit radiation, not as an endless stream like water from a faucet, but rather as leaky packets termed quanta, which he referred to as 'energy elements.' He assumed that the energy E of these packets was inversely correlated with their wavelength (λ), with the maximum energy being associated with the shortest wavelengths. The wavelength of these quanta and their energy were related analytically, and this relationship, known as the Plank relation, began to be understood [18, 21].

Therefore, the criterion of zero electric fields at the wall must be satisfied by a mode for an electromagnetic wave in a cavity. If the wavelength of the mode is shorter, there are various possibilities to accommodate it into the cavity to match that criterion. In this regard, Rayleigh and Jeans' careful investigation revealed that the number of modes was related to the frequency squared. In this view, we would enumerate all the potential ways that the integer n values could be combined to determine how many modes can satisfy this requirement. The volume of a three-dimensional grid containing the values of n, or a "n-space," can be exploited to approximate the number of combinations [22]. When three "n" axes' coordinates are specified by the n values in the relationship for a sphere's volume, the result is

$$\text{"Volume" of } n\text{'s} = \frac{4\pi}{3}(n_1^2 + n_2^2 + n_3^2)^{3/2} \tag{3.49}$$

There are a few issues with this, though. In contrast to the wave equation solution, which only employs positive definite values, we have used both positive and negative values of n when employing a sphere. As a consequence, we must subtract 1/8 of the volume above. Another technical issue is that waves can be polarized in two perpendicular planes; to take that into account, we must multiply by two. When the size of the cavity is substantially larger than the wavelength, as it is in the case of electromagnetic waves in finite cavities, the volume can then be regarded to be a measure of the number of modes (Figure 3.16), becoming a very reasonable approximation [18-21]. Using the relationship identified for n-values, this turns out to be:

$$\text{Number of modes} = N = \frac{\pi}{3}(n_1^2 + n_2^2 + n_3^2)^{3/2} = \frac{8\pi L^3}{3\lambda^3} \tag{3.50}$$

Equation 3.50 reveals that the requirement must always be fulfilled for equilibrium standing wave electromagnetic radiation in a cubical cavity of dimensional L. So, how many modes are there in a wavelength? Since deriving an expression for the number of standing wave modes present in a cavity, it is required to determine their distribution concerning wavelength. By taking the derivative of the number of modes for wavelength, this can be achieved by:

$$\frac{dN}{d\lambda}=\frac{d}{d\lambda}\left[\frac{8\pi L^3}{3\lambda^3}\right]=-\frac{8\pi L^3}{\lambda^4} \tag{3.51}$$

The fact that this symbol is negative shows that as wavelength increases, the number of modes drops. Let's divide the left side of equation 3.52 by the cubical cavity volume L^3 to obtain the number of modes per unit volume per unit wavelength.

$$\frac{dN}{d\lambda}\left[L^{-3}\right]=\frac{d}{d\lambda}\left[\frac{8\pi L^3}{3\lambda^3}\right]=-\frac{8\pi}{\lambda^4} \tag{3.52}$$

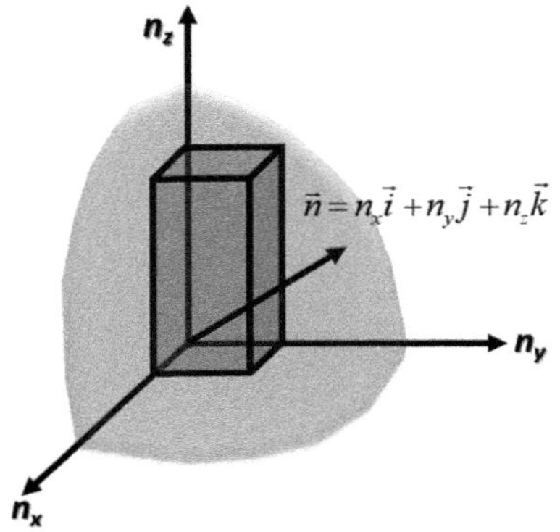

Figure 3.16: Determination number of modes.

Keep in mind that this does not entail rough sphere-to-cube conversion! We were able to count the number of potential modes since the sphere we used to calculate the number of modes was one in "*n*-space." Additionally, using a cubical cavity for the calculation only allows us to simplify the development's shape; yet, the outcome is unaffected by the geometry of the cavity.

Planck developed a radiation formula based on the presumption that the energy of the electromagnetic modes in a cavity was quantized, with the quantum energy being equal to Planck's constant times the frequency [21-23]. The average energy per "mode" or "quantum" is the quantum's energy multiplied by the likelihood that it would be formed as:

$$\langle E\rangle = hv\left[e^{\frac{hv}{kT}}-1\right]^{-1} \tag{3.53}$$

Here T is the absolute temperature in kelvin and k is Boltzmann's constant (1.3807×10^{-23} joule s per kelvin ($J \cdot K^{-1}$)). When stated in terms of frequency or wavelength, this average energy times the density of such states yields:

$$\rho(v)=\frac{8\pi}{c^3}v^2, \qquad \rho(\lambda)=\frac{8\pi}{\lambda^4} \tag{3.54}$$

Let's say that we give the energy density, then the Planck radiation formula is:

$$S_v=\frac{8\pi h}{c^3}v^3\left[e^{\frac{hv}{kT}}-1\right]^{-1} \tag{3.55}$$

Equation 3.55 demonstrates the Plank quanta energy per unit volume per unit frequency S_v. Therefore, the plank quanta energy per unit volume per unit wavelength S_λ is casted as:

$$S_{\lambda} = \frac{8\pi hc}{\lambda^5}\left[e^{\frac{hv}{kT}} - 1\right]^{-1} \tag{3.56}$$

The Bose-Einstein statistics of energy distribution are illustrated by the Planck radiation formula. The aforementioned equations are produced by adding the photon energy, the photon energy times the Bose-Einstein distribution function, and the normalization constant $A = 1$ to the density of states expressed in terms of frequency or wavelength. Therefore, multiplying the energy density by 0.25 c will yield the radiated power per unit area from a surface at this temperature. Setting inward-outward results in a factor of 0.5 for the radiated power outward since the density in the example above is for thermal equilibrium. To obtain a factor of 0.5 for the angular dependency, which is the square of the cosine, one must then average over all angles [18-23].

For low frequencies and long wavelengths, the Rayleigh-Jeans curve and Planck radiation formula are compatible as demonstrated in Figures 3.17 and 3.18 respectively. Despite not being a true representation of nature, the Rayleigh-Jeans Law was a significant stride in our understanding of the equilibrium radiation from a heated object. The Rayleigh-Jeans law, which was carefully constructed, served as the foundation for the quantum understanding embodied in the Planck radiation formula (Figure 3.19).

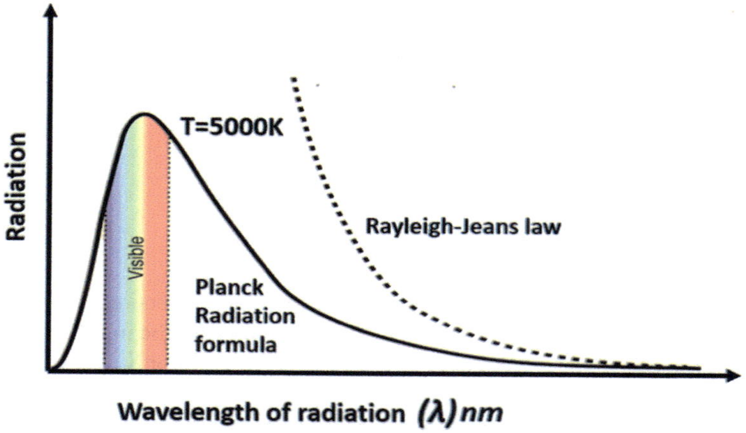

Figure 3.17: Radiated intensity as a function of frequency.

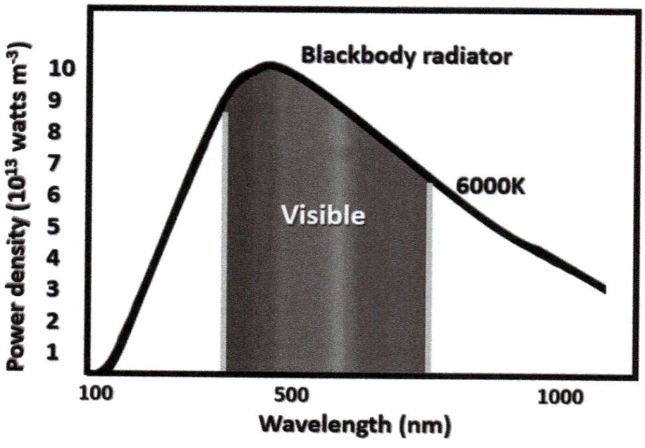

Figure 3.18: Radiated intensity as a function of wavelength.

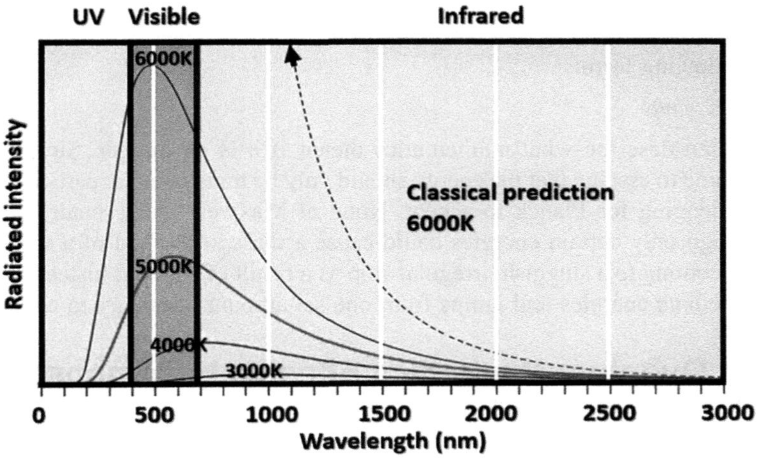

Figure 3.19: Rayleigh-Jeans equilibrium radiation from a heated object.

Therefore, when compared to the blackbody curve at 6000 K, the irradiance outside the atmosphere can be said to be similar. It is possible to verify that the sun irradiation is identical to that of a blackbody radiator by looking at the 400-700 nm nominal range of visible light. If white is defined as having the same amount of energy per wavelength, the curve at the surface is substantially flatter, or more "white."

Needless to say, the terms "blackbody radiation" and "cavity radiation" refer to an object or system that completely absorbs all radiation that strikes it before reradiating energy that is unique to this radiating system alone and is not influenced by the amount of energy that strikes it. The emitting cavity can be conceived of as releasing the energy as standing waves or resonant modes [15, 19, 21].

However, Planck was unable to articulate the rationale behind the radical notion he put out when energy should be quantized. Blackbody radiation was the only set of experimental data that his idea initially attempted to explain. Quantization would become a law if it were found to occur for a lot of diverse phenomena. A theory to explain how legislation might be created in the future. In the end, Planck's hypothesis served as the basis for modern physics. On the other hand, the spectrum distribution of blackbody radiation was explained by Max Planck as the outcome of electron oscillations. Radio waves are also created by the oscillation of electrons in an antenna. Max Planck focused on simulating the oscillating charges that must be present in the oven walls since they must radiate heat inwards while also being driven by the radiation field while in thermodynamic equilibrium. He discovered that if he disallowed these oscillators from radiating energy continuously, as the classical theory would require, and instead allowed them to only lose or gain energy in chunks, known as quanta, of size hv for an oscillator of frequency, he could explain the observed curve.

The radiation energy density inside the oven was determined by Planck using the following formula:

$$dS_v(v,T) = \frac{2hv^3}{c^2}\left[e^{\frac{hv}{kT}} - 1\right]^{-1} dv \tag{3.57}$$

It is also possible to express Planck's radiation energy density in terms of wavelength (Equation 3.57) as:

$$S_v(\lambda,T)\,d\lambda = \frac{2hc^2}{\lambda^5}\left[e^{\frac{hc}{\lambda kT}} - 1\right]^{-1} d\lambda \tag{3.58}$$

The computations suggested that a blackbody's energy was not continuous but instead released in predictable bursts at a limited range of wavelengths. If Planck believed that blackbody radiation's energy took the following form:

$$E = nh\nu \tag{3.59}$$

Plank could then describe what mathematics meant if n is an integer. Since there was no justification at the time to assume that the energy should only be transmitted at particular frequencies, this was very challenging for Planck to accept. None of Maxwell's laws made this suggestion. It appeared as though only certain energies could cause a mass at the end of a spring to vibrate. Consider the mass coming to a sluggish, irregular stop as a result of friction. Instead, the mass skips through the intermediate energies and jumps from one set amount of energy to another [21-24].

3.7 Why can Quantum Mechanics Effectively Demonstrate Black Body Radiation?

Blackbody radiation has a distinct, continuous frequency spectrum that, in experiments, is hardly influenced by the body's temperature. Scientists might be more concise: a body emits radiation at a certain temperature and frequency in a similar manner that it absorbs radiation at a specific temperature and frequency. In this view, Gustav Kirchhoff proved this statement: if we instead assume that a specific body can absorb more effectively than it emits, then in a room filled with objects with a similar temperature, it will absorb photons from other bodies more effectually than it radiates energy back to them. This implies it will become warmer while the rest of the space will become colder, breaching the second law of thermodynamics. To avoid violating the second rule of thermodynamics, a substance must release radiation just as well as it absorbs the same radiation at a particular temperature and frequency.

Consistent with the above perspective, anybody would radiate to some degree at any temperature above absolute zero, with the amount and distribution of the radiation's frequencies depending on the specifics of the body's structure. To start evaluating heat radiation, we must be clear about the body that is emitting it. The simplest situation is an idealized entity that is an ideal emitter as well as a perfect absorber (according to the previous argument). So, how do we create the ideal absorber in the lab? In 1859 Kirchhoff discovered that a little hole in the side of a huge box makes an efficient absorber because any radiation that passes through it bounces around inside the box, absorbing a significant amount with each bounce and having little chance of ever leaving.

The energy distribution of blackbody radiation could be measured quite precisely by the 1890s because of advancements in experimental methods. In 1895, Wien and Lummer at the University of Berlin drilled a tiny hole in the side of an otherwise perfectly closed oven and started measuring the radiation that emerged. A diffraction grating was used to send the various wavelengths and frequencies of the beam exiting the hole in various directions, all of which were directed toward a screen. To determine the amount of radiant energy being emitted at each frequency band, a detector was moved up and down the screen [15, 21].

They were also able to develop Stefan-Law Boltzmann's and Wien's Displacement Laws, two significant phenomenological Laws (i.e., established from experimental data, not from fundamental laws of nature), by monitoring the blackbody emission curves at various temperatures (Figure 3.20).

The Stefan-Boltzmann Law (1879), which asserts that the total power radiated from a square meter of the black surface goes as the fourth power of the absolute temperature (Figure 3.21), was the first quantitative hypothesis based on experimental findings:

$$E = \sigma T^4 \tag{3.60}$$

Equation 3.60 demonstrates that the total amount of radiation emitted by the object per square meter (Watts m^{-2}), therefore, the Stefan-Boltzmann constant is σ (5.67×10^{-8} Watts m^{-2} K^{-4}) and T(K) is the absolute temperature. Comparing the integral value, or area under the curves, of the

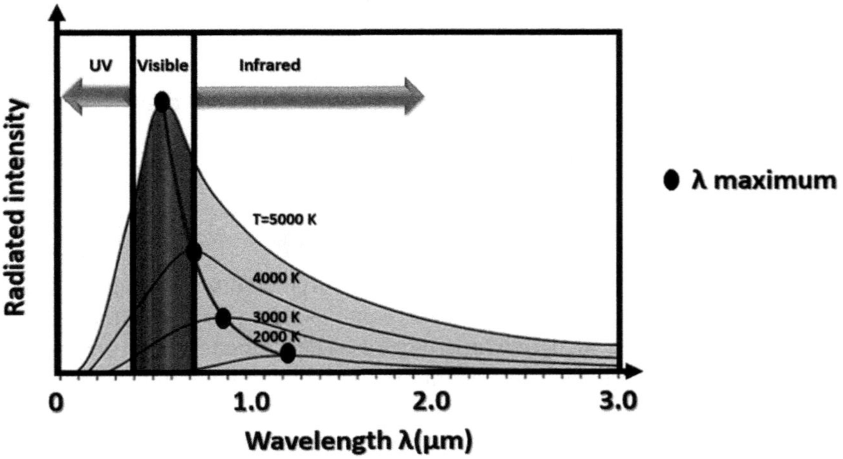

Figure 3.20: Blackbody radiation's spectral distribution at various temperatures is based on Stefan-Boltzmann's Law and Wien's Displacement Laws.

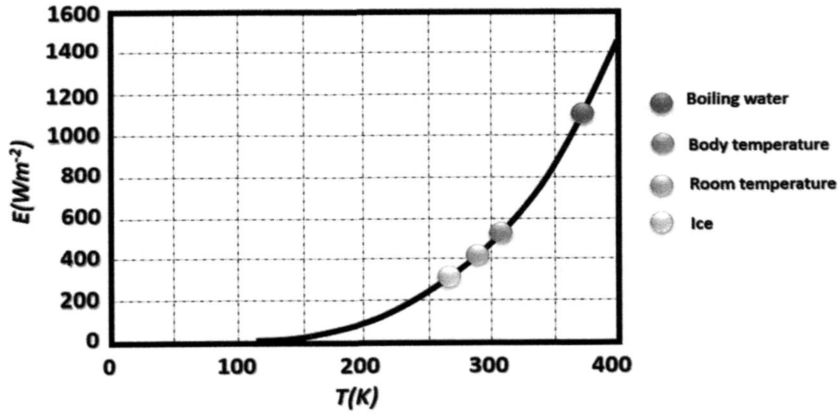

Figure 3.21: Stefan-Boltzmann Law of the fourth power of the absolute temperature.

experimental black-body radiation distribution at various temperatures makes it simple to detect the Stefan-Boltzmann Law. By applying classical thermodynamic reasoning to a box filled with electromagnetic radiation and utilizing Maxwell's equations to connect pressure to energy density, Boltzmann deduced this T^4 behavior from theory in 1884. In other words, the minimal energy emitted from the hole would naturally be temperature-dependent, such as the radiation intensity within it.

Wien's Displacement Law was the second phenomenological conclusion drawn from the experiment. The dominating (peak) wavelength or color of light coming from a body at a specific temperature is determined by Wien's law. The frequency at which the emitted radiation is most strong varies with the oven temperature. In actuality, the relationship between frequency and absolute temperature is direct:

$$v_{\max} \alpha T \tag{3.61}$$

In equation 3.61, 5.87910^{10} Hz/K serves as the proportionality constant. In 1893, Wien explicitly derived this law theoretically by applying Boltzmann's thermodynamic justification. It had previously been observed, at least semi-quantitatively, by an American astronomer, Langley. Everyone is familiar with this increase in v_{max} with T because the first visible radiation that is produced when the iron is heated in a fire (at about 900 K) is deep red, the lowest frequency of visible light. At extremely high temperatures (10,000 K or more) where the peak in radiation intensity has migrated beyond the visible into the ultraviolet, the hue changes to orange, then yellow, and eventually blue as T increases [20, 22, 24].

In terms of the peak wavelength of light, the following is another illustration of Wien's Law (Equation 3.61):

$$\lambda_{max} = bT^{-1} \tag{3.62}$$

Equation 3.62 says T is the absolute temperature in kelvin and b is a proportionality constant known as the Wien's displacement constant, which is equal to 2.89×10^{-3} mK or, more conveniently, $b \approx 2900$ μm.K to obtain the wavelength in micrometres. The relationship between wavelength and temperature is inverse. Therefore, the wavelength of thermal radiation will be shorter or smaller as temperature increases. The wavelength of thermal radiation is longer or bigger at lower temperatures. Hot things emit bluer light than cool ones do for visible radiation.

It's important to keep in mind that thermal radiation always has a wide range of wavelengths, and Equation 3.62 only identifies the peak wavelength. The Sun, therefore, seems yellowish-white, but when sunlight is split using a prism, all the hues of the rainbow are revealed. Yellow only denotes the emission's distinctive wavelength.

A black body seen in the dark at the lowest just barely observable temperature would subjectively appear grey even though its objective physical spectrum has its peak in the infrared range since the human eye cannot perceive light waves at lower frequencies. It appears dull red when it becomes slightly hotter. In this understanding, The equipartition of energy, a concept from classical physics, states that energy is continuous and equally distributed among a system's multiple degrees of freedom. This was precisely the concept behind Rayleigh-Jeans' Blackbody Radiation Equation. However, it was discovered that this equation contradicts actual experimental findings by predicting that the energy of black body radiation would become infinite at very high radiation frequencies. Therefore, conventional physics is erroneous. Max Planck later demonstrated that the findings produced are in perfect agreement with the experimental results if we avoid traditional equipartition and think of energy as discrete. Thus, quantum physics was created. In other words, According to the classic theory, a warm object would radiate heat at all wavelengths, even those with arbitrary short wavelengths. The thermal radiation from a light bulb, a heater, or even your own body would be released as fatal gamma rays.

Planck found a solution (See section 3.6) to this problem by assuming that radiation is released in discrete units, or "quanta," rather than constantly. Little can change for long wavelengths (low energy), but for short wavelengths (high energy), there merely wouldn't be enough energy to emit numerous of these high energy quanta, drastically altering the emission characteristics. Later interpretations of Planck's hypothesis claim that the heated objects' radiation is produced by the atoms inside of them. Since an atom can only contain a given amount of energy, it follows that it can only output a certain amount of energy. As a consequence, rather than being continuous, an atom's energy is quantized, existing only in particular defined numbers (like a wave). The addition or subtraction of one or more "packets" of discrete energy amounts causes each change in the atom's energy. The energy of each of these energy packets, known as a quantum, is denoted by the symbol hv as have been discussed in section 3.6. The energy of the radiation emitted or absorbed is equal to the difference in the atoms' states i.e. ΔE atom = E emitted or absorbed radiation = nhv, and consequently, an atom changes its energy state by emitting or absorbing one or more quanta (plural of quantum). In this understanding, the smallest change arises when an atom transitions from a given n energy state to an adjacent state, or when $n = 1$: $E = hv$, because the atom can only alter its energy by integer multiples of hv. This is why black body radiation cannot be explained by the classical paradigm [19, 23].

Since there is no relationship between wavelength and energy in classical theory, all of these modes have the same energy, and there is no reason why they should all be excited differently when the blackbody cavity is heated. The anticipated energy transported by each of these modes at a given temperature T is kT, where k is the Boltzmann constant as discussed earlier. The issue is that there are countless numbers of them! Therefore, the total energy held in these modes is limitless at a finite temperature. In this sense, The fact that each mode can only be excited by an integer multiple of hc/λ, where λ is the wavelength, solves the issue. Shorter wavelength modes are more difficult to excite, which resolves the UV disaster. Shorter wavelength modes have a very high energy requirement to excite them. Needless to say, this is inevitably one of quantum mechanics' notable mysteries: the correlation between length scale and energy can only be achieved through Planck's constant.

3.8 The Photoelectric Effect

Nature seems to have been quantified (non-continuous, or discrete). If this was the scenario, how could Maxwell's equations precisely predict the consequences of the blackbody radiator? To no avail, Planck spent a significant amount of time attempting to reconcile the behavior of electromagnetic waves with the discrete nature of blackbody radiation. It was not until 1905, when Albert Einstein published yet another work, that the wave nature of light was enlarged to incorporate the particle explanation of light, which empirically verified Planck's equation.

Heinrich Hertz, a German physicist, originally described the photoelectric effect in 1887, and it is thus commonly referred to as the Hertz effect. Hertz found that substances emit a visible spark when they absorb specific frequencies of light while working with a spark-gap transmitter (a rudimentary radio-broadcasting device). J.J. Thomson identified this spark as light-excited electrons (entitled "photoelectrons") leaving the metal's surface in 1899 (Figure 3.22).

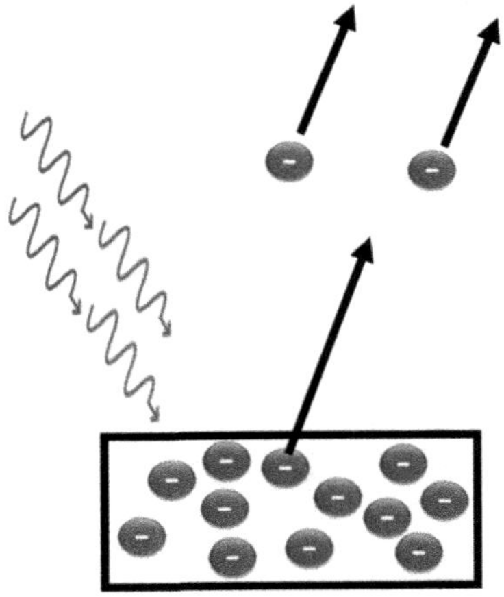

Figure 3.22: Photoelectrons concept.

The conventional explanation for the photoelectron effect was that the atoms in the metal encapsulated electrons that were traumatised and exacerbated to oscillate by the incident radiation's excitation voltage. Some of them would inevitably be shaken loose and ejected from the cathode. Consider how the amount and speed of electrons emitted would be predicted to change with the intensity and color of the incident radiation, as well as the time required to view the photoelectrons. In this perspective, increasing the intensity of radiation would cause the electrons to vibrate extremely vigorously, causing more energy to be expelled and, overall, faster discharge out [25-27]. Therefore, the electrons would be shaken extremely violently with a boost in radiation frequency, eventually resulting in a rapid expulsion of the electrons. An electron would need some time to build up adequate vibrational amplitude to shake loose in very weak light.

3.8.1 Intensity Dependence Speculation

Philipp Lenard, Hertz's student, explored how the energy of released photoelectrons varied with light intensity in 1902. He employed a carbon arc light that had a thousand-fold increase in intensity. The expelled electrons collided with another metal plate, the collector, which was linked to the cathode by a wire equipped with a sensitive ammeter to measure the current generated by the lighting (Figures 3.23 and 3.24). Lenard charged the collection plate negatively to reject electrons coming toward it to detect the energy of the ejected electrons. Only electrons ejected with sufficient kinetic energy to climb this potential peak would thus cause the electric current [25]. In other words, Lenard's photoelectric experiment demonstrated that the high blue light intensity increases photocurrent (Figure 3.23). On the other hand, the low blue light intensity has reduced photocurrent (Figure 3.24).

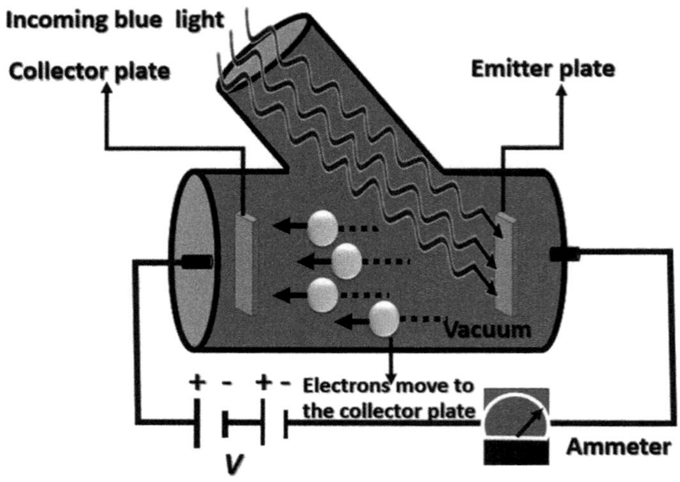

Figure 3.23: Lenard's photoelectric experiment with high-intensity blue light.

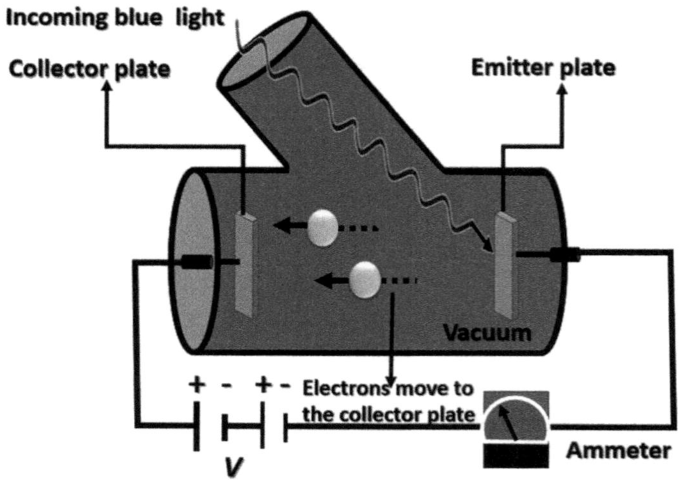

Figure 3.24: Lenard's photoelectric experiment with low-intensity blue light.

3.8.2 Wavelength Dependence Speculation

Following up on Lenard's research (Figure 3.23), American experimental physicist Robert Millikan was able to generate sufficient light intensity using a strong arc lamp to separate the hues and test the photoelectric effect using light of different colors. He discovered that the maximal energy of the ejected electrons was color-dependent, with a shorter wavelength, higher frequency light ejecting photoelectrons with higher kinetic energy (Figure 3.23). In other words, The battery depicts the source of variable voltage that Lenard could have used to charge the collecting plate negatively. The voltage provided by the battery is less than V_{stop} for blue light because the electrons emitted by the blue light are reaching the collecting plate, a low-energy red light that is ominous [26, 28]. The voltage provided by the battery exceeds V_{stop} for a red light because the electrons emitted by the red light cannot reach the collection plate (Figure 3.25).

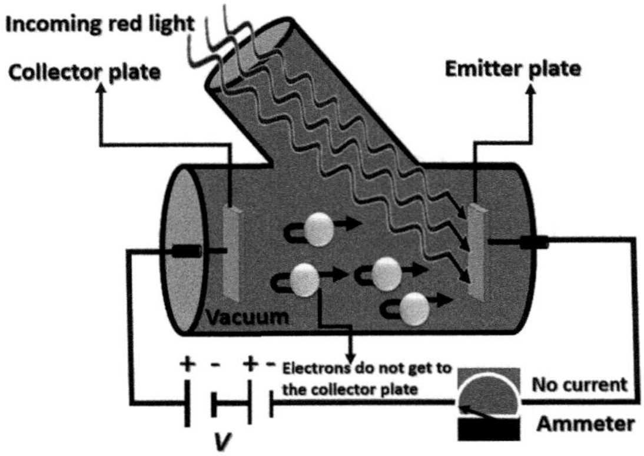

Figure 3.25: Electrons emitted by red light fail to reach the collection plate.

Since no current is detected below v_0, the kinetic energy of the electrons is linearly proportional to the frequency of the incident radiation above that value. The kinetic energy is also unaffected by the radiation's intensity. However, above the threshold value of v_0 (i.e., no current is seen below v_0), the number of electrons (i.e., the electric current) is proportional to the intensity and independent of the frequency of the incident radiation.

Lenard and Millikan's results show that there is a wavelength dependence, which is inconsistent with the predictions of classical theory, which states that the energy carried by light is proportional to its amplitude regardless of its frequency. Therefore, Figures 3.26 to 3.29 address the characteristics of the photoelectric effect from Lenard's and Millikan's experiments. Consistent with this view, Figure 3.26 shows that the kinetic energy of any single electron that is emitted rises linearly as the

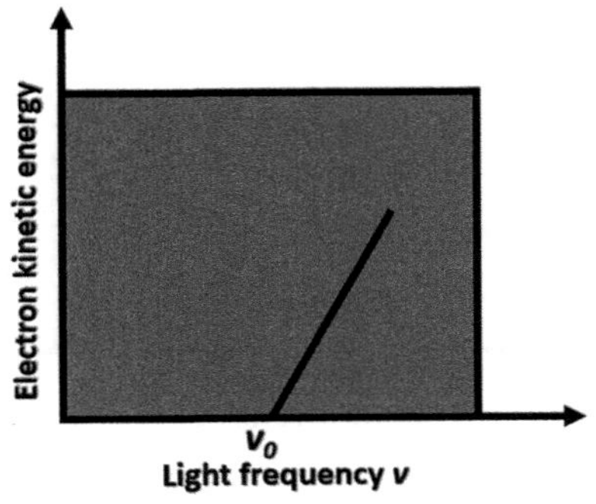

Figure 3.26: The linear relationship between emitted kinetic energy and frequency.

frequency rises above a certain threshold. However, Figure 3.27 depicts that light intensity above the threshold frequency and zero below do not affect the electron kinetic energy. Thus electric current, or the amount of electrons released each second, is independent of light frequency above the threshold frequency and zero below as illustrated in Figure 3.28. On the other hand, Figure 3.29 reveals that the quantity of electrons rises linearly with the intensity of the light [25-28].

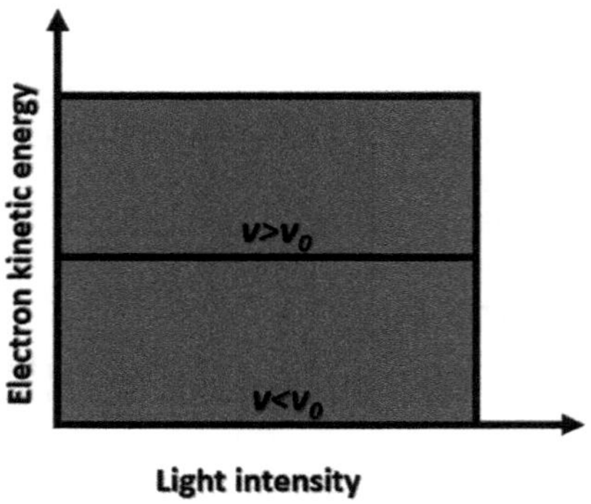

Figure 3.27: Independent relationship between emitted kinetic energy and light intensity.

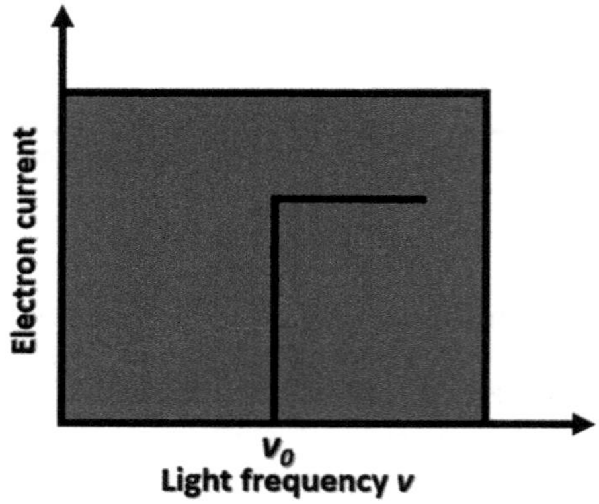

Figure 3.28: Independent relationship between electron current and frequency.

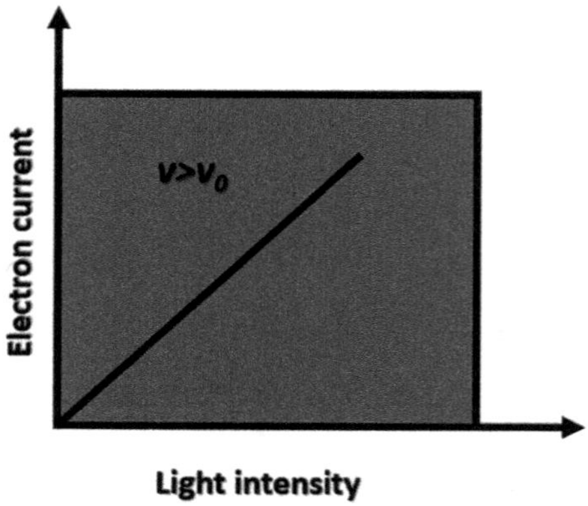

Figure 3.29: The linear relationship between electron current and light intensity.

3.8.3 The Quantum Perspective of Einstein

Assuming that the incoming radiation should be conceived of as quanta of energy hv, with a frequency v, Einstein provided a relatively straightforward explanation of Lenard's findings in 1905. Einstein did this by borrowing Planck's concept about quantized energy from his blackbody research. One such quantum is absorbed by one electron during photoemission [10, 17, 25]. The electron will lose some energy as it goes toward the surface if it is buried deep within the cathode material. As the electron departs the surface, there will always be some electrostatic penalty; this is the workfunction Φ. The electrons that are released with the maximum kinetic energy (KE) are those that are very close to the surface:

$$KE = hv - \Phi \tag{3.63}$$

The maximum kinetic energy electrons (KEe) must have generated an energy eV_{stop} upon leaving the cathode as evidenced by turning up the negative voltage on the collector plate until the current simply stops, that is, to V_{stop}. Hence,

$$eV_{stop} = hv - \Phi \tag{3.64}$$

In this way, Einstein's theory delivers a very explicit quantitative prediction: if the incident light's frequency is changed and V_{stop} is plotted as a function of frequency, the slope of the line should be hc^{-1} (Figure 3.26). It is also evident that a given metal v_0 has a minimum light frequency, which is the frequency at which the quantum of energy is equivalent to Φ (Equation 3.63). In other words, the value of the work function Φ, which is dependent on the metal, is the minimal amount of energy needed to cause photoemission of electrons from a metal surface matter how powerful, light below that frequency won't release electrons. Planck and Einstein both claimed that light's energy is related to its frequency rather than its amplitude, hence an electron must be ejected at a minimal frequency, v_0, to do so without losing any energy [19, 25, 27].

Since every photon with adequate energy can only excite one electron, increasing the light's intensity (i.e., the number of photons/sec) only leads to an increase in the quantity of liberated electrons rather than their kinetic energy. Additionally, since the atom does not need to be heated to a threshold temperature, the release of the electron upon absorption of the light occurs almost instantly. Finally, a threshold frequency exists below which no photoelectrons are seen since the photons must have more energy than a specific threshold to meet the workfunction. In honor of the person who discovered the photoelectric effect, this frequency is expressed in Hertz (1/second) units.

The quantitative characteristics of the photoelectric effect are explained by Einstein's equation 3.63. This experiment has the unusual conclusion that light can operate as a sort of massless "particle" currently known as a photon, whose energy $E = hv$ can be transferred to a real particle (an electron), giving it kinetic energy, just like in an elastic collision between two hefty particles like billiard balls.

Robert Millikan worked on the photoelectric effect for ten years, until 1916, but first rejected Einstein's hypothesis because he considered it an assault on the notion of light waves. Even methods for cleaning the metal surfaces inside the vacuum tube were developed by him. Despite his best efforts, he was met with disappointing outcomes: after ten years, he confirmed Einstein's theory. In his published research, Millikan continues to valiantly attempt to elude this conclusion. Though, He had dramatically changed his views on the occasion of his Nobel Prize victory speech!

When electromagnetic radiation (photons) strikes a material object's surface, the photoelectric effect is said to be the emission of photoelectrons from that object. When a photon approaches or above the threshold energy, the photoelectric effect only emits electrons [14, 17, 19, 27].

3.9 Photovoltaic Effect

What is meant by the photovoltaic effect, then? The appearance of a potential difference (voltage) between two layers of a semiconductor slice whose conductivities are opposing, or between a semiconductor and a metal, when a light stream is applied, is what is known as the photovoltaic effect. This brings up the crucial question: What distinguishes the photovoltaic effect from the photoelectric effect? The key distinction is that the term "photovoltaic effect" is now commonly used when the excited charge carrier is still contained within the material and "photoelectric effect" is used when the electron is discharged out of the material (primarily into a vacuum) [25, 28, 30]. In other words, the photovoltaic effect, in contrast to photoelectric effects, occurs when photons from a light source simply displace electrons from their atomic orbitals while maintaining them within the material. This permits electrons to move freely within the material. The photoelectric effect, on the other hand, involves light photons confiscating electrons from a substance.

Recalling the notion of the quantum wave function $\vec{\Psi}_T(\vec{r},t)$ (section 3.5) is necessary to comprehend the nature of the photovoltaic effect. We may derive formulations for the physical attributes of the electron, such as energy, momentum, and location; thus, the wave function is said to describe the electron's state. The potential energy function E_p determines the function $E_p(\vec{r},t)$. The Schrödinger equation is a second-order differential equation that can be solved for a given $E_p(\vec{r},t)$ and $\vec{\Psi}_T(\vec{r},t)$ as:

$$i\hbar\frac{\partial}{\partial t}\vec{\Psi}_T(\vec{r},t)=\frac{-\hbar}{2m}\nabla^2\vec{\Psi}_T(\vec{r},t)+E_p(\vec{r},t)\vec{\Psi}_T(\vec{r},t) \tag{3.65}$$

The Schrödinger equation demonstrates that the states of an electron orbiting a nucleus are quantized, which means that there are defined states that the electron can and cannot be in. An electron, for example, can only have an energy of x, $x + 1$, $x + 2$, etc., but not $x + 0.5$. The current energy levels (or orbitals), states of specified energy in which electrons can exist, are the result of this. An electron can "jump" up to the next highest energy level if it has enough energy. Nevertheless, an electron must get there all at once. An electron cannot jump halfway there twice in this situation. There are four sorts of energy levels, indicated as *s*, *p*, *d*, and *f*, each with many potential states. This can help us comprehend how electromagnetic radiation reflected from air turbulence particles can easily discriminate between different sorts of air turbulences. When electromagnetic radiation strikes atmospheric gas concentrations, their atoms fill up with electrons, starting with the state closest to the nucleus and working outward [29-31]. According to this viewpoint, if a gas atom has many electrons, only the outermost valence electrons alter the state into distinct orbitals, because the inner electrons are "trapped" in between the nucleus and the valence electrons and so are unable to move. Since valence electrons cannot fill states already occupied by inner electrons, they can only hold to higher levels.

On the word of the above perspective, if the wave function of one of the photons is identified, the position of brightness variations across pixels in remote sensing data can be applied to extrapolate the pattern of the second.

3.10 De Broglie Spectulation of Electromagentic Spectra

Relativity and the photoelectric effect, both had an impact on Louis DeBroglie as a young student at the University of Paris. The photoelectric effect revealed that light, which was previously thought to be a wave phenomenon, actually has particle qualities. He questioned whether electrons and other "particles" might have wave characteristics. The combination of these two novel concepts suggested an intriguing prospect. He correlated wavelength (λ), and momentum (P_{EM}): $\lambda = \hbar p^{-1}$. He stated that the momentum of a photon is given by $P_{EM} = Ec^{-1}$ and the wavelength (in a vacuum) by $\lambda = cv^{-1}$, where c is the speed of light in a vacuum. In this view, the relationship between the wavelength (λ) and momentum ($m*v$) for DeBroglie's "particle-wave" is obtained by:

$$\lambda=\frac{\hbar}{m*v} \tag{3.66}$$

Therefore, the relationship between energy (E) and momentum ($m*v$) as:

$$\frac{\hbar}{m*v} = \hbar\frac{c}{E} \tag{3.67}$$

Abridge, and solve for E:

$$E = mvc \tag{3.68}$$

In this view, the maximum velocity (v) reasonable by matter is the speed of light (c); consequently, the maximum energy can be given by:

$$E = mcc \tag{3.69}$$

or

$$E = mc^2 \tag{3.70}$$

The De Broglie equation, which has photons act like matter particles and apply to all particles, is expressed in equation 3.70. (Figure 3.30). Thus, the wave-like behavior of matter is related to its momentum by Equation 3.68.

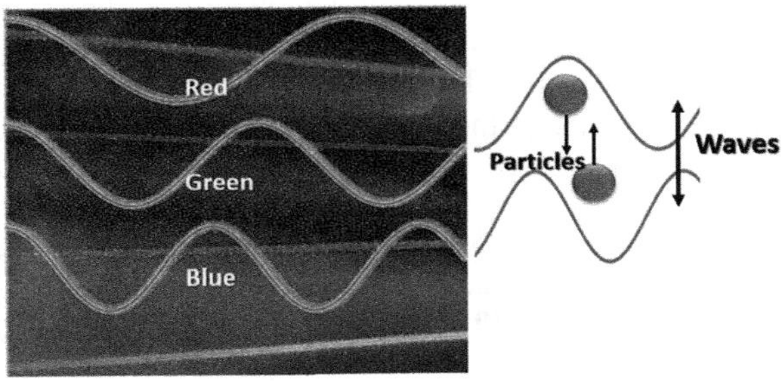

Figure 3.30: Photons behave like wave-particle.

De Broglie's notion demonstrated how far the understanding of electromagnetic beams in quantum physics has advanced the understanding of earlier hypotheses. While photons were insubstantial electromagnetic beams, electrons were parts of matter or substance. Nevertheless, they act differently depending on the situation, sometimes like waves and sometimes like particles. This insight led to a great fit between this postulation and Bohr's atomic model. An electron could only occupy certain orbits around the nucleus in this circumstance as if it were moving along tracks, and it made quantum leaps between them as it received or lost energy in the form of photons [26, 29, 31].

But how were the permissible and unacceptable orbits chosen? Only certain orbitals, with jumps in between, were possible for electrons to occupy, according to Bohar. If they could keep an even number of wavelengths, only orbitals would be possible (Figure 3.31). In the word of this theory, the electron is an energetically stable standing wave that is occupied inside the atom.

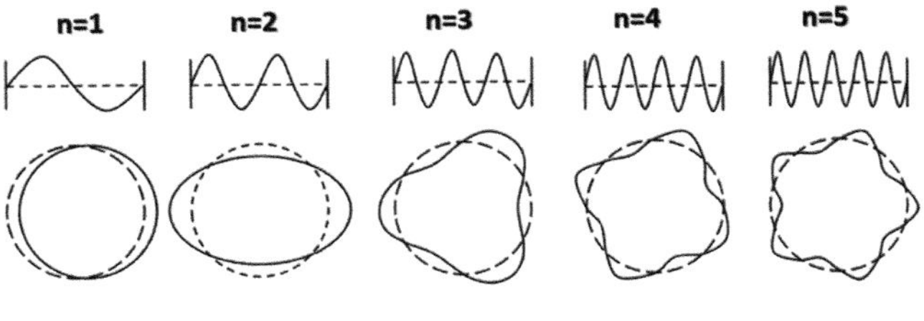

Figure 3.31: De Broglie's wavelength notion.

The principles of EM energy spectra are, therefore, summarized in Figure 3.32. In particular, the wavelengths of visible light fall within the 0.38–0.75 m, 2–3 eV range.

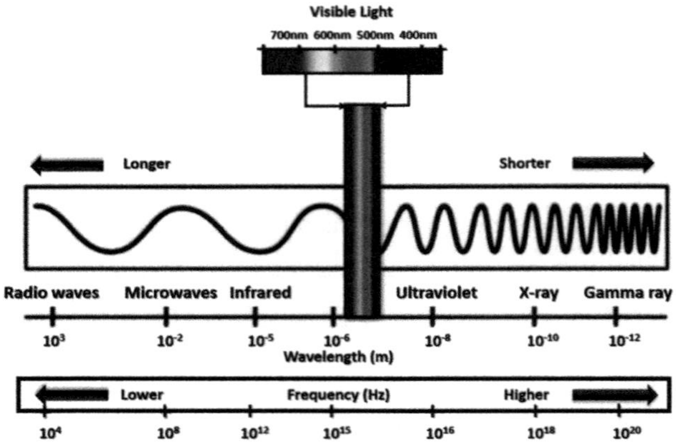

Figure 3.32: Electromagnetic spectra.

According to this perspective, an object must be larger than the wavelength of the electromagnetic radiation being exercised to be imaged. This culminates in a maximum resolution of about 400 nm for visible light microscopes. The summary of the electromagnetic spectra properties is summarized in Table 3.1. Additionally, the units for wavelength and frequency are provided in Tables 3.2 and 3.3, respectively.

Table 3.1: Summary of the essentials of electromagnetic spectra

Electromagnetic spectra	Wavelength	Frequency
Gamma-ray	< 10 pm	> 30 EHz
X-rays (Hard)	1 pm–100 pm	300,000-3000 (PHz)
X-rays (Soft)	100 pm–10,000 pm	3000-30 (PHz)
Ultraviolet (UV)	0.30 μm–0.38 μm	750–30,000 (THz)
Visible Spectrum	0.4 μm–0.7 μm	379–769 (THz)
Infrared (IR) Spectrum	0.7 μm–100 μm	0.3–430(THZ)
Microwave Region	1 mm–1 m	1–110 (GHz)
Radio Waves	(>1 m)	< 3 Hz–3000 (GHz)

Table 3.2: Units of wavelength

Units and Symbols	Values
Millimeter (mm)	10^{-6} m
Micrometer (μm)	10^{-6} m
Nanometer (nm)	10^{-9} m
Picometer (pm)	10^{-12} m

Table 3.3: Units of frequency

Units and Symbols	Values
Kilohertz (KHz)	10^{3} Hz
Megahertz (MHz)	10^{6} Hz
Gigahertz (GHz)	10^{9} Hz
Terahertz (THz)	10^{12} Hz
Petahertz (PHz)	10^{15} Hz
Exahertz (EHz)	10^{18} Hz

Along with Table 3.1, radio waves have a longer wavelength (>1 m), whereas gamma rays have the shortest wavelength (10 pm). Exahertz (EHz), which equals 1018 Hz, has the highest frequency, and Kilohertz (kHz), which equals 103 Hz, has the lowest frequency. The Gamma-ray has the greatest frequency of >30 EHz, as seen in Tables 3.2 and 3.3 respectively [4, 17, 30].

Notwithstanding, the electromagnetic spectrum can be separated into an invisible and visible half. Within a specific region of the electromagnetic spectrum, light is electromagnetic radiation. Infrared light, which is at the lower end of the visible light spectrum, is invisible to humans because it no longer contains photons with sufficient individual energy to permanently alter the conformation of the visual molecule retina, which is responsible for the perception of vision. In this understanding, a very narrow band of visible light is the only electromagnetic energy that is visible. Commonly, visible light is described as having wavelengths between 400 and 700 nanometers (nm). In the early discussion, quanta were addressed as photons at the lower end of the energy spectrum that is capable of inducing electronic excitation within molecules, which results in changes in the bonding or chemistry of the molecule and forms electromagnetic radiation in the visible light range. The reason why electromagnetic radiation at the lower end of the visible light spectrum, known as infrared, is invisible to humans is that its photons no longer have enough individual energy to permanently alter the conformation of the visual molecule retina in the human retina, which alteration is what causes the sensation of vision [6, 10, 17, 20, 30].

In other words, the term "infrared" describes a wide spectrum of frequencies, starting at the upper end of those utilized for communication and extending up to the low frequency (red) end of the visible spectrum. The wavelength range is approximately 1 mm to 750 nm. The term "far infrared" refers to the region of the spectrum with longer wavelengths that are outside of the visible spectrum. Therefore, infrared mostly causes molecules to vibrate when it interacts with matter. The vibrational spectra of molecules are frequently investigated using infrared spectrometers.

The near ultraviolet is the range of wavelengths immediately below the visible spectrum. Most solid materials absorb it quite strongly, and air even absorbs it somewhat. The far ultraviolet carries some of the risks associated with other ionizing radiation since the shorter wavelengths can ionize numerous molecules. Sunburn is one of the tissue effects of UV radiation, although it can also have some therapeutic effects. Although the sun is a powerful source of ultraviolet radiation, most of the shorter wavelengths are blocked by the atmosphere. UV radiation has a high potential for harming the eyes. Because welding arcs' ultraviolet content can induce eye inflammation, welders are required to wear protective eye shields.

From low-energy radio waves and infrared radiation, through visible light, to high-energy X-rays and gamma rays, photons are classified depending on their energies. Radio waves, infrared, ultraviolet, microwaves, and gamma radiation are all examples of invisible radiation. Alpha, beta, and "cathode rays," which are all streams of particles, are additionally invisible [1, 6, 11, 18, 27, 30].

While some radar bands exist between 1,300 and 1,600 MHz, most microwave applications operate between 3,000 and 30,000 MHz (3-30 GHz). The 3-30 GHz range is also used for amateur radio and radio navigation. Microwave radiation primarily causes molecular rotation and torsion during interactions with materials, while microwave absorption results in heat. The most accurate method for determining the lengths and angles of a molecule's bonds is to analyze its molecular rotational spectra, which can provide information on its molecular structure. Additionally, electron spin resonance spectroscopy makes use of microwave radiation. Consequently, the 535–1605 kHz range corresponds to the Amplitude Modulated (AM) radio carrier frequencies. For both aircraft navigation and maritime communication, the range of frequencies employed is 30-535 kHz. 10 kHz intervals are used to assign carrier frequencies, which range from 540 to 1600 kHz. Subsequently, the L-Band, which is utilized for a variety of satellite communication uses, is a band in the ultrahigh radio frequency range that spans 390 to 1550 MHz (Figure 3.32 and Table 3.1).

High-energy electromagnetic radiation (without rest mass or charge) is referred to as X-rays or X-radiation. High-energy photons called X-rays have short wavelengths and a high frequency because of this. Since it controls a photon's energy, the radiation frequency is the fundamental characteristic of all photons. From low-energy radio waves and infrared radiation, through visible light, to high-energy X-rays and gamma rays, photons are categorized based on their energies. The majority of X-rays have wavelengths between 0.01 and 10 nanometers (31016 Hz and 31019 Hz), which correspond to energies between 100 eV and 100 keV. Wavelengths of X-rays are generally longer than those of gamma rays but shorter than those of UV rays.

Gamma radiation commonly referred to as gamma rays, is a tremendously energetic form of electromagnetic radiation that has neither rest mass nor charge. High-energy photons known as gamma rays have extremely small wavelengths and high frequencies. Gamma rays are particularly invasive and dangerous to biological systems since they are made up entirely of very high-energy photons. Gamma rays can penetrate the human body with ease and can travel thousands of feet through the air. Therefore, at low gamma-ray energy, the photoelectric effect predominates. When light (photons) shines upon the matter, the photoelectric effect causes the emission of photoelectrons from that material [1, 4, 9, 19, 23, 27, 30, 32].

3.11 Wave-Particle Duality

The fundamental characteristic of matter known as "wave-particle duality" describes how it can operate both like a particle and like a wave simultaneously. It is worthwhile to consider the distinctions between particles and waves to comprehend wave-particle duality.

Marble,for instance, can be used to demonstrate the characteristics of particles. The marble is a sphere-shaped piece of glass that is suspended in space. The marble gains kinetic energy when our finger is flicked across it; as the marble moves, it absorbs this energy. Thrown into the air, a handful of marbles tumble to the ground, each one transferring energy to the surface it hits. Waves, on the other hand, are dispersed. Large ocean waves, pond ripples, sound waves, and light waves are all forms of waves. Like the ripples that form when we drop a pebble in a pond, a wave that is localized at one point would spread rapidly out over a wide area. The wave carries motion-related energy [33, 35]. In contrast to a particle, a wave's energy is dispersed over space.

Why are waves and particles such diverse things is undoubtedly a highly important topic? In contrast to colliding particles, which will bounce off one another, clashing waves will pass through one another and remain unaltered. The wave perhaps entirely vanishes where a dip crosses a crest while overlapping waves can interfere.

This can be observed in Young's double slit experiment when a wave's components pass through holes that are closely spaced. There are areas in space where the wave vanishes and other areas where it intensifies as a result of the wave's dispersion and interference in all directions (Figure 3.33). Therefore, this occurrence is known as diffraction. A marble, however, either bounces

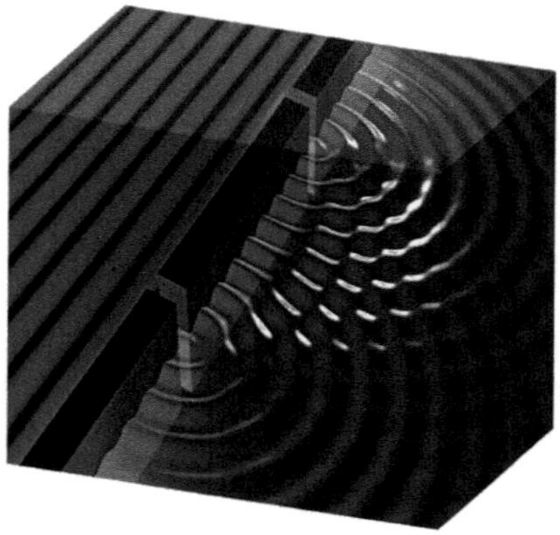

Figure 3.33: Young's double-slit experiment shows wave diffraction.

off the screen or passes straight through one of the holes when it is hurled at it [5, 9, 33, 35]. The marble will be spotted moving in one of two ways on the opposite side of the screen, depending on which hole it passes through.

Could particles like electrons and atoms, which are assumed to be wave-like but behave like particles, behave like waves of light, which we previously believed to be wave-like, also exhibit particle-like behavior? It was deemed necessary to assume that particles have wave-like qualities to explain the structure and behavior of atoms. If this is the case, a particle should behave exactly like a wave when passing through two closely spaced holes. Research has shown that atomic particles behave just like waves. In this scenario, a complete diffraction pattern can be observed precisely, when electrons are fired at one side of a screen with two closely spaced holes and measure the distribution of electrons on the opposite side. Another illustration of Young's slit experiment using electron waves as opposed to light, which is previously demonstrated in Figure 3.33. These ideas serve as the foundation for quantum theory, which is arguably the best scientific theory ever created [33-35].

Assume a light source emits light onto a screen through two tiny openings (Figure 3.34). One may see the typical dark and light stripes of interference fringes with both apertures open. Consequently, we might say that light is a wave. However, by sufficiently dimming the light at the specific location, the intensity drops so low that individual discrete photons can pass through the equipment. A detector in this case can detect the flashes as they approach the screen. The photons start scattering again into the striped interference array as a result.

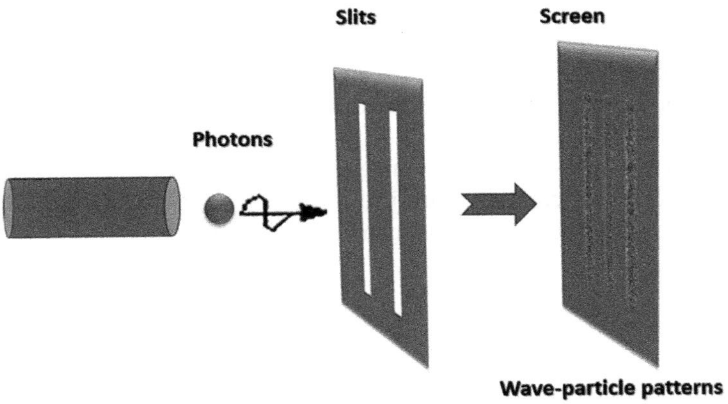

Figure 3.34: Photon acts as wave and particle in Young double slits experiment.

How does a single photon identify which slit to cross through to create the striped interference pattern, then? If a detector is inserted into one of the slits, the interference pattern vanishes. As a result, there are no interference stripes and just a basic pile-up of photons on the screen. The photons behave neither as waves nor as particles in this situation.

The peculiarity of the diffraction experiment is that contrary to what one might anticipate from a wave crashing on the shore, the electron wave does not distribute its energy evenly throughout the detector's surface. The electron's energy is concentrated in a single place, just as if it were a particle. Consequently, the electron interacts at a place like a particle even while it moves through space like a wave. This is referred to as wave-particle duality.

The critical question is now: what transpires to the remaining portion of the wave if an electron or photon propagates as a wave but deposits its energy at a single location? It vanishes from every position of space, never to be seen again! Those portions of the wave that are further away from the point of interaction somehow become aware of the energy loss and instantly vanish. If this were to occur with ocean waves, one of the surfers would receive all the energy and the ocean wave would instantly vanish down the beach. The other surfers would be sitting calmly on the water's surface while one surfer would be shooting through it. Photons, electrons, and even atom waves all behave in this manner. Many scientists, including Einstein, were outraged by this paradox. In most cases, it is glossed over and casually referred to as "the collapse of the wavefunction" on measurement [25, 29, 33, 36].

In quantum physics, wave function collapse happens when a wave function, which was initially superposed in several eigenstates, drops to only one eigenstate as a result of interaction with the outside environment. This interaction, which links the wave function with traditional observables like position and momentum, is referred to as an observation and is the essence of measurement in quantum physics. One of the two ways that quantum systems evolve is through collapse; the other way is through continuous evolution driven by Schrödinger as shown in equation (3.65).

3.12 The Uncertainty Principle

According to the above perspective, it is impossible to simultaneously measure a particle's position and momentum with arbitrarily high accuracy. The sum of the uncertainty from these two measurements has a minimal value. The sum of the uncertainty in time and energy has a minimum as well. Or, more specifically, where is the particle as the wave moves forward? Well, we cannot be certain. It can be identified in a region of space that has a dimension corresponding to the distribution of wavelengths that characterize its wave. This scenario is well known as Heisenberg's uncertainty principle. The wavelengths of consistent daily particles like marbles, salt, and sand are so minuscule that their location might well be precisely detected. On the contrary, this seems harder evident for atoms and electrons [23, 26, 30, 33, 36].

Since the electron wavelength is large in the diffraction experiment, the location of the electron is quite ambiguous. The electron relocates through both slits at the same time, much like a wave. In this scenario, this is impossible to fully imagine in terms of particles since it contradicts our predictable experience. As a corollary, physical quantities such as a particle's position, x, and momentum, p, can be predicted given initial conditions as follows:

$$\Delta x \Delta p \geq 0.5\hbar \tag{3.71}$$

It is worth noticing that equation $\Delta p = h[\Delta x]^{-1}$ was approximated. This was done to obtain a qualitative relationship that demonstrates the contribution of the Planck constant to the relationship between Δx and Δp and, consequently, the contribution of h to the determination of the confinement energy. The other reason was to compare the energy required to confine an electron in a nucleus to an electron confinement energy that is similar to that observed in nature.

Wave-particle duality and the DeBroglie hypothesis are critical stages in comprehending the uncertainty principle. When you go down to atomic dimensions, it's no longer valid to think of a particle as a hard sphere, because the smaller the dimension, the more wave-like it becomes. It no longer makes sense to claim that you have accurately determined both the position and momentum of such a particle. In this understanding, the confinement energy for the electron in the atom is only 0.06 eV if one truly implements the constraint condition made possible by the uncertainty principle, $\Delta p = \hbar[2\Delta x]^{-1}$. This is because this technique only restrains the electron in one dimension, leaving it free to move in all other directions [33-36].

When one remarks that the electron behaves like a wave,it refers to the quantum mechanical wavefunction. In this view, it is related to the chance of locating the electron at any point in space. A perfect sine wave for the electron wave disperses that probability over space, and the electron's "position" is uncertain.

In this view, the uncertainty principle has implications for the amount of energy required to hold a particle in a given volume. The energy required to contain particles originates from the fundamental forces, specifically the electromagnetic force. In this understanding, it provides the attraction required to contain electrons within the atom and the strong nuclear force. Therefore, it derives the attraction force required to contain particles within the nucleus. However, Planck's constant, which appears in the uncertainty principle, indicates the degree of restriction that these forces can generate. Another way to describe it is that the scales of the atom and nucleus are determined by the strengths of the nuclear and electromagnetic forces, as well as the restriction contained in the value of Planck's constant [5, 20, 29, 33, 35].

The author's point of view is that has nothing to do with the measurement precision of remote sensing devices or the efficacy of their methods; rather, it results from the wave qualities that are part and parcel of the quantum mechanical explanation of nature. Even with ideal sensors and techniques for remote sensing, there will always be some degree of uncertainty.

3.13 Marghany's Pioneering Concept in Remote Sensing

The significant query indeed is: can quantum mechanics deliver the novel concept of remote sensing? Can quantum mechanics fulfill the novel concept of remote sensing? It is a crucial question. Therefore, Marghany in his best-seller book titled "Remote Sensing and Image Processing in Mineralogy" which was published by CRC in 2022 innovated a novel concept of remote sensing. For a more precise definition of remote sensing, Marghany [37] stated and proved mathematically that it entails looking for the quantum information of the atomic physical characteristics of each element. Due to the interaction between photons and any atomic element in the universe through absorption or emission, this quantum data exists everywhere in the universe. Based on this theory, the spectral signature is nothing more than the signature of photon interactions with the physical characteristics of the atoms of any object. In other words, the quantization of object atom electron energy-remote sensing photon interaction is the name given to Marghany's speculation. This understanding addresses such a revolutionary notion of remote sensing techniques. According to this innovative concept, remote sensing is the mobilization of the spectral signature of valence electron energy level fluctuations that vary from one element to another and is based on the interaction of various photon spectra with various levels of electron energy in each element. The quantization of the spectral signature of various components or objects based on the quantization of the black body radiation and wave-particle duality can be used to explain this innovative definition.

In other words, it can be said that remote sensing is the quantum information amassed from any object on the ground or space owing to the reflection or backscattering of photons, which is a function of the quantum mechanical properties of every object's atoms. In this understanding, different object atoms would interact with photons, i.e., absorption and emission as a function of their different wavelengths and frequencies under circumstances of the uncertain principle of Heisbenger.

The author asserts that the uncertain principle phenomenon is triggered by wave properties that are fundamental to the quantum mechanical explanation of nature rather than the measurement accuracy or effectiveness of distant sensing devices. There would constantly be a degree of ambiguity in remote sensing, even with the perfect sensors and techniques.

This new definition of remote sensing delivers a new era of quantum image processing to obtain accurate information about the universe. The uncertainty of conventional remote sensing image processing could be issued owing to conventional computing techniques which are based on the classical computer system i.e. bit. It is also understanding quantum remote sensing technology based on Marghany's [33] definition, many mysteries in atmospheric turbulence can be tackled. In this regard, the next chapter will deliver answers to the mysteries of the High-Frequency Active Auroral Research Program (HAARP) impacts on the ionospheric layer.

References

[1] Huray, Paul G. (2009). *Maxwell's Equations*. John Wiley & Sons.
[2] Fleisch, D. (2008). *A Student's Guide to Maxwell's Equations*. Cambridge University Press.
[3] Frankel, T. (1974). Maxwell's equations. *The American Mathematical Monthly*, 81(4), 343–349.
[4] Kirsch, A. and Hettlich, F. (2009). *The Mathematical Theory of Maxwell's Equations*. Lecture notes.
[5] Bartsch, M., Dehler, M., Dohlus, M., Ebeling, F., Hahne, P., Klatt, R., ... & Wolter, H. (1992). Solution of Maxwell's equations. *Computer Physics Communications*, 73(1-3), 22–39.
[6] Romanov, V.V.G. and Kabanikhin, S.I. (1994). Inverse problems for Maxwell's equations (Vol. 2). VSP.
[7] Cochran, J.F. and Heinrich, B. (2020). *Applications of Maxwell's Equations*. John F. Cochran, Bretislav Heinrich. Simon Fraser University.
[8] Kirsch, A. and Hettlich, F. (2016). *Mathematical Theory of Time-harmonic Maxwell's Equations*. Springer International Pu.
[9] Idemen, M. (1973). The Maxwell's equations in the sense of distributions. *IEEE Transactions on Antennas and Propagation*, 21(5), 736–738.
[10] Arbab, A.I. (2013). Complex Maxwell's equations. *Chinese Physics B*, 22(3), 030301.
[11] Strocchi, F. (1970). Gauge problem in quantum field theory. III. Quantization of Maxwell equations and weak local commutativity. *Physical Review D*, 2(10), 2334.
[12] Gersten, A. and Moalem, A. (2015, May). Consistent quantization of massless fields of any spin and the generalized Maxwell's equations. *Journal of Physics: Conference Series*, 615(1), 012011. IOP Publishing.
[13] Dappiaggi, C. and Lang, B. (2012). Quantization of Maxwell's equations on curved backgrounds and general local covariance. *Letters in Mathematical Physics*, 101, 265–287.
[14] Pfenning, M.J. (2009). Quantization of the Maxwell field in curved space times of arbitrary dimension. *Classical and Quantum Gravity*, 26(13), 135017.
[15] Simulik, V.M. and Krivsky, I.Y. (2004). The Maxwell equations with gradient-type sources, their applications and quantization. *Developments in Quantum Physics*, 143–165.
[16] Gersten, A. and Moalem, A. (2011, December). Maxwell's equations, quantum physics and the quantum graviton. *Journal of Physics: Conference Series*, 330(1), 012010. IOP Publishing.
[17] Bennett, R., Barlow, T.M. and Beige, A. (2015). A physically motivated quantization of the electromagnetic field. *European Journal of Physics*, 37(1), 014001.
[18] Dung, H.T., Knöll, L. and Welsch, D.G. (1998). Three-dimensional quantization of the electromagnetic field in dispersive and absorbing inhomogeneous dielectrics. *Physical Review A*, 57(5), 3931.
[19] Matloob, R., Loudon, R., Barnett, S.M. and Jeffers, J. (1995). Electromagnetic field quantization in absorbing dielectrics. *Physical Review A*, 52(6), 4823.
[20] Huttner, B. and Barnett, S.M. (1992). Quantization of the electromagnetic field in dielectrics. *Physical Review A*, 46(7), 4306.
[21] Barut, A.O. and Barut, A.O. (Eds.). (1980). *Foundations of Radiation Theory and Quantum Electrodynamics*. (Vol. 165). New York: Plenum Press.
[22] Karlovets, D.V. and Pupasov-Maksimov, A.M. (2021). Nonlinear quantum effects in electromagnetic radiation of a vortex electron. *Physical Review A*, 103(1), 012214.
[23] Wilson, W. (1923). The quantum theory and electromagnetic phenomena. Proceedings of the Royal Society of London. Series A, Containing Papers of *Mathematical and Physical Character*, 102(717), 478–483.
[24] Bohr, N., Slater, J.C. and Kramers, H.A. (1924). *The Quantum Theory of Radiation* (pp. 785–802). Taylor & Francis.
[25] Klassen, S. (2011). The photoelectric effect: Reconstructing the story for the physics classroom. *Science & Education*, 20, 719–731.
[26] Sorokin, A.A., Bobashev, S.V., Feigl, T., Tiedtke, K., Wabnitz, H. and Richter, M. (2007). Photoelectric effect at ultrahigh intensities. *Physical Review Letters*, 99(21), 213002.
[27] Adawi, I. (1964). Theory of the surface photoelectric effect for one and two photons. *Physical Review*, 134(3A), A788.
[28] Pratt, R.H., Ron, A. and Tseng, H.K. (1973). Atomic photoelectric effect above 10 keV. *Reviews of Modern Physics*, 45(2), 273.

[29] Copeland, A.W., Black, O.D. and Garrett, A.B. (1942). The photovoltaic effect. *Chemical Reviews*, 31(1), 177–226.
[30] Bockris, J. O. M., and Khan, S. U. (2012). Quantum electrochemistry. Springer Science & Business Media.
[31] von Baltz, R. and Kraut, W. (1981). Theory of the bulk photovoltaic effect in pure crystals. *Physical Review B*, 23(10), 5590.
[32] Goldstein, B. and Pensak, L. (1959). High-voltage photovoltaic effect. *Journal of Applied Physics*, 30(2), 155–161.
[33] Angelo, R.M. and Ribeiro, A.D. (2015). Wave-particle duality: An information-based approach. *Foundations of Physics*, 45, 1407–1420.
[34] Dimitrova, T.L. and Weis, A. (2008). The wave-particle duality of light: A demonstration experiment. *American Journal of Physics*, 76(2), 137–142.
[35] Selleri, F. (Ed.). (1992). *Wave-particle Duality*. New York: Plenum Press.
[36] Yang, C.D. (2005). Wave-particle duality in complex space. *Annals of Physics*, 319(2), 444–470.
[37] Marghany, M. (2022). *Remote Sensing and Image Processing in Mineralogy*. CRC Press.

CHAPTER

4

Quantization of Haarp-Ionosphere Disturbance: Generating Turbulence in Ionospheric Plasma

We previously investigated turbulence and remote sensing quantization theories. This chapter speculates on possible explanations for the heatwave that hit Europe in July 2022. In this scenario, the ionosphere is the keystone of the Earth's weather fluctuations. Any disturbance of the ionosphere layer leads to dramatic climate changes. One factor that has received less study is the High-Frequency Active Auroral Research Program (HAARP). This chapter is devoted to fully understanding the possible impact of HAARP on the disturbance of the ionosphere. On this occasion, the quantum mechanics' view of the entanglement concept is implemented to explain the association of ionosphere plasma turbulence with HAARP signal energy impacts. As a result, the consequences of ionosphere disturbances will be discussed in the following chapter. In this view, the next chapter will investigate HAARP's impacts on generating heat wave turbulence across European atmospheric layers in July 2022. The effects of these heat waves will be studied in detail to assess whether or not there is a direct correlation between HAARP's energy signals and the disturbance of the ionosphere.

4.1 What is Ionosphere?

What exactly is the ionosphere and its relevance for fluctuations in air turbulence the questioning of this chapter? Since it receives energy from the Sun, the ionosphere, a very active component of the atmosphere, expands and contracts. The gases in these layers are energized by solar radiation to generate ions, which have an electrical charge, hence the name ionosphere. Put differently,the mesosphere, thermosphere, and exosphere are covered by an intriguing layer known as the ionosphere (Figure 4.1). It is a particularly dynamic region of the atmosphere, expanding and

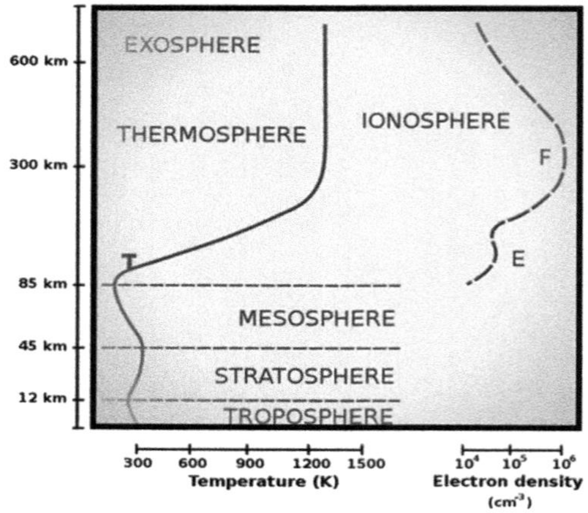

Figure 4.1: Ionospheric layer.

contracting in response to the solar energy it absorbs. Its name is derived from the fact that solar radiation excites the gases in these layers to generate "ions," which have an electrical charge [1].

The question is now: how does the ionosphere's temperature change? The sun's radiation is so strong in the ionosphere that it ionizes, or releases electrons from various atmospheric elements. Ionosphere temperatures range from 200 Kelvin (or –37.2222 °C) to 500K (or 226.667 °C) as a result of fluxes in solar radiation (or 226.667 °C) [2].

Therefore, what impact does the ionosphere have on Earth's life? There are numerous benefits of the ionosphere. It shields Earth's living things by absorbing those dangerous ultraviolet radiations. A portion of the waves arriving from Earth is also reflected by the electrically charged particles in the ionosphere. The ionosphere in particular reflects radio waves.

4.2 Mechanism of Ionospheric Layer Ionization

The ionosphere surrounds the Earth and ranges in height from about 50 km (30 mi) to more than 1,000 km. It is composed of electrons and electrically charged atoms and molecules (600 mi). UV rays from the Sun are principally responsible for their existence. In this understanding, UV, X-ray, and shorter solar radiation wavelengths ionize because they comprise photons with adequate energy to bounce an electron loss from a neutral gas atom or molecule upon absorption. In this process, the energetic electron accelerates to a high velocity, creating an electronic gas with a temperature thousands of times higher than that of ions and neutrals. Recombination, in which a free electron is "captured" by a positive ion, is the opposite of ionization. Since the gas molecules and ions are closer together as gas density increases at lower altitudes, the recombination process takes precedence. The concentration of ionization generated is determined by the balance between these two processes. In particular, ionization is essentially reliant on the Sun, whose Extreme Ultraviolet (EUV) and X-ray irradiation is very variable with solar activity. There are more sunspot active zones on the Sun at any given time as its magnetic activity increases. Particularly during episodic magnetic eruptions that include solar flares that increase ionization on the Earth's sunlit side and solar energetic particle events that can increase ionization in the polar regions, sunspot active regions are the source of increased coronal heating and accompanying increases in EUV and X-ray irradiance [1–3].

As a consequence, the ionosphere's quantity of ionization exhibits both an 11-year solar cycle and a diurnal (daytime) cycle. Since the local winter hemisphere is inclined away from the Sun and receives less solar radiation, there is also a seasonal dependence on the degree of ionization. Geographical location also affects the radiation received (polar, auroral zones, mid-latitudes, and equatorial regions). Additionally, some factors diminish ionization and destabilize the ionosphere [2–4].

4.3 Ionization-related Layers

According to the above perspective, from 60 to 300 kilometres (37 to 190 miles) above the surface of the earth, there is the ionosphere. The F-Region, E-Layer, and D-Layer are the three regions or layers that make up this structure. In this regard, the ionization in the E and D layers is incredibly low at night, exposing only the F layer with any discernible ionization. The D, E, and F layers all see significant increases in ionization during the day, and the F layer also develops a new, weaker region of ionization known as the F1 layer (Figure 4.2). The primary zone for the refraction and reflection of radio waves is the F2 layer, which is permanent day and night [1,3].

4.3.1 D-layer

At 48 km (30 mi) to 90 km (56 mi) above the Earth's surface, the D layer is the deepest. Hence, nitric oxide is ionized by Lyman series-alpha hydrogen radiation with a wavelength of 121.6

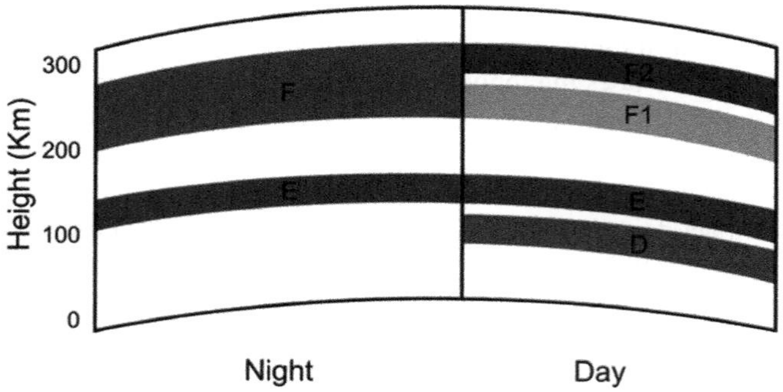

Figure 4.2: Atmospheric D-layer.

nanometers (nm) (NO) (Figure 4.3). Indeed, Theodore Lyman, the series' discoverer, is honoured with a name for it. Therefore, when an electron transitions from the lowest energy level of n = 1 to n = 2 (where n is the primary quantum number), known as the Lyman series, it emits ultraviolet light from the hydrogen atom in a sequence of changes. In this view, the energy of the electromagnetic emission increases with the size of the primary quantum number difference. Put differently, an electron orbits the nucleus of an atom of hydrogen. The electron can exist in a variety of quantum states, each with its energy, due to the electromagnetic force between it and the nuclear proton [1, 3, 5].

The Rydberg R_H formula that yielded the Lyman series (Figure 4.2) would be as follows:

$$\frac{1}{\lambda} = R_H\left(1 - \frac{1}{n^2}\right) \quad \left(R_H \approx 1.0968 \times 10^7\,\mathrm{m}^{-1} \approx \frac{13.6\mathrm{eV}}{hc}\right) \tag{4.1}$$

here n is a natural integer that is either higher than or equal to 2 (e.g., $n = 2, 3, 4$). Since $n = 2$ is on the right and $n = \infty$ is on the left, the lines in the illustration above correspond to those wavelengths. There are an infinite number of spectral lines, but as they become closer to $n = \infty$ (the Lyman limit), they get very dense and only part of the early lines and the last one are visible. In this sense, the Lyman series has only ultraviolet wavelengths. Moreover, solar flares can produce hard X-rays with wavelengths of less than 1 nm that ionize N_2 and O_2. In the D layer, where recombination rates are high, there are significantly more neutral air molecules than ions [3,5].

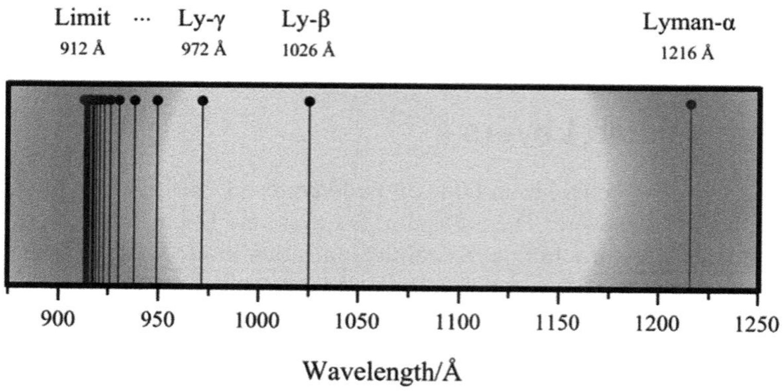

Figure 4.3: UV Lyman series of hydrogen atom spectral lines.

The rationale behind why hydrogen spectral lines fit Rydberg's formula was revealed in 1914 when Niels Bohr developed his Bohr model concept [5]. The electron attached to the hydrogen atom, according to Bohr, must have quantized energy levels that are represented by the following formula:

$$E_n = -\frac{m_e e^4}{2(4\pi\varepsilon_0\hbar)^2}\frac{1}{n^2} = -\frac{13.6\text{ eV}}{n^2}. \tag{4.2}$$

Bohr's third assumption states that every time an electron travels from an initial energy level E_i to an ending energy level E_f, the atom would release radiation with the following wavelength [5, 7]:

$$\lambda = \frac{hc}{E_i - E_f}. \tag{4.3}$$

Equation 4.3 says that the transfer of an electron from a higher energy state to a lower energy level, or the "jump," results in spectral emission. Commonly, the lower energy state is abbreviated as n', and the higher energy state is abbreviated as n, to distinguish the two states. The difference in energy between the two states is represented by the energy of a photon that is released. Every transition would always result in a photon with the same energy because each state's energy is set, as is the energy difference between them. In this regard, the spectral lines are categorized into series consistent with n'. Therefore, lines are termed consecutively preliminary from the longest wavelength/lowest frequency of the series, exploiting Greek letters within separate series. For instance, the 2 → 1 line is called "Lyman-alpha" (Ly-α), while the 7 → 3 line is called "Paschen-delta" (Pa-δ) (Figure 4.4).

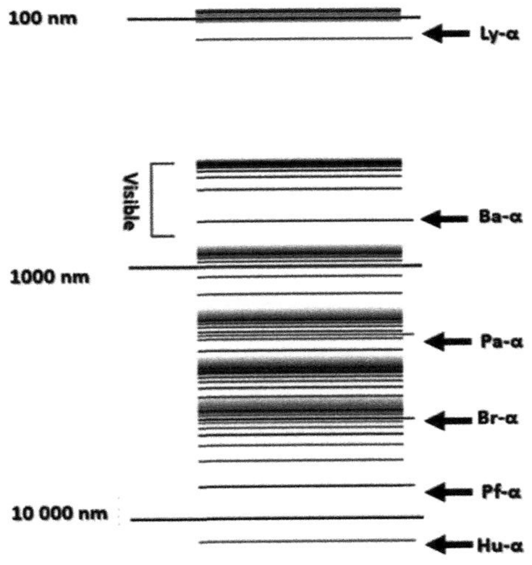

Figure 4.4: The logarithmic range of the hydrogen spectral series.

When dealing with energy in electronvolts eV and wavelengths in angstroms Å, there is also a more convenient notation.

$$\lambda = \frac{12398.4eV}{E_i - E_f}\text{Å}. \tag{4.4}$$

By substituting the expression for the energy in the hydrogen atom, where the initial energy corresponds to energy level n and the final energy corresponds to energy level m, for the energy in the aforementioned formula,

$$\frac{1}{\lambda} = \frac{E_i - E_f}{12398.4\text{eV Å}} = R_H\left(\frac{1}{m^2} - \frac{1}{n^2}\right) \tag{4.5}$$

R_H stands for hydrogen in Rydberg's well-known formula, which is the same Rydberg constant. This also implies that the Lyman limit is equal to the inverse of the Rydberg constant. Therefore, equation 4.1 says each wavelength of the emission lines corresponds to an electron dropping from a certain energy level (greater than 1) to the first energy level. Consequently, in the following aspects, joules and electronvolts are comparable to the Rydberg unit of energy:

$$1\,Ry \equiv hcR_\infty = \frac{m_e e^4}{8\varepsilon_0^2 h^2} = \frac{e^2}{8\pi\varepsilon_0 a_0} = 2.179\,872\,361\,1035(42)\times 10^{-18}\ \mathrm{J} \quad (4.6)$$
$$= 13.605\,693\,122\,994(26)\ \mathrm{eV.1}.$$

Here h is the Planck constant; c is the speed of light in vacuum; ε_0 is the permittivity of free space; e is the elementary charge; and m_e [5–7].

4.3.2 E-Layer

The E layer, which is located between 90 km (56 mi) and 150 km (93 mi) above the Earth's surface, is the middle layer. Molecular oxygen is ionized by far ultraviolet (UV) solar radiation and soft X-rays (1–10 nm) that are present in the environment (O_2). Typically, with an oblique incidence, this layer can only reflect radio waves with frequencies below about 10 MHz and may slightly aid in the absorption of frequencies above. The E_s layer, however, can reflect frequencies as high as 50 MHz when there are powerful occasional E occurrences. In this regard, signals from relatively small "clouds" in the lower E region, which is located at altitudes of roughly 95–150 km (50–100 miles), are reflected by sporadic E propagation (Figure 4.5). Ionized metals that have been abated off of micrometeoroids make up these "clouds." The more common types of skywave propagation in the higher F region of the ionosphere refract off layers of electrons knocked off of gasses by intense UV light, which are renewed on a fairly regular daily cycle, in contrast to E layer propagation, which depends on the transient abundance of metallic meteor dust. In both situations, the ionized material, when present, bends or "refracts" radio waves such that they travel via a "bent pipe" back toward the Earth's surface [1, 3, 5, 7].

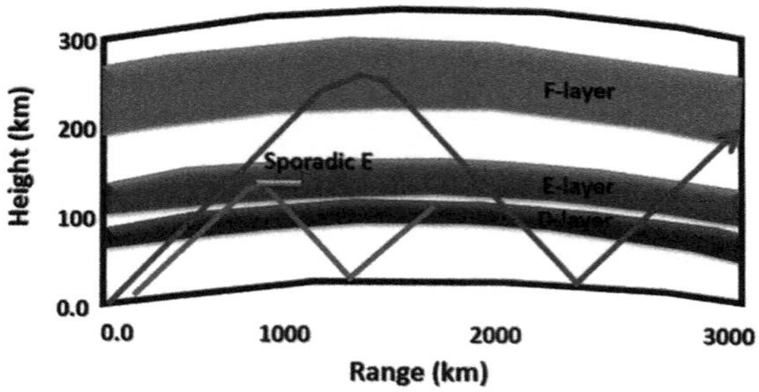

Figure 4.5: Ray curve for a sporadic E event.

Consequently, the conflicting effects of ionization and recombination essentially dictate the vertical structure of the E layer. Because of the absence of the prime component of ionization during the night, the E layer becomes weaker. The range to which radio signals can travel by reflection from the layer grows as the height of the E layer maximum increases after sunset.

4.3.2.1 Occurrence of Sporadic E

Sporadic E is extremely challenging to predict. Yet, considerable statistical information has been gathered about its occurrence. It has been discovered that sporadic E occurs in several regions of the world.

Temperate regions: It is discovered to occur primarily in summer in temperate regions, i.e. those in the mid-latitudes between the equatorial regions. The months of May to August in the northern hemisphere produce the most openings, with June being the busiest month. December sees a slight peak as well. In the southern hemisphere's in the months of November to February, a similar trend may also be seen. Typically, only during the middle of the irregular E season, primarily in June and July in the northern hemisphere, are frequencies well into the VHF region of the spectrum affected [1, 5, 7].

Polar regions: The phenomenon, known as auroral sporadic *E*, happens here. Again, there is little seasonal variation, with it commonly appearing in the morning.

Areas around the equator: The incidence of sporadic *E* is predominantly a daylight event in tropical locations, and as may be predicted by the location, there is little seasonal variation. Since it occurs more frequently than in temperate climates, it is thought that a distinct process may be at play in its formation.

4.3.2.2 What Causes Sporadic E

Uncertainty surrounds the sporadic *E* mechanism. It is believed that several phenomena could contribute to its formation:

The entry of meteors into the atmosphere is thought to be one phenomenon that causes occasional *E*, according to some evidence. There might be some relation because the *E* region is where meteorites generally burn up.

Electrical storms: These have electrical effects that are much above the clouds and may reach high altitudes. These are thought to be potential sources of energy for the intermittent development of *E* clouds [6, 8].

Auroral activity has also been connected to the prevalence of Sporadic *E* throughout the wintertime at night. This is unquestionably the case for auroral sporadic *E*, which is caused by kinetic electrons leaving the magnetosphere and entering the atmosphere [5, 7, 10].

Fast-moving winds in the upper atmosphere: According to certain concepts, these dense clouds of ionization, particularly in temperature ranges, may be created by shearing forces caused by the fast-moving winds in the upper atmosphere [7, 9, 11].

The process of their development is the subject of numerous theories. It is very conceivable that a variety of different physical processes contribute to the infrequent high levels of ionization in the *E* area in ways that are strikingly similar. Therefore, different sporadic ionization events may fall within the Sporadic *E* category together with other forms of sporadic ionization phenomena. The fact that Sporadic *E* at the equator is more stable than Sporadic *E* at higher latitudes lends credence to this notion. There are further differences as well. More information about its recurrence is being gathered, which will probably deepen our comprehension of the event and help us make more precise predictions [1, 6, 8, 12].

It seems that the sunspot cycle has some impact on temperate zone sporadic *E*, which is one intriguing connection that has been made. It has been observed that there are more openings during the sunspot minimum. Radio communications propagation known as sporadic *E* is particularly fascinating. Since Sporadic *E* is sporadic, it is more challenging to research and comprehend. Because it is difficult to connect the cause to the effect, it is shrouded in mystery. It is not because it is sporadic.

4.3.3 E_S-Layer

The E_s layer, also known as the sporadic *E*-layer, is distinguished by tiny, delicate clouds of high ionization that can allow radio wave reflection, commonly up to 50 MHz and infrequently up to 450 MHz. Sporadic *E* occurrences could last anywhere between a few minutes and several hours. When long-distance propagation pathways that are ordinarily unreachable "open up" to two-way communication, sporadic *E* propagation makes VHF operating by radio amateurs exceedingly thrilling. Researchers are continuously looking at a variety of reasons for sporadic *E*. The mid-latitudes of the northern hemisphere see this propagation every day in June and July when high signal levels are frequently attained [1, 8, 13].

Therefore, the skip distances are often approximately 1,640 kilometres (1,020 mi). One hop propagation can cover distances of up to 2,500 km (900 mi) (1,600 mi). Additionally typical are multi-hop transmissions of 3,500 km (2,200 mi), occasionally to distances of 15,000 km (9,300 mi) or more [10–14].

4.3.4 F-Layer

The Appleton-Barnett layer sometimes referred to as the F layer or region, is located between 150 km (93 mi) and 500 km (310 mi) above the surface of the Earth. Since it has the highest electron density of any layer, any signals that pass through it will likely escape into space. Extreme ultraviolet (UV, 10-100 nm) radiation that ionizes atomic oxygen is the main source of electron generation. In this view, High-energy ultraviolet radiation (HEUV), also known as extreme ultraviolet radiation (EUV or XUV), is electromagnetic radiation in the region of the electromagnetic spectrum with wavelengths between 124 and 10 nm and, as defined by the Planck-Einstein equation, photons with energies between 10 and 124 eV. The solar corona generates EUV naturally, and plasma, high harmonic generation sources, and synchrotron light sources produce it artificially. There is some overlap between the phrases because UVC goes up to 100 nm. It is worth understanding that similar to what occurs when X-rays or electron beams are absorbed by materials when an EUV photon is absorbed, photoelectrons and secondary electrons are released by ionization [1, 10, 15].

Similar to other ionizing radiation forms, EUV and the electrons it releases, either directly or indirectly, are likely sources of device deterioration. Damage could occur as a result of trapped charge after ionization or oxide desorption. Additionally, the Malter effect's indefinite positive charge can cause damage. Positive ion desorption is the sole option to regain neutrality if free electrons are unable to come back and cancel out the net positive charge. Desorption, however, essentially entails that the surface deteriorates during exposure, and additionally, the desorbed atoms pollute any exposed optics. The Extreme UV Imaging Telescope's CCD radiation aging has already shown evidence of EUV damage (EIT). Consequently, a well-known problem that has been researched concerning plasma processing damage is radiation damage. Indeed, surface charge at wavelengths below 200 nm is quantifiable [4, 7, 16]. While VUV (Vacuum Ultraviolet) radiation demonstrated positive charging centimeters inside the limit values, EUV radiation demonstrated positive charging centimeters outside the permissible levels (Figure 4.6). By these principles, any abrupt changes in the ionosphere layer would culminate in weather catastrophes like hurricanes, extremely heavy rain that causes flooding, and flash disasters, which will be covered in more detail later in the following chapters.

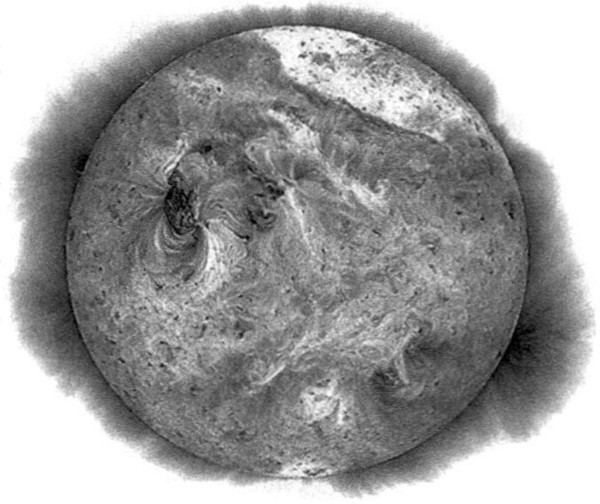

Figure 4.6: Composite image of the sun in extreme ultraviolet light.

At night, the F layer is just one layer (F_2), while during the day, the electron density profile frequently creates a secondary peak (designated F_1). The majority of radio wave skywave propagation and long-distance high frequency (HF, or shortwave) radio communications occur in the F_2 layer since it is present day and night. Lastly, the concentration of oxygen ions diminishes above the F layer, and lighter ions such as hydrogen and helium take over. The topside ionosphere refers to this zone that lies below the plasmasphere and above the F-layer peak [1, 7, 11, 16].

4.4 Signal Attenuation in Ionospheric Layer

The signals are weaker as they move through this area. The frequency determines the attenuation level. Higher frequencies are less muted than lower ones. In actuality, it has been discovered that attenuation varies as the inverse square of the frequency, meaning that doubling frequency reduces attenuation by a factor of four. This means that, except during the night when the zone vanishes, low-frequency signals are frequently prohibited from reaching the upper regions.

$$\text{attenuation} = k\,f^{-2} \tag{4.7}$$

here k is constant, and f presents the frequency of operation (Hz). The radio waves in the D region force the free electrons in the specific area to vibrate sympathetically, attenuating the coherent radiation. A small amount of energy is yielded each time an electron collides with a molecule while it is vibrating. A considerable fall in the total signal level results from the energy loss, which becomes apparent when there are countless millions of mobile electrons vibrating. Numerous variables affect how much signal is instantly lost.

The number of gas molecules present: The quantity of gas molecules present is one component. There are more collisions and hence more attenuation when there are more gas molecules present. There are still a lot of gas molecules at the height where the D area is located, which means that under many conditions there are plenty of ion-molecule collisions to absorb a lot of energy.

Ionization level: Ionization level is also highly significant. The more electrons that vibrate and crash into molecules, the higher the level of ionization.

Signal frequency: The third key element is the signal's frequency. The wavelength of the vibration shortens with increasing frequency, and there are fewer collisions between free electrons and gas molecules. Ones at lower frequencies in the radio frequency spectrum are therefore muted far more than signals at higher frequencies. High-frequency signals nevertheless have some signal strength loss [17–19].

In real-world applications, it has been discovered that the level of attenuation is adequate to block signals from the MF section of the spectrum from reaching the higher layers. They can access the higher layers at night when the ionization in the D zone decreases, and messages from further away may be audible. This is apparent at higher frequencies and in the medium wave range, where the D region absorbs the impulses [1, 11, 15, 19].

The amount of attenuation, therefore, would vary according to the frequency for signals at higher frequencies that are "reflected" by higher locations in the ionosphere. It is crucial to bear in mind that the signal would need to pass through the D region twice, losing strength each time, for each reflection. As a consequence, signals that are reflected more than once can experience considerable attenuation [14–18].

4.5 Ionospheric Oscillations

As the amount of ionization decreases at night, it has previously been shown that the time of day influences the ionosphere's status in very significant ways. Nevertheless, the ionosphere is influenced by a wide range of additional elements. The Sun itself is the most important one, but there are others, such as the time of year and the location on the planet.

The quantity of emission that the ionosphere receives varies with the distinct seasons in the same way as the reasonable amount of specific heat that various specific locations on Earth experience invariably does. This is owing to the Earth's surface being closer to being at right angles to the rays' direction during the summer, which causes the radiation received to spread across a more compact region. Winter causes the radiation to spread over a more spacious area since the Earth's surface is at a more considerable angle. The ionosphere, therefore, receives less energetic radiation in the typical winter than in the summer [20, 22].

Lower levels of ionization in the winter than in the summer are the expected response in the D and E regions, and the F1 region also exhibits this trend. The F2 area, on the other hand, is affected by additional elements and reacts differently.

The heating action of the Sun has a significant impact on how the F_2 region reacts. Because the sun is lower in the sky during the winter, more of the heat from it is dispersed across a broader area than it is during the summer. Summertime increases air activity and the temperature of the gas in the F_2 zone, which causes more molecules to ascend higher into the atmosphere. In winter, the heavier molecules sink as the temperature drops, allowing the lighter atoms to ascend to the top [5, 12, 20].

In other words, the F_2 region's greater altitude has a higher percentage of atoms throughout the winter. Since atoms are simpler to ionize than gas molecules, there are more potential targets for the radiation to ionize. Ionization levels throughout the day are thus higher in the winter than they are in the summer. Since the Sun's energy is present for a smaller percentage of the day in winter than it is in summer, the overall impact is that the peak daytime ionization levels rise higher in winter than they do in summer. However, they decline to a lower level as a consequence [11, 19, 21].

Geographical variations: The location on the planet has an impact on ionization levels. Some fluctuations naturally occur according to latitude, with polar regions receiving less radiation and equatorial parts receiving significantly higher levels. In general, this causes the D, E, and F_1 zones to experience higher amounts of ionization at the equator than at the poles.

In conjunction with the Earth's magnetic field and other sources of ionization, the F_2 zone is subject to a variety of other factors that influence its level of ionization. These factors led to the discovery that the ionization levels are higher in Asia and Australia than in the western hemisphere, which encompasses Africa, Europe, and North America [1, 5, 22].

The crucial query posed was, "Can the ionosphere be easily manipulated?" Since radio waves move in straight lines but the Earth isn't flat, it might be challenging to transfer radio signals across the globe. By causing ionosphere-wide abnormalities that would cause radio signals to bounce across vast distances, HAARP's discoveries might assist us in expanding the range of radio communications. Can the radio wave growths in this scenario destabilize the ionosphere layer and create appreciable changes in the weather? This knowledge acquisition will be explored in the chapter's following sections.

4.6 What Renders HAARP Particularly Incredible?

What is meant by Haarp is the key question. Presently, the High-Frequency Active Auroral Research Program (HAARP) is the key facility for synthesizing extremely low frequency (ELF) electromagnetic radiation in the ionosphere.

Construction on the HAARP facility began in 1993 and was finalized in 2007 for a total expected cost of $290 million. The transmitter array at the facility, which consists of 180 crossed dipoles spaced over a 30-acre area, is the main tool for studying the ionosphere. Together, the dipoles can generate up to 3,600 kW of radiated power in a band from 2.8 to 10 MHz, which is within what is commonly known as the HF band (3–30 MHz) of electromagnetic radiation. The dipoles are arranged in a 12 by 15 rectangular grid and are phased to continue providing steering [23].

On a 40-acre spot, HAARP's ionospheric research instrument is typically comprised of 180 aluminium antenna towers. The towers operate simultaneously completely to transfer radio waves into the ionosphere, which begins at a height of about 50 miles. There, sunlight briefly ejects the electrons from gas molecules, resulting in charged particles. The continuous signal from HAARP is merely modified by modern scientists to trigger processes in the lower ionosphere that result in complex phenomena like radiating auroral currents, also known as "virtual antennas," which reflect incredibly low-frequency waves to Earth. The continuous waves' unique ability to penetrate the considerable depths of the limitless ocean may typically enhance submarine communication. The nighttime absence of direct sunlight intentionally causes the ionosphere's operational base to briefly vanish. This merely enables HAARP to carry out research that might result in improved methods for typically utilizing a standard procedure known as "skywave propagation" [23, 24, 26].

Because HAARP can produce waves with frequencies that are similar to those of human brain waves, the author is concerned that it may be able to control people's minds. The author also thinks that the facility has the power to alter the weather. In fact, according to The Ultimate Weapon of the Conspiracy, the HAARP project was hurriedly finished after the 2005 hurricane season, which included Katrina, to prevent the storms from making landfall. The author also thinks it was to blame for the tsunami that hit in 2004 and the loss of the space shuttle Columbia in 2003. Conforming to this principle, HAARP could force electrons upward in the atmosphere, causing particles below to move and evolving Earth's weather. This concept of HAARP is backed up by the widely accepted scientific principles that state that like charges repel and opposites attract [23–26].

4.7 How does HAARP Operate?

The precise location of HAARP is 62.39° N and 145.15° W (Figure 4.7). In other words, it is located precisely in the north magnetic latitude of 63.09° N, which coincides exactly with the west magnetic longitude of 92.44°W. Uniquely situated in the Alaska Scale or domain, HAARP is conveniently located at the operational base of two mountain ridges. This remoteness and distinct lack of light pollution efficiency are two of the most significant advantages that naturally make HAARP ideal for a space observatory. Moreover, the mountainous terrain around the secure facility adequately provides instinctive shielding from radio-frequency interference, universally allowing for better detection of faint signals. As a consistent result of these advantageous factors, HAARP is admirably suited for conducting astronomical observations and studying the complex interaction between Earth's unique atmosphere and the considered Sun's plasma. Additionally, the geographic location of HAARP naturally makes it ideal for monitoring satellites and observing transient astronomical phenomena such as gamma-ray bursts [24–26].

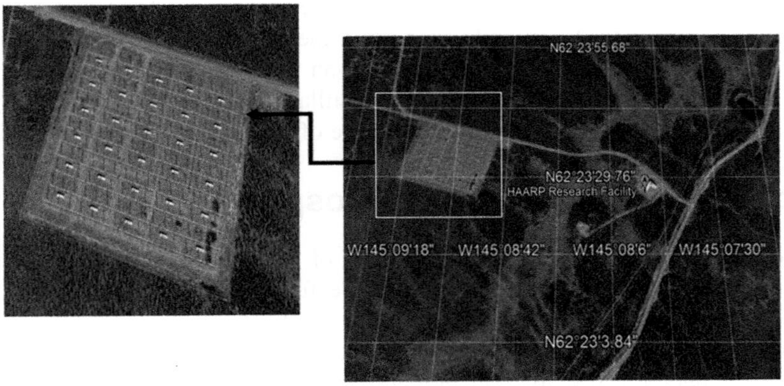

Figure 4.7: The geographical location of Haarp Research Facility, Alaska.

The transmitted power can be facilitated as high as 10 MHz ensuring that the facility can probe into the F-region even under high plasma density conditions. Such a broad range allows operation throughout a completed solar cycle. The low end of the chosen band is just below twice the electron gyro frequency in the ionosphere. The effective range of transmission frequencies from 0.5 MHz to 10 MHz in the notable HF band typically allows for a comprehensive assessment of ionospheric conditions. In this regard, the total transmitted power can range from 3,600 W to 3.6 MW while maintaining a consistent antenna pattern because each HAARP transmitter can produce between 10 W and 10 kW (Figure 4.8). The HAARP antenna, as previously mentioned, is made up of 180 crossed dipoles arranged in a grid of 12 by 15 squares that are phased to provide steering. The array's main lobe beam width is approximately 15° at the lower end of the frequency band (2.8 MHz), and approximately 5° at the upper end of the band (10 MHz) [23, 25, 27].

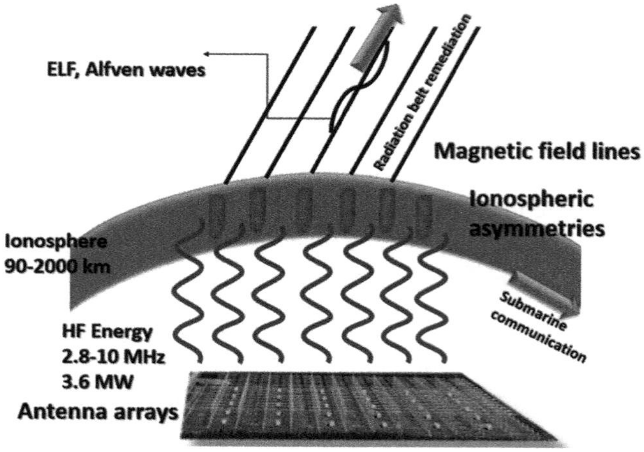

Figure 4.8: Operation of HAARP as HF ionospheric heating.

Consequently, the antenna beam can be directed anywhere in the sky below 5.8 MHz without the risk of grating lobes. The beam can be moved in a new location in roughly 15 μs, allowing it to be swept across a space almost continuously or to move quickly from one location to another. Almost any signal that can be represented as a ".wav" file, including AM, FM, phase, and pulse, can be modulated by the transmitter with great flexibility.

The HAARP site has several diagnostic tools in addition to the IRI, as well as the infrastructure to support more tools. The HAARP program owns some diagnostics (some of which have an associated principal investigator [PI]) while supporting the ownership of other diagnostics by PI institutions. A seismometer, magnetometers, riometers, an ionosonde, ultrahigh-frequency (UHF) and very-high-frequency (VHF) radars, optics, GPS scintillation receivers, ELF and VLF receivers, an HF receive antenna, and spectrum monitors are some of the diagnostic tools used [26, 28, 30].

4.8 How does HAARP Evoke the Ionosphere?

The ionosphere in the F region can undergo a variety of changes when exposed to high-power radio waves in the 2.6–10 MHz frequency range. The fundamental physical processes behind the ionosphere's artificial changes. Specific thresholds in HF power are required to start various degrees of ionosphere disturbances. These disturbances can affect different physical properties such as electron temperature, density, and pressure. The electron temperature is raised by the pump's electromagnetic (EM) wave at its lowest power, which also creates localized areas of enhanced pressure that are transmitted along magnetic field lines via thermal conduction and plasma diffusion. Above a critical HF power threshold, electron density in the ionosphere increases due to enhanced ionization from both electron and ion attachment to oxygen molecules [1, 5, 29].

A thermal parametric instability that channels high-frequency electrostatic waves (Langmuir and upper hybrid) into field-aligned cavities, where the waves' pressure intensifies them, leads to the production of field-aligned plasma irregularities with increases in HF power. Either by (i) direct mode conversion, in which the frequency of the wave from electromagnetic to electrostatic does not change, or by (ii) parametric decay, in which two electrostatic waves are simultaneously excited and their combined frequencies are equal to the frequency of the driving wave Electrostatic or electromagnetic forces can act as the driving wave [5, 26, 30].

High-frequency electrostatic waves can be converted back into an EM wave by changing their mode, which results in stimulated electromagnetic emissions that can travel to ground receivers and be picked up as a sideband of the original EM pump wave. If the high-frequency waves' phase velocity and electron velocity are in phase, then the electrons can be accelerated to energies high enough to create an artificial aurora or breakdown ionize the background neutral species [1, 26, 29].

Currently, the HAARP facility in Alaska has an exceptional ability: the generation of artificial enhancements of the electron plasma density by high-power radio waves. In light of this, the crucial query is: How can an antenna generate electrons in plasma beams? The conductor's electrons oscillate back and forth, which causes the antenna to radiate. The general theory of relativity has an intriguing explanation for radiation. Nothing, including fields, can move faster than the speed of light, according to the general theory of relativity. which implies that fields cannot be altered immediately. The amount of time it takes for the field to change is finite. It's called "retarded potential" when this occurs. Think about a dipole antenna (Figure 4.9). The AC voltage applied across the antenna causes electrons to accelerate and decelerate inside the conductor [27–30].

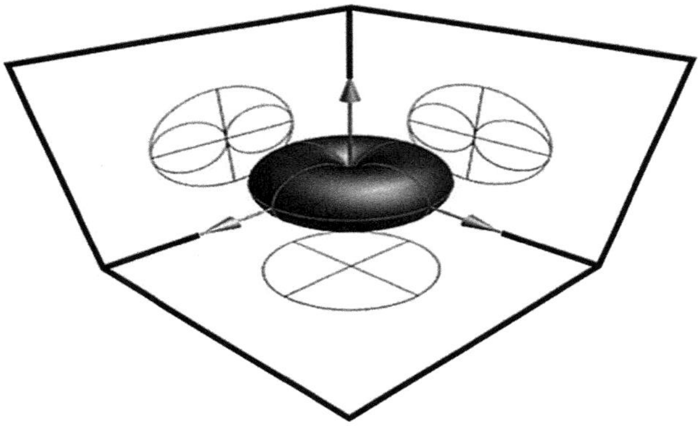

Figure 4.9: Dipole antenna radiation.

As seen in the image above, the F-field begins on a positive charge and ends on a negative charge. The field must adjust as the charges move. The E-field lines must slightly bend because the field cannot change instantly. The bent E-field eventually disconnects from its source. Because the E-field is time-varying, a time-varying magnetic field is created. This produces a time-varying electric field. An antenna produces a transverse electromagnetic wave as a result. Note that static charge cannot create a magnetic field; it can only create an electric field. Consequently, a charge moving at a constant speed can create an electric field and a magnetic field [15, 26, 30].

The main reasons for this are (i) the transmitter's ability to operate continuously (3.6 MW total); (ii) the array's 12 × 15 elements, which have the highest gain (30 dB at 10 MHz); and (iii) the HAARP system's full range frequency agility (2.6 to 10 MHz). Hence, the HAARP array's capability for beam-pointing and beam-forming is crucial for creating artificial plasma clouds with HAARP. Parenthetically, A pencil beam cannot produce a stable plasma cloud because the geometry of the beam prevents the plasma from forming on the cloud's bottom side. The use of a structured beam is the only method currently known for creating a long-lasting patch of artificial ionization. Structured beams can produce a stable plasma cloud by introducing a radial symmetry, which prevents any one side of the beam from becoming too dense or too dispersed.

Nevertheless, with the proper phasing of the HAARP array transmissions, a "twisted beam" can be formed into an annulus pattern with minimum power at the center, in which the HAARP twisted beam can form regions of artificial ionization. The peak electric field in this wider-angle beam is approximately 5 dB less than the power of a pencil beam at the same frequency. This makes it possible to develop a "directed energy weapon" that can control ionosphere plasma for a variety of purposes, including communication disruption, the delivery of chemical and biological weapons, and even the manipulation of weather patterns. The utilization of HAARP array technology for the development of directed energy weapons is a reality that has been extensively studied over the past few decades [1, 7, 27, 30].

Figures 4.10 and 4.11 flaunt antenna gain patterns that mimic the pencil and twisted beams for HAARP. With a gain of 24 dB, the zero-order L = 0 mode forms a single maximum. A ring is formed by the first-order L = 1 mode, with a maximum gain of 19 dB. This is an incredible increase in gain compared to the first-order L = 1 mode and demonstrates the effectiveness of HAARP's pencil (Figure 4.10) and twisted beams. The L = 2 and higher order modes also show an increase in gain (Figure 4.11), but with a much smaller maximum than the first two. This shows that the HAARP pencil and twisted beams are highly effective at concentrating energy, with the L = 0 modes being able to form a single maximum with a gain of 24 dB, compared to the 19 dB maximum of the L = 1 mode [31, 35, 37, 46].

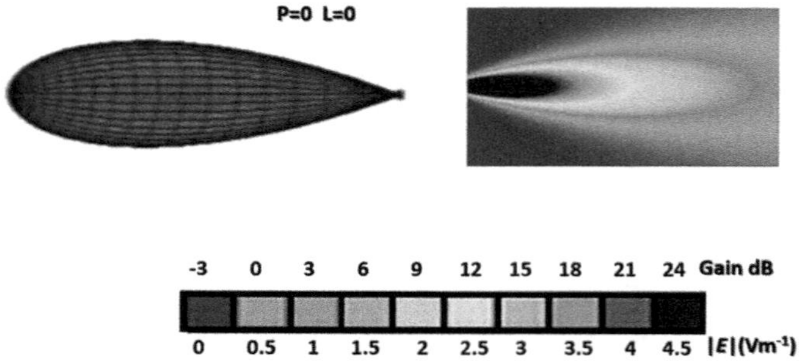

Figure 4.10: The antenna gain pattern of a pencil beam.

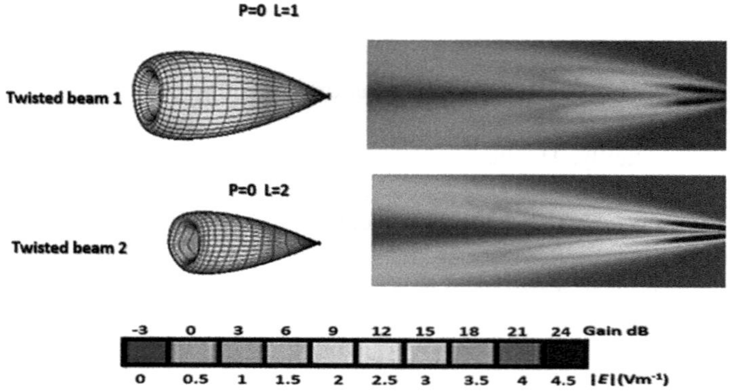

Figure 4.11: Antenna gain pattern of a twisted beam.

The significant query is: How can strong-field ionization's stronger heating be described by quantum mechanics? A critical aspect of HAARP's strong-field ionization is that the emitted electrons have very high kinetic energies compared to the energies of the originally bound electrons.

4.9 Quantization of HAARP-Ionosphere Turbulence Plasma Generation

What is "quantum ionization," exactly, and how does it work? Quantum ionization is a physical process that occurs when an electron absorbs energy from external sources, such as an electric field, and gains enough energy to become completely detached from its parent atom or molecule as demonstrated in the previous section. This is a significant process because it leads to the formation of charged particles known as ions, which are capable of carrying an electric current to the ionosphere layer. In quantum ionization, the HF electron can gain enough energy to escape the confines of its parent atom or molecule at HAARP devices, leaving it in a free state. This process is a cornerstone of disturbance of the ionosphere layer, as it forms a cloud plasma, which has a direct effect on the energy levels of particles in the ionosphere layer [32, 34, 36, 47].

A HAARP mechanical operation system can be described by quantum mechanics as a two-body (or bipartite) problem consisting of the electron-ion-core system (e-ion-core system) in the single active electron approximation, driven by a strong laser pulse. Since the relevant EM field wavelengths are several orders of magnitude longer than the system's size, we consider their interaction with the laser pulse in the dipole approximation, or as an external time-dependent electric field. The Hamiltonian for the Haarp atom system is written in the length gauge and takes the form of a two-body problem consisting of an electron, an ion core, and a strong laser pulse interacting with each other. The Hamiltonian for the Haarp system is as follows, using the length gauge [37, 47]:

$$H_{ec} = \frac{\vec{P}_c^2}{2m_e} + \frac{\vec{P}_c^2}{2m_c} - [\vec{r}_e - \vec{r}_c] + E(t)(\vec{r}_e - \vec{r}_c), \tag{4.8}$$

where the electron and core masses, respectively, are m_e (=1) and m_c. Therefore, the Hamiltonian is obtained by also using the reduced mass, $\mu = m_e m_c [m_e + m_c]^{-1} = m_e m_c [M]^{-1}$:

$$H_{ec} = \frac{\vec{P}_0^2}{2M} + \frac{\vec{P}_0^2}{2\mu} - |\vec{r}|^{-1} + \vec{E}(t)\vec{r}, \tag{4.9}$$

In equation 4.9, the coordinate transformation to the centre of mass (r_0; P_0) and relative coordinates (r; P) as follows $\vec{P}_0 = \vec{P}_e + \vec{P}_c$, and $\vec{r} = \vec{r}_e - \vec{r}_c$; that is used to simplify equation 4.8 into equation 4.9. Consequently, equation 4.9 can be carried into the entanglement of the individual particles Ψ_{ec} as follows:

$$\vec{\Psi}_{ec}(\vec{r}_e; \vec{r}_c; t) = \vec{\Psi}(\vec{r}, t)\vec{\Psi}_0(\vec{r}_0, t), \tag{4.10}$$

The time-dependent Schrödinger equation (TDSE), which is a component of the Hamiltonian's centre of mass, describes a free-particle propagation of HAARP plasma wave as follows:

$$i\frac{\partial}{\partial t}\vec{\Psi}_0 = \vec{P}_0^2 \otimes [2M]^{-1} \left[\frac{\frac{\sigma}{\sqrt{\pi}}}{\sigma^2 + i\frac{t}{M}} \right] e^{\left(-\vec{r}_0^2 \left[2\sigma^2 + i\frac{t}{M} \right]^{-1} \right)} \tag{4.11}$$

A Bohr radius, denoted by the parameter $\sigma = 1$, was set. This is a well-known free wave packet with root mean square deviations in each direction of the coordinates of the centre of mass, spreading as:

$$\Delta x_0 = \Delta y_0 = \Delta z_0 = \sqrt{\sigma^2 + t^2(M^{-2}\sigma^{-2})} \tag{4.12}$$

Evaluating equation 4.12 for the period indicated by the exciting pulse's maximum duration T_{max}. Strong field experiments typically use T_{max} = 300 a.u or a few femtoseconds. Put differently, heating the ionosphere by HAARP-free plasma wave particles occurs in a few femtoseconds. This causes an increment in the quantum entropies of ionization particles because the spreading during the interaction is so negligible (about 1.3% of the original width) due to the large value of M ~ 1837. The interaction of HAARP-free plasma wave particles with the ionosphere causes heating in a matter of femtoseconds. This heating increases ionization particle quantum entropies, due to the interaction being so brief that it is unable to spread significantly, resulting in less energy expenditure. Despite the tiny amount of energy expended, the entropies produced from this interaction are greatly increased [33–38]. In this understanding, the entangled wavefunction Ψ_{ec} of the HAARP heating ionosphere can be uniquely decomposed into the sum shown below:

$$\vec{\Psi}_{ec}\left(\vec{r}_e,\vec{r}_c,t\right)=\sum_k \lambda_k(t)\phi_k(\vec{r}_c,t)\psi_k(\vec{r}_e,t) \tag{4.13}$$

Equation 4.13 demonstrates that the HAARP wave plasma $\phi_k(\vec{r}_c,t)$ influences the stability of the ionosphere $\psi_k(\vec{r}_e,t)$ ionization system through the entanglement of two free plasma wave particles. In this view, the author demonstrates how the HAARP electrons can ionize the ionosphere layer using the Feynman diagram (Figure 4.12). In this sense, the wavy lines represent the incoming and scattered photons with momenta k_1 and k_2, respectively. The electron-electron interaction is represented by the dashed line and serves to redistribute momentum between the two electrons.

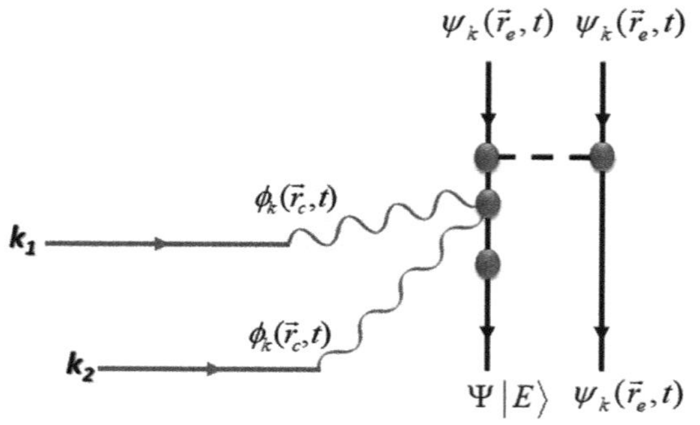

Figure 4.12: The author's view of HAARP ionizes the ionosphere layer using the Feynman diagram.

That is, measuring one particle changes the quantum state of the other particle in a nonlocal way. In this circumstance, by introducing quantum entropies, the eigenvalues $\lambda_k(t)$ enable one to quantify the entanglement of the particles between HAARP and the ionosphere [47]. The quantum entropies and eigenvalues $\lambda_k(t)$ provide valuable information about the degree of entanglement of both plasma wave particles, as they can be used to measure how much each wave-particle is correlated with each other. in other words, quantum entanglement between $\phi_k(\vec{r}_c,t)$ and $\psi_k(\vec{r}_e,t)$ can be given by:

$$\min\left[S\left[\psi_k(\vec{r}_e,t),\phi_k(\vec{r}_c,t)\right]\right]\leq S\left[\psi_k(\vec{r}_e,t):\phi_k(\vec{r}_c,t)\right]\Rightarrow \ell_{ec}^{quant} \tag{4.14}$$

In this view, ℓ_{ec}^{quant} involves both classical and quantum correlations. In this understanding, heating the ionosphere layer through HAARP occurs due to quantum entanglement between the interaction of plasma wave particles induced by HAARP and ionosphere particles. However, the temperature is completely irrelevant to equation 4.14. This equation is concerned solely with the behaviour of particles, and how they interact with each other from the point of view of entanglement quantum. Therefore, the temperature is an expression of the average kinetic energy of particles within the ionosphere. Thus this is where the concept of temperature is relevant. In this view, assume that all ℓ_{ec}^{quant} quantum states have the same energy, denoted by ε_i, and that the population of atoms at each of these energy levels is given by N_i, where *i* is just an index used to identify the energy levels [32, 35, 38, 40, 47].

$$N_i = \frac{N}{\sum \ell_{ec}^{quant} e^{\frac{\varepsilon_i}{K_B T}}} \ell_{ec}^{quant} e^{-\frac{\varepsilon_i}{K_B T}} \tag{4.15}$$

The Boltzmann constant, temperature and total population across all energy states are denoted by K_B, *T*, and *N*, respectively. Now let's examine Boltzmann's well-known equation for the entropy of a macro-state with *W* micro-states:

$$S[\psi_k(\vec{r}_e,t):\phi_k(\vec{r}_c,t)] = K_B \ln W \tag{4.16}$$

This number of micro-states *W of* ionosphere ionization due to HAARP interaction can be used to calculate the entropy of the ionosphere ionization, and therefore, its thermodynamic state:

$$W = \Pi\left[\ell_{ec}^{quant}\right]^{N_i}\left[N_i!\right]^{-1} \tag{4.16.1}$$

Using equation 4.16.1, the entropy can be mathematically expressed as:

$$S[\psi_k(\vec{r}_e,t):\phi_k(\vec{r}_c,t)] = K_B \sum N_i \ln \frac{N_i}{\ell_{ec}^{quant}} + K_B N \tag{4.17}$$

By including internal energy term of $NK_BT^2\left(\frac{\partial \ln \sum \ell_{ec}^{quant} e^{\frac{\varepsilon_i}{K_B T}}}{\partial T}\right)$ in Equation 4.17, the relationship between entropy and temperature can be obtained as follows:

$$S[\psi_k(\vec{r}_e,t):\phi_k(\vec{r}_c,t)] = NK_BT \ln \frac{\sum \ell_{ec}^{quant} e^{\frac{\varepsilon_i}{K_B T}}}{N} + NK_BT^2\left(\frac{\partial \ln \sum \ell_{ec}^{quant} e^{\frac{\varepsilon_i}{K_B T}}}{\partial T}\right)_V + K_B N \tag{4.18}$$

Equation 4.18, which links entropy's absolute value to temperature *T*, takes a rather general form and includes terms whose values depend on both temperature and volume/pressure. The sum of four individual entropies, or four different sources of missing information, is used to represent the overall entropy [41–46]. These include the indistinguishability of the particles, positional uncertainty, momenta uncertainty, and the uncertainty principle of quantum mechanics. After adding up the four parts, the Sackur-Tetrode equation is as follows:

$$S[\psi_k(\vec{r}_e,t):\phi_k(\vec{r}_c,t)] = NK_B\left[\frac{3}{2}\ln T + \ln\frac{V}{N} + \frac{3}{2}\ln\frac{2\pi m K_B}{h^2} + \frac{5}{2}\right] \tag{4.19}$$

Where *V* is volume; *h* is the Planck constant and *m* is gas mass. The Sackur-Tetrode equation is a thermodynamic equation of state that describes the entropy of an ideal gas in terms of its temperature, pressure, and number of particles. The Sackur-Tetrode equation is a fundamental expression for entropy that provides insight into the nature of ideal gases and their properties [54]. Another way to express the equation is in terms of the thermal wavelength λ:

$$S[\psi_k(\vec{r}_e,t):\phi_k(\vec{r}_c,t)] = NK_B\left[\ln\left(\frac{V}{N\lambda^3}\right) + \frac{5}{2}\right], \tag{4.20}$$

Equations 4.18 to 4.20 show that the increase in ionospheric warming is directly proportional to the increase in entropy due to the effect of HAARP. Under these circumstances, the heating of the ionosphere by HAARP causes more kinetic energy in the electron dynamic fluctuations. This leads to a turbulent plasma wave flow. This turbulence further increases the entropy, allowing for an even greater exchange of energy between particles in the ionosphere and therefore a larger increase in the ionospheric temperature. As the kinetic energy of electrons in the ionosphere increases, this causes an increase in temperature, leading to an increase in entropy due to the increased energy exchange between particles [43, 47, 49].

According to the above perspective, "Turbulence", a flow with vortices of different sizes, occurs in a high-temperature plasma trapped by the magnetic field. The heat of the enclosed plasma is forced outwards by the turbulence, disturbing the plasma and lowering its temperature. If enough turbulence is created, the phenomenon known as thermal conduction can be greatly enhanced. The heat of the trapped plasma is pushed outwards by the turbulence, disturbing the plasma and lowering its temperature. If the turbulence intensity is high enough, this phenomenon—known as heat conduction—can be significantly enhanced. This improves the efficiency of heat conduction and enables more efficient heat transfer over longer distances [40–52].

The authors suggest that a plasma power spectra $E(k)$ gain can be produced when there is an increase in entropy due to HAARP entanglement in the ionosphere layer as follows:

$$S\left[\psi_k(\vec{r}_e,t):\phi_k(\vec{r}_c,t)\right]\approx E(k)\ln E(k) \tag{4.21}$$

According to equation 4.21, by substituting the Kolmogorov-Obukhov spectrum, as stated in section 1.15 of chapter 1 and given by equation 1.41, the modified equation can be written as follows:

$$S\left[\psi_k(\vec{r}_e,t):\phi_k(\vec{r}_c,t)\right]\approx C_k\epsilon^{\frac{2}{3}}k^{\frac{-5}{3}}\ln C_k\epsilon^{\frac{2}{3}}k^{\frac{-5}{3}} \tag{4.22}$$

Equation 4.22 demonstrates that the entropy-generating plasma turbulence flow is entangled to the Kolmogorov constant C_k and energy dissipation rate ϵ. In this view, the Kolmogorov constant is a universal measure of energy dissipation rate per unit mass in plasma turbulent flows.

The turbulent plasma energy spectra (i.e., the sum of kinetic and magnetic energy), in the form of a power-law scaling, can be used to measure the amount of energy available for heating. By looking at the plasma energy spectra, we can understand how much of that energy is dissipating, and how much of it can be effectively used to heat the ionosphere. Using the plasma energy spectra to measure the available heating energy can give us insight into how much of that entanglement entropy is being used to heat the ionosphere:

$$S\left[\psi_k(\vec{r}_e,t):\phi_k(\vec{r}_c,t)\right]\approx\sqrt{\epsilon V_A}\,k^{-\frac{3}{2}}\ln\sqrt{\epsilon V_A}\,k^{-\frac{3}{2}} \tag{4.23}$$

here is the Alfvén velocity. Since the importance of collisions between charged and neutral particles, the ionospheric E- and F-regions of the Earth have unique conditions. It has been speculated that some low-frequency ionospheric waves in these areas may also produce strong turbulence with spectra of the universal power law. This is a significant phenomenon because it raises the possibility that the properties of the Earth's ionosphere could be significantly altered by strong ionospheric turbulence. The atmosphere, ionosphere, and magnetosphere system may then experience significant changes as a result of this.

Therefore, the integral of the mean-square of the electric field over all frequencies, for example, can be used to define both scalar and vector spectra, resulting in the total power as given by:

$$\left\langle\phi_k(r_c,t)\right\rangle=\int^{\infty}E(k)dk \tag{4.24}$$

The plasma entropy in the F and E regions of the ionosphere is increased by entanglement entropy, leading to an increase in the turbulence plasma spectra:

$$S\left[\phi_k(\vec{r}_c,t)\right]=\left\langle\phi_k^2(\vec{r}_c,t)\right\rangle\ln\left\langle\phi_k^2(\vec{r}_c,t)\right\rangle \tag{4.25}$$

HAARP increases turbulence plasma spectra, but the possibilities for generating an ionosphere disturbance are vast. The ionosphere disturbance can be used to control weather, alter global temperatures, and even interfere with communication signals. In this regard, a critical question is how remote sensing technology tracks ionosphere disturbances and their effect on air temperature. Detailing the answer would be covered in the following sections.

4.10 How does Turbulence Plasma Wave Track from Space?

It is possible to study the turbulent plasma waves brought on by HAARP using satellite beacon signals. In this sense, satellite beacon signals passing through the altered ionosphere offer a potent method for detecting irregularities in artificially created plasma. Testing the susceptibility of navigation, communications, and radar systems to natural ionospheric disturbances requires the creation of artificial irregularities that can cause radio scintillations. High-power radio waves have not yet been able to replicate the intense scintillation environment of the natural ionosphere associated with auroral arcs, strong substorms, and polar cap patches. In this regard, HAARP is a powerful tool to allow scientists to generate large-scale plasma irregularities and measure the effects of scintillation on satellite beacon signals passing through the altered ionosphere.

The National Research Laboratory's (NRL) TACSat4 was tasked with transmitting 253 MHz through the ionosphere over HAARP to track the scintillations from artificial ionization regions. The NRL communications experiment (COMMX) is aboard TACSat4, which is in an orbit with a 63° inclination, to showcase the capabilities of VHF/UHF satellite communications (SATCOM) (Figure 4.13). The repeating orbit of the TACSat4 satellite passes directly over the HAARP transmitter. The TACSat4 COMMX receiver picked up the 401.25 MHz beacon transmissions used by satellite stations in Cold Bay, Alaska, and Yellow Knife, Canada, for Doppler orbitography and radio positioning integration for the HAARP experiments [51].

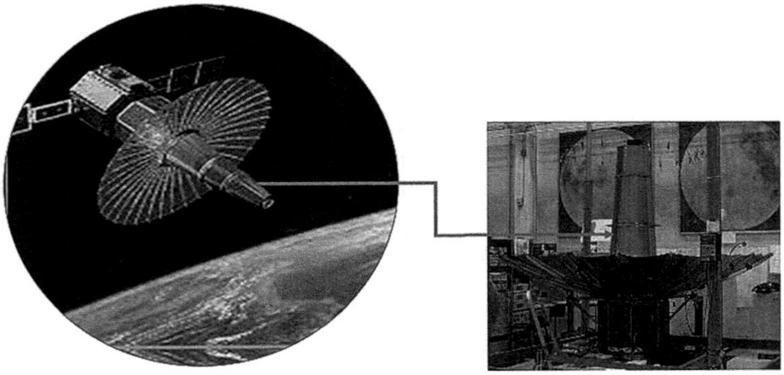

Figure 4.13: TACSat4 with UHF antenna.

Therefore, these signals are then converted by COMMX to 253 MHz for rebroadcast via the satellite's high-gain parabolic antenna. This antenna was permanently aimed at the ground receiver, which was situated directly below the HAARP-modified ionosphere. The VHF/UHF signals were translated to 10.7 MHz by the NRL ground receiver system for TACSat4 before being digitalized and subjected to further processing. The final output of the system was then transmitted back to the ground receiver via the satellites' high gain parabolic antenna, where it was subjected to a final level of processing before being converted into its original form for further use. The TACSat4 system worked by receiving signals from the HAARP-modified ionosphere through a special antenna. The antenna was connected to a 10.7 MHz receiver, which was able to process the signals and convert them into digital data [52].

The comparative S4 index, which accurately compares the standard deviation of signal power to the average signal intensity, is a key metric for amplitude scintillations. Enthusiastic reasonable anomalies correctly are those that typically occur at high latitudes during the solar maxima and produce S4 indices above 0.6. These apparent anomalies can adversely affect UHF satellite communications and the outstanding performance of other UHF systems. Even during the solar minimum, strong phase scintillations have been carefully observed at high latitudes [53].

Most prevailing UHF systems won't be negatively impacted by radio scintillations that contain an effective S4 index of 0.3 or less, which are considered weak scintillations from unique experiments at EISCAT and HAARP. The controlled experiments used signals with 240 MW ERP, 4 dB fluctuations in the 250 MHz band, and reliable S4 indices between 0.1 and 0.35 at specific frequencies between 3.85 and 5.56 MHz. At HAARP, ongoing research aims to intensify irregularities and efficiently generate UHF radio scintillations that are close to essential unity [51–53].

4.11 Quantum Algorithm for Retrieving Ionosphere Turbulent Plasma Wave Due to HAARP

Let's assume that the spectra density of turbulent plasma can be initiated in the form of quantum Langmuir dispersion relation as follows:

$$|\omega\rangle = \sqrt{\left|\omega_p^2\right\rangle + \frac{3K_B T_{0\|} k^2}{m} + \frac{\hbar^2 k^4}{4m^2}}. \tag{4.26}$$

here K_B is Boltzmann's constant and plane wave perturbations proportional to $e^{(ikz-i\omega t)}$ are assumed, without loss of generality. Therefore, ω_p is the plasma frequency and $T_{0\|}$ represents the equilibrium parallels to the wave propagation. In this view, for a zero-temperature Fermi gas, one has $T_{0\|} = (2/5)T_F$ where T_F is fermi temperature. Consequently, m is the mass of the charge carriers and $\hbar = \frac{h}{2\pi}$.

The key question right now is how does quantum computing simulate the turbulent plasma wave in the ionosphere caused by the HAARP impact?. Understanding the principles of quantum computing is necessary to respond to this query. The revolutionary field of computing known as quantum computing processes information much more quickly than traditional computing methods by making use of quantum-mechanical phenomena like entanglement and superposition. This technology, when applied to the phenomenon of HAARP-induced turbulence in the ionosphere, could help explain why this phenomenon is occurring and how it can be predicted or manipulated [41, 43, 47].

The wave function's temporal evolution is thought to be the cause of the disturbance field in ionosphere plasma spectra, which can be expressed by phase shift ϕ as:

$$e^{(-iHt)}|\Psi\rangle = e^{(-iEt)}|\Psi\rangle = e^{(-i2\pi\phi)}|\Psi\rangle \tag{4.27}$$

Equation 4.27 demonstrates the phase shift retrieved by the inverse of the quantum Fourier Transform (IQFT). The eigenvalue problem of the shifted Hamiltonian H' of ionosphere entropy, which consists of the Hamiltonian of the system H and the entanglement between HAARP plasma and ionospheric plasma, is used for direct calculations of spin state energy disturbances. Put differently, the eigenvalue problem of the shifted Hamiltonian H' allows for the examination of entanglement dynamics between HAARP plasma and ionospheric plasma in a way that was previously impossible. This can be achieved through a quantum algorithm $|\Re\rangle \times |\Re\rangle$. In this understanding, the shifted Hamiltonian H' of the energy phase can be formulated as:

$$H'\left|\Psi_{S_H \neq S_I}\right\rangle = E'_{S_H \neq S_I}\left|\Psi_{S_H \neq S_I}\right\rangle = E_{S_H \neq S_I} + j\left\{S\left[\phi_k(\vec{r}_c,t)\right] \times S\left[\phi_k(\vec{r}_c,t)\right] + 1\right\}\left|\Psi_{S_H \neq S_I}\right\rangle \tag{4.28}$$

Under the shifted Hamiltonian H', the energy gap between the two spin states of HAARP $\left|\Psi_{S_H}\right\rangle$ and ionosphere $\left|\Psi_{S_I}\right\rangle$ is calculated as follows:

$$\Delta E' = E_{S_I} - E_{S_H} + j\left\{\left[S[\phi_k(\vec{r}_c,t)]\times S[\phi_k(\vec{r}_c,t)]+1\right]-\left[S\left[\psi_k(\vec{r}_e,t)\right]\times S\left[\psi_k(\vec{r}_e,t)\right]+1\right]\right\} \tag{4.29}$$

Under the circumstances of the shifted Hamiltonian H', the HAARP plasma wave turbulence would spin with the ionospheric plasma wave having a similar eigenvalue E'. In this sense, the wave function $|\Psi_0\rangle$ would demonstrate the superposition of the two spin states, which is also considered an eigenfunction of H′. As such, the superposition of these two states would be an eigenstate of the new Hamiltonian H′ and would manifest in terms of the plasma wave turbulence as follows:

$$|\Psi_0\rangle = c_{S_H}\left|\Psi_{S_H}\right\rangle + c_{S_I}\left|\Psi_{S_I}\right\rangle \tag{4.30}$$

Therefore, the square overlap $|\langle\Psi_0|U(H',j,t)|\Psi_0\rangle|^2$, where $U(H',j,t)$, can be used to estimate the deviation of $|\Psi_0\rangle$ from the eigenfunction of H' as:

$$U(H';j,t) = e^{\{-i(H+jS[\phi_k(\vec{r}_c,t)]\times S[\phi_k(\vec{r}_c,t)]+1\}} \tag{4.31}$$

Consequently, a SWAP test can be used to evaluate the square overlap $|\langle\Psi_0|U(H',j,t)|\Psi_0\rangle|^2$ effectively. In quantum computing, the SWAP test has been extensively used, particularly for entanglement detection. Two qubits that are suppositionally entangled have been measured using the SWAP test. In this view, the SWAP analysis comprises two Hadamard gates and a controlled-SWAP gate, which swaps the quantum states of $|\varphi\rangle$ and $|\psi\rangle$ when the constraint qubit is in the $|1\rangle$ state, and the quantity of a qubit cast-off as the regulator. Therefore, two qubits are used in the SWAP gate. The SWAP gate switches the states of the two qubits involved in the operation when expressed in basis states:

$$SWAP = \begin{pmatrix} 1 & 0 & 0 & 0 \\ 0 & 0 & 1 & 0 \\ 0 & 1 & 0 & 0 \\ 0 & 0 & 0 & 1 \end{pmatrix} \tag{4.32}$$

Subsequently, a single-qubit operation called the Hadamard gate converts the basis states from $|0\rangle$ to $\frac{|0\rangle+|1\rangle}{\sqrt{2}}$ and from $|1\rangle$ to $\frac{|0\rangle-|1\rangle}{\sqrt{2}}$. Resulting in a superposition of the two basis states that is equal. The Hadamard gate can be described as:

$$H_d = \frac{1}{\sqrt{2}}\begin{pmatrix} 1 & 1 \\ 1 & -1 \end{pmatrix} \tag{4.33}$$

Figure 4.14 depicts the modified quantum circuit for the direct calculation of spin state energy gaps. In this view, The N_{DOR} denotes the number of doubly populated orbitals, the N_{SOR} indicates the number of singly congested orbitals and N_{UOR} denotes the number of vacated orbitals. Thus, the Hartree–Fock wave function, along with the quantum algorithm, allows for a computationally-generated model of ionosphere ionization energies caused by (HAARP) pulses:

$$\hat{F}[\{\phi_j\}](1) = -\frac{1}{2}\nabla_1^2 - \sum_{\alpha}\frac{Z_\alpha}{r_{1\alpha}} + \sum_{j=1}^{N/2}[2\hat{J}_j(1) - \hat{K}_j(1)], \hat{F}[\{\phi_j\}] \tag{4.34}$$

Equation 4.34 says the electron exchange energy resulting from the total N-electron wave function's antisymmetry is defined by the exchange operator $\hat{K}_j(1)$. The electron-electron repulsion energy caused by each of the two electrons in the *j-th* orbital is defined by the Coulomb operator, or $\hat{J}_j(1)$. Simply put, this "exchange energy" operator $\hat{K}_j$ is a Slater determinant artefact. Therefore, the orbitals ϕ_j deliver the one-electron Fock operator $\hat{F}[\{\phi_j\}](1)$ as a function of the level Z_α

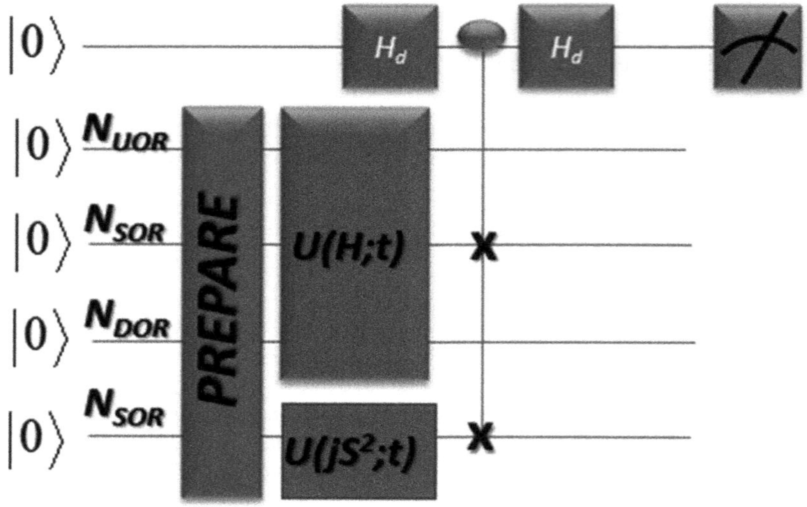

Figure 4.14: Quantum circuit gate to model HAARP ionizing ionosphere layer.

and orbital radius $r_{1\alpha}$ respectively. Simply put, the "(1)" that appears after each operator symbol denotes an operator with a single electron.

In light of this, we used the wave function |0 as follows to apply the quantum algorithm to the direct calculations of ionization energies:

$$|\Psi_0\rangle = \frac{1}{\sqrt{2}}\left(\left|\hat{F}[\{\phi_j\}](1)\right\rangle - a_i\left|\hat{F}[\{\phi_j\}](1)\right\rangle\right) \tag{4.35}$$

Here, a_i denotes an annihilation operator acting on the electron to be ionized, and $\left|\hat{F}[\{\phi_j\}](1)\right\rangle$ is a Hartree-Fockwave function of the neutral state. Consequently, $a_i\left|\hat{F}[\{\phi_j\}](1)\right\rangle$ is a rough representation of the ionized state's wave function. Needless to say, the re-expression of the molecular Hamiltonian allows for easier computation and manipulation of the system, while still providing a reasonable approximation to the exact Hamiltonian. We performed numerical quantum circuit simulations for the ionization of atoms (He, Li, Be, B, C, and N) and small molecules to demonstrate the quantum algorithm (HF, BF, CF, CO, O_2, NO, CN, F_2, H_2O, and NH_3).

According to the above perspective, an initial growth in electron density can be formed at the point where the pump frequency matches the existing plasma frequency profile. This indicates that the formation of artificial plasma clouds requires an ambient ionosphere with a density greater than the critical density for the reflection of the HF pump wave. The statement suggests that an artificial plasma cloud is formed when the ionosphere's density is increased through the pump wave and exceeds the critical density of reflection. A further understanding implies that the artificial plasma clouds reflect the pump wave, allowing for greater levels of interaction between the electromagnetic field and charged particles (Figure 4.15).

Consequently, the heating region produced in the twisted beam mode is different from the more conventional "solid spot" region of a pencil beam in that it is shaped like a ring. The horizontal structure of the ring may make the ring-heating pattern more favourable for the development of stable artificial airglow layers. The formation of layers that are more durable than pencil beam-induced layers may be possible as a result of this structure, which may produce a more even and uniform horizontal temperature profile. This is especially advantageous because these layers

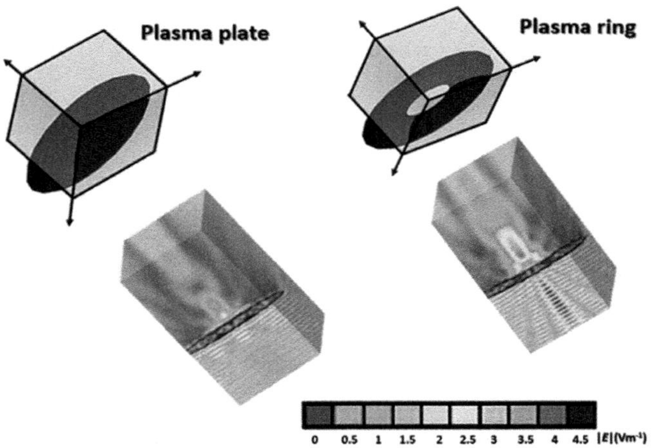

Figure 4.15: Simulation of plasma cloud intensities generated by plasma plate and plasma ring.

can last for long periods and may be more favourable for the development of other atmospheric phenomena like dust devils.

It was later found that the use of twisted beams reduces energy consumption because the wider beam coverage leads to a stronger heating effect over a larger area and more energy can be absorbed than with pencil beams. As this energy efficiency allows more energy to be stored in the atmosphere over a longer period, this is particularly helpful in creating longer-lasting airglow layers. This makes it possible to create a temperature profile with improved energy efficiency and increased potential for atmospheric phenomena over longer periods.

The third electron gyroharmonic, or about 4.335 MHz, is where an artificial ionization cloud formed, based on the aforementioned viewpoint. At a latitude of 150–280 km above Arctica, this peak plasma begins in less than one second (Figure 4.16a). As a result, after 60 seconds, the artificial ionization cloud has a range of 200 km to 320 km and a peak plasma frequency of less than 5 MHz (Figure 4.16b). As a result, the plasma frequency peak rises to 9 MHz at 120 seconds and creates turbulence ionization clouds, which are primarily extended from 80 km to 500 km (Figure 4.16c). Then, from 80 km to 700 km, plasma frequency varied between 4 MHz and 9 MHz, forming turbulence divergence ionization clouds that descended to 80 km above Arctica.

According to the above perspective, an artificial ionization cloud formed with peak plasma frequency at the third electron gyroharmonic around 4.335 MHz. This peak plasma is initiated within less than one second at a latitude between 150 km and 280 km (Figure 4.16a) above Arctica. Therefore, the artificial ionization cloud extends between 200 km to 320 km with a peak plasma frequency of fewer than 5 MHz after 60 sec (Figure 4.16b). Consequently, at 120 sec, the plasma frequency peak increases to 9 MHz and forms turbulence ionization clouds, which are mainly extended from 80 km to 500 km (Figure 4.16c). Subsequently, plasma frequency fluctuated between 4 MHz to 9 MHz from 80 km to 700 km forming turbulence divergence ionization clouds descending to 80 km above Arctica. However, a portion of the turbulence divergence ionization clouds rises to 600 km in 180 sec (Figure 4.16d).

The plasma turbulence peak frequency range is between 4 MHz and 9 MHz as these turbulence ionization clouds then spread above Europe (Figure 4.17). This arrangement worked best for producing a potent artificial ionization in the atmosphere. Observations showed that the use of this technique resulted in a significant concentration of electrons, allowing for an increase in plasma density relative to the environment.

The particle density of electrons per m^{-3} is represented by a color map. With each color denoting a specific number of electrons per m^{-3}, the color map is a useful tool for identifying regions of

(a)

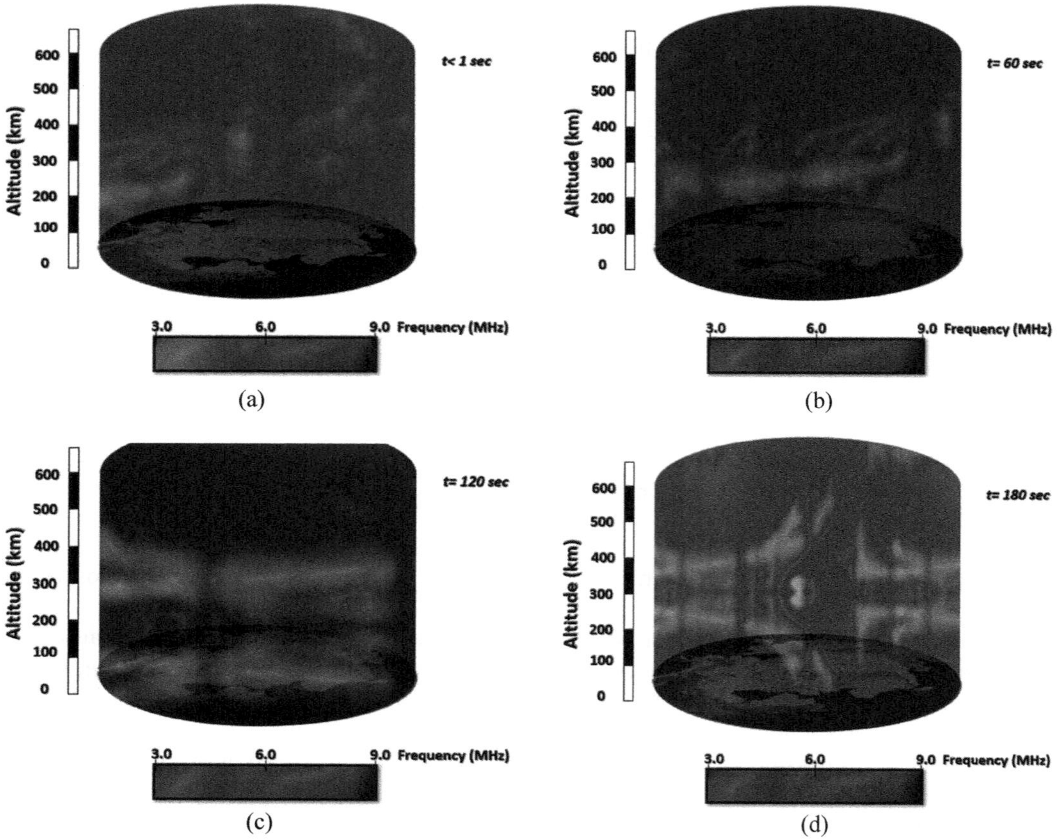

Figure 4.16: Artificial ionization turbulence cloud patterns within (a) less than a second; (b) 60 sec; (c) 120 sec; and (d) 180 sec.

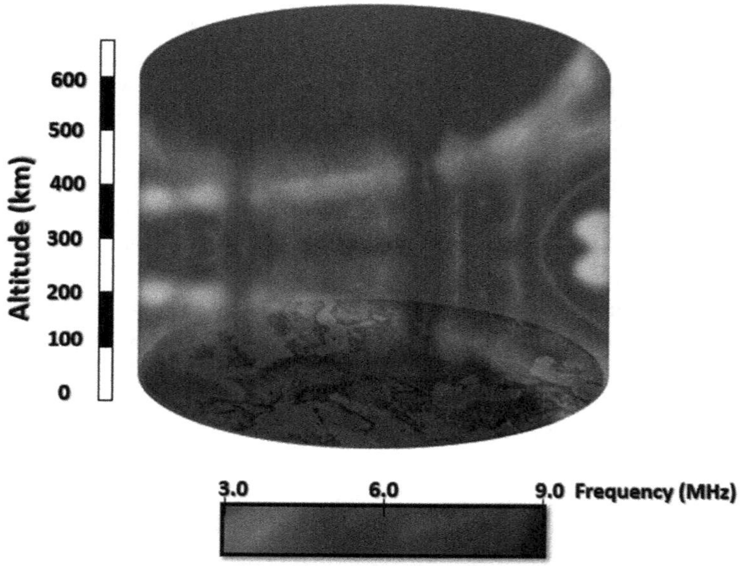

Figure 4.17: Artificial ionization turbulence cloud patterns above Europe.

high and low electron density. It can be used to identify the disturbance of the ionosphere layer. The particle density map can also be used to identify areas of high and low temperatures, allowing scientists to gain insight into the chemical and physical properties of various ionization regions, especially arctic and Western European countries. In this sense, there is the tongue of the density of electrons ranging from 3.1×10^4 to 5.9×10^4 m^{-3} extending from Artica to Europe atmosphere layers of 80 km to higher than 600 km. The tongue of the density of electrons is dominated by a maximum frequency of 9 MHz (Figures 4.18 and 4.19).

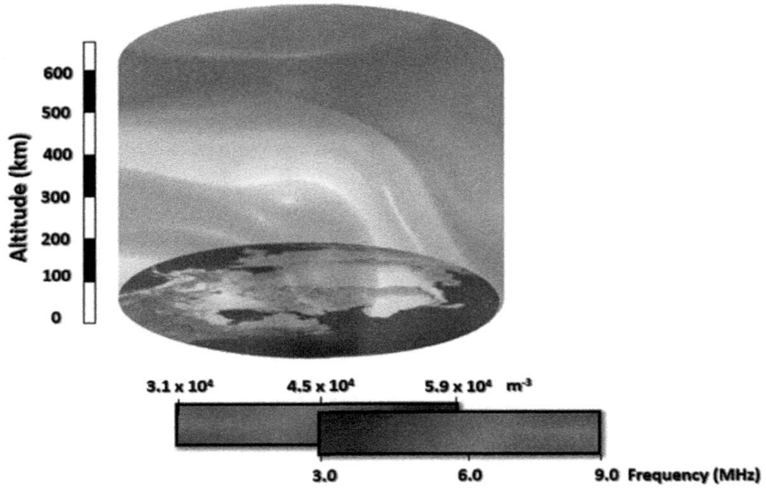

Figure 4.18: The tongue of the density of electrons in Arctica's atmosphere.

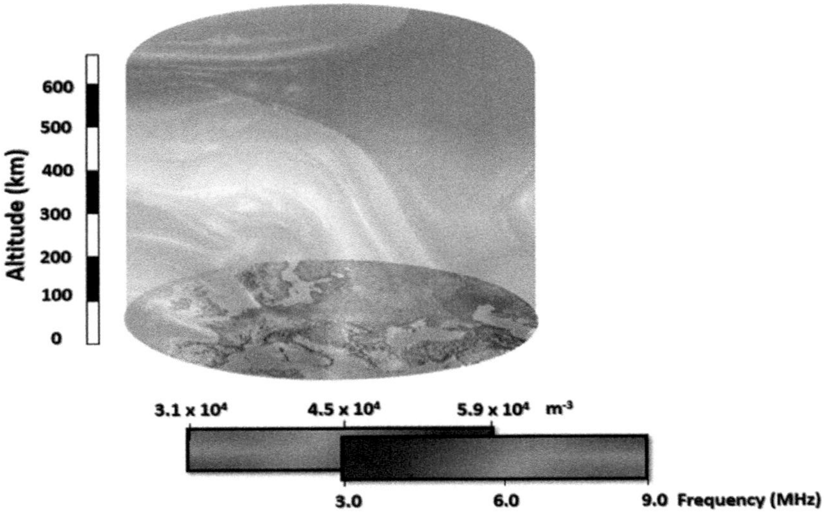

Figure 4.19: The tongue of the density of electrons in Europe's atmosphere.

Therefore, the tongue of the density of electrons grows from Actica to the European atmosphere to form turbulence plasma energy of low-frequency zone of 15 kHz and highest ionization energy of 24 $20\log_{10}|E|$ at approximately high altitude of 500 km. It is worth noting that this turbulence plasma energy forms mesoscale eddy patterns above Arctica and the European atmosphere (Figures 4.20 and 4.21).

As a result, from Arctica to Europe, an electron density tongue grows, creating turbulence plasma with a low-frequency zone of 15 kHz and a maximum ionization energy of 24 20log10 |E| at a height of about 500 km. It is important to note that the mesoscale eddy patterns formed by this turbulence plasma energy are found above Arctica and the European atmosphere (Figures 4.20 and 4.21). These eddy patterns, driven by winds from the lower atmosphere, can cause a redistribution of energy among various frequency bands and significantly affect ionospheric processes. In particular, these eddies can create perturbations in the total electron content, thus creating instabilities which affect both high-frequency and low-frequency radio wave propagation. As a result, these eddies can cause a redistribution of energy among various frequency bands that has a strong impact on the ionosphere, allowing the turbulence plasma energy to significantly affect the total electron content.

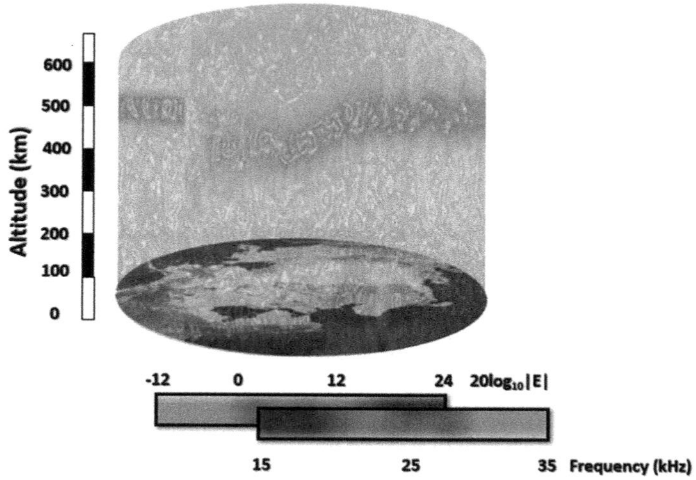

Figure 4.20: Turbulence plasma energy in Arctica's atmosphere.

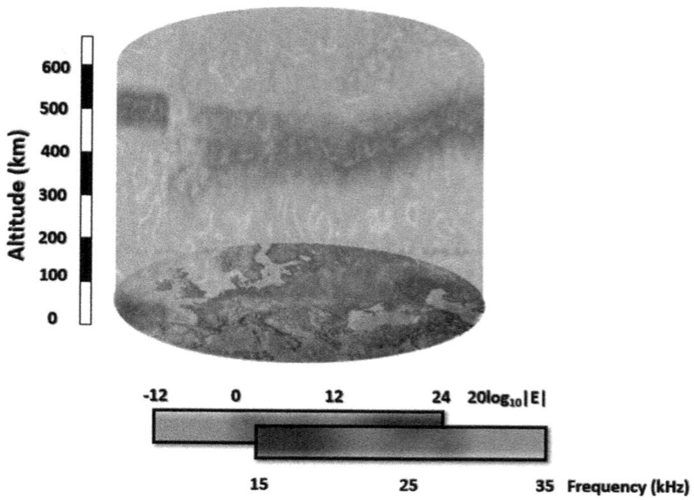

Figure 4.21: Turbulence plasma energy in Europe's atmosphere.

In fact, during the day, scintillation is more intense near the poles due to the increased density of plasma structures in those regions. At night, however, the intensity of scintillation is greater at lower latitudes due to the increased ionization of plasma near the equator. This phenomenon is a result of the Earth's magnetic field, which causes the electrons and protons to be concentrated in particular regions of the ionosphere. Therefore, these bursts are most frequently explained by the current generation of ion-sound or ion-cyclotron oscillations or by the crossing of small-scale spatial disturbances by satellite. However, in our case, these disturbances are observed at medium latitudes (32°–47° N), making it difficult to associate them with a geomagnetic storm because similar fluctuations were also seen for other half orbits under calm geomagnetic conditions. This suggests that these oscillations are not caused by a geomagnetic storm but could be due to interactions between the HAARP signal and the ionospheric plasma, possibly generated by its passage through an unstable layer or boundary in the ionosphere. These interactions could be due to the ionosphere reacting to the high-frequency electric field generated by HAARP, thereby creating a 'bump' in the medium latitudes of the ionosphere. This theory of ionospheric reaction to the HAARP signal is further supported by the fact that these disturbances were observed only in medium latitudes (32°–47° N) and did not occur under calm geomagnetic conditions [48–51].

Needless to properly say, the quantum algorithmic rule is capable of significantly reducing the computational cost and typically increasing the reasonable accuracy of ionization predictions compared to existing classical methods. Additionally, these accurate simulations merely revealed that the quantum algorithm exhibits a considerable degree of scalability for more complex systems, undoubtedly meaning that it can accurately predict the ionization of increasingly larger molecules with higher precision and at faster rates. This typically indicates that a quantum algorithm remains a promising tool for scientifically studying ionization in a comprehensive range of efficient systems, from unstable atoms to small molecules and beyond [50–54].

Ionospheric scintillation can be observed at all frequencies and across all latitudes, though the intensity of the signal is highly dependent on frequency and location. For example, during the day, scintillation is more intense near the poles due to the increased density of plasma structures in those regions. At night, however, the intensity of scintillation is greater at lower latitudes due to the increased ionization of plasma near the equator. This phenomenon is a result of the Earth's magnetic field, which causes the electrons and protons to be concentrated in particular regions of the ionosphere. This concentration of electrons and protons creates a higher level of scintillation in the corresponding regions, resulting in greater signal intensity. Thus, the intensity of scintillation varies based on the region of the ionosphere and can be more intense in areas with higher plasma density and increased ionization due to the presence of Earth's magnetic field. The Earth's magnetic field is therefore an important factor in the phenomenon of scintillation, as it provides an opportunity for electrons and protons to be concentrated in certain areas of the ionosphere, leading to higher levels of scintillation and increased signal [1, 14, 30, 46, 50, 53].

This process is the foundation of ionosphere layer disturbances because it produces cloud plasma, which has a direct effect on particle energy levels in the ionosphere layer. By exciting or ionizing particles within the layer, this cloud plasma allows for energy to be exchanged between atmospheric particles and radiation, altering the energetic state of the ionosphere layer. In this scenario, this cloud plasma allows for energy to be exchanged between atmospheric particles and radiation, altering the energetic state of the ionosphere layer. The energetic changes that occur within the ionosphere layer, when excited or ionized by the cloud plasma, allow for different types of electromagnetic radiation to be transferred into or out of the atmosphere. This transfer of radiation has direct implications for Earth's climate as it affects the temperatures within the lower atmosphere, allowing energy to be absorbed or reflected [24, 37, 48, 53].

Ultimately, the ionosphere layer disturbances caused by cloud plasma can have a dramatic effect on Earth's climate. As a result, the ionosphere layer can be considered an important regulator of climate on Earth due to its ability to absorb and reflect electromagnetic radiation. Additionally, these energetic changes can also affect the composition of the atmosphere and can even lead to changes in the concentrations of certain gases, such as carbon dioxide. This change in atmospheric composition can, in turn, cause variations in the global climate system. These variations in climate can range from short-term alterations, such as changes in air temperature and pressure patterns, to longer-term shifts in global temperatures over time. The ionosphere, therefore, acts as an important mechanism for regulating the Earth's climate by controlling the amount of radiation that reaches the surface and by modulating atmospheric gas concentrations. Consequently, changes in the ionospheric layer can drastically influence the climate of our planet, as shown by the European drought and the dramatic floods in Saudi Arabia in 2022. In particular, extreme weather events such as floods and droughts are thought to be partially driven by ionospheric alterations.

The following chapter will continue to investigate how HAARP's ionosphere disturbance is causing dramatic heat waves across the European continental during the summer of 2022. The investigation will focus on how the ionosphere disturbances caused by HAARP could have created air pressure imbalances that, when combined with other environmental factors, triggered these unusually extreme weather patterns.

References

[1] Kelley, M.C. (2009). *The Earth's Ionosphere: Plasma Physics and Electrodynamics*. Academic Press.
[2] Ratcliffe, J.A. (1956). Some aspects of diffraction theory and their application to the ionosphere. *Reports on Progress in Physics*, 19(1), 188.
[3] Axford, W.I. (1961). Note on a mechanism for the vertical transport of ionization in the ionosphere. *Canadian Journal of Physics*, 39(9), 1393–1396.
[4] Yonezawa, T. (1966). Theory of formation of the ionosphere. *Space Science Reviews*, 5(1), 3–56.
[5] Beyer, A., Maisenbacher, L., Matveev, A., Pohl, R., Khabarova, K., Grinin, A. et al. (2017). The Rydberg constant and proton size from atomic hydrogen. *Science*, 358(6359), 79–85.
[6] Zhao, P., Lichten, W., Layer, H.P. and Bergquist, J.C. (1986). Remeasurement of the Rydberg constant. *Physical Review A*, 34(6), 5138.
[7] Series, G.W. (2009). The Rydberg constant. *Contemporary Physics*, 50(1), 131–150.
[8] Kirkwood, S. and Nilsson, H. (2000). High-latitude sporadic-E and other thin layers – The role of magnetospheric electric fields. *Space Science Reviews*, 91(3–4), 579–613.
[9] Bauer, S.J. (2012). *Physics of Planetary Ionospheres*. Springer Science & Business Media.
[10] Cosgrove, R.B. and Tsunoda, R.T. (2002). A direction-dependent instability of sporadic-E layers in the nighttime midlatitude ionosphere. *Geophysical Research Letters*, 29(18), 11–1.
[11] Zhou, C., Tang, Q., Song, X., Qing, H., Liu, Y., Wang, X. et al. (2017). A statistical analysis of sporadic E layer occurrence in the midlatitude China region. *Journal of Geophysical Research: Space Physics*, 122(3), 3617–3631.
[12] Tang, Q., Zhou, C., Liu, H., Liu, Y., Zhao, J., Yu, Z. et al. (2021). The possible role of turbopause on sporadic-E layer formation at middle and low latitudes. *Space Weather*, 19(12), e2021SW002883.
[13] Rishbeth, H. (1997). The ionospheric E-layer and F-layer dynamos—A tutorial review. *Journal of Atmospheric and Solar-Terrestrial Physics*, 59(15), 1873–1880.
[14] Heelis, R.A. (2004). Electrodynamics in the low and middle latitude ionosphere: A tutorial. *Journal of Atmospheric and Solar-Terrestrial Physics*, 66(10), 825–838.
[15] Maute, A. and Richmond, A.D. (2017). F-region dynamo simulations at low and mid-latitude. *Space Science Reviews*, 206(1–4), 471–493.
[16] Immel, T.J., Sagawa, E., England, S.L., Henderson, S.B., Hagan, M.E., Mende, S.B. et al. (2006). Control of equatorial ionospheric morphology by atmospheric tides. *Geophysical Research Letters*, 33(15).
[17] Withers, P. (2011). Attenuation of radio signals by the ionosphere of Mars: Theoretical development and application to MARSIS observations. *Radio Science*, 46(02), 1–6.
[18] Kuverova, V.V., Adamson, S.O., Berlin, A.A., Bychkov, V.L., Dmitriev, A.V., Dyakov, Y.A. et al. (2019). Chemical physics of D and E layers of the ionosphere. *Advances in Space Research*, 64(10), 1876–1886.
[19] Angling, M.J., Cannon, P.S. and Bradley, P. (2012). Ionospheric propagation. *In:* Propagation of Radiowaves, 3rd Edition (pp. 199–233). Institution of Engineering and Technology.
[20] Artru, J., Lognonné, P. and Blanc, E. (2001). Normal modes modelling of post-seismic ionospheric oscillations. *Geophysical Research Letters*, 28(4), 697–700.
[21] Lei, J., Burns, A.G., Tsugawa, T., Wang, W., Solomon, S.C. and Wiltberger, M. (2008). Observations and simulations of quasiperiodic ionospheric oscillations and large-scale traveling ionospheric disturbances

during the December 2006 geomagnetic storm. *Journal of Geophysical Research: Space Physics*, 113(A6).

[22] Rishbeth, H. and Garriott, O.K. (1964). Relationship between simultaneous geomagnetic and ionospheric oscillations. *Radio Sci. D.*, 68, 339–343.

[23] Rosenberg, T.J., Weatherwax, A.T., Detrick, D.L. and Lutz, L. (1999). High frequency active auroral research program imaging riometer diagnostic. *Radio Science*, 34(5), 1207–1215.

[24] Lance, C. and Eather, R. (1993). High Frequency Active Auroral Research Program (HAARP) Imager. Philips Lab Hanscom AFB MA.

[25] Pilipenko, V.A., Kozyreva, O.V., Engebretson, M.J., Detrick, D.L. and Samsonov, S.N. (2002). Dynamics of long-period magnetic activity and energetic particle precipitation during the May 15, 1997 storm. *Journal of Atmospheric and Solar-Terrestrial Physics*, 64(7), 831–843.

[26] James, M.K., Yeoman, T.K., Mager, P.N. and Klimushkin, D.Y. (2013). The spatio-temporal characteristics of ULF waves driven by substorm injected particles. *Journal of Geophysical Research: Space Physics*, 118(4), 1737–1749.

[27] Piddyachiy, D., Bell, T.F., Berthelier, J.J., Inan, U.S. and Parrot, M. (2011). DEMETER observations of the ionospheric trough over HAARP in relation to HF heating experiments. *Journal of Geophysical Research: Space Physics*, 116(A6).

[28] Smith, J.E. (1998). *HAARP: The Ultimate Weapon of the Conspiracy*. Adventures Unlimited Press.

[29] Su, F., Wang, W., Burns, A.G., Yue, X. and Zhu, F. (2015). The correlation between electron temperature and density in the topside ionosphere during 2006–2009. *Journal of Geophysical Research: Space Physics*, 120(12), 10–724.

[30] Korobko, Y. and Musa, M. (2014). *The Shifting Global Balance of Power: Perils of a World War and Preventive Measures*. Xlibris Corporation.

[31] Basu, S., MacKenzie, E. and Basu, S. (1988). Ionospheric constraints on VHF/UHF communications links during solar maximum and minimum periods. *Radio Science*, 23(03), 363–378.

[32] Grach, S., Sergeev, E., Shindin, A., Mishin, E. and Watkins, B. (2014). Artificial ionosphere layers for pumping-wave frequencies near the fourth electron gyroharmonic in experiments at the HAARP facility. *Doklady Physics*, 59(2).

[33] Li, M., Geng, J.W., Liu, H., Deng, Y., Wu, C., Peng, L.Y. et al. (2014). Classical-quantum correspondence for above-threshold ionization. *Physical Review Letters*, 112(11), 113002.

[34] Nozik, A.J., Beard, M.C., Luther, J.M., Law, M., Ellingson, R.J. and Johnson, J.C. (2010). Semiconductor quantum dots and quantum dot arrays and applications of multiple exciton generation to third-generation photovoltaic solar cells. *Chemical Reviews*, 110(11), 6873–6890.

[35] Sahu, A., Garg, A. and Dixit, A. (2020). A review on quantum dot sensitized solar cells: Past, present and future towards carrier multiplication with a possibility for higher efficiency. *Solar Energy*, 203, 210–239.

[36] Herath, T., Yan, L., Lee, S.K. and Li, W. (2012). Strong-field ionization rate depends on the sign of the magnetic quantum number. *Physical Review Letters*, 109(4), 043004.

[37] Jensen, R.V., Susskind, S.M. and Sanders, M.M. (1991). Chaotic ionization of highly excited hydrogen atoms: Comparison of classical and quantum theory with experiment. *Physics Reports*, 201(1), 1–56.

[38] Grach, S.M., Sergeev, E.N., Mishin, E.V. and Shindin, A.V. (2016). Dynamic properties of ionospheric plasma turbulence driven by high-power high-frequency radiowaves. *Physics-Uspekhi*, 59(11), 1091.

[39] Borisova, T.D., Blagoveshchenskaya, N.F., Kalishin, A.S., Rietveld, M.T., Yeoman, T.K. and Hägström, I. (2016). Modification of the high-latitude ionospheric F region by high-power HF radio waves at frequencies near the fifth and sixth electron gyroharmonics. *Radiophysics and Quantum Electronics*, 58, 561–585.

[40] Sergeev, E.N., Shindin, A.V., Grach, S.M., Milikh, G.M., Mishin, E.V., Bernhardt, P.A. et al. (2016). Exploring HF-induced ionospheric turbulence by Doppler sounding and stimulated electromagnetic emissions at the High Frequency Active Auroral Research Program heating facility. *Radio Science*, 51(7), 1118–1130.

[41] Bernhardt, P.A., Siefring, C.L., Briczinski, S.J., McCarrick, M. and Michell, R.G. (2016). Large ionospheric disturbances produced by the HAARP HF facility. *Radio Science*, 51(7), 1081–1093.

[42] Briczinski, S.J., Bernhardt, P.A., Siefring, C.L., Han, S.M., Pedersen, T.R. and Scales, W.A. (2015). "Twisted beam" SEE observations of ionospheric heating from HAARP. *Earth, Moon, and Planets*, 116, 55–66.

[43] Ivanov, I.A., Nam, C.H. and Kim, K.T. (2019). Entropy-based view of the strong field ionization process. *Journal of Physics B: Atomic, Molecular and Optical Physics*, 52(8), 085601.

[44] Briczinski, S.J., Bernhardt, P.A., Pedersen, T.R., Rodriguez, S. and SanAntonio, G. (2012). "Twisted Beam" SEE observations of ionospheric heating from HAARP. *In:* AGU Fall Meeting Abstracts (Vol. 2012, SA13A–2156).

[45] Haas, F. and Mahmood, S. (2022). Linear and nonlinear waves in quantum plasmas with arbitrary degeneracy of electrons. *Reviews of Modern Plasma Physics*, 6(1), 7.

[46] Briczinski, S.J., Bernhardt, P.A., Siefring, C.L., Han, S.M., Pedersen, T.R. and Scales, W.A. (2015). "Twisted beam" SEE observations of ionospheric heating from HAARP. *Earth, Moon, and Planets*, 116, 55–66.

[47] Majorosi, S., Benedict, M.G. and Czirják, A. (2017). Quantum entanglement in strong-field ionization. *Physical Review A*, 96(4), 043412.

[48] Maxwell, A.S., Madsen, L.B. and Lewenstein, M. (2022). Entanglement of orbital angular momentum in non-sequential double ionization. *Nature Communications*, 13(1), 4706.

[49] Rivera-Dean, J., Stammer, P., Maxwell, A.S., Lamprou, T., Tzallas, P., Lewenstein, M. and Ciappina, M.F. (2022). Light-matter entanglement after above-threshold ionization processes in atoms. *Physical Review A*, 106(6), 063705.

[50] Nishi, T., Lötstedt, E. and Yamanouchi, K. (2019). Entanglement and coherence in photoionization of h 2 by an ultrashort XUV laser pulse. *Physical Review A*, 100(1), 013421.

[51] Jenkins, P.P., Bentz, D.C., Barnds, J., Binz, C.R., Messenger, S.R., Warner, J.H. et al. (2013). Initial results from the TacSat-4 solar cell experiment. *In:* 2013 IEEE 39th Photovoltaic Specialists Conference (PVSC) (pp. 3108–3111). IEEE.

[52] Hoang, T., Armiger, W., Baldauff, R., Nguyen, B., Mahony, D. and Robinson, W. (2012). Performance of COMMx loop heat pipe on TacSat 4 spacecraft. *In:* 42nd International Conference on Environmental Systems (p. 3498).

[53] Duffey, T., Hurley, M., Raynor, B., Specht, T., Weldy, K., Bradley, E. et al. (2012). TACSAT-4 Early Flight Operations Including Lessons from Integration, Test, and Launch Processing. Naval Research Lab Washington DC.

[54] David, C.W. (1988). On the Legendre transformation and the Sackur-Tetrode equation. *Journal of Chemical Education*, 65(10), 876.

CHAPTER

5

Quantization of HAARP-Inducing Turbulence Plasma Extreme Heatwaves

5.1 Heatwave: What is it?

A heatwave is a stretch of exceptionally hot weather (Figure 5.1) that generally takes two days or longer. The temperature should be below the long-term average temperature of the region. For instance, a few summer days in Maine with a high of 95° might be regarded as a heatwave, but a few summer days in Death Valley with a high of 95° would hardly be note worthy. Meteorologists consider not only the actual temperatures but also the humidity and other elements that affect the apparent temperature when determining the start of a heatwave [1–3].

Figure 5.1: Extreme temperature is an indicator of a heatwave.

Figure 5.2: Death Valley has high temperatures.

Hence, a "heatwave" is a period of inordinately or exceptionally hot weather. However, less severe heatwaves can also be found in spring and early autumn. Extreme events are most frequent during the middle of summer. Slow-moving synoptic-scale events that permit hot air masses to develop continuously over vast areas for days and, in rare circumstances, weeks could be the cause of heatwaves. These occurrences have been linked to a variety of atmospheric factors, including an unusually effective high-pressure system over the continent and increased humidity, which reduces the difference in temperature between day and night [2, 4, 6].

5.2 Meteorology of Heatwave

A heatwave has no inclusive definition that is standardized. The World Meteorological Organization defines it as five or more days in succession, where the daily maximum temperature is 5 °C (9 °F) or higher than the average maximum temperature. Some nations have standards that they use. The U.S. The National Weather Service defines a heatwave as a period of "abnormally and uncomfortably hot and unusually humid weather" lasting two days or more, in contrast to the India Meteorological Department's requirement that temperatures rise 5–6 °C (9–10.8 °F) or more than the average temperature [1, 5, 7].

Particularly in the middle latitudes, where many people, including the very young, the very old, and those with health problems, might be more susceptible to heat stress, oppressively hot and humid air masses lingering over populated areas can result in many fatalities. The European heatwave of 2003, which claimed the lives of more than 30,000 people; the U.S. heatwave and drought of 1988, which claimed the lives of more than 4,000 people; the Russian heatwave of 2010 (which covered 1,036,000 square kilometers [400,000 square miles]); and the Indian heatwave of 2015 are notable recent episodes (which killed more than 2,500 people) [2–5].

The mechanisms for heat buildup include advection from lower latitudes, large-scale subsidence transporting higher potential temperature air from upper levels, surface heating, development of the diurnal mixed layer, and replacement from below by the new mixed layer for the succeeding day. Therefore, this evidence supports surface heating as the dominant potential contributing factor. This is because, with surface heating, the air rises and cools adiabatically; the warm air can then be replaced by cooler air from the lower levels of the atmosphere, resulting in a build-up of heat [1, 5, 7].

When the latent heat flux (evaporation) decreases owing to drought, the sensible heat component of the land surface radiation budget increases. For example, if the maximum temperature rise by 1–3 degrees Celsius somewhere, it is typically associated with high temperatures following periods of drought [1, 4, 7].

Accordingly, the morphology of severe heatwaves is characterized by diurnal variation in the boundary-layer depth. As the amount of surface sensible heat increased, the daytime mixing increased. Radiation balance studies of the 2003 European extreme heatwave noted that in situations where high surface temperatures result from dry soils, sensible heating into the shallow nocturnal boundary layer persists and contributes to anomalously high overnight temperatures. The depth of the daytime boundary layer is limited in these circumstances, which limits the mixing of cool air [1–4].

The accumulation of tropospheric heat is caused by daytime heating. A stationary or slow-moving Rossby wave pattern in the mid-troposphere is another characteristic of heatwaves (Figure 5.3). Polarward side of the anticyclonically curved mid-troposphere (500 hPa) jet stream. The "warm anticyclone" is a property of this supporting long-wave ridge. For instance, depending on the severity of the event, the anticyclone tilts westward from the surface to 700 or 500 hPa. Unlike stationary Rossby wave patterns during the cooler months, there is no typical split-jet structure to support a traditional "blocking pattern" [2–6].

The predictability of these Rossby waves and whether they are likely to become stationary have an impact on the seasonality, onset, and duration of heatwaves. Recent studies have linked an increase in the frequency of weather extremes in the Northern Hemisphere to a weakening of the

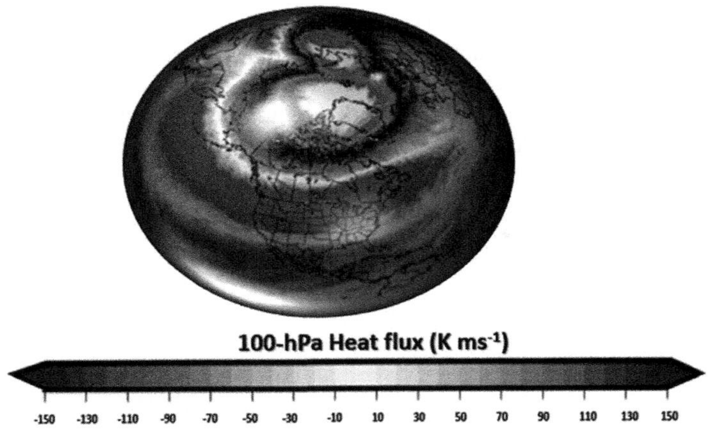

Figure 5.3: Rossby wave pattern is associated with the heat wave.

mid-tropospheric temperature gradients, which favors standing Rossby waves. Reduced sea ice is largely responsible for Arctic amplification or enhanced Arctic warming, which has significantly increased the heat flux into the polar sea [3, 5, 7].

In contrast to the Northern Hemisphere, the Antarctic continent maintains relatively stable surface flux. A stable wave train is excited by anomalous mid- to low-latitude heat fluxes, which appear to be the source of the Rossby wave structure in the Southern Hemisphere. There is a modality in the ocean temperature structure that produces statistically significant anomalous mid- to low-latitude meridional sea surface heat fluxes during intense heatwaves. Anomalously cool (warm) SST can be linked to long-wave ridge (trough) positions that are conducive to a steady, stationary Rossby wave train [1–4].

5.3 Mechanisms that Cause Heatwaves

When high pressure develops in the atmosphere (between 3,000 and 7,600 m) and persists over a region for several days to several weeks, heatwaves are generated. As the jet stream "follows the sun," this occurs frequently in the summer (in both the Northern and Southern Hemispheres). The high-pressure region was in the upper layers of the atmosphere on the equator side of the jet stream [3, 5, 8].

Weather patterns tended to change more slowly in summer than in winter. This upper level high-pressure also moves slowly. The high pressure causes the air to sink (subside) towards the surface, where it warms and dries adiabatically, preventing convection and cloud formation. Reduced cloud cover increases the amount of shortwave radiation reaching the surface. An increase in warming is caused by low surface pressure that causes surface winds from lower latitudes to bring warm air. Alternatively, the surface winds might originate from the warm interior of the continent and blow towards the coast, causing heatwaves there, or they might originate from a high elevation and blow towards a low elevation, causing subsidence and thereby enhancing adiabatic warming [4–7].

In this view, a high-pressure system from the Gulf of Mexico that becomes stationary just off the Atlantic Coast (commonly referred to as a Bermuda high) can cause a heatwave in the Eastern United States. While hot, dry air masses form over the desert southwest and northern Mexico, hot, humid air masses form over the Caribbean Sea and Gulf of Mexico. As a result of the southwest winds on the back side of the High continuing to pump hot, humid Gulf air northeastward, many of the Eastern States are experiencing hot and humid weather. The crucial question is: What contribution does the Bermuda High make to the development of such a dramatic heatwave? The Bermuda High is a region of elevated atmospheric pressure that develops over the western Atlantic Ocean during the summer and has a significant influence on North American weather patterns [1, 5, 8].

In this connection, the Azores High also called the Bermuda-Azores High or the North Atlantic (subtropical) High/Anticyclone, is a sizable subtropical, semi-permanent center of high atmospheric pressure situated in the Atlantic Ocean south of the Azores at Horse latitudes (Figure 5.4).The Icelandic Low serves as the other pole of the North Atlantic Oscillation. Large portions of North Africa, Southern Europe, and, to a lesser extent, eastern North America are affected by the system's effects on weather and climatic patterns [2, 4, 8].

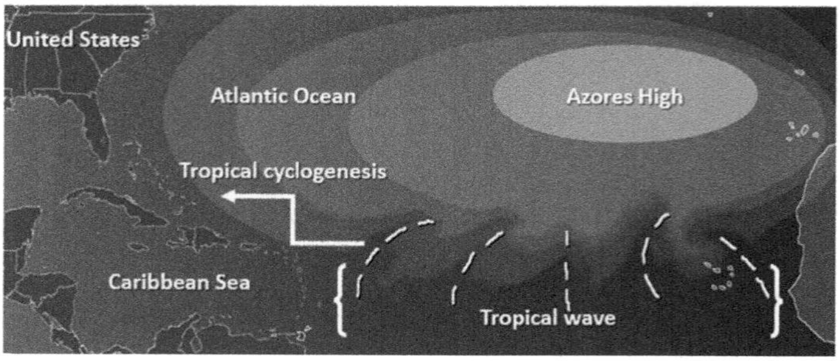

Figure 5.4: Azores high pattern.

The large-scale subsidence and sinking motion of the air in the system is responsible for the aridity of the Sahara Desert and summer drought in the Mediterranean Basin. A southwest flow of warm tropical air is produced by the high in its summer position, which is centered near Bermuda and moves towards the East Coast of the United States. The Azores-Bermuda High is strongest in the summer. At approximately 1024 mbar, the central pressure was constant (hPa) [3, 6, 8].

Therefore, Bergwind, which is created when high-pressure inland air and low-pressure offshore combine, can cause a heatwave in South Africa's Western Cape Province. The temperature increases by approximately 10 °C from the interior to the coast as the air warms and descends from the Karoo interior. In summer, temperatures can reach over 40 °C, and humidity is typically very low. One summer, a bergwind along the Eastern Cape coastline, caused the highest official temperatures ever recorded in South Africa (51.5 °C) [4–7].

In Europe, soil moisture may also play a role in the intensification of heatwaves. Low soil moisture triggers a variety of intricate feedback mechanisms, some of which can increase surface temperatures. Reduced-atmospheric evaporative cooling is one of the main mechanisms. Water evaporation uses energy, which immediately lowers the temperature. When the soil is extremely dry, sunlight heats the air without much or any cooling from moisture evaporating from the ground.

5.4 Heatwave Index

The question is now: What is the heat-wave index? The heat index can be investigated to fully understand the dynamic heat fluctuation, which is used to determine the pattern of heatwaves and evaluate the heatwave. The heat index also referred to as the "apparent temperature," is a gauge of how hot it feels when combined with relative humidity. To make it simpler for people to understand how hot or cold the weather is, it considers the combined effects of temperature and humidity [6, 7, 9].

In other words, the heat index (HI) is a measure of how hot it would feel if the relative humidity in the shade was another value. It combines air temperature and relative humidity in shaded areas to produce a temperature that is perceived by people as being hot and relative humidity in shaded areas to produce a temperature that is perceived by people as being hot. The consequence is additionally referred to as the "felt air temperature," the "apparent temperature," the "real feel," or "feels like." For instance, the heat index was 41 °C (106 °F) when the temperature was 32 °C (90 °F) and the relative humidity was 70% (Figure 5.5). It is customary to omit the humidity level at which the heat index feels. Only when the humidity is 21% is the heat index example; in this case, 41°C, feels like 41°C (Table 5.1) [1, 5, 8, 10].

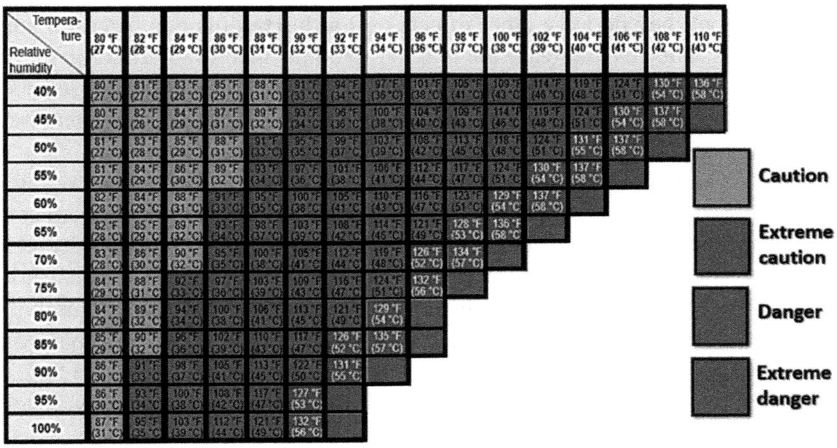

Figure 5.5: Heat index patterns.

Table 5.1: Heat index patterns

Temperature (°C)	Heat Index
27–32	Caution: Fatigue is possible with prolonged exposure and activity. Continuing activity could result in heat cramps.
32–41	Extreme caution: Heat cramps and heat exhaustion are possible. Continuing activity could result in heat stroke.
41–54	Danger: Heat cramps and heat exhaustion are likely; heat stroke is probable with continued activity.
Over 54	Extreme danger: Heat stroke is imminent.

However, supersaturation of air, which occurs when the air is more than 100% saturated with water, is an extreme condition where the heat index does not perform well. The heat index significantly underestimated the severity of intense heatwaves, such as those that hit Chicago in 1995. Other problems with the heat index include the assumption that the person is healthy and has easy access to water and shade, as well as a lack of precise humidity data in many geographic areas.

5.5 Heatwaves on Earth: Are Solar Storms Responsible?

Most of the energy emitted by solar flares and their associated coronal mass ejections is reflected in space by the Earth's magnetic field, even though they can bombard Earth's outermost atmosphere with enormous amounts of energy. Energy has no measurable impact on the surface temperature of our planet because it never reaches the surface.

From this perspective, it is reasonable to wonder if the heatwaves that hit the eastern and central United States in March 2012 and the spate of solar eruptions are connected. The warmth of the Earth originates from the energy of the sun. However, unlike visible and ultraviolet light, which can pass through Earth's atmosphere and warm the surface, most of the energy released by solar storms, such as those on March 8–10, is similar. Electrically charged particle bursts are launched into space by solar storms, where they collide with the thermosphere and the magnetic field of the Earth [11, 13].

Thus, the thermosphere was warmed by a stream of energetic particles. Coolants in the thermosphere carbon dioxide and nitrogen oxide absorb energy and then radiate heat back into space.

Layers of the atmosphere below the thermosphere receive a small amount of extra heat from solar flares, but this amount pales in comparison to the regular heating that these layers of the atmosphere already receive from incoming visible and ultraviolet sunlight. Nevertheless, heatwaves are not a result of solar flares, but they do have other effects on Earth. Hazards and beautiful auroras have negative consequences. They can increase the drag on satellites in a low-Earth orbit and rain down additional radiation on them. The increased electromagnetic activity caused by solar storms can also interfere with radio communication and power grids. Commercial jet passengers travelling on polar routes may be exposed to more electromagnetic radiation [11–13].

According to this understanding, longer-term variations in solar output may have an impact on Earth's climate, but transient solar explosions did not affect weather phenomena such as heatwaves in March 2012. The Maunder Minimum, which lasted for several decades in the second half of the seventeenth century, is thought by many scientists to have been the catalyst for the Little Ice Age, which chilled the Northern Hemisphere from roughly 1650 to 1850. However, multiple records reveal that the amount of energy the Earth receives from the Sun is relatively stable over the long term. Since the Scientific Revolution, astronomers have pointed to telescopes at the Sun, and recent research has been able to reconstruct solar activity over the past three centuries. Satellites have been monitoring the sun since 1978. They discovered that solar activity varies by approximately one-tenth of one percent on an approximately 11-year cycle [4, 10, 14].

Regarding the solar storm that occurred at the beginning of March 2012, although it released a sizable amount of energy, almost all of it radiated back into space, and only a small amount reached the lower atmosphere. In actuality, the additional energy from this storm is 100,000 times lower than the energy typically received at the Earth's surface. It is so small that you would not notice it, but even a small amount is enough to disrupt Earth's magnetic field and electromagnetic radiation. This disruption can cause satellite, power grid, and radio-communication system problems [1, 9, 11, 13].

5.6 Are the Ocean Heat Contents Blamed for the Earth's Heatwave?

What is the memory of the ocean? This is a critical question that needs to be answered. The heat content of the ocean is the primary component of both the ocean and the atmosphere. Ocean heat content is not only a measure of the energy balance between the ocean and the atmosphere, but it also acts as a long-term memory that stores heat over many decades. Therefore, ocean heat content is responsible for the dynamic behavior of the ocean. The ocean heat content is an important factor in global climate change, as it plays a crucial role in climate regulation. Increasing ocean temperatures increase the amount of energy available to fuel powerful storms, extreme weather events, and other climate-related disasters. Furthermore, rising ocean temperatures can cause sea levels to rise because of the thermal expansion of water. As a result, ocean heat content is not only an important factor in global climate change but can also have drastic consequences for coastal regions due to the increased risk of flooding and other climate-related disasters [5, 9, 13].

To accurately predict weather and climate change, it is crucial to understand the relationship between ocean heat content and atmospheric circulation. In the past, ships had to lower sensors or sample collectors in the ocean to measure the temperature. Only a portion of the vast oceans on the surface of the planet can have temperatures determined using this laborious method. Scientists have used satellites to gauge the height of the ocean surface to obtain global coverage. Sea surface heights can be used to estimate ocean temperature because warm water expands as it cools. Scientists now have a better understanding of ocean temperatures around the world owing to the use of satellites, which allow them to measure the temperature of a much larger area [10–14].

Scientists and engineers use a variety of in-situ temperature-sensing instruments to obtain a more comprehensive view of the heat content of the ocean at various depths. A fleet of more than

3,000 robotic "floats" that track ocean temperatures globally is one of these. These sensors, also referred to as "Argo floats," float through the ocean at various depths. They rise through the water every approximately 10 days, as directed by their programming, measuring the temperature (and salinity) along the way. A float that has reached the surface sends its location and other data to scientists via satellite before returning to depth. In other words, instruments for measuring ocean temperatures include Argo floats, expendable bathythermographs, and conductivity–temperature–depth (CTD) instruments. To measure temperatures in hard-to-reach places, seals have been fitted with an instrument (Figure 5.6).

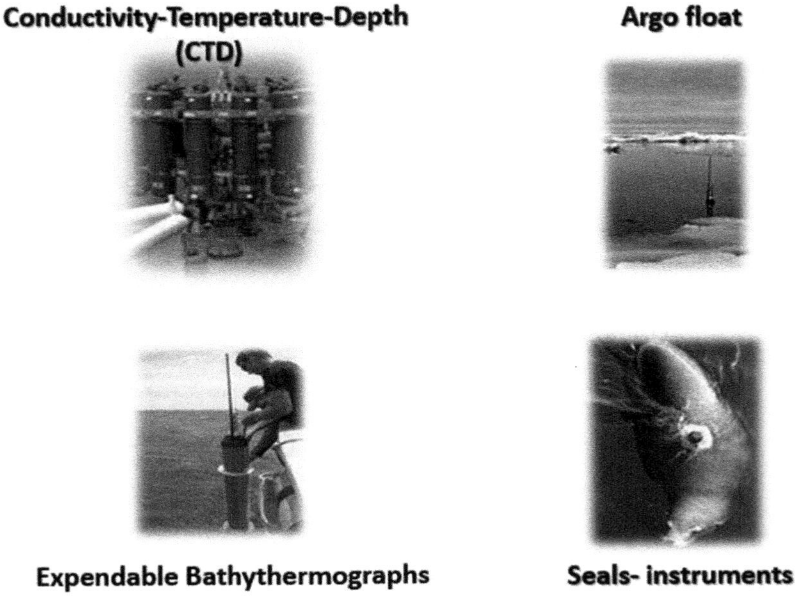

Figure 5.6: Instruments used for the heat measurements.

How does heat move? The ocean is the largest solar energy device on Earth. Water not only compensates for more than 70% of the surface of our planet, but it can also absorb a significant amount of heat without substantially increasing the temperature. The ocean plays a crucial role in maintaining Earth's climate system because of its extraordinary capacity to store and release heat over extended periods. Sunlight is the primary source of heat in the ocean. In addition, heat absorbed by clouds, water vapor, and greenhouse gases is released, and some of the heat energy enters the ocean. Heat is constantly transferred from warmer to cooler latitudes and deeper ocean levels by waves, tides, and currents.

In this sense, the heat absorbed by the ocean is transferred from one location to another but does not move away. The heat energy eventually returns to the rest of Earth's system by melting ice shelves, evaporating water, or directly reheating the atmosphere. As a result, decades after it is absorbed, the heat energy from the oceans can continue to warm the planet. The ocean's heat content increases if it absorbs more heat than it expels. Understanding and simulating the global climate requires an understanding of the amount of heat energy the oceans absorb and release [11–15].

From 1990 till 2022, the heat content anomaly has increased continuously to 16 × 10 Joules (Figure 5.7). Contingent on which research group's analysis you consult, the average heat-gain rates from 1993 to 2021 for depths between 0 and 700 meters (down to 0.4 miles) ranged from 0.37 (±0.05) to 0.44 (±0.12) Watts per square meter (Wm^{-2}). For depths between 700 and 2,000 meters, heat gain rates ranged from 0.17 (±0.03) to 0.29 (±0.03) Wm^{-2} (0.4–1.2 miles). Between September 1992 and January 2012, the estimated increase was 0.07 (±0.03) Wm^{-2} for depths between 2000 and 6000 meters (1.2 and 3.7 miles). The full-depth ocean heat gain rate ranges from 0.64 to 0.80 Wm^{-2} applied to the entire surface of the Earth (Figure 5.8) [12–17].

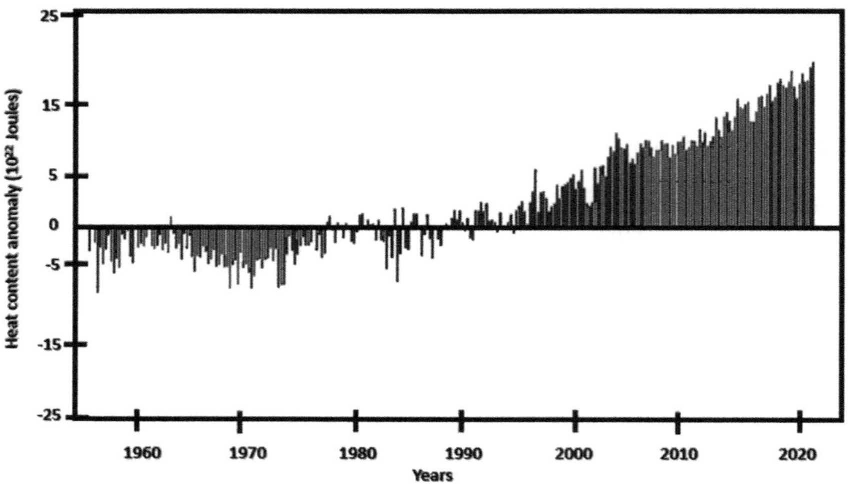

Figure 5.7: Heat content anomaly over eight decades.

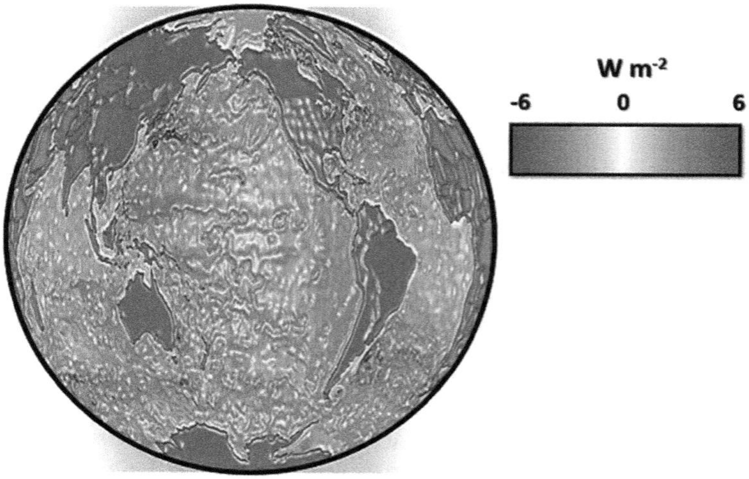

Figure 5.8: Average heat fluctuations from 1993 to 2021.

The following section examines the possibility of heatwaves occurring because of ionospheric plasma disruption caused by HAARP.

5.7 Quantized Marghany-Heatwaves

Conduction, convection, and radiation are the three main pathways through which heat is transferred. In contrast to convection, which results from the motion of molecules in gases or liquids, such as hot air rising, conduction pronounces heat transfer through direct contact with the materials. Empty spaces were not covered by these two. Sun warming the Earth is an excellent illustration of how radiation—heat transfer by electromagnetic waves—can occur in a vacuum. Radiation transmits heat in a vacuum, as opposed to conduction and convection, which require a medium to do so. As a result, radiation is the most efficient pathway for heat transfer from the sun to the Earth [18–20].

The novel theory and experience underpinning this natural phenomenon is a crucial question. According to one theory, the movement of air masses, which in turn creates wind patterns, is caused by the differential heating of the air near the equator and poles. However, other variables, such as temperature, humidity, and Earth's rotation, also have an impact on atmospheric pressure. According to this viewpoint, air masses are simply composed of tiny particles that are entangled with one another to create warm or cold air masses. To move or diffuse from one location to another, these particles require a specific quantity of energy [18, 20, 21].

Consequently, the quantity of heat can be expressed in the form of $k^2\pi^2 \frac{T}{3h}$, where k is the Boltzmann constant, T is the temperature, and h is the Planck constant. The most well-known instance of the electrical conductance quantum, denoted by the symbol $\frac{e^2}{h}$ (where e is the electron charge), is in the quantum Hall effect [20–22]. As a result, heat can be transferred over distances of the order of nanometers using quantum fluctuations, a type of churning of transient particles, and fields that occur even in empty spaces. In this regard, heat can travel through a vacuum, although it only has a significant impact over very short distances, as shown in the HAARP turbulence plasma in the previous chapter.

The air masses of the particles vibrated like drumheads because of this heat; the warmer the particles, the more ferociously they did so. The temperatures of the two air masses equalized when they were brought within a few hundred nanometers of one another, separated only by empty space, indicating that heat had been transferred between them.

Therefore, what is the significant role of quantum fluctuations in heat-wave transfer? It is reasonable to assume that the two air gas molecules are physically pushed together to form air masses by fleeting gravity waves in a vacuum. The gravitational Casimir effect is a gravitational force that attracts matter together. Detecting this would demonstrate the existence of gravitons, which are hypothetical particles that transmit gravitational forces. Gravity plays a significant role in daily life; however, studying it on exceedingly small scales is challenging. The author suggests that although gravity is an essential force, its study at the quantum level presents significant challenges. The detection of gravitons, which are the theoretical particles responsible for transmitting gravitational force, is currently beyond the scope of our technological capabilities.

In accordance with this perspective, gravity and the Casimir effect have long served as examples of how electromagnetism is fundamentally a quantum phenomenon. Two thin mirrors are spaced micrometres apart in a vacuum in the traditional experiment. The electromagnetic force's carrying particles, known as photons, constantly pop in and out of existence in a vacuum, even though it appears to be space. These transient photons behave similarly to waves and move the mirrors by reflecting off them [19, 21]. Some waves cannot pass between the mirrors because of their proximity. The electromagnetic waves on either side of the mirrors push in a direction that is stronger than the waves in between, and the mirrors begin to move in that direction (Figure 5.9).

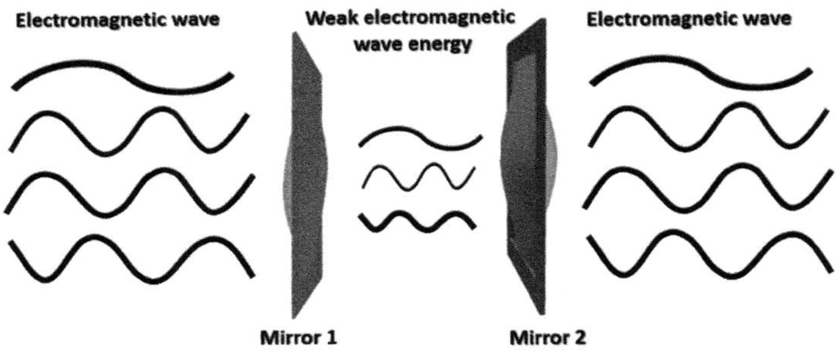

Figure 5.9: Simplification of Casimir effect in dual mirrors experimental.

However, the quantum fluctuation produces an attractive force between different atmospheric surface zones separated by different pressures, leading to a new type of heat transfer in the form of the wave propagation. In this sense, a heatwave can be defined as the stream flow of an ionized wave particle through the atmosphere. This wave-particle stream is responsible for the increased temperature of the atmosphere, which can be observed in the form of a heatwave. Therefore, the quantized heatwave energy can be simply described by:

$$E = \frac{1}{2}\hbar\omega. \tag{5.1}$$

Let us consider the distance between isobars is d_i and the heat flow would transfer in the form of the longitudinal standing wave for a temporary period through the atmosphere. This can be described in the following mathematical form:

$$E_n(x, y, z; t) = e^{-i\omega_n t} e^{ik_x x + ik_y y} \sin(k_n z), \tag{5.2}$$

here E_n stands for two perpendicular components of heatwave energy propagation in xy-plane with wave number of k_x and k_y, respectively. Therefore, the frequency of this quantized heatwave can be given by:

$$\omega_n = c\sqrt{k_x^2 + k_y^2 + \frac{n^2\pi^2}{d_i^2}}, \tag{5.3}$$

where n is an integer and the quantized heatwave spread over a sizable area as a function of the speed of light, as shown in Equation 5.3. Or, to put it another way, the electromagnetic radiation from the sun, which travels at the speed of light, is the primary source of atmospheric heat. This implies that variations in the sun's radiation can affect the amount of heat energy in a particular area. This may influence regional weather patterns and climatic conditions. Therefore, by using a regulator, equation 5.3 can compute the sum without encountering infinite values, making it valuable for modeling complex physical processes [18, 22]. However, it's important to carefully choose the regulator and set appropriate limits to ensure accurate results." In this perspective, the sum that regulates the quantized heatwave energy is cast as:

$$\langle E(t)\rangle = \frac{1}{2}\sum_n \hbar \mid \omega_n \mid e^{\left(-t^2|\omega_n|^2\right)} \tag{5.4}$$

A quantum heatwave propagates in the form of longitudinal energy waves. As opposed to a classical heatwave that spreads outwards from its source, a quantum heatwave remains localized in a specific region. This unique behavior is due to the wave-particle duality that characterizes quantum mechanics. In this view, high-pressure impacts cause warm air particles to spin downward while the pressure spins up. The hot air particles are trapped near the ground in the form of longitudinal waves because the high pressure acts as an entanglement lock, preventing it from rising. In other words, this heatwave is made up of warm gas molecules or particles that entangle with one another to form a heat-wave jet. This heat-wave jet moves along the ground because the particles are pushed by the high-pressure zone. This can cause heatwaves or temperature increases in affected areas. In this regard, heatwaves occur more frequently in summer when high pressure spreads over a region. High-pressure systems move slowly and can remain in a small-scale region for several days or weeks [2–5].

Considering the aforementioned viewpoint, a heatwave can be quantized as two repeated zones moving in the same direction as the particle-wave duality. Air particles form a compression zone with a high concentration of warm air particles when they are superimposed on one another. By contrast, as the air molecules spin upward, this compression zone lines up with the high-pressure zone. On the other hand, a lower concentration of air-warm particles forms an extension (or rarefaction) zone that joins the low-pressure zone as the air particles spin downward (Figure 5.10).

Therefore, when the air molecules spin up and down in the same direction, the compression and rarefaction zones converge. On the other hand, divergence occurs if air molecules spin up and down in different directions. From this perspective, the heatwave occurs and propagates owing to the compression and rarefaction of warm-air particles. This novel mechanism is called the quantized Marghany heatwave mechanism (Figure 5.10).

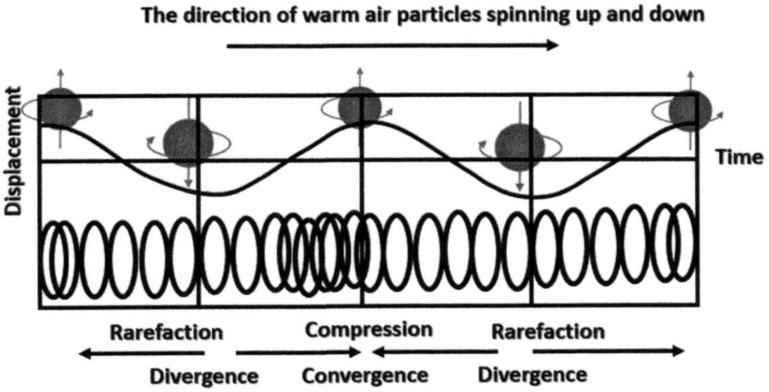

Figure 5.10: Simplification of the quantized Marghany heatwave.

The compression and rarefaction zones were split and joined together. The divergence and convergence zones are not superimposed in this quantized form. The compression and rarefaction zones in this quantized pattern are the causes of the heatwaves. Heatwaves are produced by a series of high- and low-pressure zones that are created because of the back-and-forth motion of air molecules. These zones of high and low pressure are also responsible for the formation of sound waves, which similarly travel through the air. The quantized pattern of the compression and rarefaction zones allows us to hear sounds and feel heatwaves.

The quantized Marghany heatwave concept can be defined as a temporary, random change in the amount of thermal energy in the atmosphere owing to quantum fluctuations. This occurs in discrete Gaussian form. This concept has significant implications for weather and climate modelling. Additionally, studying quantized Marghany heatwaves can provide valuable insights into the behaviour of the atmosphere at the quantum scale.

Consequently, the quantized Marghany heatwave is thermodynamically equivalent, demonstrating the quantum regime between different air masses. In other words, the quantized Marghany heatwave can be described as a quantum-thermal signature that accounts for wave-like properties or quantum coherence. This understanding can have potential applications in various fields such as renewable energy and quantum computing. They also highlight the intricate connection between quantum mechanics and thermodynamics, which can lead to further research and development. The discovery of quantized Marghany heatwaves presents exciting opportunities for scientific advancement. This knowledge not only impacts renewable energy and quantum computing but also sheds light on a deeper understanding of the relationship between quantum mechanics and atmospheric thermodynamics.

5.8 Satellite Data are Used for Heatwave Simulation

The Moderate Resolution Imaging Spectrometer (MODIS) is a sensor-board platform of Terra and Aqua. This sensor has 36 spectral bands at a moderate resolution (0.25–1 km). The spatial is 250 m for channels 1 and 2, 500 m for channels 3 to 7, and 1000 m for channels 8 to 36. The bandwidths of each spectral band were optimized for imaging specific surface and atmospheric features. A description of each spectral band is presented in Table 5.2 [23–25].

Table 5.2: MODIS spectral bands

Primary Use	**Band**	**Bandwidth**
Land/Cloud/Aerosols Boundaries	1	620–670 nm
	2	841–876 nm*
Land/Cloud/Aerosols Properties	3	459–479 nm
	4	545–565 nm
	5	1230–1250 nm
	6	1628–1652 nm*
	7	2105–2155 nm
	8	405–420 nm
	9	438–448 nm
	10	483–493 nm
Ocean Color Phytoplankton Biogeochemistry	11	526–536 nm
	12	546–556 nm
	13	662–672 nm
	14	673–683 nm
	15	743–753 nm
	16	862–877 nm*
Atmospheric Water Vapour	17	890–920 nm
	18	931–941 nm
	19	915–965 nm*
Surface/Cloud Temperature	20	3.660–3.840 µm
	21	3.929–3.989 µm
	22	3.929–3.989 µm
	23	4.020–4.080 µm*
Atmospheric Temperature	24	4.433–4.498 µm
	25	4.482–4.549 µm*
Cirrus Clouds Water Vapor	26	1.360–1.390 µm
	27	6.535–6.895 µm
	28	7.175–7.475 µm*
Cloud Properties	29	8.400–8.700 µm*
Ozone	30	9.580–9.880 µm*
Surface/Cloud Temperature	31	10.780–11.280 µm
	32	11.770–12.270 µm*
Cloud Top Attitude	33	13.185–13.485 µm
	34	13.485–13.785 µm
	35	13.785–14.085 µm
	36	14.085–14.385 µm

MODIS can view the surface of Earth every 1–2 days making observations over land, ocean surfaces, and clouds. MODIS is also able to measure various parameters regarding the surface of the Earth, such as surface temperature (land and ocean) and fire detection, ocean colour (sediment and phytoplankton), global vegetation maps and change detection, cloud characteristics, aerosol concentrations and properties, temperature and moisture soundings, snow cover, and characteristics and ocean currents [23, 25].

Satellite remote sensing observations from the NASA Moderate Resolution Imaging Spectroradiometer (MODIS) can be used to derive chlorophyll. MODIS is a moderate-resolution multispectral sensor currently flying on two NASA satellites, AQUA and TERRA. MODIS uses mid- and thermal-IR to measure surface emissivity. MODIS chlorophyll products were corrected for atmospheric disturbances. The AQUA satellite was designed to observe the Earth's water cycle. Its sun-synchronous polar orbit (south to north) passes over the equator in the afternoon (1:30 p.m.), and it acquires data for nearly the entire Earth each day. The TERRA satellite was designed to collect data relating to Earth's biogeochemical and energy systems. TERRA is also in a sun-synchronous orbit and crosses the equator at approximately 10:30 a.m. local time [24, 26].

This study is based on the standard AQUA and TERRA MODIS Level 3, 4.63 km gridded 3-day and 8-day composite products generated through surface emissions. The gridded data were generated by binning and averaging the nominal 1 km swath observations, yielding a ≈ 4 km gridded global data. The data were distributed in HDF-EOS format. The use of MODIS data is limited to cloud-free conditions. Because MODIS is installed on both TERRA and AQUA satellites, four data points are available per day. MODIS data can be freely obtained through direct broadcasting, which requires an X-band antenna and its control equipment, or from the NASA MODIS website. The data were distributed in the HDF5-EOS format [24–26].

5.9 Quantum Simulation of Heatwave

The evolution of a multipartite quantum system is used in quantum computation to effectively carry out computations that are thought to be insurmountable with a conventional computer. For instance, Shor's quantum algorithm can be used to efficiently decompose a large number into its prime factors, which is exponentially faster than with any known classical algorithm, provided several significant obstacles are overcome. This is due to the ability of quantum systems to exist in multiple states simultaneously, allowing for parallel processing and increased computational power. However, the fragility of quantum states and the difficulty in controlling them pose significant challenges to the development of practical quantum computers.

Considering brightening temperature as the keystone index of the heatwave in optical satellite data such as MODIS. In this sense, thermal bands of MODIS data (bands 31 and 32) are converted to radiance then brightness temperature is computed by [27–29]:

$$T_B = \frac{1.439 \times 10^4 \, k \, \mu\text{m}}{\lambda \times \ln\left(\frac{1.1911 \times 10^8 \, \text{WM}^{-2}\text{sr}^{-1} \left(\mu\text{m}^{-1}\right)^{-4}}{(\lambda_5 \times R) + 1} \right)} \tag{5.5}$$

Let us assume that the brightness temperature can be described by a two-dimensional Hilbert space $\mathcal{H}_2$. In this view, let us consider $|\psi\rangle_B \in \mathcal{H}_2^{\otimes n}$ the quantum brightness temperature and heatwave which can be described mathematically by:

$$|\psi\rangle_B = \sum_{i_1=0}^{1} \ldots\ldots \sum_{i_n=0}^{1} T_{B_i} \ldots\ldots i_n \, |i_i\rangle \otimes \ldots\ldots \otimes |i_n\rangle, \tag{5.6}$$

In equation 5.6, a single-spin orthonormal basis is denoted by $\{|0\rangle, |0\rangle \in \mathcal{H}_z\}$. As a result, it is effective to simulate any quantum evolution of an n-qubit system when its current state always factors into a set of states that each contain, at most, n qubits. This is a significant result in the field of quantum computing as it shows that certain quantum evolutions can be efficiently simulated, which has important implications for the development of quantum algorithms and applications. However, it is important to note that not all quantum evolutions can be efficiently simulated using this approach [30].

The author, therefore, demonstrates how to effectively simulate the pure-state quantum dynamics of n entangled qubits using a classical computer whenever only a small amount of entanglement is present in the system. Consequently, entanglement is a crucial resource for speeding up (pure state) quantum computation. More generally, the author establishes an upper bound on the maximum speedup that quantum computation can achieve in terms of the degree of entanglement. Quantum computations with mixed states also have an analogous upper bound but in terms of correlations (either classical or quantum). This upper bound is important for understanding the limitations of quantum computing and designing quantum algorithms that can take advantage of entanglement. The author's findings have implications for the development of practical quantum computers and the study of fundamental questions in quantum information theory.

To achieve this, let us consider a discretized evolution of n qubits, starting in state $|0\rangle^{\otimes n}$, by a succession of single-qubit and two-qubit gates with poly(n) (i.e., a number polynomial in n). However, the exploration of the circuit model can efficiently approximate any evolution of n qubits based on the most recent single-qubit and two-qubit Hamiltonians with arbitrary accuracy; therefore, the present results also apply to this more general scenario [30,32].

Let us subset the n qubit by sequences of the kernel's partition $\Im:\gamma$ to efficiently deal with big data such as MODIS, which covers a large-scale region of the western side of Europe. Consequently, the partition $\Im:\gamma$ can be mathematically expressed as:

$$|\psi\rangle_B = \sum_{\alpha=1}^{\ell_\Im} \lambda_\alpha \left|\vec{B}_\alpha^{[\Im]}\right\rangle \otimes \left|\vec{B}_\alpha^{[\gamma]}\right\rangle, \tag{5.7}$$

Equation 5.7 demonstrates that $\left|\vec{B}_\alpha^{[\Im]}\right\rangle$ and $\left|\vec{B}_\alpha^{[\gamma]}\right\rangle$ are dual eigenvectors with eigenvalue $|\lambda_\alpha|^2 > 0$ of the compact density matrix $\rho^{[\Im]}$ and $\rho^{[\gamma]}$ for qubit partition $\Im:\gamma$. Therefore, the entanglement between the qubits in $\Im$ and those in γ can be quantified by ℓ:

$$\ell \equiv \max_{\Im} \ell_\Im, \tag{5.8}$$

This notion of "smallness" for ℓ is important in the study of computational complexity, as it allows us to distinguish between problems that can be efficiently solved and those that are likely to be computationally intractable. Specifically, problems with small ℓ are often considered to be "easy" or "tractable," while those with large ℓ are typically much more difficult to solve. In this view, let us decompose the quantum pure state of MODIS brightness temperature $|\psi\rangle_B \in \mathcal{H}_2^{\otimes n}$ in terms of n-tensors $\left\{\mathcal{T}^{[\Im]}\right\}_{\Im=1}^{n}$ and $n-1$ vectors $\left\{\vec{\lambda}^{[\Im]}\right\}_{\Im=1}^{n}$, which is cast as:

$$|\psi\rangle_B \in \mathcal{H}_2^{\otimes n} \leftrightarrow \mathcal{T}^{[1]}\lambda^{[1]}\ldots\ldots\mathcal{T}^{[\ell]}\ldots\ldots\lambda^{[n-1]}\mathcal{T}^{[n]}. \tag{5.9}$$

Formally, a thermal expectation value $|\psi\rangle_B$ can be defined as follows for some observable O:

$$\langle O\rangle_B = \left\langle O e^{-B\mathcal{H}}\right\rangle_B \tag{5.10}$$

where $\langle\ \rangle$ is the standard quantum mechanical expectation value and B is the inverse temperature (this definition should be properly normalized, which we will ignore for now) [31–33]. We can expand the expectation value in the energy eigen basis to gain a more intuitive understanding of what this means.

$$\langle O\rangle_B = \langle n|O|n\rangle_B\, e^{BE_n} \tag{5.11}$$

What is meant by equation 5.11? e^{BE_n} penalizes higher energy contributions and the lower-energy states contribute more to the thermal expectation value. This means that in a system with multiple energies, such as HAARP, ionosphere turbulence plasma, and heatwave states, the high-energy states are more likely to be occupied and contribute more to the overall behavior of the interaction of different systems. In addition, this preference for high-energy states would create unstable heatwaves. In this understanding, the thermal expectation value is the expectation value of a system in the state:

$$\rho = e^{BE_n} \tag{5.12}$$

How are the temperature and quantum states related? We can define the temperature from the energy for any quantum state with an average energy E by solving:

$$\left| e^{BE_n} \right\rangle = \sum_{\alpha_1,\ldots,\alpha_n} \mathscr{T}_{\alpha_1}^{[1]i_1} \lambda_{\alpha_1}^{[\ell]} \ldots\ldots \mathscr{T}_{\alpha_{n-1}}^{[n]i_n}. \tag{5.13}$$

This process of swapping qubits can be time-consuming and may lead to errors in the computation. Therefore, it is important to carefully design the gate operations to minimize the number of swaps required. When we have two qubits, things become more interesting. If the two wavefunctions overlap significantly (an example of the superposition of waves), they will be coherent and entangled, that is, connected (another example of the superposition of waves) [30, 32, 34]. As a result, they act in a single mixed state:

$$\begin{aligned} |\psi\rangle &= (\alpha|1\rangle + \beta|0\rangle).|1\rangle \\ &= \alpha|1\rangle|1\rangle + \beta|0\rangle|1\rangle \end{aligned} \tag{5.14}$$

Because of their entangled origins, the two qubits enable the conscious modification of only one, known as the control qubit, whereas the target qubit is subject to computations that would cause it to change in a predetermined manner. In a quantum OR gate, for instance, the second qubit can be flipped (inverted) if the first qubit is 1 and remains unchanged in all other cases.

$$\begin{aligned} &\text{Quantum OR Gate} \\ &|\text{control\_bit, target\_bit}\rangle \\ &U|01\rangle \rightarrow |01\rangle \\ &U|00\rangle \rightarrow |00\rangle \\ &U|10\rangle \rightarrow |11\rangle \\ &U|11\rangle \rightarrow |10\rangle \end{aligned} \tag{5.15}$$

Quantum or gate is used to determine the heatwave as a function of the brightness temperature in the MODIS data. This application of the quantum or gate is particularly useful for monitoring and predicting heatwaves, which can have significant impacts on human health and the environment. By analyzing sequences of MODIS data, researchers can gain a better understanding of how heatwaves form and develop over time [30–35].

5.10 Quantized Heatwave Owing to Ionospheric Turbulence Plasma

This section explores the multi-channel representation for quantum images MCQI provides a more comprehensive and visually appealing representation of satellite images by utilizing the RGB channels, which allows for a more accurate interpretation of the data. This mathematical formula enables researchers to easily analyze and manipulate the colour information encoded in MCQI images [30,35]." In this sense, the brightness temperature in equations 5.6 to 5.8 can be encoded by MCQI algorithm as follows:

$$|\psi(n)\rangle_B = \frac{1}{2^{n+1}} \sum_{i=0}^{2^{2n}-1} \left| T_{RGB\alpha}^{i} \right\rangle_B \otimes |y\rangle|x\rangle, \tag{5.16}$$

where $\left|T^i_{RGB\alpha}\right\rangle_B$ is 2-D quantum brightness temperature retrieved in MODIS data along column y-axis and raw x-axis, respectively. The raw MODIS Terra data that were collected on June 20, 2022, along the western edge of Europe are presented in Figure 5.11. Consequently, Figure 5.12 reveals how the raw data is encoded into qubits. Following that, Figure 5.13 shows the atmospheric MODIS Terra temperature and quantum elevated brightness land surface, which are rooted on bands 24: 9.580–9.880 μm; 25: 4.482–4.549 μm; 31: 10.780–11.280 μm; and 32: 11.770–12.270 μm, respectively (Table 5.2). These figures provide valuable information for analyzing the land surface temperature and atmospheric conditions along the western edge of Europe on June 20, 2022. The data can be used to identify patterns and trends in the region's climate and inform future research on climate change [36–38].

Figure 5.11: MODIS Terra data used in simulation heatwave using MCQI.

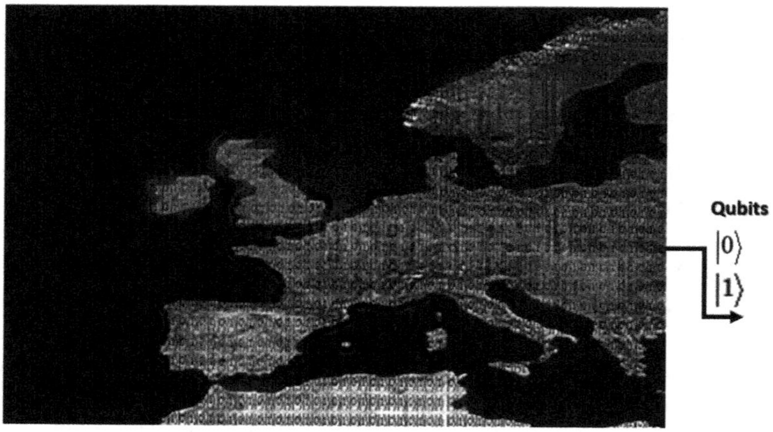

Figure 5.12: Converted MODIS Terra data into qubits.

In mid-June 2022, the Iberian Peninsula was already experiencing extreme heat, with a low 40s in Spain, Portugal, Morocco, and Algeria during the warmest afternoon hours. In addition, the heat was dispersed northeastward into France and England. South-central France is experiencing high- to mid-40 °C temperatures, and Spain is experiencing even higher temperatures that could

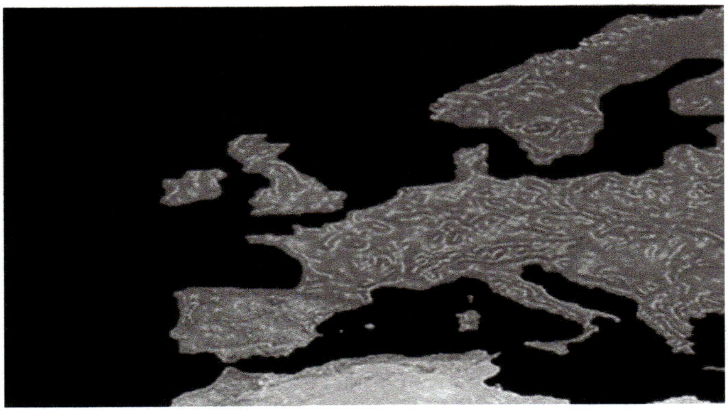

Figure 5.13: MODIS Terra brightness temperature is simulated by MCQI.

Figure 5.14: Heatwave in western Europe in mid-June 2022 using MCQI.

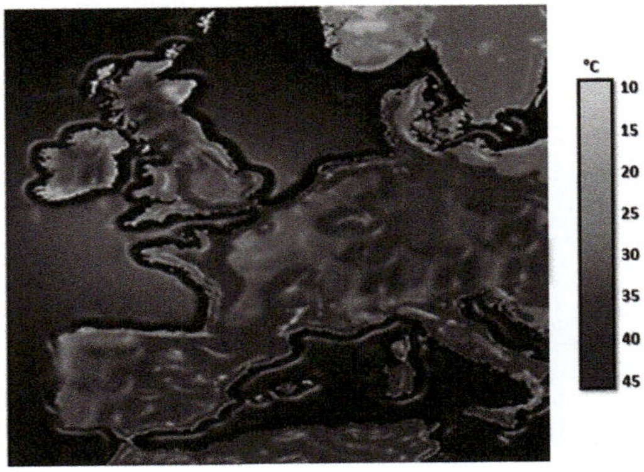

Figure 5.15: Heatwave in western Europe in July 2022 using MCQI.

reach 45 °C. For June and July 2022, both nations could break some old heat records (Figures 5.14 and 5.15). For the middle of June, July, and August 2022, these temperatures would be "extremely" high in western and southwestern Europe.

With a maximum temperature of 45 °C, these heatwaves will strengthen across Europe beginning in mid-June and peaking in August 2023. (Figure 5.16). Mediterranean drought is cited as the primary cause of the heatwave that has influenced the entirety of Europe. As shown in Figure 5.17, numerous places experienced extreme temperatures that were far above or below the July average. Due to the cut-off low west of Portugal, this heat was generated in the Sahara and moved north (Figure 5.17). This dynamic pattern is comparable to the June 2019 heatwave, which also saw record-breaking temperatures in several southern French regions.

After June 20th, the cut-off low stretched and moved inland, leading to severe thunderstorms. The high-pressure system stretched northeast, bringing the heat north of the Alps. A week later, there was yet another extended ridge structure (Figure 5.17), leading to heat extremes stretching from southern Italy to northern Norway.

In this view, around the low and high contour centres in the Northern Hemisphere, the air rotates anticlockwise and clockwise. The air rotates counterclockwise around high-contour centres and clockwise around low-contour centres in the southern hemisphere. The contours accurately depict the primary tropospheric waves that "control" our weather: low heights denote middle tropospheric cyclones and troughs, whereas high heights denote anticyclones and ridges. Because wind speeds are inversely correlated with the space between contours, densely packed contours indicate strong wind (Figure 5.17).

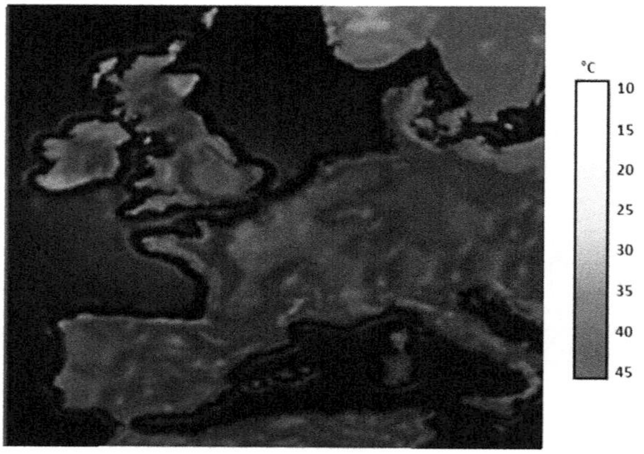

Figure 5.16: Heatwave in western Europe in August 2022 using MCQI.

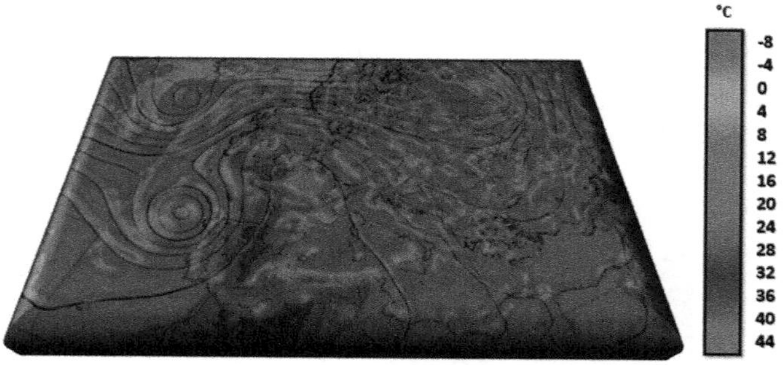

Figure 5.17: 500 hPa geopotential height (contours) and 850 hPa temperature on 18 July 2022.

In this context, 500 hPa geopotential height contours (in the dam) were measured at four dam intervals. These provide an overview of how high in the atmosphere one should ascend before the pressure drops to 500 hPa. This level is commonly 5.5 km above sea level (ASL). Because the underlying weather systems generally move in the same direction as the winds at the 500 hPa level, it is frequently referred to as the steering level. This information is useful for weather forecasting, as it helps meteorologists track the movement of weather systems and predict their future paths. Additionally, changes in the 500 hPa height contours can indicate shifts in the jet stream, which can have a significant impact on weather patterns. Nevertheless, there are circumstances, particularly during the winter months, where this approach is not appropriate. As it frequently lies close to the level of maximum moisture transport in the atmosphere, the 850 hPa level is also significant for predicting precipitation. Furthermore, wind strength and direction at this altitude can be useful indicators of future weather patterns and storm tracks [35, 38].

The very warm air mass over the larger Iberian Peninsula is visible in the Meteosat-11 simulated air mass using MCQI beginning in July, which demonstrates pulsating (with diurnal heating) red to dark red shades (Figure 5.18). Data from the SEVIRI water vapour: 6.2-µm, water vapour: 7.3-µm, infrared: 9.7-µm, and infrared: 10.8-µm channels were amalgamated to generate the Airmass RGB in this view. Since 1977, Meteosat satellites have been delivering vital information for weather forecasting. Currently, Meteosat-9, -10, and -11 are operated by EUMETSAT in geostationary orbit (36,000 km) above Europe, Africa, and the Indian Ocean.

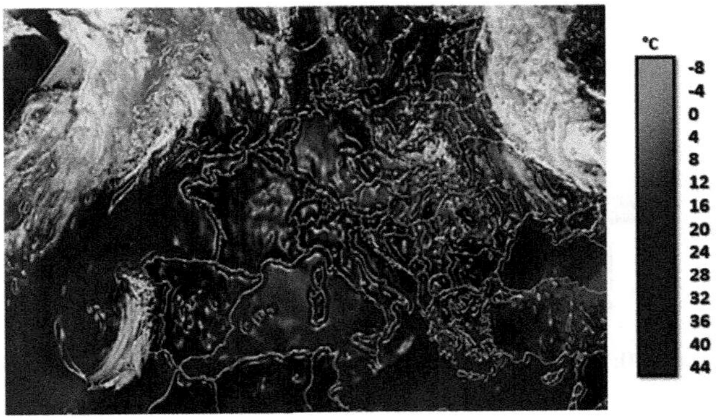

Figure 5.18: Heatwaves flux simulated by MCQI from Meteosat-11 on July 2022.

Intuitively, Meteosat-11, based on MODIS data, confirmed the occurrence of heatwaves. In this regard, maximum temperatures on July 18 exceeded 40 °C in western and southern France and central Spain, with western European nations recording temperatures of approximately 42 °C. The hottest day in the UK was July 19, when at least 34 stations broke the previous record of 38.7° set in 2019 at several stations in central and eastern England (provisional maximum of 40.3 °C, measured at Coningsby, Lincolnshire). The heatwaves in Europe have been attributed to a high-pressure system that has caused a domino effect of hot air being pushed up from the Sahara Desert towards the continent. Heatwaves have led to multiple wildfires and health warnings across Europe [35–38].

5.11 How Could HAARP Lead to Worldwide Heatwaves?

How can the HAARP heat the world? This is crucial. Europe appears to have evolved into a tropical zone. As the heat flux covers London's land surface, the land temperature simulated by the MCQI reached 44 °C (Figure 5.19). Therefore, as the temperature rises to 50 °C, Madrid's land surface appears worse than that of London. This occurred in Madrid because a large amount of heat was trapped along the side of the narrow road (Figure 5.20).In addition to western Europe, heatwaves spread to China, Saudi Arabia, the centre of Africa, and India. Figure 5.21 reveals that the mid-June 2022 heatwaves occurred at a high radio wave frequency of 9 MHz and an extremely high concentration of electronic density flux of 5.9×10^4 m^{-3} (Figure 5.21).

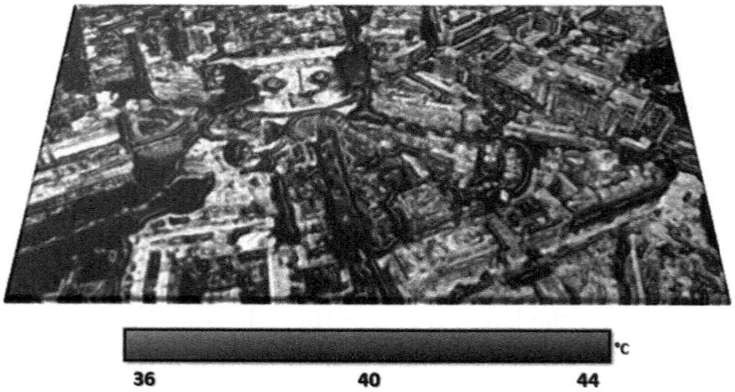

Figure 5.19: London infrastructure during the heatwave.

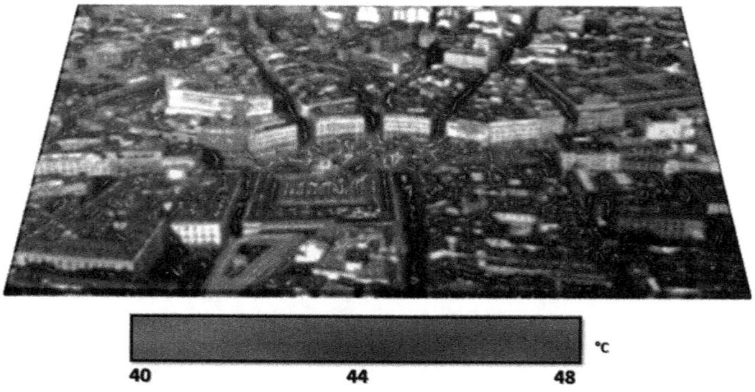

Figure 5.20: Land surface and buildings during the Madrid heatwave.

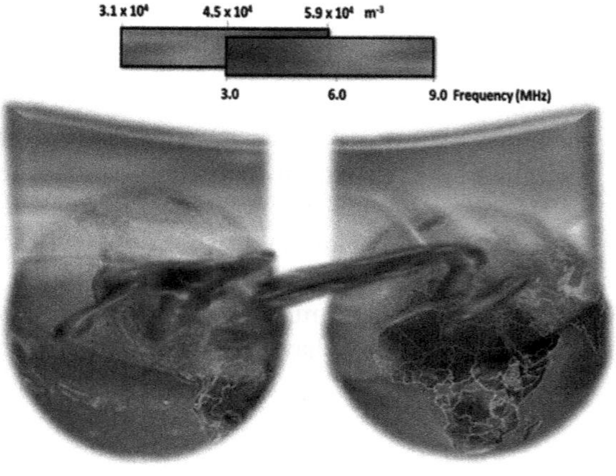

Figure 5.21: Heatwaves were brought on by high electronic density flux in the northern hemisphere.

Indeed, a heavy electrical discharge of 100 million volts leaves the air heated to approximately 30,000 °C in a few milliseconds. This generates shock waves, whose propagation is heard as thunder. Therefore, turbulent heatwave generation can also be obtained from plasma spectra. In other words, turbulence in plasma can generate heatwaves due to energy being transferred from low-frequency to high-frequency fluctuations. Therefore, the heatwave generated by the plasma spectra, combined with the turbulent environment, created an opportunity to increase the efficiency of thermal energy. This heatwave was initiated across North America and covered many parts of South America with a maximum temperature of 44 °C (Figure 5.22). Consequently, these heatwaves propagated across western Europe until India and China caused the highest rate of atmospheric temperatures of less than 45 °C (Figure 5.23). The HAARP turbulence plasma has been attributed to climate change and the increase in atmospheric temperatures [37–40].

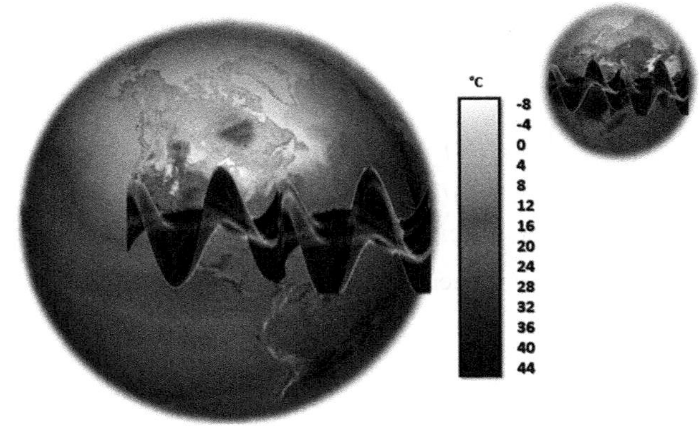

Figure 5.22: Extreme heatwave due to HAARP influences across North and South America.

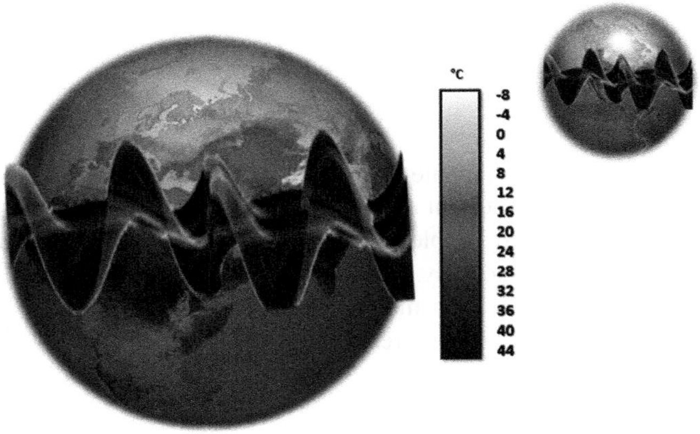

Figure 5.23: Extreme heatwave propagation across Europe to India and China due to HAARP impacts.

Thermal energy can cause droughts in many locations on the planet. The increased efficiency of thermal energy has had devastating consequences for people living in these places. During heatwaves, the sun's energy is trapped in the atmosphere owing to a combination of increased air temperature and decreased air pressure, creating an unstable situation for ground temperatures. These temperature spikes, compounded by a lack of precipitation and an increase in evaporation rates, have caused an extreme reduction in the amount of available water in many regions of the world. This could be implemented in the middle of 2022 to help alleviate drought in Europe, particularly in France. Such dire consequences can be seen in the recent droughts experienced by Europe and France, particularly during the summer of 2022 [40–42].

The results showed that the HAARP turbulence plasma that caused heatwaves had a high testing accuracy and performed well in terms of recall, precision, and F1 score. These metrics indicate that the model is reliable and can be used to predict heatwave events caused by HAARP turbulence plasma:

$$|\Psi(P)\rangle = \frac{(\psi|p|1\rangle + \psi|p|0\rangle)}{(\psi|p|1\rangle + \psi|p|0\rangle + \psi|n|0\rangle + \psi|n|1\rangle)} \tag{5.17}$$

The accuracy is examined from the perspective of quantum mechanics as a function of the wave function $|\Psi(P)\rangle$ of the likelihood that heatwaves would be created by the HAARP turbulence plasma. In this understanding, the HAARP signal is expected to have a positive probability impact $\psi|p|1\rangle$ based on the number of false-positive samples $\psi|n|1\rangle$. Consequently, the true negative HAARP signal probability $\psi|n|0\rangle$ is addressed along with the number of false-negative events $\psi|p|0\rangle$.

$$|\Psi(C)\rangle = \frac{(\psi|p|1\rangle)}{(\psi|p|1\rangle + \psi|n|1\rangle)} \tag{5.18}$$

Equation 5.18 demonstrates the concise wave function $|\Psi(C)\rangle$ of the impacts of the HAARP signal on developing heatwaves. Subsequently, the quantum memory function $|\Psi(M)\rangle$ then can be casted of:

$$|\Psi(M)\rangle = \frac{(\psi|p|1\rangle)}{(\psi|p|1\rangle + \psi|n|0\rangle)} \tag{5.19}$$

A combination of equations 5.18 and 5.19 can lead to a quantum score function equation as follows:

$$|\Psi(\Upsilon)\rangle = 2\otimes\left[\frac{|\Psi(C)\rangle \otimes |\Psi(M)\rangle}{(|\Psi(C)\rangle \oplus |\Psi(M)\rangle)}\right] \tag{5.20}$$

This equation can be used to calculate the probability of a particular measurement outcome in a quantum system and is an important tool in quantum information processing and quantum computing [44]. It is based on the principles of quantum mechanics and allows for the prediction of experimental results with high accuracy. In this sense, the quantum score function provided an unexpected precious effect of HAARP plasma on generating extreme heatwaves. The achievement accuracy of the quantum score function is presented as a higher true positive rate (TPR) in ROC curves of HAARP plasma and extreme heatwave. In terms of the ROC area, the impacts of HAARP plasma on forming extreme heatwaves is proved by 95% by using the quantum score function (Figure 5.24).

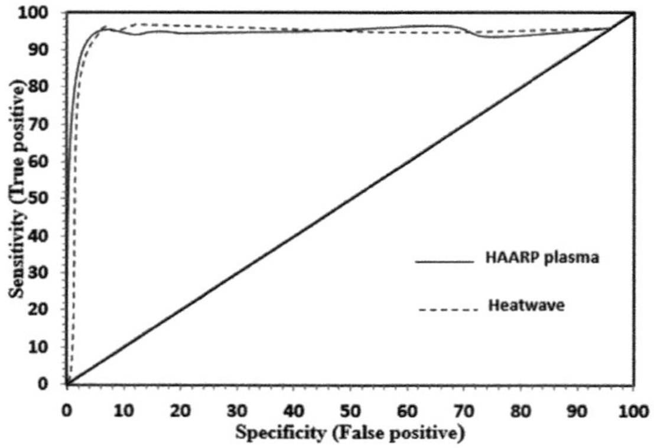

Figure 5.24: Accuracy of HAARP plasma induced extreme heatwave using quantum score function.

When readers ask the initially quoted question "How hot is plasma?" They frequently refer to the thermal phenomena occurring on the surface of the substrate being treated, rather than the actual temperature of the plasma. In industrial applications, the maximum temperatures are typically around 1 eV, although naturally occurring plasmas can reach temperatures of up to 106 eV (1 eV = 11600 K). As a result, plasmas are in a high-energy state, with temperatures determined by the energies of their species (neutral atoms, electrons, and ions), and are significantly influenced by the level of ionization in each plasma. As a result, it is possible to divide plasmas into two main groups based on their temperature: thermal plasmas and nonthermal plasmas. Thermal plasmas, which occur at temperatures ranging from 1 eV to 106 eV, contain neutral atoms, electrons, and ions that have an equal temperature. On the other hand, non-thermal plasmas occur at temperatures above 106 eV and contain species that have different temperatures. For example, nonthermal plasmas often contain electrons that are much hotter than neutral atoms and ions. Nonthermal plasmas are more complex than thermal plasmas and often involve additional energy sources, such as an externally applied electric field. These additional energy sources can cause an increase in the electron temperature beyond that of neutral atoms and ions, leading to a situation where non-thermal plasmas become electrically conductive. Such a situation can cause an electric current to flow through the plasma, leading to further heating and acceleration of the electrons. This process can create an unstable state in which the plasma temperature rises, leading to instabilities, such as shocks and turbulence. The instability associated with nonthermal plasmas can produce a wide variety of interesting effects, such as changes in electrical resistance, increased ionization of the gas, and waves that can propagate through the plasma. Such instability can cause several phenomena, including the generation of Alfvén waves, which are oscillations in the magnetic field caused by the interaction between charged particles and neutral atoms. These Alfvén waves can cause further heating of the plasma and induce changes in its electrical properties, leading to phenomena such as turbulent flows and shocks. This process can create a highly energetic and turbulent state, making it difficult to predict the behavior of the plasma accurately. This phenomenon is known as "plasma turbulence," and has been studied for many decades to better understand how plasmas interact with their surroundings. Plasma turbulence is a complex process that affects many physical and chemical properties of plasma, such as electrical conductivity, heating rates, and transport of charged particles [39–41].

When plasma is fully ionized and all species are at the same temperature, as in the solar corona or fusion plasma, we refer to it as being "thermal." These plasmas contain ions, neutral particles, and electrons of various tensile strengths. As a result, while most gas particles become significantly less hot or remain at room temperature, electrons can reach temperatures of 10,000 K. However, static measurement of the plasma flame during operation with dry compressed air as the plasma gas would produce temperatures below 1000 °C. The plasma flame, which is the most noticeable portion of the plasma jet, is also referred to as "remote plasma" and is typically where the substrate surfaces are treated [39, 41, 43].

In other words, rapid ionization of ionosphere plasma particles is achieved in a few femtoseconds by the quantum entanglement of HAARP plasma wave particles. This ionization process leads to the generation of an electric field, which produces a transfer of energy from the HAARP to the ionosphere. This energy transfer is further intensified by the powerful ionosphere-magnetosphere coupling, which is believed to cause an increase in the temperature of the ionosphere layer. This increased temperature, in turn, causes an increase in the electrical conductivity of the ionosphere layer, leading to more energy being applied to the magnetosphere. This process can be used to manipulate the atmosphere to enhance radio communication; theoretically, extreme weather phenomena can be created by focusing large amounts of energy on certain areas of the atmosphere [40, 43].

Ionospheric scintillation can be observed at all frequencies and across all latitudes, although the intensity of the signal is highly dependent on frequency and location. For example, during the day, scintillation is more intense near the poles because of the increased density of the plasma structures in these regions. At night, however, the intensity of scintillation is greater at lower latitudes because of the increased ionization of the plasma near the equator. This phenomenon is a result of the Earth's magnetic field, which causes electrons and protons to be concentrated regions of the ionosphere. This concentration of electrons and protons creates a higher level of scintillation in the corresponding regions, resulting in greater signal intensity. Thus, the intensity of scintillation varies based on the region of the ionosphere and can be more intense in areas with higher plasma density and increased ionization owing to the presence of the Earth's magnetic field. The Earth's magnetic field is therefore an important factor in the phenomenon of scintillation, as it provides an opportunity for electrons and protons to be concentrated in certain areas of the ionosphere, leading to higher levels of scintillation and an increased signal [39–43].

This chapter has illustrated a novel method for producing extreme heatwaves using HAARP turbulence plasma, which was discussed in the fourth chapter. It is undeniably true that during the summer of 2022, HAARP typically caused extreme heatwaves that spread from North America to China through Europe. The novel theory presented and mathematically proven by the author, known as the quantized Marghany heatwave mechanism in Section 5.7, demonstrates the comprehensive understanding of heatwave generation by HAARP. This new understanding of HAARP's role in heatwave generation could have significant implications for climate change research and mitigation efforts. Further studies are needed to fully explore the potential impacts of this novel method on global weather patterns and the environment.

References

[1] Oliver, E.C., Benthuysen, J.A., Darmaraki, S., Donat, M.G., Hobday, A.J., Holbrook, N.J. et al. (2021). Marine heatwaves. *Annual Review of Marine Science*, 13, 313–342.

[2] Holbrook, N.J., Sen Gupta, A., Oliver, E.C., Hobday, A.J., Benthuysen, J.A., Scannell, H.A. et al. (2020). Keeping pace with marine heatwaves. *Nature Reviews Earth & Environment*, 1(9), 482–493.

[3] Oliver, E.C., Donat, M.G., Burrows, M.T., Moore, P.J., Smale, D.A., Alexander, L.V. et al. (2018). Longer and more frequent marine heatwaves over the past century. *Nature Communications*, 9(1), 1–2.

[4] Van Oldenborgh, G.J., Krikken, F., Lewis, S., Leach, N.J., Lehner, F., Saunders, K.R. et al. (2021). Attribution of the Australian bushfire risk to anthropogenic climate change. *Natural Hazards and Earth System Sciences*, 21(3), 941–960.

[5] Straughan, B. (2011). *Heatwaves*. Springer Science & Business Media.

[6] Meehl, G.A. and Tebaldi C. (2004). More intense, more frequent, and longer lasting heatwaves in the 21st century. *Science*, 305(5686), 994–997.

[7] Perkins, S.E. and Alexander, L.V. (2013). On the measurement of heatwaves. *Journal of Climate*, 26(13), 4500–4517.

[8] Gershunov, A. and Guirguis, K. (2012). California heatwaves in the present and future. *Geophysical Research Letters*, 39(18).

[9] Russo, S., Dosio, A., Graversen, R.G., Sillmann, J., Carrao, H., Dunbar, M.B. et al. (2014). Magnitude of extreme heatwaves in present climate and their projection in a warming world. *Journal of Geophysical Research: Atmospheres*, 119(22), 12–500.

[10] Kuchcik, M.A. (2006). Defining heatwaves – Different approaches. *Geographia Polonica*, 79(2), 47.

[11] Foster, G. and Rahmstorf, S. (2011).Global temperature evolution 1979–2010. *Environmental Research Letters*, 6(4), 044022.

[12] Kulmala, M., Riipinen, I., Nieminen, T., Hulkkonen, M., Sogacheva, L., Manninen, H.E. et al. (2010). Atmospheric data over a solar cycle: No connection between galactic cosmic rays and new particle formation. *Atmospheric Chemistry and Physics*, 10(4), 1885–1898.

[13] White, R.H., Anderson, S., Booth, J.F., Braich, G., Draeger, C., Fei, C. et al. (2023). The unprecedented pacific northwest heatwave of June 2021. *Nature Communications*, 14(1), 727.

[14] Hesketh, A.V. and Harley, C.D. (2023). Extreme heatwave drives topography-dependent patterns of mortality in a bed-forming intertidal barnacle, with implications for associated community structure. *Global Change Biology*, 29(1), 165–178.

[15] Pearce, A.F. and Feng. M. (2013). The rise and fall of the "marine heatwave" off Western Australia during the summer of 2010/2011. *Journal of Marine Systems*, 111, 139–156.

[16] Jung, S., Kim, Y.J., Park, S. and Im, J. (2020). Prediction of sea surface temperature and detection of ocean heatwave in the South Sea of Korea using time-series deep-learning approaches. *Korean Journal of Remote Sensing*, 36(5_3), 1077–1093.

[17] Bryden, H.L. and Imawaki, S. (2001). Ocean heat transport. *International Geophysics*, 77, 455–474.

[18] Wang, J.S. (2007). Quantum thermal transport from classical molecular dynamics. *Physical Review Letters*, 99(16), 160601.

[19] Wang, J.S., Wang, J. and Lü, J.T. (2008). Quantum thermal transport in nanostructures. *The European Physical Journal B*, 62, 381–404.

[20] Kane, C.L. and Fisher, M.P. (1997). Quantized thermal transport in the fractional quantum Hall effect. *Physical Review B*, 55(23), 15832.

[21] Vinjanampathy, S. and Anders, J. (2016). Quantum thermodynamics. *Contemporary Physics*, 57(4), 545–579.

[22] Kosloff, R. (2013). Quantum thermodynamics: A dynamical viewpoint. *Entropy*, 15(6), 2100–2128.

[23] Agathangelidis, I., Cartalis, C., Polydoros, A., Mavrakou, T. and Philippopoulos, K. (2022). Can satellite-based thermal anomalies be indicative of heatwaves? An investigation for MODIS land surface temperatures in the Mediterranean region. *Remote Sensing*, 14(13), 3139.

[24] Arshad, S., Kazmi, J.H., Shaikh, S., Fatima, M., Faheem, Z., Asif, M. and Arshad, W. (2022). Geospatial assessment of early summer heatwaves, droughts, and their relationship with vegetation and soil moisture in the arid region of Southern Punjab, Pakistan. *Journal of Water and Climate Change*, 13(11), 4105–4129.

[25] Teuling, A.J., Seneviratne, S.I., Stöckli, R., Reichstein, M., Moors, E., Ciais, P. et al. (2010). Contrasting response of European forest and grassland energy exchange to heatwaves. *Nature Geoscience*, 3(10), 722–727.

[26] Chen, W., Yao, T., Zhang, G., Woolway, R.I., Yang, W., Xu, F. and Zhou T. (2023). Glacier surface heatwaves over the Tibetan Plateau. *Geophysical Research Letters*, 50(6), e2022GL101115.

[27] Song, C. and Jia, L. and Menenti, M. (2013). Retrieving high-resolution surface soil moisture by downscaling AMSR-E brightness temperature using MODIS LST and NDVI data. *IEEE Journal of Selected Topics in Applied Earth Observations and Remote Sensing*, 7(3), 935–942.

[28] Yue, H., He, C., Zhao, Y., Ma, Q. and Zhang, Q. (2017). The brightness temperature adjusted dust index: An improved approach to detect dust storms using MODIS imagery. *International Journal of Applied Earth Observation and Geoinformation*, 57, 166–176.

[29] Gong, A., Li, J. and Chen, Y. (2021). A spatio-temporal brightness temperature prediction method for forest fire detection with MODIS data: A case study in San Diego. *Remote Sensing*, 13(15), 2900.

[30] Marghany, M. (2022). *Remote Sensing and Image Processing in Mineralogy*. CRC Press.

[31] Su, J., Guo, X., Liu, C. and Li, L. (2020). A new trend of quantum image representations. *IEEE Access*, 8, 214520–37.

[32] Haque, M.E., Paul, M., Ulhaq, A. and Debnath, T. (2023). Advanced quantum image representation and compression using a DCT-EFRQI approach. *Scientific Reports*, 13(1), 4129.

[33] Lisnichenko, M. and Protasov, S. (2023). Quantum image representation: A review. Quantum Machine Intelligence, 5(1), 2.

[34] Liu, K., Wei, Y. and Li, H.S. (2023). The quantum realization of image linear gray enhancement. *Quantum Machine Intelligence*, 5(1), 15.

[35] Felsche, E., Böhnisch, A. and Ludwig, R. (2023). Inter-seasonal connection of typical European heatwave patterns to soil moisture. *NPJ Climate and Atmospheric Science*, 6(1), 1–1.

[36] Domeisen, D.I., Eltahir, E.A., Fischer, E.M., Knutti, R., Perkins-Kirkpatrick, S.E., Schär, C. et al. (2023). Prediction and projection of heatwaves. *Nature Reviews Earth & Environment*, 4(1), 36–50.

[37] Rouges, E., Ferranti, L., Kantz, H. and Pappenberger, F. (2023). European heatwaves: Link to large-scale circulation patterns and intraseasonal drivers. *International Journal of Climatology*. 43, 3189–3209.

[38] Faranda, D., Messori, G., Jezequel, A., Vrac, M. and Yiou, P. (2023). Atmospheric circulation compounds anthropogenic warming and impacts of climate extremes in Europe. *Proceedings of the National Academy of Sciences*, 120(13), e2214525120.

[39] Collins, K.V. (2023). Development of a scalable, low-cost meta-instrument for distributed observations of ionospheric variability. Doctoral dissertation, Case Western Reserve University.

[40] Kalishin, A.S., Blagoveshchenskaya, N.F., Borisova, T.D. and Egorov, I.M. (2022). Comparison of spectral features of narrowband stimulated electromagnetic emission excited by an extraordinary pump wave in the high-latitude ionospheric F region at frequencies below and above the F2 layer X-component critical frequency. *Russian Meteorology and Hydrology*, 47(12), 921–930.

[41] Sivokon', V. and Cherneva, N. (2022). Dynamic characteristics of field-aligned ionospheric irregularities under the conditions of ionosphere modification. *In:* Problems of Geocosmos–2020: Proceedings of the XIII International Conference and School (pp. 473–479). Cham: Springer International Publishing.

[42] Liu, M., Zhou, C. and Feng, T. (2022). Electron acceleration by Langmuir turbulence in ionospheric heating. *Earth and Planetary Physics*, 6(6), 529–535.

[43] Petrukovich, A., Mogilevskii, M., Kozlov, I., Pulinetz, S., Dobrolenskii, Y., Anufreichik, K. et al. (2021). Monitoring of physical processes in upper atmosphere, ionosphere and magnetosphere in ionosphere space missions. *In:* EPJ Web of Conferences (Vol. 254, p. 02010). EDP Sciences.

[44] Wu, S.L., Chan, J., Guan, W., Sun, S., Wang, A., Zhou, C. et al. (2021). Application of quantum machine learning using the quantum variational classifier method to high energy physics analysis at the LHC on IBM quantum computer simulator and hardware with 10 qubits. *Journal of Physics G: Nuclear and Particle Physics*, 48(12), 125003.

CHAPTER

6

Utilizing Quantum Spectral Energy Signatures for Identifying Cloud Patterns

In recent years, the exploration of quantum principles in various scientific domains has yielded novel insights and approaches. Within the realm of atmospheric sciences and remote sensing, the application of quantum spectral energy signatures has emerged as a promising avenue for detecting and identifying intricate cloud patterns. By harnessing the unique characteristics of quantum interactions, this chapter seeks to unlock new dimensions of cloud analysis, offering the potential to enhance our understanding of cloud dynamics and their implications for weather prediction, climate studies, and beyond. This chapter delves into the utilization of quantum spectral energy signatures as a tool for unravelling the complexities of cloud patterns and their underlying physical phenomena.

6.1 Types of Clouds

Until the early nineteenth century, clouds were not formally distinguished and categorized despite ancient astronomers nominating the major stellar constellations approximately 2,000 years ago. Although the French environmentalist J-B Lamarck (1744–1829) insinuated the initial classification for identifying the sort of clouds in 1802; his work did not receive an inclusive ovation. One year later, an English naturalist pronounced as, Luke Howard established a cloud classification approach that achieved universal recognition. Essentially, Howard's pioneering scheme devoted Latin words to designate clouds as they seem to a ground viewer. In this sense, a sheetlike cloud *stratus* (Latin for "layer"); a puffy cloud *cumulus* ("heap"); a wispy cloud *cirrus* ("hair's curl"); and a rain cloud *nimbus* ("violent rain") are pronounced universally after him. In Howard's scheme, these were the four rudimentary cloud varieties. Other clouds could be described by combining the basic types. For instance, nimbostratus is a rain cloud that demonstrates layering, although cumulonimbus is a rain cloud having pontificated upright expansion [1, 8, 10].

Based on their visual attributes, such as height in the atmosphere, shape, and behavior, clouds are categorized. Based on their altitude, the World Meteorological Organization (WMO) divides clouds into four main groups: (i) low clouds; (ii) middle clouds; (iii) high clouds; and (iv) clouds with Extensive vertical development (Table 6.1) [12]. In this view, low clouds form at low altitudes, generally below 6,500 feet (2,000 meters). Low clouds are typically composed of water droplets and can sometimes contain supercooled water. Stratus clouds and cumulus clouds are the common sorts of low cloud types. Stratus clouds appear as a uniform gray layer with a smooth, featureless base. They often bring overcast skies and can lead to drizzle or light rain. Thus cumulus clouds are puffy, white clouds with a flat base. They form through convection and are generally associated with fair weather.

Middle clouds, therefore, form at intermediate altitudes, between 6,500 and 20,000 feet (2,000 to 6,000 meters) (Table 6.2). These clouds can consist of a mix of water droplets and ice crystals. Common middle cloud types include (i) altocumulus clouds; and (ii) altostratus clouds

Figure 6.1: Low cloud types.

(Figure 6.2). In this perspective, altocumulus clouds appear as white or gray patches with a wavy or globular appearance. They can indicate changes in weather patterns. Consequently, altostratus clouds are gray or blue-gray clouds that cover the sky with a uniform layer. They often precede rain or snowfall [13].

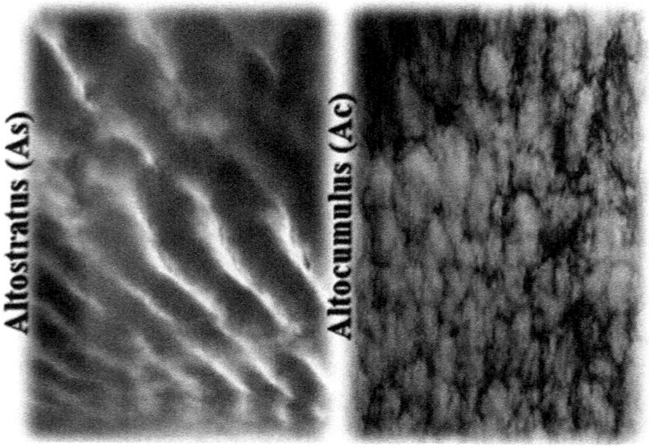

Figure 6.2: Middle cloud types.

High clouds, on the other hand, form at high altitudes, usually above 20,000 feet (6,000 meters) (Table 6.2). These clouds are composed mostly of ice crystals. Common high cloud types are mainly known as cirrus clouds; and cirrostratus clouds. Cirrostratus clouds are thin, whitish clouds that cover the sky and can produce halos around the sun or moon [11–13]. Therefore, cirrus clouds are thin, wispy clouds with a feathery appearance (Figure 6.3). They are made up of ice crystals and often indicate fair weather.

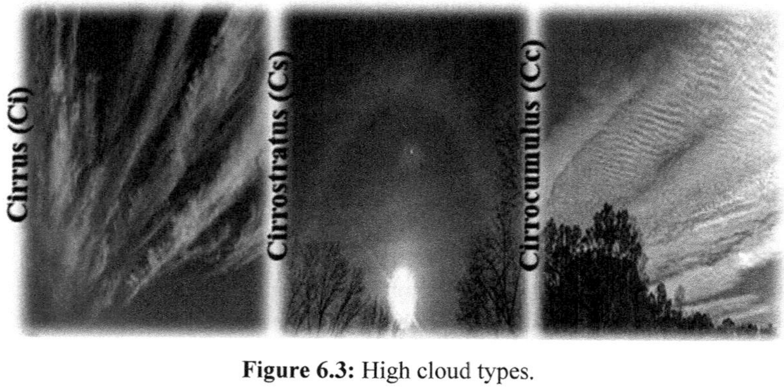

Figure 6.3: High cloud types.

Lastly, clouds with extensive vertical development (Figure 6.4) are associated with strong convective activity and can extend through multiple cloud levels (Figure 6.5). The most prominent cloud type in this category is cumulonimbus clouds. In this view, cumulonimbus clouds are large and vertically developed clouds associated with thunderstorms (Figure 6.6). They can reach great heights and produce heavy rain, lightning, and even severe weather. Understanding cloud classification is essential for meteorologists, weather forecasting, and aviation, as it helps in predicting weather patterns and associated phenomena [12–14].

Figure 6.4: Clouds with vertical development cloud types.

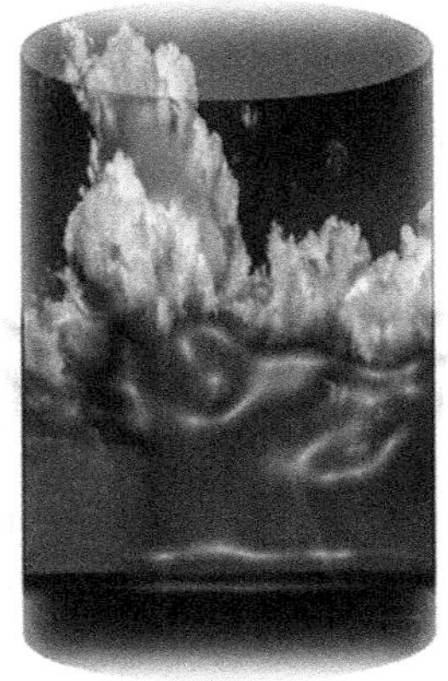

Figure 6.5: Vertical convective flow.

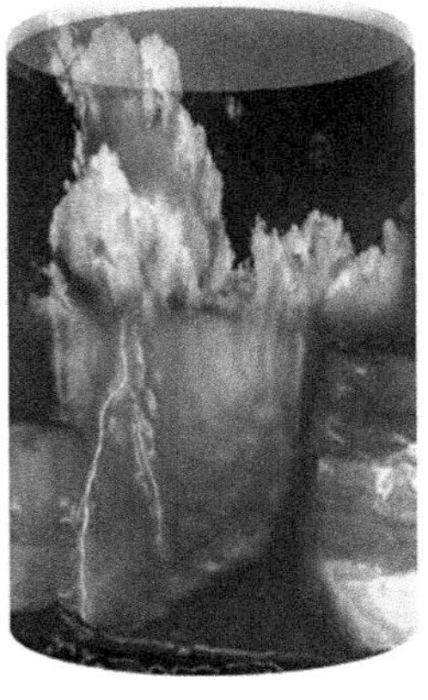

Figure 6.6: Vertically developed clouds associated with thunderstorms.

Table 6.1: Types of clouds and their specific Latin names

Types of Clouds	Specific Cloud Latin Names and Symbols
Low clouds	Stratus (St) Stratocumulus (Sc) Nimbostratus (Ns)
Middle clouds	Altostratus (As) Altocumulus (Ac)
High clouds	Cirrus (Ci) Cirrostratus (Cs) Cirrocumulus (Cc)
Clouds with vertical development	Cumulus (Cu) Cumulonimbus (Cb)

Table 6.2: Approximate elevation of cloud bases in different world zones

Cloud Groups	Polar Zone	Mid-latitude Zone	Tropical Zone
Low clouds	Surface to 2000 m	Surface to 2000 m	Surface to 2000 m
Middle clouds	2000 to 4000 m	2000 to 7000 m	2000 to 8000 m
High clouds	3000 to 8000 m	5000 to 13,000 m	6000 to 18,000 m

6.2 What is the Magical Association between Clouds and Turbulence?

Variable clouds and atmospheric turbulence preserve an intimate relationship because turbulence is frequently brought on by the continuous movement of unstable air masses with varying tremendously seasonable temperatures and humidity levels, which can be positively influenced by the possible presence of clouds. The considerable following typically represents a few responsible mechanisms that cloud and atmospheric turbulence are properly connected:

6.2.1 Microphysics of Clouds and Turbulence

In winter clouds, turbulence is characterized by a regular alternation of updrafts and downdrafts, in which air containing cloud particles rises or falls. Consistent with Petrov et al. [1], the size and concentration of cloud particles and other microphysical aspects of clouds can influence turbulence's intensity and duration.

6.2.2 Cloud Microphysics and Precipitation in Turbulence

According to this perspective, Zhu et al. [2] looked into how turbulence affected the microphysics and precipitation of warm marine boundary layer clouds. In this sense, turbulence is crucial to the formation of precipitation and cloud microphysics. To be more specific, turbulence can promote cloud droplet collision and coalescence, resulting in the formation of larger droplets that are more likely to fall as precipitation. Therefore, smaller cloud droplets can collide with each other and merge, forming larger droplets. The larger the droplets become, the more effectively they can overcome the upward air currents, increasing the likelihood of them falling as precipitation, such as rain. In this context, turbulence indeed plays a significant role in the formation of precipitation. When air is turbulent, it causes the mixing of different air masses with varying properties, including temperature, humidity, and aerosol concentrations. As warm, moist air rises and encounters turbulent conditions, it can lead to the formation of clouds through condensation and subsequent droplet growth.

6.2.3 Cloud Ice and Turbulence

An important development in our understanding of cloud physics is the discovery that vertical air motions promote the formation of ice in mixed-phase clouds. It is difficult to understand the behavior of mixed-phase clouds, which are clouds that contain both liquid water droplets and ice crystals Scientists believed that vertical air currents present in mixed-phase clouds could increase the collision and merging of supercooled liquid droplets and ice crystals. This could result in more effective ice formation mechanisms [3].

It is challenging to describe the numerous pathways by which ice forms in clouds because the pertinent processes occur in a turbulent environment at various scales. Altocumulus and stratocumulus shallow supercooled cloud layers have been studied in ice-formation processes frequently in the last ten years. Bourgeois et al. [4] recently demonstrated that these mid-level clouds can significantly cool Earth's climate. As demonstrated by Ansmann et al. [5] Westbrook and Illingworth [6] and de Boer et al. [7] immersion freezing may be the predominant ice-formation process at temperatures above 27°C. Shallow cloud layers also offer narrow constraints on temperature, pressure, and humidity [3].

Ice formation primarily takes place in the liquid phase below this temperature. As a result, shallow cloud layers that form ice serve as an excellent natural laboratory for research on aerosols, dynamics, ice formation, and their interactions. Even though shallow cloud layers have straightforward conditions, it is still challenging to gauge the vertical air velocity within these layers. Since the air's velocity cannot be directly measured, tracers like aerosol particles or cloud droplets must exist to serve as alternative sources for wind speed [3–7].

Since almost no precipitation would fall from clouds in the middle latitudes of the Earth without the presence of ice, it is pointless to declare that the formation of ice in clouds is an important process in the atmosphere. Despite how extensive these processes may be, many of the specifics are still poorly understood and are not included in the weather and climate models.

6.3 Can Turbulence be Caused by Clouds?

We discussed the fact that turbulence is a common concern for travelers in Chapters 1 and 2, particularly for frequent flyers who fear flying. Some of these concerns can be addressed by being aware of the causes of turbulence such as clouds. It can be useful to understand how these mechanisms interact because clouds and weather patterns contribute significantly to the formation of turbulence during flight.

Clouds are among the many potential causes of turbulence. For instance, cumulus clouds are more likely to produce turbulence than other cloud formations. Convective processes in the atmosphere, where warm air rises and creates updrafts, are the common cause of cumulus convection. These updrafts may result in the development of clouds and ultimately thunderstorms. When flying through clouds, turbulence may be caused by the wind shear within the cloud. An abrupt change in wind direction or speed over a short distance is known as wind shear, which has the potential to produce turbulent air masses.

The aircraft may experience abrupt movements along its flight path owing to updrafts or downdrafts as a result of this turbulence within the clouds. Turbulence can also be caused by factors other than cumulus clouds and thunderstorms, such as irregularities on the Earth's surface or changes in air pressure. Owing to changes in air temperature and humidity, the cloud base, where the bottom of the clouds meets the atmosphere, can also be a source of turbulence.

6.4 What Sorts of Clouds Yield the Most Turbulence?

Strong convective activity and atmospheric instability are essentially connected to the types of clouds that yield the most turbulence. These types of clouds, known as cumulonimbus clouds, are often associated with thunderstorms and can reach great heights in the atmosphere. The strong updrafts within cumulonimbus clouds create an environment of instability, leading to intense convective activity and the potential for severe weather phenomena such as lightning, heavy rain, and even tornadoes.

Even though high clouds like cirrus, cirrostratus, and cirrocumulus generally cause less turbulence compared to low-level and convective clouds, there are still some factors that can contribute to turbulence in high clouds [8–10]. However, it is crucial to bear in mind that weather-related factors such as even high clouds can cause some turbulence. There are nevertheless a few factors that may trigger turbulence in high clouds, including jet stream; wind shear; weather fronts; mountain waves; and weather systems [3, 4, 9].

According to the above perspective, high-altitude winds, such as jet streams, can be a source of turbulence, even in the absence of visible clouds. Cirrus clouds located near jet streams can indicate areas of strong winds, and flying through these regions can result in turbulence for aircraft.

Wind shear refers; therefore, to a change in wind speed or direction with altitude. Even in stable atmospheric conditions, abrupt changes in wind speed or direction between layers of high clouds can create turbulence. In this regard, high clouds can be associated with the presence of weather fronts, such as warm fronts or cold fronts, which can lead to atmospheric instability and turbulence in the surrounding air [4, 7, 9].

Mountain Waves: High clouds downwind of mountain ranges can be a sign of mountain wave activity. While these clouds themselves might not cause turbulence, the mountain waves that generate them can lead to areas of turbulence near the mountains. Consequently, high clouds can be part of larger weather systems, and depending on the dynamics of the system, there may be areas of turbulence associated with them [9–11].

In general, pilots and meteorologists consider high clouds to be less likely to produce turbulence compared to low-level and convective clouds. However, weather conditions can vary, and turbulence can occur in any part of the atmosphere, so aviation professionals must stay informed about weather patterns and turbulence forecasts to ensure flight safety [8–11].

6.5 Have any Recent Developments in Cloud Types Involved Turbulence?

The WMO has added a few new cloud types to the International Cloud Atlas for the first time in more than 30 years. They may share some characteristics with the clouds that have been previously studied, but they may also have some characteristics that are a little different. These new cloud types contain volutes, asperities, and cavum. Volutus clouds are long, tube-shaped clouds that resemble a rolling wave. Asperitas clouds have a turbulent and chaotic appearance, often resembling the underside of a rough sea. Cavum clouds are characterized by large, circular holes or gaps in the cloud layer. These additions to the International Cloud Atlas provide meteorologists with a more comprehensive understanding of cloud formations and their impact on weather patterns.

6.5.1 Volutus

Volutus clouds are relatively rare cloud formations that have a distinct appearance. They are characterized by their low-level, horizontal, and tube-shaped structure, resembling a cylindrical roll stretching across the sky. Meteorologists commonly associate Volutus clouds with thunderstorms and occasionally with cold fronts. These clouds often form along the leading edge of advancing thunderstorms or on the outflow boundary of a thunderstorm system [14, 16]. They can also be associated with the gust front, which is the leading edge of a cooler and drier air mass that moves ahead of a thunderstorm (Figure 6.7).

Figure 6.7: Volutus clouds.

Roll clouds typically appear as well-defined and smooth cylindrical rolls, and they can extend for several miles in length. The rolling motion of these clouds is what gives them their name. Despite their unique appearance, volutus clouds are not associated with tornadoes or other severe weather phenomena. However, they can be an indication of atmospheric instability and the presence of turbulent weather conditions [14–16].

As with all cloud types, Volutus clouds add to the diversity and wonder of the Earth's atmosphere. The study and observation of these cloud formations help meteorologists better understand weather patterns and atmospheric dynamics.

6.5.2 Asperitas

When viewed from below, asperitas, a relatively uncommon cloud formation, looks like the rough surface of an ocean during a storm. The undulating waves and rough, wavy patterns of these clouds give them a distinctive and chaotic appearance (Figure 6.8).

Figure 6.8: Asperitas clouds with chaotic appearance.

Asperitas clouds are frequently linked to convective weather, especially thunderstorms. The dynamic atmospheric conditions that can result in the development of thunderstorms are reflected in the turbulent and unsettled appearance of these clouds. It's crucial to keep in mind, though, that the presence of Asperitas clouds does not automatically herald the arrival of a thunderstorm. They may be a sign of the potential for convective activity and atmospheric instability [15, 18].

It's interesting to note that Asperitas clouds can be seen in calmer settings. Although they are frequently linked to convective weather systems, they can develop under different atmospheric circumstances. The exact mechanisms behind the formation of Asperitas clouds are still not entirely understood, and research in this area continues to shed light on the processes responsible for their appearance [16–18].

Interest among meteorologists and cloud enthusiasts alike has increased as a result of the Asperitas clouds' inclusion in the International Cloud Atlas in 2017. The discovery of this cloud type has improved our knowledge of cloud classification and the wide variety of cloud formations that can take place in the atmosphere of the Earth.

6.5.3 Fluctus Clouds

A wave-like cloud formation called a fluctus can be seen on the top surface of some cloud types. The short duration and curls or breaking wave patterns (Figure 6.9) of these cloud waves add to the cloud deck's dynamic reliability [14, 19].

Figure 6.9: Wavey pattern of Fluctus clouds.

Cirrus, Altocumulus, Stratocumulus, Stratus, and occasionally Cumulus clouds are among the many cloud types that frequently coexist with flutus clouds. These wave formations can both improve the cloud layer's aesthetic appeal and offer useful information about the atmosphere's state at that particular level.

The interaction of air currents and turbulence within the cloud layer is one of the mechanisms that causes Fluctus clouds to form. On the cloud tops, these interactions result in wave-like patterns that can be seen. Fluctus clouds are more likely to form when there are different wind directions and speeds at various cloud layer altitudes [15,17,19].

Fluctus clouds, like many other cloud types, provide meteorologists and weather enthusiasts with useful data. The appearance and prevalence of Fluctus clouds in various cloud types can be studied to help scientists better understand the dynamics and processes of the atmosphere. These findings support ongoing efforts to enhance climate modeling and weather forecasting.

6.5.4 Cavum Clouds

Cavum clouds, also known as "fallstreak holes" or "hole punch clouds." Cavum is a distinct cloud formation that typically exhibits a well-defined circular shape when viewed from below (Figure 6.10). However, when seen from a distance, it may appear oval-shaped. When formed due to an aircraft's interaction with the cloud layer, Cavum clouds can appear linear in the form of a dissipating trail.

These cloud formations are particularly associated with Altocumulus and Cirrocumulus clouds. They are less commonly observed with Stratocumulus clouds. Cavum clouds are relatively rare but are easily recognizable and can captivate onlookers due to their unique appearance [15–19].

Figure 6.10: The circular shape of Cavum clouds.

The formation of Cavum clouds occurs through a fascinating process involving the interaction between supercooled water droplets and ice crystals in the cloud. When the cloud layer contains supercooled water droplets (water that remains liquid below freezing temperature), it can be destabilized by disturbances like aircraft passing through the cloud or changes in air pressure or temperature. As a result, the supercooled droplets freeze and fall from the cloud, creating a hole or gap in the cloud layer. This hole is what gives Cavum clouds their distinct appearance [2, 15, 18].

The presence of Cavum clouds offers valuable insights into atmospheric conditions and the interaction between aircraft and cloud layers. Understanding these cloud formations helps improve our understanding of cloud physics and atmospheric processes, contributing to advancements in weather forecasting and aviation safety.

6.5.5 Murus Clouds

Indeed, Murus clouds, commonly known as "wall clouds," are an essential and fascinating feature associated with severe thunderstorms. These clouds are characterized by a localized, persistent, and abrupt lowering of the cloud base from the main body of the Cumulonimbus cloud [17–20].

Figure 6.11: Wall clouds of Murus clouds.

The term "Murus" is derived from Latin, meaning "wall," which aptly describes the cloud's appearance. When observed from a distance, a Murus cloud often appears as a distinct, dark, and ominous wall-like structure extending downward from the base of the Cumulonimbus cloud [14, 18, 20].

Murus clouds are typically associated with severe thunderstorms and are known to develop in the rain-free portion of the Cumulonimbus cloud. They are often located at the leading edge of the storm, known as the gust front or the outflow boundary. The presence of a Murus cloud signals a high potential for severe weather, including the possibility of tornado formation [16–20].

These wall clouds are formed due to the strong updrafts within the Cumulonimbus cloud, which cause air to rise and cool rapidly. As the air cools, it may reach its dew point, leading to condensation and cloud formation. This process creates the localized lowering and distinct appearance of the Murus cloud [18, 20].

Meteorologists closely monitor Murus clouds as they can be an indication of the storm's intensity and potential for severe weather events. They are important for severe weather forecasting and tornado detection, as they often precede tornado formation and indicate areas of strong convective activity within the thunderstorm.

Due to the potential hazards associated with severe thunderstorms and tornadoes, people in affected areas must pay close attention to weather advisories and take appropriate safety measures when Murus clouds or other signs of severe weather are present.

6.5.6 Cauda Clouds

"Tail clouds," also referred to as cauda clouds (Figure 6.12). At low altitudes, Cauda is a horizontal, tail-shaped supplementary cloud formation that extends from a Cumulonimbus cloud's main precipitation region to the Murus clouds.

Figure 6.12: A pattern of Cauda clouds.

These clouds are typically attached to the Murus cloud, with both their bases found at the same height. The term "Cauda" is derived from Latin, meaning "tail," which accurately describes the cloud's appearance as a tail-like extension from the main Cumulonimbus cloud. Cauda clouds, therefore, are relatively uncommon and are most frequently associated with severe thunderstorms. They are often observed in the vicinity of the Murus cloud, which is a characteristic feature of severe thunderstorms. The presence of Cauda clouds adds to the dynamic and complex nature of these intense weather systems [14, 17, 20].

The formation of Cauda clouds is closely related to the convective processes within the Cumulonimbus cloud. As the Cumulonimbus cloud grows vertically due to strong updrafts, it may develop an overshooting top, which is an area where the cloud extends beyond the typical anvil-shaped top. From this overshooting top, the Cauda cloud extends horizontally as a supplementary cloud feature [16, 21].

Meteorologists pay close attention to Cauda clouds as they can be an indication of the storm's intensity and potential for severe weather. The presence of Cauda clouds, along with Murus clouds, can suggest strong convective activity and atmospheric instability within the thunderstorm system. These conditions increase the risk of severe weather phenomena, such as tornadoes and damaging winds [18, 20].

As with any severe weather event, individuals in affected areas need to stay informed about weather updates and follow safety guidelines to protect themselves from potential hazards associated with severe thunderstorms.

6.5.7 Flumen Clouds

It appears that the term "Flumen" is used to describe bands of low clouds associated with supercell thunderstorms, specifically Cumulonimbus clouds. These bands of clouds are arranged parallel to the low-level winds and move towards the supercell thunderstorm [16, 20, 23].

Unlike the wall cloud, which is typically associated with supercells and located at low levels, "Flumen" clouds are described as accessory clouds that are not attached to the wall cloud. Additionally, their cloud base is much higher than that of the wall cloud.

It's important to note that the term "Flumen" may not be an officially recognized cloud classification by the World Meteorological Organization (WMO) or in the International Cloud Atlas. As it was mentioned earlier, cloud classification is a rigorous process, and any new cloud types or features must undergo scientific scrutiny and validation before being officially recognized [19–22].

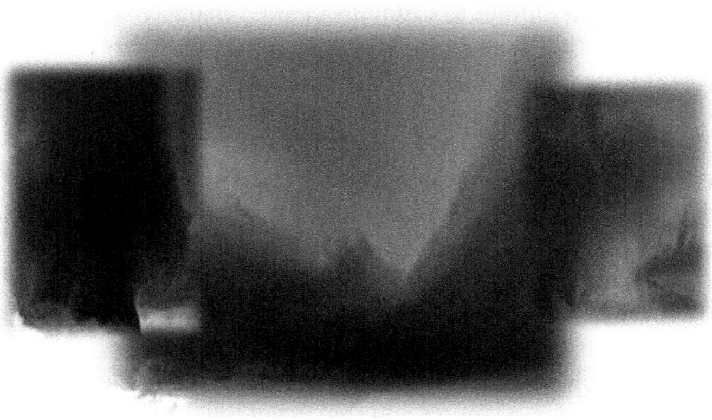

Figure 6.13: Flumen bands of low clouds.

If "Flumen" is a term used informally or within specific meteorological communities, it's essential to be aware that it might not have widespread acceptance or recognition in the broader meteorological field. Supercell thunderstorms are complex and powerful weather systems that can produce severe weather phenomena such as tornadoes, large hail, and damaging winds. The presence of accessory clouds like "Flumen" can provide valuable insights into the atmospheric conditions and dynamics within these intense storms [19–23].

6.5.8 Flammagenitus Clouds

Indeed, Flammagenitus clouds, also known as "pyrocumulus" or "fire clouds," are a fascinating and distinctive type of cloud formation that originates from convection initiated by localized forest fires, wildfires, or volcanic eruptions. These unique clouds are a result of the intense heat and updrafts generated by fire or volcanic activity.

Figure 6.14: Flammagenitus clouds or fire clouds.

Flammagenitus clouds are typically formed when the heat from the fire or volcanic eruption warms the air near the ground. As the air becomes significantly warmer than the surrounding air, it starts to rise rapidly, creating a strong updraft. As this warm air ascends, it cools adiabatically (due to expansion) and reaches its dew point temperature, leading to the condensation of water vapor into cloud droplets. The rapid ascent of the air and the presence of water vapor from fire or volcanic emissions contribute to the cloud's formation [20, 22, 24].

These clouds may appear similar to Cumulus clouds, characterized by their puffy, cauliflower-like shape, but they are formed under entirely different conditions. Flammagenitus clouds can have a significant vertical extent, reaching into the middle and upper troposphere, and may even develop into more massive and taller pyrocumulonimbus clouds (pyroCb) under extreme conditions [16, 20, 22].

Flammagenitus clouds are an important meteorological phenomenon as they can have significant effects on weather and fire behavior. They can intensify fire activity by producing stronger updrafts, which may lead to more rapid fire spread. The clouds can also influence local weather by affecting wind patterns, temperature, and even precipitation in some cases [14, 21, 23].

These cloud formations are intriguing to researchers and meteorologists, and their study helps improve our understanding of the complex interactions between wildfires, volcanic eruptions, and the atmosphere. As with other cloud types, the recognition and classification of Flammagenitus clouds are important for accurately documenting and understanding cloud formations around the world.

6.5.9 Homogenitus Clouds

Homogenitus clouds are cloud formations that are primarily the result of human actions rather than normal meteorological processes. These clouds are anthropogenic, which means that human activities, especially emissions from industry and cities, are to blame for their formation [23, 25, 27].

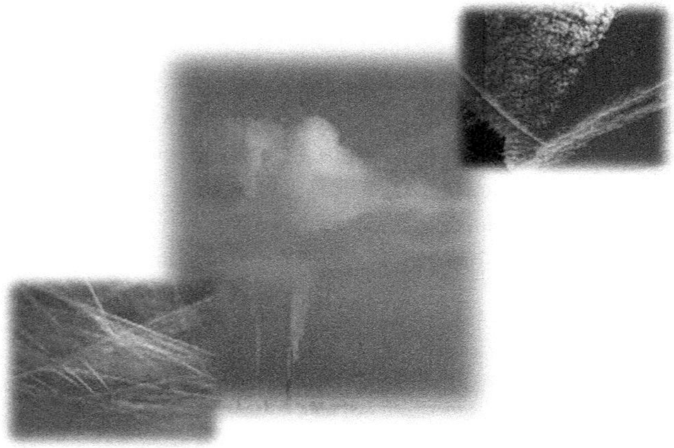

Figure 6.15: Homogenitus clouds are caused by human activities.

One of the most common examples of Homogenitus clouds is the contrail (short for "condensation trail") formed by aircraft. Contrails are long, narrow, and linear clouds that appear behind jet aircraft flying at high altitudes. They are created when hot engine exhaust gases mix with the cold, humid upper atmosphere. The moisture from the exhaust gases condenses into tiny ice crystals, forming the visible contrail [26, 28].

Other sources of Homogenitus clouds include emissions from thermal power plants, industrial processes, and urban pollution. These clouds are formed when the warm exhaust gases from industrial or urban sources interact with the surrounding atmospheric conditions, leading to condensation and cloud formation[24,28].

Homogenitus clouds are an essential component of the study of anthropogenic impacts on the atmosphere and climate. Their presence reflects human activities and their influence on local and regional weather patterns and cloud cover. Researchers and climate scientists analyze Homogenitus clouds to better understand the effects of human emissions on cloud formation and the Earth's climate system [26, 27, 29].

It's worth noting that Homogenitus clouds are distinct from Homomutatus clouds, which are cloud formations that are altered or modified by human activities, such as cloud seeding experiments.

As our understanding of cloud formation and climate science continues to evolve, studying the impact of human activities on cloud formations and their subsequent effects on weather and climate remains an important area of research.

6.5.10 Homomutatus Clouds

Homomutatus clouds are formed as a result of persistent contrails that endure for an extended period under the influence of strong winds. These contrails continue to spread and transform, eventually taking on the appearance of natural cirriform clouds, which are high-altitude clouds with a wispy and fibrous appearance (Figure 6.16).

Figure 6.16: Wispy and Fibrous appearance of Homomutatus clouds.

Homomutatus clouds are a type of anthropogenic cloud formation, just like Homogenitus clouds, which are clouds formed due to human activities. In the case of Homomutatus, the persistent contrails left by aircraft can linger in the upper atmosphere for an extended period, particularly under conditions of strong winds. The contrails spread out and mix with the surrounding air, undergoing internal transformations due to atmospheric processes [27–30].

Over time, these altered contrails lose their initial linear appearance and evolve into cloud formations that resemble cirriform clouds. Cirriform clouds typically occur at high altitudes and are composed of ice crystals. They appear wispy, thin, and fibrous, and they often have a feathery or filaments-like structure [26, 28, 30].

Homomutatus clouds serve as an example of how human activities can influence cloud formations and alter the natural appearance of the sky. The presence of persistent contrails and their transformation into cirriform clouds can have implications for atmospheric conditions, including potential impacts on weather and climate [27, 29].

As researchers continue to investigate the effects of human activities on the atmosphere and cloud formations, the study of Homomutatus clouds contributes to a broader understanding of anthropogenic influences on weather patterns and the Earth's climate system.

6.5.11 Cataractagenitus Clouds

Cataractagenitus clouds are a unique type of cloud formation that occurs locally in the vicinity of large waterfalls. These clouds are generated when the powerful force of the waterfall causes water to break into spray, which then rises into the air and forms clouds. The name "Cataractagenitus" is derived from "cataract," which refers to a large, and powerful waterfall [18, 27, 29].

The process of Cataractagenitus cloud formation involves the interaction between the ascending spray from the waterfall and the surrounding air. As the water plunges downward, it breaks into

Figure 6.17: Cataractagenitus clouds occur in large waterfalls.

tiny droplets, creating a mist or spray. The motion of the locally ascending air, possibly due to the heating of the moist air near the falls or other local atmospheric conditions, compensates for the downdraft created by the water current moving downward [26, 30].

The rising spray, combined with the upward motion of the local air, leads to the formation of clouds near the waterfall. These clouds can appear as a localized and temporary phenomenon, usually dissipating shortly after they are formed. Cataractagenitus clouds are an intriguing weather spectacle, adding to the scenic beauty of large waterfalls. They are a natural consequence of the interplay between the powerful forces of nature and the atmospheric conditions in their vicinity [27, 30].

It's important to note that Cataractagenitus clouds are a relatively localized phenomenon and are not included in the official cloud classification of the International Cloud Atlas. As such, they are considered a cloud accessory rather than a distinct cloud type. Nevertheless, they are a captivating example of how various natural elements can interact to create unique cloud formations in specific environments [18, 28, 30].

6.5.12 Silvagenitus Clouds

Silvagenitus clouds are cloud formations that can develop locally over forests. The name "Silvagenitus" is derived from "silva," which means "forest" in Latin. These clouds are formed as a result of increased humidity in the air, which is primarily caused by evaporation and transpiration from the trees in the forest.

Figure 6.18: Silvagenitus clouds are formed over the forest.

The process of Silvagenitus cloud formation involves the release of moisture into the atmosphere from the leaves of trees through a process known as transpiration. Transpiration is the natural process by which trees and other plants release water vapor into the air through tiny pores called stomata on their leaves. Additionally, evaporation from the moist forest floor and bodies of water within the forest can also contribute to increased humidity in the surrounding air [18, 29, 31].

As the humidity levels rise due to these processes, the air may become saturated, and cloud formation can occur. Silvagenitus clouds can appear as a localized and temporary phenomenon, typically concentrated over the forested area. Silvagenitus clouds,therefore, are not officially recognized in the International Cloud Atlas as a distinct cloud type, but they are considered a cloud accessory formed under specific localized conditions. These clouds can add to the serene and scenic atmosphere of a forested landscape, creating a unique and tranquil experience for those observing them [28, 31].

As with other types of localized cloud formations, the presence of Silvagenitus clouds showcases the fascinating interactions between the natural environment and the atmosphere. The study of such clouds contributes to a deeper understanding of the role of forests and vegetation in the water cycle and the overall climate system [30–33].

6.6 Satellite Observation of Clouds

Now the question is: what are spectral regions that are relevant for cloud identifications?

In the context of imaging clouds, spectral remote sensing relies on capturing the interaction of various wavelengths of light with cloud particles and droplets to gain insights into cloud properties and characteristics. Clouds interact with sunlight and emit radiation, and these interactions vary based on cloud composition, phase (liquid, ice, or mixed), particle size distribution, and other factors. By using sensors that detect multiple wavelengths across the electromagnetic spectrum, remote sensing can provide valuable information about clouds.

Clouds can be detected in different spectral regions of the electromagnetic spectrum, depending on the properties of the clouds and the remote sensing instrument used. In this regard, remote sensing of the Earth in the S/MWIR spectral region can be greatly obscured by the presence of clouds. However, some researchers have explored cloud detection using cloud optical properties, such as spectral content, viz, near-infrared (NIR). Therefore, there is no single spectral region that is most commonly used for cloud detection in remote sensing. Different studies have used different spectral regions depending on the remote sensing instrument and the properties of the clouds. In this view, the cloud detection algorithm-generating method for remote sensing data can be implemented at visible to SWIR wavelengths.

Certain IR wavelengths are particularly useful for observing thin clouds. These are often referred to as "infrared window channels" because they allow the transmission of IR radiation through the atmosphere and are not strongly absorbed by water vapor or other atmospheric constituents. Examples of such channels include the 8-12 μm and 10.5-12.5 μm bands.

Infrared (IR) emission from thin clouds is an important aspect of remote sensing and atmospheric studies. Clouds emit and interact with infrared radiation in different ways, depending on their properties, thickness, and wavelength of the emitted radiation. In some cases, thick clouds can create thermal anomalies on the Earth's surface. For example, convective thunderstorms can lead to cold cloud tops and relatively warm areas on the surface, which can be detected using IR window channels. In this view, the clouds, like all objects with a temperature above absolute zero, emit thermal radiation in the infrared spectrum. The Earth's surface and the atmosphere emit infrared radiation, and clouds act as both emitters and absorbers of IR radiation. The intensity of IR emission; therefore, from clouds is directly related to their temperature. Clouds at higher altitudes are generally colder and emit less intense IR radiation compared to lower, warmer clouds. The IR emission spectrum of clouds typically peaks in the far-infrared region.

For thin clouds, which have lower water content and are less optically dense, the infrared radiation emitted by the underlying Earth's surface can pass through the cloud relatively easily. Consequently, the IR emission from the cloud may be partially absorbed or transmitted, depending on the cloud's composition and thickness. Additionally, the radiative properties of thin clouds are crucial for understanding their impact on the Earth's energy balance. Thin clouds can have a cooling effect on the planet during the day by reflecting sunlight into space (albedo effect) and a warming effect during the night by trapping outgoing longwave radiation (greenhouse effect). On the other hand, since dense clouds have a higher albedo (reflectivity) than refined clouds, they are seen brighter on a visible satellite image. Nevertheless, similar albedo is associated with middle and low clouds, so it is challenging to discriminate between them simply by experiencing them in visible light.

Lastly, IR window channels can help classify cloud types, such as low-level clouds, middle-level clouds, and high-level clouds. Different cloud types have distinct thermal signatures, allowing researchers to differentiate them based on their IR emissions.

The majority of emission techniques rely on measurements made in different atmospheric window regions of the Earth's atmospheric absorption spectrum. Nevertheless, a few approaches attempt to use readings of emission in absorbing regions originally intended for temperature sounding. To map out thin cirrus clouds, for instance, the CO_2 slicing method has been used with IR emission measurements [37]. The use of specific MW sounding channels in addition to MW window channels has recently been suggested by Bauer et al. [38] for retrieving profiles of precipitation over both water and land.

Emission methods applied to optically thin media are commonly formulated using very simplistic expressions of radiative transfer, often posed in terms of the transfer through a single layer in the form of the following equation:

$$I_{out} = I_{in} \times \varepsilon \tag{6.1}$$

In this formulation, the outgoing radiant intensity I_{out} is directly proportional to the incident radiant intensity I_{in} and the emissivity ε of the medium. In this understanding, for optically thin media, the radiative transfer processes are relatively simple because there is minimal absorption and scattering of radiation within the medium. As a result, they often assume the radiant intensity of the outgoing radiation to be directly related to the emissivity of the medium and the incoming radiation. Therefore, it is essential to note that this formulation is appropriate for optically thin media and may not be valid for optically thick media, where more complex radiative transfer processes involving multiple layers and interactions become significant. In those cases, more sophisticated models and equations are required to accurately describe the radiative transfer.

In this view, what is the role of a blackbody in IR emission formulation? The emission equation using the concept of a blackbody is known as Planck's law. Planck's law describes the spectral radiance (also known as spectral intensity or spectral radiative flux density) emitted by a blackbody at a specific wavelength and temperature. The equation is given by:

$$B_\lambda(T) = \frac{2hc^2}{\lambda^5} \times \frac{1}{e^{\left(\frac{hc}{\lambda KT}\right)} - 1} \tag{6.2}$$

In equation 6.2, $B_\lambda(T)$ is the spectral radiance (emission) at wavelength λ and temperature T (in Kelvin). Therefore, h is the Planck constant (approximately 6.626×10^{-34} J·s); c is the speed of light in a vacuum (almost 3.00×10^8 m/s); k is the Boltzmann constant (roughly 1.3806×10^{-23} J/K); and T is the temperature of the blackbody (in Kelvin). Accordingly, Planck's law represents the spectral distribution of thermal radiation emitted by a blackbody, and is a fundamental concept for understanding the behavior of thermal radiation across the electromagnetic spectrum. When plotted as a function of wavelength, the curve obtained from Planck's law shows how the emission intensity changes with temperature and wavelength.

It is important to note that Planck's law describes the emission from an idealized blackbody, which absorbs all the radiation incident on it and emits radiation purely because of its temperature. Real-world objects may not behave precisely as blackbodies; however, Planck's law serves as a useful approximation for understanding the emission characteristics of many objects in nature.

6.7 Quantization of IR Emission from Clouds

The author thoroughly explained the quantization of blackbody radiation in sections 3.6 and 3.7 of Chapter 3. Therefore, the potential use of ultimate cloud atoms and molecules as immutable quantum systems for radiative temperature measurements. In this sense, cloud atoms and molecules are fundamental constituents of matter and are described by quantum mechanics. In the context of this passage, "immutable" means that the properties of atoms and molecules are well-defined and unchanging, at least under certain conditions. These quantum systems have well-characterized interactions with electromagnetic radiation, meaning they can absorb and emit photons of specific energies, leading to quantized transitions between energy levels. Therefore, utilizing the quantized energy transitions of specific atoms or molecules, it would be possible to create a fully quantum-based realization of temperature measurements. This method would provide a direct and accurate link between temperature measurements of clouds and fundamental constants of nature, such as Planck's constant (*h*), which is used in the description of quantum transitions of cloud phases. In this perspective, the passage points out that the blackbody radiation-induced Stark shift is currently the largest uncompensated systematic error in optical clocks. This means that the influence of blackbody radiation on the quantum states of the cloud atoms or ions used in optical clocks is not entirely accounted for, leading to an error that limits the clocks' precision.

Pointless to reveal, a quantum blackbody is a concept that arises from applying the principles of quantum mechanics to blackbody radiation. It involves the quantization of energy levels in the form of photons and explains the emission and absorption of radiation by a blackbody at different temperatures and frequencies without encountering the issues presented by classical physics. Planck's law successfully describes the energy density of radiation emitted by a blackbody and provides a crucial insight into the behavior of electromagnetic radiation at the quantum level. In this understanding, the energy of a quantum state cloud state $\left|i_{cloud}\right\rangle$ is also shifted by a certain amount by blackbody radiation as formulated by:

$$\Psi\left|\Delta E_i\right\rangle = \left[6\pi^3\varepsilon_0 c^3\right]^{-1}\left(\frac{k_B T}{\hbar}\right)^3 \sum_j \left|\langle i|d|j\rangle\right|^2 \left(-\frac{\pi^2}{3}y\right)\left(\frac{\hbar\omega_{ij}}{k_B T}\right). \tag{6.3}$$

Considering two quantum states $\left|j_{cloud}\right\rangle$ and $\left|i_{cloud}\right\rangle$ of the blackbody radiation with frequency ω_{ij}. In this scenario, the impulsive decay rate $\Psi\left|\Gamma_{ij}\right\rangle$ is mathematically expressed as:

$$\Psi\left|\Gamma_{ij}\right\rangle = \frac{\left|\langle i|d|j\rangle\right|^2 \left|\omega_{ij}^3\right\rangle}{3\varepsilon_0 \hbar\pi c^3}\left\langle n(\omega)\right\rangle \tag{6.4}$$

In equation 6.4, the term $\langle n(\omega)\rangle$ represents the mean photon number with frequency ω Therefore, at room temperature, a molecule's vibrational transition frequencies are frequently much nearer to the peak of the blackbody spectrum. A molecule can go through a variety of transitions when it absorbs or emits radiation. Vibrational transitions, in which the molecule modifies its vibrational energy level, are one of the most prevalent types of transitions. The corresponding frequencies for these transitions, which involve the quantized vibrational modes of the molecule, are typically in the infrared (IR) portion of the electromagnetic spectrum. These IR frequencies correspond to the energy differences between different vibrational states of the molecule. Accordingly, they can provide valuable information about molecular structure and bonding. Additionally, vibrational transitions are often accompanied by rotational transitions, further expanding the range of frequencies observed in the IR spectrum. Thus, at room temperatures, the thermal energy of the molecules causes them to populate higher vibrational energy levels. As a result, the probability of a molecule undergoing a vibrational transition and interacting with infrared radiation is relatively high. This leads to a significant absorption and emission of radiation in the infrared region, which aligns with the peak of the blackbody spectrum at room temperatures.

According to the above perspective, the concept proposed by Prabhakara et al. [41] involves using the differential absorption and emission of infrared (IR) radiation by ice crystals in cirrus clouds to determine cloud optical thickness and particle size. This technique has been applied extensively to satellite IR radiance data collected at or near wavelengths of 10.8 and 12 micrometers (µm) in the atmospheric IR window spectral region. In this sense, cirrus clouds consist of ice crystals, and they are often thin and high-altitude clouds composed of ice particles rather than water droplets. The IR window spectral region refers to the range of wavelengths in the infrared spectrum where atmospheric gases have low absorption, allowing radiation to pass through relatively unimpeded. In this context, the principle behind this technique is that different wavelengths of infrared radiation are absorbed and emitted differently by the ice crystals in the cirrus cloud. By measuring the radiance of the atmosphere at two specific wavelengths (10.8 µm and 12 µm), which are chosen to be in the atmospheric IR window, it is possible to analyze how much radiation is absorbed and emitted by the ice crystals [35–40].

In accordance with the above technique of perspective, what are the advantages of infrared images in sensing clouds? Infrared (IR) imaging is a technique of identifying and depicting infrared radiation emitted by entities as a function of their temperature. In this perspective, an enhanced image of the real radiating surface: Infrared images acquire the thermal radiation generated by clouds rather than the reflected visible light. This enables to disclose variations in temperature on the surface, which could prove valuable for numerous applications like thermal inspections, monitoring cloud developments, or even identifying kinds of clouds.

From the perspective of quantized blackbody speculations, warmer clouds radiate more energy in the form of infrared radiation compared to cold ones. By employing infrared imaging of equation 6.3, one can discriminate between clouds or areas with varying temperatures, making it valuable for applications like identifying temperature anomalies in equipment and locating thermal leaks in cloud structures. In other words, infrared imaging may be applied to examine and categorize clouds based on their temperature. Low clouds tend to be warmer than high clouds owing to their varying heights and compositions. As a consequence, infrared measurements may discriminate between warm low clouds, which seem darker in the image, and cool high clouds, which appear lighter (Figure 6.19).

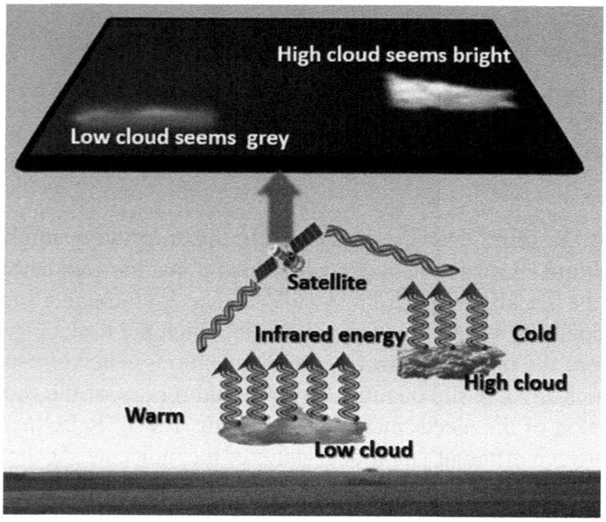

Figure 6.19: Distinguish between low cloud and high cloud in infrared energy.

How we can distinguish cloud kinds in visible and infrared bands? During the daylight, the satellite can capture images from orbit, termed a visible satellite image. This makes it easy to identifying where the larger cloud is, but it is tougher to discern a narrow fragile cirrus cloud.

However, low clouds could be seen in a variety of grayscale tones or even deeper hues in infrared images. The infrared sensor's observation of the temperature differential between the clouds' surroundings and those of the clouds themselves caused this kind of appearance (Figure 6.20).

Figure 6.20: Low clouds appear in grayscale tone across infrared image.

Nevertheless, clouds generally show dazzling white or light-colored patches in the visible band of an optical satellite picture (recorded at the same time as the infrared band). This is because the cloud particles disperse sunlight (Figure 6.21).

Figure 6.21: Bright cloud appearance in the visible band.

Sunlight has a wide spectrum of colours (ROYGBIV) in the visible wavelength range, and clouds are effective light scatterers. Sunlight scatters in all directions as it interacts with cloud particles, including toward the observer (in this example, the satellite sensor). As a consequence of reflecting a considerable percentage of the incoming sunlight, clouds look dazzling and white in visual images. Therefore, clouds are more visible in the visible spectrum than other things on the Earth's surface due to the scattering mechanism. Clouds, unlike the ground or other terrestrial features, disperse sunlight in all directions, making them very visible from space. The above perspective entails radiative transfer theory. So, how would this theoretical framework enable us to comprehend the remote sensing technique of imaging clouds?

6.8 Radiative Transfer Theory for Cloud Imaging Mechanism

Radiative transfer theory is a fundamental concept in understanding how electromagnetic radiation interacts with the atmosphere and objects within it. By applying this theory to remote sensing, we can analyze how clouds interact with different wavelengths of light and ultimately interpret the information captured by imaging sensors. This allows us to gain insights into cloud properties such as their composition, height, and optical thickness, which are crucial for various applications including weather forecasting and climate studies.

The radiative transfer theory, particularly when applied to plane-parallel geometry, provides a general functional form for spectral reflectance from cloud layers. In the context of cloud remote sensing, spectral reflectance refers to the ratio of the reflected electromagnetic radiation (light) to the incident radiation at various wavelengths. The general functional form of spectral reflectance from cloud layers, as predicted by radiative transfer theory, is often described by the bidirectional reflectance distribution function (BRDF) or bidirectional reflectance factor (BRF). The BRDF/BRF quantifies how the reflectance of a surface or cloud layer varies with the direction of incident and reflected light. In this understanding, for plane-parallel clouds, the spectral reflectance is influenced by several factors, including the cloud's optical properties, the incident angle of sunlight, the cloud's thickness, and the underlying surface's reflectance. It is typically expressed as a function of the following formula:

$$\begin{aligned}\Psi\left|\mathcal{R}_\lambda\right\rangle &= \Psi\left|R_\lambda\right\rangle \\ &= \Psi\left|\tau\right\rangle,\Psi\left|P\right\rangle,\Psi\left|\varpi_0\right\rangle,\Psi\left|\alpha_{sfc}\right\rangle,\Psi\left|\xi\right\rangle,\Psi\left|\xi_\odot\right\rangle\end{aligned} \tag{6.5}$$

Equation 6.5, demonstrates that the surface reflection $\Psi\left|\mathcal{R}_\lambda\right\rangle$ is a function of numerous quantum states that shape the spectral reflectance of clouds. In this perspective, the reflectance can vary with the wavelength of the incident light. Different cloud particles may exhibit distinct quantized scattering and absorption state properties at different wavelengths. Therefore, the quantized spectral reflectance is a function of the quantum state between the incident solar radiation $\Psi\left|\xi_\odot\right\rangle$ and the sensor view direction $\Psi\left|\xi\right\rangle$. In this view, different view angles $\cos\Theta = \Psi\left|\xi\right\rangle.\Psi\left|\xi_\odot\right\rangle$ can create a relativity mechanism in reflectance values. As $\Psi\left|\mathcal{R}_\lambda\right\rangle$ would be varied as Θ is varied. Therefore, optical depth can create another quantum state $\Psi\left|\tau\right\rangle$ of the measure of the extinction of light as it passes through the cloud layer. This quantum state depends on the cloud's thickness and its scattering and absorption properties. Consequently, a quantum state of single scattering albedo $\Psi\left|\varpi_0\right\rangle$ is created by cloud particles in a single scattering event. Subsequently, the phase function $\Psi\left|P\right\rangle$ describes the angular distribution of scattered light by cloud particles [40–44].

6.9 Spectral Regions Relevant to Clouds

To fully understand the image processing algorithm, it is crucial to ascertain its specific development for spectral regions pertinent to clouds. Identifying clouds in satellite imagery requires the use of certain spectral regions or wavelength bands. These distinct wavelengths offer important insights into cloud properties and aid in differentiating clouds from other elements. The spectral regions that would be discussed are crucial for identifying clouds. In this regard, the visible Spectrum (0.4–0.7 micrometres) imagines clouds in bright white or light-coloured due to the scattering of sunlight by cloud particles (Figure 6.21). This wavelength range allows for the identification of clouds based on their high reflectivity. Near-Infrared (NIR) Spectrum (0.7–1.3 micrometres); therefore, is sensitive to cloud particle size and cloud thickness. Thicker and denser clouds tend to absorb more NIR radiation, making them appear darker in this band compared to thinner clouds. The Shortwave Infrared (SWIR) Spectrum, ranging from 1.3 to 3.0 micrometres, can effectively distinguish between high and low clouds. Generally, low clouds tend to be warmer and emit more SWIR radiation, which makes them appear darker than higher, colder clouds [35, 40, 42, 44].

Thermal Infrared Spectrum (3.0–14 micrometres) (Figure 6.22); consequently, provides information about cloud-top temperatures. Colder clouds emit less thermal infrared radiation and appear brighter, while warmer clouds appear darker. On the other hand, microwave spectra are used to detect and study clouds, especially in situations where visible and infrared bands are not effective due to atmospheric interference, nighttime, or heavy cloud cover. Without a doubt, the collection of data from relevant areas is made possible by a range of satellite instruments and sensors that operate in different spectral bands. The combination of information from multiple bands not only enhances cloud identification but also provides crucial details about cloud types, altitudes, temperatures, and properties [40, 43, 45].

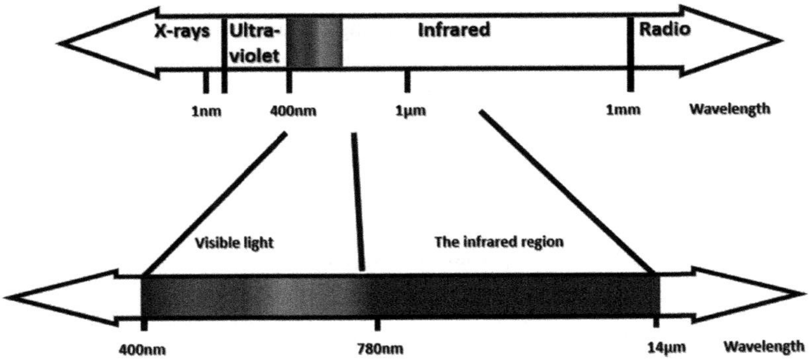

Figure 6.22: The spectral region of cloud identifications.

6.10 Entanglement of Cloud Spectral Absorption

According to the aforementioned viewpoint, a crucial inquiry is: what causality exists between photons that are emitted and those that are absorbed? It is essential to know the relationship between photons that are emitted and those that are absorbed to comprehend how light behaves in diverse systems. Scientists may learn more about phenomena like fluorescence, the photoelectric effect, and even quantum entanglement by investigating this link. One may learn more about the basic properties of light and how it interacts with clouds by investigating the complex relationship between photons that are released and those that are absorbed. Entanglement has the power to alter the effects of photon-matter interactions, and conversely, interactions have the power to alter the non-classical correlations already present in the spectroscopic system. Following this notion, a single photon pulse is completely absorbed in optically dense clouds. The atoms and the field become entangled as a result of the pulse's absorption by the medium; as soon as the pulse is absorbed, the atoms leave in an entangled state engaged in a single excitation. Conversely, the energy that the field is constantly losing is used to excite the mobility of the cloud atoms.

Therefore, entanglement by Absorption refers to the phenomenon where a particle becomes entangled with its environment through absorption interactions. This occurs when a particle interacts with another system, causing their quantum states to become correlated and entangled. The absorbed system then carries information about the original particle's state, leading to entanglement between the two systems; which could be demonstrated in equation 6.5. In this regard, when a single photon is absorbed in an optically dense medium, the process can lead to the creation of entangled states between the photon (the electromagnetic field) and the atoms or molecules of the medium. This phenomenon is a fundamental aspect of quantum optics and is known as "quantum memory" or "quantum state mapping" [46–48].

Accordingly, the absorption of the photon by clouds can lead to the creation of an entangled state between the cloud molecules and the remaining electromagnetic field. This entangled state is a superposition of different energy states, involving both the cloud molecules and the field. Let $|\Psi\rangle$ be the entangled state involving both the cloud molecules and the field. This state can be written as a superposition of energy eigenstates $|E_i\rangle$, where i represents the different energy states:

$$|\Psi\rangle = \sum_i c_i |E_i\rangle \tag{6.6}$$

Here c_i are the complex coefficients representing the probability amplitudes of the system being in the corresponding energy eigenstate $|E_i\rangle$. The sum is taken over all possible energy states involved in the entangled state. The specific form of the energy eigenstates $|E_i\rangle$ and the coefficients c_i; therefore, will depend on the particular physical system and the interactions between the cloud molecules and the field. To fully determine the entangled state, one would need to know the details of the quantum system and the Hamiltonian governing its evolution. In this view, the eigenvalue equation of a Hamiltonian operator represents the energy eigenstates of a system. In the context of supersymmetric quantum mechanics, the eigenvalue equation of the Hamiltonian operator can be used to derive entangled states of absorption and remission by clouds [46, 48, 50]. The eigenvalue equation of the Hamiltonian is given by:

$$\hat{H}|\Psi_E\rangle = E_i|\Psi_E\rangle \tag{6.7}$$

Consequently, the mathematical expression of equation 6.7 after absorption is given by:

$$|\varphi_j > B\ |g > B \rightarrow \gamma|\bar{\varphi}_j > B|e > B + \delta|\varphi_j > B|g > B \tag{6.8}$$

Here $|e>$ denotes the exciting internal state and $\bar{\phi}$ and $\bar{\varphi}$ are the wave functions for emission and absorption states; respectively. In this view, $|\alpha|^2 + |\beta|^2 = 1$ and $|\gamma|^2 + |\delta|^2 = 1$ Therefore, the probability of multi-absorption processes is extremely low and can be ignored. In the presence of multi-photon absorptions, a non-linear regime would be considered [47–49]. Subsequently, the mathematical expression of the final state after multi-photon interaction with different cloud molecules is expressed by:

$$|\psi_f > = \frac{1}{\sqrt{2}}\left(\alpha\gamma\,|\bar{\phi}_L > A|\bar{\varphi}_R > B + \gamma\alpha\,|\bar{\phi}_R > A|\bar{\varphi}_L > B\right)|e > A|e > B + |.... > \tag{6.9}$$

In equation 6.8, $|.... >$ comprises the relaxation terms; which do not allow double absorptions. Therefore, the probabilities of double absorptions can be computed from the expression of $|\psi_f >$ In such a manner, two options take part in the likelihood of double retention: (i) assimilation by a particle of kind A in L and by one of kind B in R and; (ii) assimilation by a particle of sort B in L and by one of sort A in R. In any case, these choices are not unmistakable (the two particles can retain at the two sides of the exhibit) and thus, reliable with the guidelines of quantum hypothesis, likelihood amplitudes should be added rather than probabilities [46–49]. In conclusion, the likelihood of twofold retention is specified a role as:

$$P_d = \left|\frac{1}{\sqrt{2}}\alpha\gamma + \frac{1}{\sqrt{2}}\gamma\alpha\right|^2 = 2|\alpha\gamma|^2 \tag{6.10}$$

Subsequently, the probability of the double mixture absorption can be given by:

$$P_d^{mix} = |\alpha\gamma|^2 \tag{6.11}$$

Dual absorption is twice as likely in the entangled state as it is in the product state. Obviously, the assimilation probabilities in multi-molecule frameworks depend on the snare. In bright recurrence trapped photons, atoms causing photochemical responses of interest regularly have electronic change energies in the bright area.

After the absorption, therefore, the imaging system is in an entangled state. In many cases, the cloud atoms will eventually re-emit the photon, leading to the re-emission of the quantum state of the absorbed photon. This re-emission can occur spontaneously or be triggered by an external control field. Consequently, the re-emission of the photon restores the quantum state of the field and, at the same time, transfers the information about the absorbed photon's quantum state to the clouds. As a result, the state of the cloud becomes entangled with the original state of the photon [47, 49, 51].

The phenomenon highlights the non-classical correlations that exist in quantum systems and how the interactions between photons and clouds can lead to the creation and manipulation of entangled states. This understanding is essential for the development of quantum technologies, including quantum computing, quantum communication, and quantum-enhanced spectroscopy.

6.11 How does Entanglement Form Spectral Discrimination of Clouds?

In quantum teleportation, the properties of the quantum trap are utilized to send a twisted state (qubit) between sensors without truly moving the elaborate molecule. Although a particle's state is destroyed on one side and extracted on the other to communicate the information it encodes, the particles themselves do not teleport (Figure 6.23). The cycle isn't momentary, because data should be conveyed conventionally between sensors as a component of the interaction with cloud molecules. The effectiveness of quantum teleportation lies in its capacity to send quantum data for arbitrary reasons over far distances without presenting quantum states to warm decoherence from the climate or other unfavourable impacts.

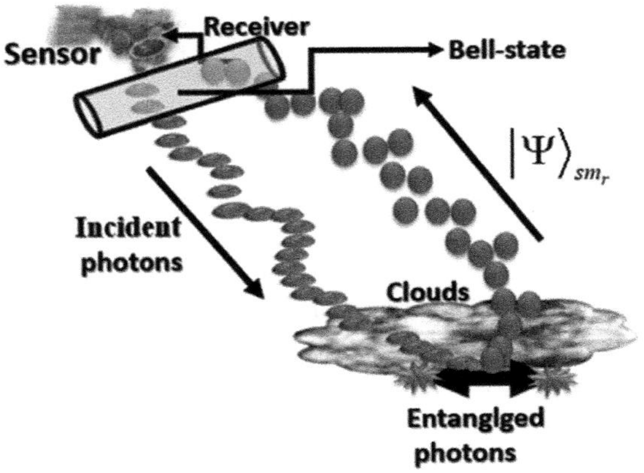

Figure 6.23: Quantum teleportation of cloud remote sensing images.

The process of quantum teleportation involves three main components: the sender, the quantum channel, and the receiver. In this perspective, the remission from cloud molecules has a qubit that they wish to teleport to the sensor. They also have another entangled qubit (the entangled pair) that is shared with the receiver or sensor. Therefore, this is the shared entangled qubit between the radiated clouds and the sensor. Quantum entanglement; obviously, ensures that the states of the absorption; and reflectance are correlated to a sensor, even when they are separated by large distances. Lastly, the sensor has one-half of the entangled pair, which was initially shared with the reflectance energy from different sorts of clouds. The sensor also has an unentangled qubit, which will receive the teleported quantum state [47, 50, 52].

Let's assume an entangled pair of electrons with spin states of sensors (spectroscopy and remote sensing) and cloud structures in a specific Bell-state, to develop a novel theory to comprehend the quantized of the clouds' spectral library:

$$|\Phi_0\rangle = \frac{1}{\sqrt{2}}\left(|\uparrow\rangle_s \otimes |\uparrow\rangle_m + |\downarrow\rangle_s \otimes |\downarrow\rangle_m\right). \tag{6.12}$$

Following equation 6.12, entangled electrons vary between sensors and cloud structures or other physical aspects. In this sense, the sensor entangles with the variant quantum state of clouds to quantify what is known as the "Bell-state" of the sensor. The sensor uses a specific, time-tested technique of photon-electron interaction to transmit the results of its reflectance measurement of the physical characteristics of clouds. The Bell states, an orthonormal foundation of entangled states for the tensor product space of spin-particles, are therefore cast off from the Pauli matrices:

$$|\Phi_0\rangle = I \otimes \sigma_0 |\Phi_0\rangle = \frac{1}{\sqrt{2}}\left(|\uparrow\rangle_s \otimes |\uparrow\rangle_m + |\downarrow\rangle_s \otimes |\downarrow\rangle_m\right). \tag{6.13}$$

$$|\Phi_1\rangle = I \otimes \sigma_1 |\Phi_0\rangle = \frac{1}{\sqrt{2}}\left(|\uparrow\rangle_s \otimes |\uparrow\rangle_m + |\downarrow\rangle_s \otimes |\uparrow\rangle_m\right). \tag{6.14}$$

$$|\Phi_2\rangle = I \otimes \sigma_2 |\Phi_0\rangle = \frac{i}{\sqrt{2}}\left(|\uparrow\rangle_s \otimes |\downarrow\rangle_m - |\downarrow\rangle_s \otimes |\uparrow\rangle_m\right). \tag{6.15}$$

$$|\Phi_3\rangle = I \otimes \sigma_3 |\Phi_0\rangle = \frac{1}{\sqrt{2}}\left(|\uparrow\rangle_s \otimes |\uparrow\rangle_m - |\downarrow\rangle_s \otimes |\downarrow\rangle_m\right). \tag{6.16}$$

Therefore, the Pauli matrices can be mathematically denoted by the following equations:

$$\sigma_0 = I = \begin{pmatrix} 1 & 0 \\ 0 & 1 \end{pmatrix}, \ \sigma_1 = \begin{pmatrix} 0 & 1 \\ 1 & 0 \end{pmatrix}, \ \sigma_2 = \begin{pmatrix} 0 & -i \\ i & 0 \end{pmatrix}, \ \sigma_3 = \begin{pmatrix} 1 & 0 \\ 0 & -1 \end{pmatrix}. \tag{6.17}$$

As a possible consequence, the spin operators are designated as $\frac{\hbar}{2}$ σ_0, σ_1, σ_2, and σ_3 for the x, y, and z axes, respectively. The quantity of the interaction between photons and cloud molecules is measured by either spectroscopy or remote sensing satellite. When cloud molecules interact, the rate of entanglement varies from one spectrum band to another. In this understanding, the speculative possibility of quantum teleportation identifies the spectral measurement of clouds by remote sensing sensors in various discrete bands. In other words, the measurement of spectral disentangles sensors from the physical characteristics of clouds and entangles reflectance into distinct bands. This approach allows for a more precise analysis of cloud properties and their interaction with photons. By measuring the reflectance in different spectral bands, scientists can gain insights into the specific molecular interactions that contribute to entanglement. This knowledge can further enhance our understanding of cloud dynamics and improve remote sensing techniques for cloud monitoring and analysis [47, 49, 51].

Depending primarily on what specific entangled state the sensor reveals, every separated spectral band will know exactly how the clouds are disentangled (m_r) and can influence clouds to assume the state that reflectance spectra have originally. Hence, the state of the distinct spectral channels or bands is "teleported" from sensors to clouds, which now has a state that appears identical to how the stored spectral channels or bands initially appeared. It is crucial to highlight that state variations of the spectral signatures that are stored binary are not retained in the processes: the no-cloning and no-deletion theorems of quantum physics prevent quantum information from becoming completely duplicated or deleted.

The state of cloud spectral signatures $|\Psi\rangle_{sm_r}$ is determined in the separated individual binary bands or channels that can be assigned by:

$$\left|\Psi\right\rangle_{sm_r} = \sum_{i=0}^{n} 0.5\left|\Phi\right\rangle_{sm_r} \otimes \begin{pmatrix} 0 & 1 \\ 1 & 0 \end{pmatrix}\left|\Psi\right\rangle_{m} \tag{6.18}$$

As a result of photon-atom interaction, cloud structures have altered the spin state of particles, which is stated as:

$$\left|\Psi\right\rangle_{m} = c_1\left|\uparrow\right\rangle_{m} + c_2\left|\downarrow\right\rangle_{m} \tag{6.19}$$

Equation 6.19 into 6.18 delivers us the specific instance of how remote sensing identifies sort of clouds as:

$$\left|\Psi\right\rangle_{sm_r} = \sum_{i=0}^{n} 0.5\left|\Phi\right\rangle_{sm_r} \otimes \begin{pmatrix} 0 & 1 \\ 1 & 0 \end{pmatrix} c_1\left|\uparrow\right\rangle_{m} + c_2\left|\downarrow\right\rangle_{m}. \tag{6.20}$$

The state to be teleported and the state of one of the entangled particles are revealed by Equation 6.20. The quantum state of cloud spectral reflections is being teleported, and photon-molecule interactions are being entangled (Figure 6.24). Consistent with this interpretation, the final spin state of the photon-atom interaction resembles the cloud spectral reflectance state [47–51].

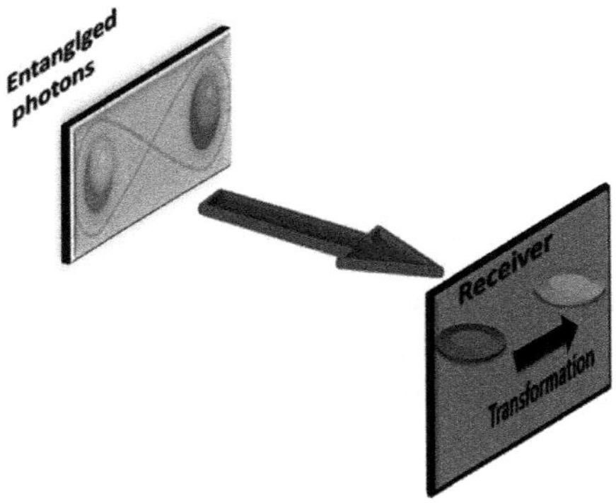

Figure 6.24: Teleported state of quantum cloud spectral entanglement.

Needless to say, this implies that the information on the cloud spectral reflectance state is transferred to the final spin state of the photon-atom interaction. This manner tolerates the teleportation of quantum information encoded in the cloud spectral reflections to be transferred to entangled particles.

6.12 Simulation of Quantum Cloud Spectral Libraries

This section introduces a novel technique to simulate the different spectra signatures of the diversity of cloud sorts. In this perspective, Density Functional Theory (DFT) is implemented to simulate the spectra energy of clouds in optical satellite images. Therefore, DFT can be used to calculate the optical properties of clouds, such as their absorption, reflection, and emission spectra. The Kohn-Sham eigenvalues obtained from DFT calculations can be used to determine the energies of photon transitions, which are directly related to the peaks and features observed in optical spectra. In this view, it is required to derive the equations of Density Functional Theory (DFT) in terms of cloud thermal energy variations, we need to consider the finite-temperature DFT formalism, known as the Mermin-Kohn-Sham (MKS) formulation. The MKS approach extends the ground-state Kohn-Sham DFT to describe cloud spectra energy at a non-zero temperature [53–56].

Consistent with the above perspective, at a non-zero temperature T, the cloud emission system is described by the canonical ensemble, characterized by a fixed number of particles N and a fixed volume V. The system's density matrix ρ is given by:

$$\hat{\rho} = e^{-[k_B T]^{-1}(\hat{H}-\mu\hat{N})} \times \left[Tr\left[e^{-[k_B T]^{-1}(\hat{H}-\mu\hat{N})} \right] \right]^{-1} \tag{6.21}$$

Equation 6.21 expresses the inverse thermal energy, with k_B being the Boltzmann constant. Therefore, $\hat{H}$ is the Hamiltonian operator for the energy spectra emission from the diversity of cloud types in satellite optical images with a total number of operators $\hat{N}$. In this sense, the energy functional for the finite-temperature DFT is obtained by taking the expectation value of the Hamiltonian operator concerning the density matrix:

$$E[\rho] = Tr\left[\hat{\rho}\hat{H}\right] \tag{6.22}$$

To express the energy functional in terms of the electron density, we use the Kohn-Sham ansatz with effective potentials:

$$E[\rho] = T_s[\rho] + E_H[\rho] + E_{xc}[\rho] + E_{ext}[\rho] \tag{6.23}$$

Equation 6.23 demonstrates that $T_s[\rho]$ is the kinetic energy of the non-interacting Kohn-Sham system, given by the sum of the Kohn-Sham orbital kinetic energies. $E_H[\rho]$ is the Hartree energy, accounting for the electron-electron Coulomb repulsion. Additionally, $E_{xc}[\rho]$ is the exchange-correlation energy, representing the quantum mechanical exchange and correlation effects, and $E_{ext}[\rho]$ is the energy arising from the external potential [55–57]. Consequently, the Kohn-Sham equations at finite temperature have the form:

$$\left(-[2\mathrm{m_e}]^{-1}\hbar^2\nabla^2 + V_{eff}(\vec{r},t) - \mu\right)\phi_i(\vec{r},t) = \varepsilon_i\phi_i(\vec{r},t) \tag{6.24}$$

The Kohn-Sham ansatz introduces a set of non-interacting fictitious electrons with time-dependent wave functions $\phi_i(\vec{r},t)$ and corresponding densities $\rho_i(\vec{r},t)$. These photons interact with an effective time-dependent potential $V_{eff}(\vec{r},t)$ of cloud physical properties to produce the same energy spectra density $\rho_i(\vec{r},t)$ as the original interacting photons. In other words, when light is incident on the diversity of cloud types, it can promote electrons from occupied Kohn-Sham orbitals to unoccupied ones. This process results in electronic transitions and absorption of photons with specific energies, corresponding to the energy differences between the occupied and unoccupied Kohn-Sham eigenstates.

The estimation of cloud spectral reflectance from optical remote sensing is assumed to be a form of the quantum linear system. If any optical remote sensing is assumed to be presented as a Hermitian matrix I^H and $DN \in I^H$. In this view, the quantum spectral radiance can be estimated as [47]:

$$E|\rho\rangle = \left(\frac{|L(\lambda)\rangle_{\max}}{254} - \frac{|L(\lambda)\rangle_{\min}}{255}\right) \otimes |DN\rangle\langle DN| + |L(\lambda)\rangle_{\min} \tag{6.25}$$

Equation 6.25 reveals the quantum state of the spectral reflectance in an optical remote sensing image; in which the quantum state of $|DN\rangle\langle DN| \in I^H$ is entangled the quantum state of spectral reflectance $E|\rho\rangle$. Moreover, the quantum gain state is presented in $\left(\frac{|L(\lambda)\rangle_{\max}}{254} - \frac{|L(\lambda)\rangle_{\min}}{255}\right)$ and quantum state of minimum spectral radiance is presented by $|L(\lambda)\rangle_{\min}$. Consequently, the spectral reflectance of a single band of the optical DNs image can be formulated as:

$$E|\rho\rangle_{sm_r} = \left(\frac{|L(\lambda)\rangle_{\max} - |L(\lambda)\rangle_{\min}}{|DN\rangle_{\max} - |DN\rangle_{\min}}\right) \otimes \left(|DN\rangle - |DN\rangle_{\min}\right) + |L(\lambda)\rangle_{\min} \pm |\varepsilon\rangle \tag{6.26}$$

In equation 6.26, the gate complexity is the number of 2-qubit gates; which is exploited. In this understanding, an algorithm is gate-efficient is if it is gate complexity is larger than its query complexity only by a logarithmic factor. Formally, an algorithm with query complexity Q is gate-efficient if its gate complexity is $O(Q.\text{poly}(\log(qN)))$. Subsequently, the runtime of the quantum algorithm for obtaining precise spectral reflectance can be demonstrated implementing $O(\kappa^2 \log N\varepsilon^{-1})$ where $\varepsilon > 0$ is the error and κ is condition number. Therefore $\kappa = \frac{\max_i |\psi\rangle_{smr}}{\min_i |\psi\rangle_{smr}}$ in which, if the condition number is not too much larger than one, the estimated spectral reflectance of the mineral is well-conditioned, and it means that its inverse can be computed with good accuracy [47, 55, 57].

When using a qubit per pixel to encode each pixel of a spectral image, the quantity of the pixel over the entire grayscale determines the grey level of each individual pixel. This could be understood in the quantum form by using polynomials to designate the code in a traditional manner. Notwithstanding, the entire technique is entirely satisfactory because in the quantum world, the |0> and |1> are present but the 0s are suppressed. The coefficient of the base state |0> is then understood to be the equivalence of polynomial zeros, and the coefficient of the base state |1> is understood to be the correspondence of polynomial ones, as regards. 0 represents "off" and 1 represents "on", then the binary code 01111111 would translate to the decimal number 127. This means that in an 8-BBC, when all bits are set to 1, it represents the maximum decimal value that can be represented with 8 bits.

Zero	**One**
0	1111111

(6.27)

For an 8-bit encoding, each value can be represented using 8 binary digits (bits). Indeed, there are 256 possible levels, and the values range from 0 to 255. These values can be used to represent a grayscale image or a color channel with 256 different intensity levels. In quantum superposition state positive scenario one can obtain:

$$(0)\cdot 128 + (1)\cdot 64 + (1)\cdot 32 + (1)\cdot 16 + (1)\cdot 8 + (1)\cdot 4 + (1)\cdot 2 + (1)\cdot 1 = 255. \quad (6.28)$$

The normalization status can then be determined using 6.27 by:

$$\begin{aligned} &(0).\frac{128}{255} + (1).\frac{64}{255} + (1).\frac{32}{255} + (1).\frac{16}{255} \\ &+(1).\frac{8}{255} + (1).\frac{4}{255} + (1).\frac{2}{255} + (1).\frac{1}{255}, \end{aligned} \quad (6.29)$$

The respective pattern results from further merging equations 6.28 and 6.29:

$$(0).\frac{128}{255} + (1).\frac{127}{255} = 0.501\ 960\ 78 \times (0) + \ 0.498\ 039\ 21 \times (1) \quad (6.30)$$

Satisfying the quantum spectral reflectance of the minerals is $E|\rho_{sm_r}|_0^2 + E|\rho_{sm_r}|_1^2 = 1$, let us consider; for instance, $E|\rho_{sm_r}|_0^2 = 0.501\ 960\ 78$ and $E|\rho_{sm_r}|_1^2 = 0.498\ 039\ 2$, then the wave function of the spectral reflectance is $E|\rho_{sm_r}\rangle = 0.501\ 960\ 78|0\rangle + 0.498\ 039\ 2|1\rangle$. Simultaneously, It has dual states of the observable reflectance spectra quantity,the positive and negative of the pixel taken [47–57].

The description provided seems to mix concepts from quantum mechanics and spectral reflectance in an unusual manner. Quantum mechanics deals with the behavior of particles at the atomic and subatomic scale, while spectral reflectance is related to the interaction of light with materials, such as minerals, and how they reflect different wavelengths of light. In quantum mechanics, the wave function represents the state of a quantum system. The wave function $E\Psi\left|\rho_{sm_r}\right\rangle$ contains information about the probabilities of different reflectance from cloud states. It can be complex-valued, and its squared modulus, $E\left|\Psi\left|\rho_{sm_r}\right\rangle\right|^2$, gives the probability density of finding the cloud in a specific state. In this view, spectral reflectance refers to the ratio of the reflected light intensity to the incident light intensity as a function of wavelength. It characterizes how a cloud interacts with light at different wavelengths. Spectral reflectance is typically represented by a continuous function $\Psi\left|R_\lambda\right\rangle$ ER(λ), where λ is the wavelength of light [53–57].

Pointless to mention, the dual images and superposition states described appear to propose the development of "dual images" connected with the quantum superposition states of $|0\rangle$ and $|1\rangle$. The sophistication of how the remote sensing image data are altered by utilizing quantum superposition states to generate these dual images might be presented in a superposition of low and high clouds in one spectral band or superimposed in a separate, distinct spectral band.

6.13 Cloud Quantum Spectral Energy Reflectance

The aforementioned concept of harnessing quantum spectral energy reflectance speculation is brought to the practical stage through the utilization of MODIS satellite data. This endeavour seeks to provide insights into the intriguing query of whether quantum spectral energy reflectance holds the potential to discern various types of clouds. In this vantage point, the initiation of quantum spectral states involves a dual procedure: the creation of the initial positive state and its counterpart, the initial negative state (Figure 6.25).

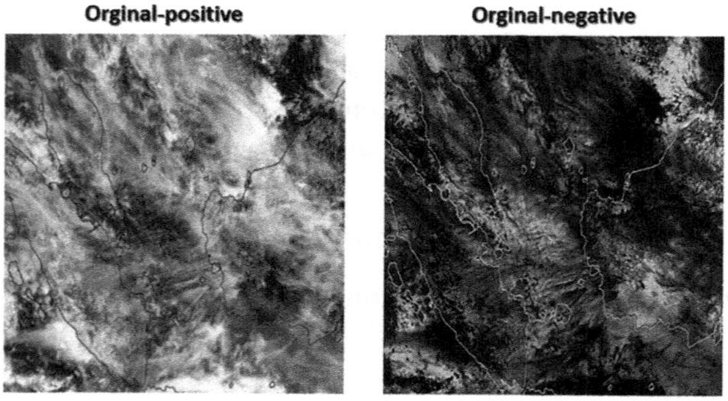

Figure 6.25: Generation of a quantum image of positive and negative states.

In this understanding, the terms "quantum image positive state" and "quantum image negative state" are used in a specialized context related to cloud physical properties, they could refer to distinct quantum states representing certain characteristics of clouds. These characteristics might include properties like cloud composition, altitude, density, or other features that can be encoded using quantum states [45, 47, 53].

Given the complex and evolving nature of quantum computing and its potential applications, it's essential to have clear context or references to accurately understand the meaning of these terms. The relationship between quantum states and cloud physical properties would depend on the specific theoretical framework or model being proposed.

Therefore, the next step in creating a quantum image involves achieving a state of superposition between the paired positive and negative states (Figure 6.26). In this view, superposition is a characteristic of quantum cloud states where a quantum system can exist in a combination of multiple states at once. Unlike classical image processing which is confined to one state, quantum systems can be in a "superposition" of different states simultaneously. This means that a quantum particle, represented by a quantum state, can be in a blend of various positions, energies, or other measurable properties which will be discussed later [50–53].

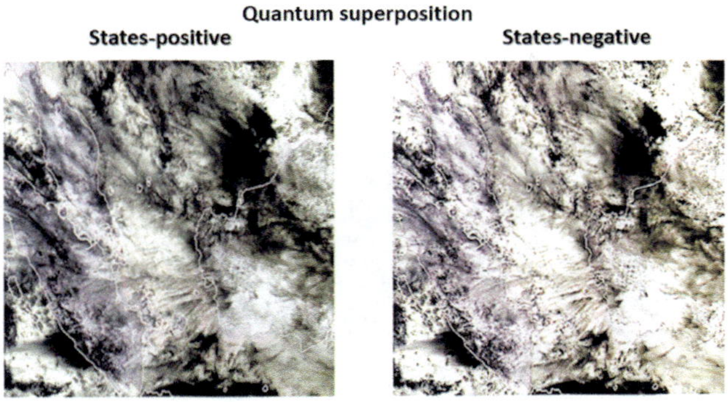

Figure 6.26: Superposition between the original positive state and the original negative state.

As a result, quantum superposition can distinguish between various types of clouds. From this standpoint, both thin and thick clouds can be accurately differentiated, as can clouds with colder and warmer temperatures (Figure 6.27). In this understanding, clouds with lower temperatures emit less thermal radiation. In MODIS data, these colder clouds appear as darker patches in thermal infrared imagery. The brightness temperature of cold clouds is relatively low compared to the surrounding background. On the other hand, clouds with higher temperatures emit more thermal radiation. These warmer clouds appear as brighter patches in thermal infrared imagery. The brightness temperature of warm clouds is relatively high compared to the surrounding background [47, 50, 53].

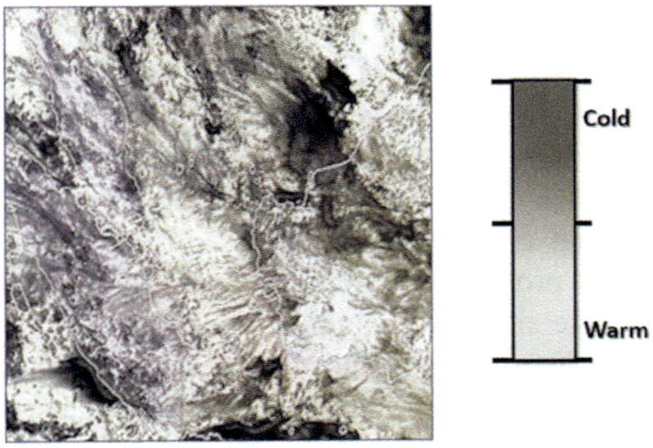

Figure 6.27: Quantum superposition is used to distinguish between cold and warm clouds in MODIS data.

Figure 6.28 demonstrates that thick clouds are characterized by strong quantum spectral signals, particularly in the visible band at 0.3 and 0.7 μm and the infrared bands at 1.12 and 1.48 μm, respectively. The spectral signature in the visible bands; therefore, might show relatively high reflectance values for cloud-covered areas. As the wavelength of the spectra increases, this spectral strength gradually diminishes, a pattern observed in both thick and thin cloud types. In this understanding, as the wavelengths of the spectra increase (move toward longer wavelengths), the strength of the quantum spectral signatures gradually decreases. In other words, the distinct features that make thick clouds stand out in the visible and lower infrared bands become less pronounced as the wavelengths become longer. Therefore, among the thickest clouds, Cumulonimbus Clouds exhibit the most pronounced high quantum spectral energy, surpassing Nimbostratus, Stratocumulonimbus, and Altostratus Clouds. Notably, Altostratus Clouds have the potential to obscure the Sun or

Moon, diffusing light and yielding lower reflectance values (Figure 6.28). Conversely, the quantum spectral signature energy of mid-level clouds falls below that of thick clouds but surpasses that of thin clouds. Mid-level clouds, exemplified by Altocumulus Castellanus, display vertical growth, a sign of atmospheric instability. These clouds can enhance reflectance values due to their denser composition [42, 44, 53, 57].

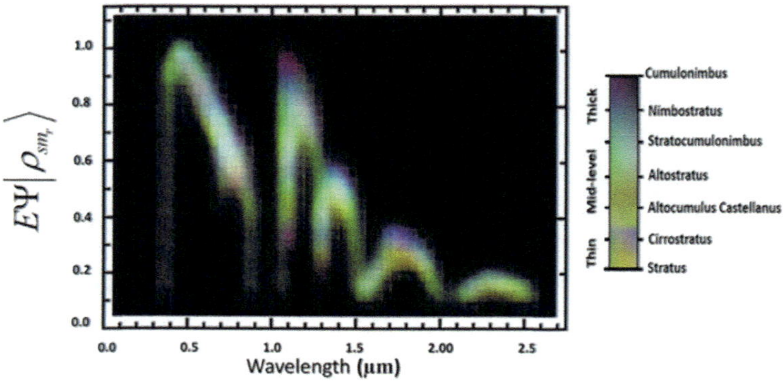

Figure 6.28: Quantum spectral signature of clouds in MODIS data.

Thin clouds like Stratus and Cirrostratus exhibit lower quantum spectral signature energy due to their composition, altitude, and transparency. Thin clouds are often composed of small ice crystals or water droplets. These particles interact with light in a way that causes less scattering and absorption compared to thicker clouds. This results in lower quantum spectral signature energy. Moreover, thin clouds are relatively transparent, allowing a significant portion of sunlight to pass through them. This transparency leads to a decrease in the scattering and reflection of light, contributing to lower quantum spectral signature energy. Thin clouds; consequently, have fewer cloud particles along their vertical extent, resulting in fewer interactions with light. In contrast, thicker clouds have more particles to scatter and reflect light, leading to higher energy signatures [35, 43, 53, 56].

The spectral signature of clouds in MODIS data refers to the distinct patterns of reflectance or brightness temperature that clouds exhibit across different wavelengths or spectral bands (Figure 6.29). These patterns provide information about cloud properties such as composition, altitude, thickness, and particle size. The combination of reflectance and brightness temperature values at various wavelengths helps identify and characterize different cloud types and atmospheric conditions. MODIS instruments on the Terra and Aqua satellites have multiple spectral bands, each sensitive to specific ranges of wavelengths [39, 41, 50, 53].

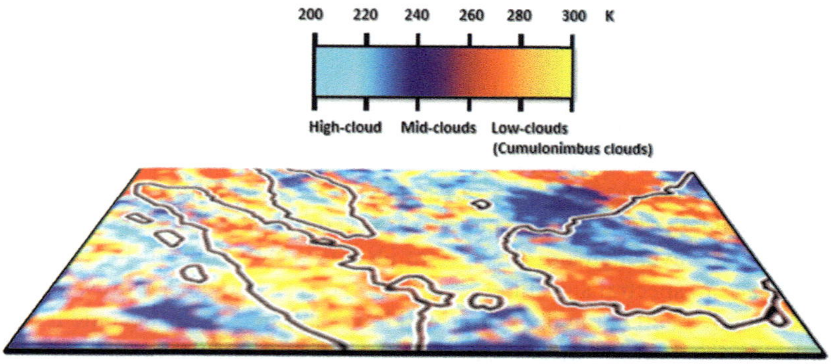

Figure 6.29: The spectral signature of clouds in MODIS data refers to their distinct reflectance or brightness temperature patterns across various wavelengths.

Accordingly, near-infrared wavelengths at 0.7 to 1.3 μm are sensitive to ice crystals and water droplets in clouds. Ice crystals strongly reflect near-infrared light, leading to high reflectance values in this range. Water droplets, on the other hand, scatter and absorb light, resulting in lower reflectance. The spectral signature in the near-infrared bands can provide information about the presence of ice crystals and liquid water in clouds. On the other hand,the thermal infrared bands beyond 1.3 μm capture the thermal radiation emitted by clouds. Cold, high-altitude clouds emit less thermal radiation, leading to higher brightness temperature values in these bands. Warmer clouds at lower altitudes emit more thermal radiation and have lower brightness temperature values. The spectral signature in the thermal infrared bands helps infer cloud-top temperatures and vertical distribution [39, 43, 45, 47, 53].

The specific shape of the spectral signature curve; consequently, can provide insights into cloud properties. For example, cirrus clouds, composed of ice crystals, might exhibit a strong peak in reflectance in the near-infrared range. Cumulus clouds, on the other hand, might have a more gradual increase in reflectance across visible and near-infrared wavelengths.

It is worth emphasizing that the spectral pattern can change depending on factors such as cloud variety, height, thickness, and how cloud particles interact with light through scattering and absorption. Through the utilization of MODIS data and similar remote sensing methods, researchers have the means to examine these patterns. This analysis provides insights into cloud characteristics, thereby contributing significantly to fields like weather prediction, climate investigation, and the study of Earth's atmosphere. Hence, the algorithm for quantum cloud spectral signatures can effectively categorize cloud types using their respective brightness temperatures. Illustrating this concept, Figure 6.29 exhibits an unconventional cloud pattern across Southeast Asia on January 24, 2023, as observed in MODIS data. In this context, the prevalence of high clouds is indicated by their lower brightness temperatures, typically ranging from 200 to 230 K, compared to the middle-level cloud coverage. Conversely, lower cloud coverage is characterized by higher brightness temperatures, falling within the range of 276 to 286 K. Notably, the primary lower cloud cover appears to be Cumulonimbus, notably distinguished by its brightness temperature of 286 K [39, 44, 47, 49, 52, 55, 57].

Utilizing quantum spectral energy for modeling cloud types and cloud brightness temperature from MODIS data leads to insightful conclusions. By employing quantum principles, researchers gain a refined understanding of the intricate relationships between cloud properties and their spectral signatures. This approach allows for enhanced classification of diverse cloud types based on their distinctive quantum manners. Furthermore, the application of quantum spectral energy provides a novel perspective on cloud brightness temperature patterns, shedding light on the underlying physics governing cloud patterns. As a result, this innovative methodology presents promising avenues for advancing cloud simulation, benefiting fields such as weather prediction, climate studies, and atmospheric research.

References

[1] Petrov, V.V., Bazanin, N.V., Kirin, D.V., Volkov, V.V. and Strunin, A.M. (2022). Relationship between microphysical characteristics and turbulence in winter clouds. *In:* International Scientific Conference "Problems of Atmospheric Physics, Climatology and Environmental Monitoring" (pp. 269–275). Cham: Springer International Publishing.

[2] Zhu, Z., Yang, F., Kollias, P. and Luke, E. (2023). Observational investigation of the effect of turbulence on microphysics and precipitation in warm marine boundary layer clouds. *Geophysical Research Letters*, 50(10), e2022GL102578.

[3] Bühl, J., Seifert, P., Engelmann, R. and Ansmann, A. (2019). Impact of vertical air motions on ice formation rate in mixed-phase cloud layers. *NPJClimate and Atmospheric Science*, 2(1), 36.

[4] Bourgeois, Q., Ekman, A.M., Igel, M.R. and Krejci, R. (2016). Ubiquity and impact of thin mid-level clouds in the tropics. *Nature Communications*, 7(1), 12432.
[5] Ansmann, A., Tesche, M., Seifert, P., Althausen, D., Engelmann, R., Fruntke, J., ... and Müller, D. (2009). Evolution of the ice phase in tropical altocumulus: SAMUM lidar observations over Cape Verde. *Journal of Geophysical Research: Atmospheres*, 114(D17).
[6] Westbrook, C.D. and Illingworth, A.J. (2013). The formation of ice in a long-lived supercooled layer cloud. *Quarterly Journal of the Royal Meteorological Society*, 139(677), 2209–2221.
[7] de Boer, G., Eloranta, E.W. and Shupe, M.D. (2009). Arctic mixed-phase stratiform cloud properties from multiple years of surface-based measurements at two high-latitude locations. *Journal of the Atmospheric Sciences*, 66(9), 2874–2887.
[8] Smith, J. (2020). *The Art of Cloud Watching*. Weather Publishing.
[9] Johnson, A. and Brown, B. (2019). Turbulence and cloud microphysics. *Journal of Atmospheric Science*, 45(3), 321–335.
[10] National Weather Service. (2022). Cloud types and turbulence. NOAA National Weather Service. https://www.weather.gov/cloudtypes.
[11] Adams, C. and Williams, D. (2021). The impact of turbulence on cloud formation. *In:* S. Johnson (Ed.), Proceedings of the International Conference on Atmospheric Sciences (pp. 123–135). Weather Publishing.
[12] World Meteorological Organization (WMO). (2017). International Cloud Atlas –Volume I: Manual on the Observation of Clouds and Other Meteors (WMO-No. 407). Geneva, Switzerland: World Meteorological Organization. [Available online: https://library.wmo.int/doc_num.php?explnum_id=4005]
[13] Ludlam, F.H. (1980). *Clouds and Storms: The Behavior, Effect, and Evolution of the Atmosphere*. New York: Houghton Mifflin Harcourt.
[14] Ahrens, C.D. (2019). *Meteorology Today: An Introduction to Weather, Climate, and the Environment*. Cengage Learning.
[15] Fitzgerald, R.J. (2017). International cloud atlas. *Physics Today*, 70(5), 76–76.
[16] Hamblyn, R. (2021). *The Cloud Book: How to Understand the Skies*. David and Charles.
[17] Lyons, W. (2017). Introducing asperitas: The newest cloud in the sky. *Weatherwise*, 70(6), 12–19.
[18] Rajini, S.A. and Tamilpavai, G. (2018). Classification of cloud/sky images based on knn and modified genetic algorithm. *In:* 2018 International Conference on Intelligent Computing and Communication for Smart World (I2C2SW) (pp. 1–8). IEEE.
[19] Dobashi, K., Uehara, H., Kandori, R., Sakurai, T., Kaiden, M., Umemoto, T. and Sato, F. (2005). Atlas and catalog of dark clouds based on digitized sky survey I. *Publications of the Astronomical Society of Japan*, 57(sp1), S1–S386.
[20] Lecours, M.J., Bernath, P.F., Sorensen, J.J., Boone, C.D., Johnson, R.M. and LaBelle, K. (2022). Atlas of ACE spectra of clouds and aerosols. *Journal of Quantitative Spectroscopy and Radiative Transfer*, 292, 108361.
[21] Lew, B.W., Apai, D., Zhou, Y., Schneider, G., Burgasser, A.J., Karalidi, T., ... and Lowrance, P.J. (2016). Cloud atlas: Discovery of patchy clouds and high-amplitude rotational modulations in a young, extremely red L-type Brown Dwarf. *The Astrophysical Journal Letters*, 829(2), L32.
[22] Tannock, M. E., Metchev, S., Heinze, A., Miles-Páez, P. A., Gagné, J., Burgasser, A., ... & Plavchan, P. (2021). Weather on Other Worlds. V. The Three Most Rapidly Rotating Ultra-cool Dwarfs. *The Astronomical Journal*, 161(5), 224.
[23] Berghaus, F., Casteels, K., Di Girolamo, A., Driemel, C., Ebert, M., Furano, F., ... and Taylor, R.P. (2018). Federating distributed storage for clouds in ATLAS. *Journal of Physics: Conference Series*, 1085(3), 032027. IOP Publishing.
[24] Kurihana, T., Moyer, E.J. and Foster, I.T. (2022). AICCA: AI-driven cloud classification atlas. *Remote Sensing*, 14(22), 5690.
[25] Lecours, M.J., Bernath, P.F., Sorensen, J.J., Boone, C.D., Johnson, R.M. and LaBelle, K. (2022). Atlas of ACE spectra of clouds and aerosols. *Journal of Quantitative Spectroscopy and Radiative Transfer*, 292, 108361.
[26] Brunke, M.A., Cutler, L., Urzua, R.D., Corral, A.F., Crosbie, E., Hair, J., ... and Ziemba, L.D. (2022). Aircraft observations of turbulence in cloudy and cloud-free boundary layers over the western north Atlantic Ocean from ACTIVATE and implications for the earth system model evaluation and development. *Journal of Geophysical Research: Atmospheres*, 127(19), e2022JD036480.

[27] Hamblyn, R. (2022). Luke Howard, namer of clouds. *Weather*, 77(11), 376–379.
[28] Di Felice, P. (2023). Infield mini radio-probes measurements of physical fluctuations inside warm clouds and environmental air. Doctoral dissertation, Politecnico di Torino.
[29] Deruelle, F. (2022). Are persistent aircraft trails a threat to the environment and health? *Reviews on Environmental Health*, 37(3), 407–421.
[30] Bryukhanov, I., Penzin, M., Kuchinskaia, O., Ni, E., Pustovalov, K., Bryukhanova, V., ... and Samokhvalov, I. (2022). Evaluation of the Applicability of the ERA5 Reanalysis for Interpretation of the Polarization Laser Sensing Data of the High-Level Clouds in Western Siberia. Preprints 2022, 2022050086.
[31] Lyons, W. (2019). Cloud forests of Costa Rica: Ecosystems in Peril. *Weatherwise*, 72(3), 32–37.
[32] Sarmiento, F.O. (2021). Dynamics of Andean treeline ecotones: Between cloud forest and páramogeocritical tropes. *The Andean Cloud Forest*, 25–42.
[33] Cristóbal-Pérez, E.J., Barrantes, G., Cascante-Marín, A., Madrigal-Brenes, R., Hanson, P. and Fuchs, E.J. (2023). Blooming plant species diversity patterns in two adjacent Costa Rican highland ecosystems. *PeerJ*, 11, e14445.
[34] Lin, J., Huang, T.Z., Zhao, X.L., Chen, Y., Zhang, Q. and Yuan, Q. (2022). Robust thick cloud removal for multitemporal remote sensing images using coupled tensor factorization. *IEEE Transactions on Geoscience and Remote Sensing*, 60, 1–16.
[35] Yuen, D.A., Scruggs, M.A., Spera, F.J., Zheng, Y., Hu, H., McNutt, S.R., ... and Tanioka, Y. (2022). Under the surface: Pressure-induced planetary-scale waves, volcanic lightning, and gaseous clouds caused by the submarine eruption of Hunga Tonga-HungaHa'apai volcano. *Earthquake Research Advances*, 2(3), 100134.
[36] Gebhardt, C., Guha, B.K., Young, R.M.B. and Wolff, M.J. (2022). A frontal dust storm in the northern hemisphere at solar longitude 97—An unusual observation by the Emirates Mars Mission. *Geophysical Research Letters*, 49(20), e2022GL099528.
[37] Wylie, D., Jackson, D.L., Menzel, W.P. and Bates, J.J. (2005). Trends in global cloud cover in two decades of HIRS observations. *Journal of Climate*, 18(15), 3021–3031.
[38] Bauer, P., Moreau, E. and Di Michele, S. (2005). Hydrometeor retrieval accuracy using microwave window and sounding channel observations. *Journal of Applied Meteorology*, 44(7), 1016–1032.
[39] Li, H., Zheng, H., Han, C., Wang, H., & Miao, M. (2018). Onboard spectral and spatial cloud detection for hyperspectral remote sensing images. *Remote Sensing*, 10(1), 152.
[40] Giuffrida, G., Diana, L., de Gioia, F., Benelli, G., Meoni, G., Donati, M. and Fanucci, L. (2020). CloudScout: A deep neural network for on-board cloud detection on hyperspectral images. *Remote Sensing*, 12(14), 2205.
[41] Prabhakara, C., Fraser, R.S., Dalu, G., Wu, M.L.C., Curran, R.J. and Styles, T. (1988). Thin cirrus clouds: Seasonal distribution over oceans deduced from Nimbus-4 IRIS. *Journal of Applied Meteorology and Climatology*, 27(4), 379–399.
[42] Bi, L., Yang, P., Liu, C., Yi, B., Baum, B.A., Van Diedenhoven, B. and Iwabuchi, H. (2014). Assessment of the accuracy of the conventional ray-tracing technique: Implications in remote sensing and radiative transfer involving ice clouds. *Journal of Quantitative Spectroscopy and Radiative Transfer*, 146, 158–174.
[43] Mishchenko, M.I. (2008). Multiple scattering, radiative transfer, and weak localization in discrete random media: Unified microphysical approach. *Reviews of Geophysics*, 46(2).
[44] O'Hirok, W. and Gautier, C. (1998). A three-dimensional radiative transfer model to investigate the solar radiation within a cloudy atmosphere. Part I: Spatial effects. *Journal of the Atmospheric Sciences*, 55(12), 2162–2179.
[45] Mayer, B. (2009). Radiative transfer in the cloudy atmosphere. *In:* EPJ web of Conferences (Vol. 1, pp. 75–99). EDP Sciences.
[46] Sancho, P. (2018). Entanglement in absorption processes. *European Journal of Physics*, 40(1), 015404.
[47] Marghany, M. (2022). *Remote Sensing and Image Processing in Mineralogy*. CRC Press.
[48] Sancho, P. (2020). Joint effects of entanglement and symmetrization: Physical properties and exclusion. *Annals of Physics*, 421, 168264.
[49] Lü, X.Y., Si, L.G., Hao, X.Y. and Yang, X. (2009). Achieving multipartite entanglement of distant atoms through selective photon emission and absorption processes. *Physical Review A*, 79(5), 052330.
[50] Song, G.Z., Tao, M.J., Qiu, J. and Wei, H.R. (2022). Quantum entanglement creation based on quantum scattering in one-dimensional waveguides. *Physical Review A*, 106(3), 032416.
[51] Chen, R.X. and Shen, L.T. (2011). Tripartite entanglement of atoms trapped in coupled cavities via quantum Zeno dynamics. *Physics Letters A*, 375(44), 3840–3844.

[52] Kurpiers, P., Magnard, P., Walter, T., Royer, B., Pechal, M., Heinsoo, J., ... and Wallraff, A. (2018). Deterministic quantum state transfer and remote entanglement using microwave photons. *Nature*, 558(7709), 264–267.

[53] de J León-Montiel, R., Svozilik, J., Salazar-Serrano, L.J. and Torres, J.P. (2013). Role of the spectral shape of quantum correlations in two-photon virtual-state spectroscopy. *New Journal of Physics*, 15(5), 053023.

[54] Laurel, C.O., Dong, S.H. and Cruz-Irisson, M. (2015). Equivalence of a bit pixel image to a quantum pixel image. *Communications in Theoretical Physics*, 64(5), 501.

[55] Bagayoko, D. (2014). Understanding density functional theory (DFT) and completing it in practice. *AIP Advances*, 4(12).

[56] Obot, I.B., Macdonald, D.D. and Gasem, Z.M. (2015). Density functional theory (DFT) as a powerful tool for designing new organic corrosion inhibitors. Part 1: An overview. *Corrosion Science*, 99, 1–30.

[57] Orio, M., Pantazis, D.A. and Neese, F. (2009). Density functional theory. *Photosynthesis Research*, 102, 443–453.

CHAPTER

7

Quantum Edge Detection Algorithm for Automated Cloud Street Identifications

7.1 What are Cloud Streets?

A cloud street refers to a meteorological phenomenon characterized by the presence of organized convection in the atmosphere. It manifests as an elongated line of cumulus clouds that typically align parallel to the prevailing direction of the wind (Figure 7.1). Cloud streets often appear in skies that are otherwise lightly clouded. The formation of these cloud streets is attributed to various sources of thermals, which are localized upward currents of warm air, distributed across the wind direction [1–3].

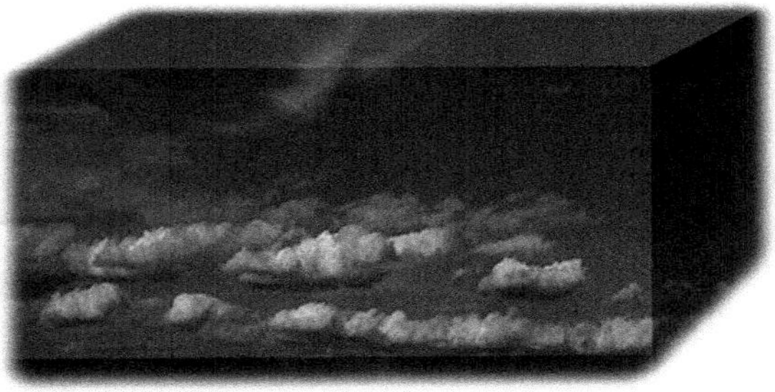

Figure 7.1: Horizontal convective rolls/cumulus mediocris radiatus.

In essence, a cloud street is a visible manifestation of the interaction between different air masses, temperature variations, and atmospheric dynamics. The parallel alignment of cumulus clouds in a cloud street is a result of the way thermals are generated and distributed within the atmosphere. These thermals create columns of rising air that trigger the formation of cumulus clouds, which then organize into the characteristic linear pattern [2–4].

Cloud streets are not only visually captivating but also provide valuable insights into the complex interactions that shape the Earth's atmosphere. Studying cloud streets aids meteorologists and atmospheric scientists in understanding the behavior of air masses, turbulence, and convection, contributing to improved weather forecasting and a deeper comprehension of atmospheric processes [1, 3, 4].

7.2 Mechanism of Cloud Street Formations

Cloud streets refer to elongated sequences of cumulus clouds that exhibit alignment with the prevailing wind direction. In technical terms, these formations are termed horizontal convective rolls (Figure 7.2). These patterns are commonly observed in satellite images. While they predominantly manifest as linear arrays, instances can arise where the clouds' progression is hindered by obstacles, causing the clouds to adopt curved configurations known as von Kármán vortex streets. This phenomenon underscores the intricate interplay of atmospheric dynamics and fluid mechanics.

Figure 7.2: Generation of cloud streets through the mechanism of horizontal convective rolls.

7.2.1 Horizontal Convective Rolls

In the intricate dance of Earth's atmosphere, there exists a phenomenon known as horizontal convective rolls, whispered by some as cloud streets. These are elongated bands of air, gracefully twirling in counter-rotating fashion, dressed in the attire of cumulus clouds (Figure 7.1). These celestial performers align themselves almost in parallel with the earthly realm, painting a captivating portrait in the planetary boundary layer.

The planetary boundary layer (PBL), often referred to as the atmospheric boundary layer (ABL) or peplosphere, constitutes the lowermost portion of the Earth's atmosphere. It is the interface where the atmosphere directly interacts with the underlying planetary surface, which can be the Earth's surface, water bodies, or even structures. The behavior of the PBL is significantly shaped by this interaction [1, 4, 5].

In this layer, the atmosphere is subject to various influences from the surface below. These influences include friction, heat exchange, moisture transfer, and momentum transfer. As air molecules near the surface experience contact with the ground, their movement is constrained by the friction generated between the air and the surface. This friction causes the air to move more sluggishly near the surface, creating a gradient of air velocity known as the surface layer.

The PBL is a dynamic region where air characteristics like temperature, humidity, and wind speed can vary considerably. During daytime, the Sun's heating of the ground leads to warm air rising and creating vertical motions that extend through the PBL. Conversely, during the night, cooling at the surface can lead to stable atmospheric conditions, limiting vertical mixing and trapping pollutants near the ground.

Understanding the behavior of the planetary boundary layer is crucial in various scientific fields, including meteorology, air quality monitoring, and climate studies. The PBL's response to changes in surface conditions plays a key role in shaping weather patterns, influencing local climate, and determining the dispersion of pollutants in the atmosphere. Thus, the planetary boundary layer serves as a critical bridge between the solid surface of the planet and the vast expanse of the atmosphere above [2–5].

As the wind's gentle touch orchestrates their movement, these horizontal roll vortices weave tales of atmospheric currents and kinematics. Their rhythmic undulations carve a space parallel to the ground, where air and cloud harmonize in a mesmerizing symphony. These cloud streets, as they're sometimes affectionately called, are a ballet of elements, performing their choreography under the vast theatre of the sky [1, 4, 6].

Indeed, the spectacle is not merely a scene in nature's gallery; it's a manifestation of the delicate equilibrium between the forces that shape our world. Their very existence echoes the whispers of the wind and the secrets of fluid dynamics. Each roll, each cloud, contributes to a story of atmospheric intricacies, a story that scientists and dreamers alike seek to decipher.

In these realms of horizontal convective rolls, insights into the dance of air and cloud unravel. And so, these cloud streets are not just wisps of water vapor; they are the brush strokes of nature's artistry, painted across the canvas of the atmosphere [2–4].

7.2.2 Kármán Vortex Street

In the realm of fluid dynamics, a Kármán vortex street, also referred to as a von Kármán vortex street, emerges as a recurring arrangement of swirling vortices. This phenomenon is a consequence of a process termed vortex shedding, which leads to the intermittent detachment of fluid flow around blunt objects [1, 6, 9, 11].

Vortex shedding occurs when a fluid, such as air or water, encounters an obstacle or body. The flow of the fluid around this object is not uniform; instead, it creates alternating regions of low and high pressure. As the fluid moves around the object, these alternating pressure zones cause the formation of vortices on both sides of the object. These vortices are the essence of the Kármán vortex street (Figure 7.3) [2, 5, 8].

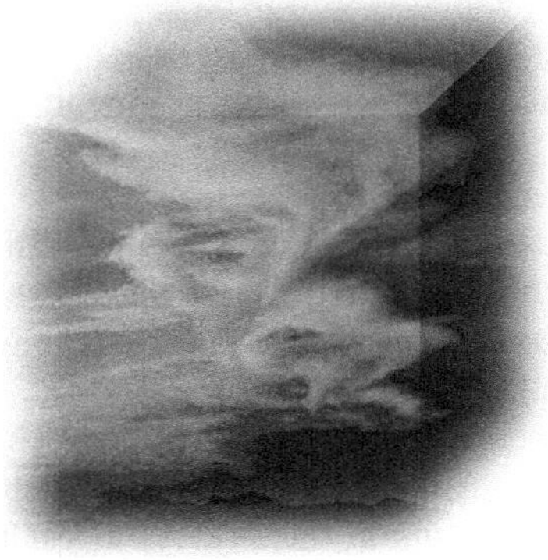

Figure 7.3: Kármán vortex cloud streets.

The term "Kármán vortex street" honors Theodore von Kármán, a Hungarian-American aerospace engineer and physicist who significantly contributed to the understanding of fluid dynamics and aerodynamics. The phenomenon is most prominently observed when the flow of the fluid reaches a certain critical speed, resulting in the consistent shedding of vortices from the object.

Kármán vortex streets have applications in various fields, including aerodynamics, civil engineering, and oceanography. Understanding the dynamics of vortex shedding is crucial in designing structures that can withstand fluid forces and optimizing the performance of vehicles moving through air or water. By unravelling the intricacies of these repeating patterns of vortices, scientists and engineers gain valuable insights into the behaviour of fluids around obstacles and bodies in motion [1, 3, 12].

7.3 Characteristics of Cloud Streets

According to the above perspective, cloud streets exhibit distinct characteristics that make them a fascinating meteorological phenomenon of scientific interest. Some of the salient features of cloud streets include linear configuration; cumulus cloud composition; regular spacing; and horizontal convective rolls.

Horizontal rolls, also known as counter-rotating vortex rolls, exhibit a distinct alignment with the mean wind direction within the Planetary Boundary Layer (PBL). These rolls are generated through convection in the presence of moderate winds or dynamic inflection point instabilities in the mean wind profile. Early theoretical models concerning these phenomena propose that the vortices can be oriented at various angles relative to the mean wind, depending on the stratification of the atmosphere. In this understanding, theoretical predictions suggest that the vortices may be aligned at an angle of up to 30° to the left of the mean wind direction for stably stratified environments. In neutral environments, however, this alignment angle is estimated to be around 18° to the left. In contrast, in unstably stratified (convective) conditions, the vortices tend to be nearly parallel to the mean wind direction [1, 3, 10].

Deriving a mathematical equation for the behavior of horizontal rolls involves complex fluid dynamics and atmospheric science principles. While a comprehensive equation requires a detailed understanding of these processes, the author can provide a simplified representation based on the general characteristics of horizontal rolls in the context of atmospheric dynamics.

Let θ represent the angle of alignment of the horizontal rolls concerning the mean wind direction. In a stably stratified environment:

$$\Theta\text{-}\theta_{stably}\text{-}30^\circ \tag{7.1}$$

However, in a neutral environment:

$$\Theta\text{-}\theta_{stably}\text{-}18^\circ \tag{7.2}$$

Thus, in an unstably stratified (convective) environment:

$$\Theta\text{-}\theta_{stably}\sim 0^\circ \tag{7.3}$$

These equations represent the approximate alignment angles of horizontal rolls based on theoretical concepts. However, it's important to note that these equations are simplified and may not accurately capture the full complexity of the atmospheric dynamics involved in the formation and behavior of horizontal rolls [1, 3, 12, 14].

Consistent with the above perspective, cloud streets are characterized by their elongated and linear arrangement. Cumulus clouds align themselves parallel to the direction of the prevailing wind, creating visually striking patterns in the sky. As a result, the cloud streets predominantly consist of a number of cumulus clouds (N_c). These clouds are characterized by their white, fluffy appearance and occur in well-defined columns within the cloud street formation. Therefore, the cumulus clouds within cloud streets often exhibit regular and consistent spacing L. This is a result of the underlying atmospheric processes that lead to the formation of cloud streets. Cloud streets, consequently, emerge due to the presence of horizontal convective rolls. These rolls are bands of rotating air generated near the interface between contrasting air masses U. The interplay of upward and downward motions within these rolls triggers cloud development [11–14]. A simplified mathematical expression for the spacing of cumulus clouds in a cloud street can be represented as:

$$L = N_c U^{-1} \tag{7.4}$$

Keep in mind that this is a highly simplified representation and doesn't encompass the full complexity of cloud street formation. The actual formation of cloud streets involves interactions between temperature gradients, wind shear, atmospheric stability, and other factors that are best captured through advanced atmospheric models and observational data. Deriving a single mathematical equation for cloud street formation based solely on temperature gradients is challenging due to the complexity of atmospheric processes involved. In this sense, let ΔT represent the temperature difference between the water surface and the air above it. Thus, a simplified equation that relates temperature gradients to cloud street formation could be:

$$L = k.\Delta T \tag{7.5}$$

Here, k is a proportionality constant that captures the relationship between temperature gradients and the resulting cloud street spacing. A larger temperature difference (ΔT) could lead to smaller cloud street spacing (L), indicating more frequent cloud formation along the path of the wind. In this perspective, the formation of horizontal convective rolls due to wind shear involves complex fluid dynamics and interactions [3, 11, 13]. While there isn't a single equation that captures the entire process, the author can provide a simplified conceptual representation:

Let $\frac{\partial u}{\partial x}$; and $\frac{\partial u}{\partial z}$ the horizontal wind shear, and the vertical wind shear, respectively, change in horizontal wind speed with distance. Additionally, assume ΔT represents the temperature difference between the surface and the atmosphere [1, 12, 13]. A simplified equation, therefore, that relates wind shear and temperature differences to the generation of horizontal convective rolls H_{cr} in the context of cloud street formation could be:

$$H_{cr} = k.\Delta T\left[\frac{\partial u}{\partial z}.\frac{\partial u}{\partial x}\right] \tag{7.6}$$

Equation 7.6 suggests that the generation of horizontal convective rolls, which contribute to cloud street formation, is influenced by a combination of vertical and horizontal wind shear, as well as temperature differences between the surface and the atmosphere [10–14].

Therefore, the stability of the lower atmosphere plays a significant role. In an unstable atmosphere, warm, buoyant air near the surface rises easily, leading to the formation of cumulus clouds. Conversely, in a stable atmosphere, vertical mixing is limited, hindering cloud development. In this regard, atmospheric stability is often quantified using the concept of potential temperature (θ) and lapse rates. One commonly used measure of atmospheric stability is the potential temperature lapse rate (Γ_θ):

$$\Gamma_\theta = -\frac{1}{\theta}\frac{\partial \theta}{\partial z} \tag{7.7}$$

The relationship between the potential temperature lapse rate (Γ_θ) and the environmental lapse rate (Γ_e) provides insights into the stability of the atmosphere. When $\Gamma_\theta > \Gamma_e$, the atmosphere is stable, and when $\Gamma_\theta < \Gamma_e$, the atmosphere is unstable. This equation represents simplified concepts of atmospheric stability. More comprehensive stability indices and parameters are used in atmospheric science to analyze and categorize stability conditions, which play a crucial role in weather patterns and phenomena [12–14].

Atmospheric stability; therefore, affects the buoyancy and vertical motion of air parcels, which in turn influences cloud street formation. Stable conditions tend to suppress vertical motion and limit cloud development, while unstable conditions promote vertical motion and enhance cloud formation. Let's consider a simplified conceptual representation that relates atmospheric stability to cloud street formation λ_{cr}:

$$\lambda_{cr} = C\frac{\partial \theta}{\partial z} \tag{7.8}$$

Here C is constant and equation 7.8 suggests that cloud street formation is influenced by the rate of change of potential temperature with height. More stable atmospheric conditions $\left(\frac{\partial \theta}{\partial z} > 0\right)$ suppress vertical motion and may hinder cloud street formation. On the other hand, more unstable conditions $\left(\frac{\partial \theta}{\partial z} < 0\right)$ encourage vertical motion and could enhance the formation of cloud streets [1, 6, 14].

Cloud streets, therefore, are commonly observed over bodies of water, such as oceans and lakes. The differential heating between the water surface and the overlying air fosters the atmospheric conditions conducive to cloud street formation. In this perspective, when cloud streets encounter terrain features like mountains or buildings, they can undergo curving and complex patterns known as von Kármán vortex streets, adding a layer of complexity to their appearance. In this understanding, Von Kármán vortex streets are a phenomenon caused by the shedding of vortices behind an obstacle in a fluid flow. While the exact equations that describe von Kármán vortex streets involve complex fluid dynamics, the author can provide a simplified representation of the phenomenon [13–15]. Consider a two-dimensional flow of a fluid past a cylindrical obstacle. The vortices formed in the wake of the obstacle led to the formation of von Kármán vortex streets. One way to represent this phenomenon is through the vorticity equation:

$$\frac{\partial \omega}{\partial t} + (\vec{u}.\nabla)\omega - \nu\nabla^2\omega = 0 \tag{7.9}$$

Here ω is the vorticity (a measure of local rotation of cloud street); $\vec{u}$ is the velocity vector; and ν is kinematic viscosity. In the context of von Kármán vortex streets, the shedding of vortices occurs due to the separation of fluid flow behind the obstacle, leading to regions of alternating high and low vorticity. These vortices form the characteristic pattern observed in von Kármán vortex streets.

7.4 Mathematical Description of Cloud Streets Causing Turbulence

The mathematical description of cloud streets causing turbulence involves considering the interactions between fluid motion, cloud formation, and turbulence. While a comprehensive model would require complex equations and numerical simulations, the author attempts to provide a conceptual representation of how cloud streets can contribute to turbulence. In this view, turbulence equations for cloud rotating movements involve describing the complex interactions between fluid motion, rotation, and turbulence [1, 16, 18]. While there isn't a single equation that captures all aspects of this phenomenon. Consider a simplified equation that represents the interplay between rotation and turbulence in the context of cloud street-rotating movements:

$$\frac{\partial \vec{u}}{\partial t} + (\vec{u}.\nabla)\vec{u} = -\frac{1}{\rho}\nabla P + \nu\nabla^2\vec{u} + \vec{F} - 2\Omega \times \vec{u} \tag{7.10}$$

This equation builds on the Navier-Stokes equations to account for the Coriolis effect due to rotation. The term $2\Omega \times \vec{u}$ represents the cross-product between the rotation vector and the velocity vector, reflecting the influence of rotation on fluid motion. Therefore, turbulence in convection is complex and typically requires additional equations to model it. Reynolds-averaged Navier-Stokes (RANS) equations are commonly used for turbulence modeling. The RANS equations involve introducing Reynolds stresses and eddy viscosity terms to account for turbulent effects [1, 15, 18]. In this view, the RANS equation for the turbulent kinetic energy (k) can be written as:

$$\frac{\partial k}{\partial t} + (u.\nabla k)\vec{u} = \frac{\partial}{\partial x_i}\left[\left(\nu + \frac{\nu_t}{\sigma_k}\right)\frac{\partial k}{\partial x_i}\right] + P_K - \varepsilon \tag{7.11}$$

Where P_K represents the production of turbulent kinetic energy; σ_k is a model constant; ν_t is the eddy viscosity; k is the turbulent kinetic energy; and ε is the dissipation rate of turbulent kinetic energy. Equations 7.10 and 7.11 are part of a more comprehensive set of equations that capture the complex interactions between turbulent motion, convection, and other atmospheric processes.

Therefore, the convergence and divergence zones between cloud street bands can lead to mixing of air masses with different properties. This mixing, combined with shear and thermal effects, can enhance turbulence within and around cloud streets. In this regard, turbulent mixing involves the transport of properties like temperature, moisture, and momentum due to random fluid motion [15–18]. In the context of cloud streets, turbulence plays a role in mixing air masses with different properties, enhancing the organization of convective rolls and cloud patterns. A simplified equation for turbulent mixing within cloud streets can be expressed as:

$$\frac{\partial M}{\partial t} + (\vec{u}.\nabla)M = \nu \nabla^2 M \tag{7.12}$$

Equation 7.12 demonstrates how the property M changes over time due to advection (transport by fluid motion) and diffusion (turbulent mixing). The term $(\vec{u}.\nabla)M$ represents the advection term and $\nu\nabla^2 M$ represents the diffusion term. In cloud streets, this equation captures how turbulent mixing contributes to the convergence and divergence zones between alternating warm and cool air bands. The mixing of air masses with different properties can lead to variations in temperature, humidity, and other parameters, affecting cloud formation and the overall structure of the cloud street pattern [1, 13, 16, 19].

Consequently, the vertical motion of clouds in the atmosphere is influenced by various factors, including buoyancy, temperature gradients, and wind patterns. The equation that describes the vertical motion of air parcels, including cloud motion, is known as the equation of motion or the vertical momentum equation.

The simplified form of the vertical momentum equation can be represented as:

$$\frac{\partial u}{\partial t} = -\frac{1}{\rho}\frac{dP}{dz} - g + \frac{Dw}{dt} + F_{\text{buoyancy}} \tag{7.13}$$

This equation captures the balance between pressure gradient $\frac{dP}{dz}$, F_{buoyancy}, and gravity acceleration g that influence vertical motion. The buoyancy of cloud streets, or any fluid parcel in the atmosphere, is described by the buoyancy force that results from the density difference between the air density ρ_{air} and its surrounding environment parcel ρ_{parcel}. The equation for buoyancy can be expressed as:

$$F_{\text{buoyancy}} = -g.(\rho_{air} - \rho_{parcel}) \tag{7.13.1}$$

In the context of cloud streets, the buoyancy force plays a crucial role in their formation. The alternating warm and cool air bands within cloud streets create variations in density, leading to differences in buoyancy forces. As warmer air rises due to lower density (positive buoyancy), it creates upward motion and contributes to the organization of horizontal convective rolls, which are characteristic of cloud streets [1, 3, 14].

Positive vertical velocity u represents upward motion, which could correspond to the upward movement of clouds. Therefore, $\frac{Dw}{dt}$ represents the rate of change of vertical velocity concerning time. In this understanding, the vertical velocity associated with convective processes can be given by Deardorff expression. The Deardorff expression is a widely-used empirical formula to estimate the convective velocity scale in the atmospheric boundary layer, which is important for understanding turbulence and vertical mixing in the boundary layer. Therefore, let us consider the virtual potential temperature difference between the surface and the mixed layer is θ_p; q_s is the saturation specific humidity at the surface temperature; q is the specific humidity of the mixed layer; and R_c is the bulk Richardson number. Thus, the Deardorff expression for convective velocity (w_{conv}) is given as:

$$w_{conv} = {}^{3}\sqrt{\left(\frac{g}{\theta_p}\right)}\sqrt{\left(\frac{q_s - q}{R_c}\right)} \tag{7.14}$$

The Deardorff expression takes into account the potential temperature difference between the surface and the mixed layer top, as well as the specific humidity difference and the bulk Richardson number. It provides an estimation of the typical vertical velocity associated with convective processes, allowing researchers to gain insights into vertical mixing and turbulence within the boundary layer. Needless to say, the Deardorff expression for convective velocity involves (i) Kinematic vertical turbulent flux of virtual potential temperature near the surface; and (ii) Convective velocity scale. The convective velocity scale is the characteristic velocity associated with convective motions. It is typically on the order of 1 ms^{-1}. On the other hand, Kinematic vertical turbulent flux represents the turbulent flux of virtual potential temperature, which is a measure of the convective activity near the surface [13–18].

7.5 Observational Mechanisms Employed by Satellites to Monitor Cloud Streets

Satellites can observe cloud streets using various remote sensing techniques and instruments. Cloud streets, which are organized rows of cumulus clouds, can be identified and studied from space through satellite imagery and data. These sensors are designed to detect and capture electromagnetic radiation in different wavelength ranges, allowing scientists to gather valuable information about cloud properties, dynamics, and atmospheric processes.

7.5.1 Visible Spectra Signature

Visible light sensors on satellites are designed to capture electromagnetic radiation within the visible light spectrum, which corresponds to the wavelengths of light that are visible to the human eye. The range of wavelengths typically covered by visible light sensors is approximately 0.38 to 0.7 μm. This range encompasses different colors, with shorter wavelengths corresponding to violet and blue light, and longer wavelengths corresponding to green, yellow, orange, and red light. Therefore, when visible bands capture imagery of cloud streets, several key factors come into play. In this view, the Sunlight illuminates the Earth during the day, and a significant portion of this sunlight is reflected by various objects on the planet's surface, such as land, water bodies, and clouds. Cumulus clouds within cloud streets have a reflective property, causing them to scatter and reflect sunlight in various directions [1–4].

Consequently, the visible light imagery obtained from these sensors provides a visual representation of the scene below. This means that the colors and brightness levels in the imagery correspond to the colors and brightness that we would perceive with our own eyes. The question now is: How do cloud streets appear in the visible band? Cumulus clouds, commonly found in cloud streets, are fluffy, white clouds with well-defined edges. They appear bright in visible light imagery due to their ability to strongly reflect sunlight. For instance, Figure 7.4 demonstrates visible image was captured by NASA's Terra satellite's Moderate Resolution Imaging Spectroradiometer (MODIS) on November 18, 2017. It depicts the presence of cloud streets, which are characterized by the occurrence of long parallel bands of cumulus clouds, situated over the Yellow Sea [4–8].

This brightness contrasts with the darker background of the sky, making the clouds stand out prominently. Subsequently, cloud streets are characterized by their organized rows of cumulus clouds. When visible light sensors capture these cloud streets, the linear arrangement of the clouds becomes evident in the imagery. This linear pattern is formed due to the organized convection processes that lead to the development of cloud streets. The repeated arrangement of bright cloud patches against the darker sky background makes the linear nature of cloud streets easily recognizable [4–6].

7.5.2 Infrared Spectral Signature

It is a widely recognized fact that the infrared spectrum is partitioned into distinct wavelength ranges: Near Infrared (NIR), which spans from 0.7 to 1.3 μm; Short-Wave Infrared (SWIR), covering the range of 1.3 to 3.0 μm; Mid-Wave Infrared (MWIR), encompassing wavelengths from 3.0 to 5.0 μm; Long-Wave Infrared (LWIR), spanning the interval of 8.0 to 15.0 μm; and

Far Infrared (FIR), extending beyond the range of 15.0 μm. The significant question at this point is: Which infrared band can be utilized for observing cloud streets?

The range of infrared wavelengths commonly used to observe cloud streets falls within the mid-wave infrared (MWIR) and long-wave infrared (LWIR) bands. These bands are selected because they allow for the detection of temperature differences between clouds and their surroundings, making cloud streets stand out in the imagery. The approximate ranges of these infrared bands are Mid-Wave Infrared (MWIR), and Long-Wave Infrared (LWIR). In these ranges, the MWIR and LWIR sensors on satellites can capture the heat radiation emitted by objects, including clouds. Cloud streets, which often have different temperatures compared to their surroundings, stand out in infrared imagery due to the temperature contrast. The linear arrangement and distinct thermal properties of cloud streets become evident in images captured within these infrared wavelength ranges. Figure 7.5 showcases the presence of cloud streets through a composite image combining Mid-Wave Infrared (MWIR) and Long-Wave Infrared (LWIR) data from MODIS. This composite image is featured in Figure 7.4 for reference [6–8].

Figure 7.4: Visible MODIS band imagined cloud streets.

Figure 7.5: The composite image of Mid-Wave Infrared (MWIR) and Long-Wave Infrared (LWIR) data derived from MODIS showcases the visual representation of cloud streets.

The key principle behind this sensing is that all objects with a temperature above absolute zero emit radiation in the form of heat. When applied to observing cloud streets, infrared sensors capture the variations in temperature between the clouds, the surface of the Earth, and the surrounding atmosphere. Clouds typically have temperatures different from their surroundings due to their interaction with solar radiation, moisture, and atmospheric conditions.

In this view, clouds, like all objects, emit radiation in the form of heat energy due to their temperature. This emitted thermal radiation is measured by the infrared sensors on the satellite. From the point of view of temperature differences, cloud streets, consisting of cumulus clouds, have a different temperature from the underlying surface and the surrounding air. Clouds are cooler due to their altitude and moisture content. This temperature difference between the clouds and their surroundings contributes to the distinctiveness of cloud streets in infrared imagery. In this perspective, warmer objects emit more infrared radiation. Because the clouds are cooler compared to the surface and the adjacent atmosphere, they emit less infrared radiation. This difference in emitted radiation based on temperature is captured by the sensors.

In these circumstances, infrared imagery represents these temperature differences using false color or grayscale schemes. Cooler areas, such as the cloud streets, are depicted in different shades than warmer areas, such as the Earth's surface. The resulting imagery creates a clear visual distinction between the clouds within the cloud streets and the surrounding environment. Infrared imagery highlights the temperature contrast between cloud streets and their surroundings. The linear arrangement of cloud streets becomes apparent as the cooler clouds stand out against the warmer background [3, 4, 6, 8].

The central inquiry pertains to the underlying reasons for the formation of these cloud streets over the regions of China and Korea. During the cold season, which marks the opportune time, the phenomenon of cloud streets becomes evident above the Yellow Sea. This occurrence is driven by the interaction between cold, dry air originating from Siberia and the comparatively warmer and moister sea surface. However, this transformative process doesn't occur instantaneously; the air mass requires a certain period over the sea to absorb the requisite heat and moisture. Consequently, the coastal region remains cloud-free due to the initial stages of this process [6–8].

As the air continues its trajectory over the sea, cloud streets begin to manifest, characterized by clouds arranged in parallel rows. These cloud streets undergo a developmental sequence, transitioning from open-cell to closed-cell cloud structures. In closed-cell clouds, the individual cloud cells feature opaque cumulus clouds at their centres (Figure 7.6). In contrast, open-cell clouds exhibit a ring of clouds encircling their cells, leaving the central areas devoid of clouds (Figure 7.7). This sequential evolution of cloud structures is intricately linked to atmospheric dynamics and underlying meteorological processes.

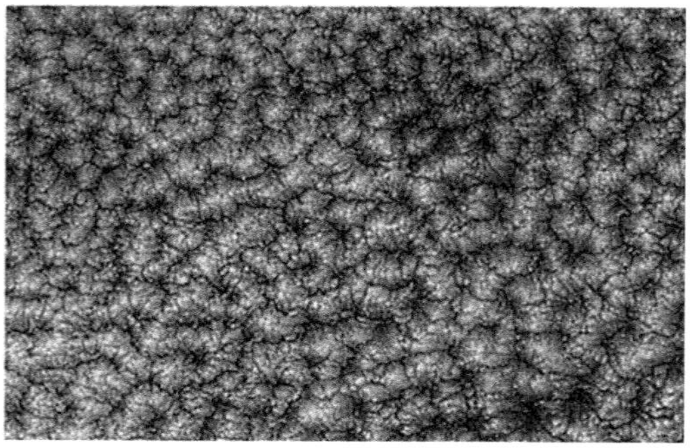

Figure 7.6: Closed-cell clouds from MODIS data.

7.6 Quantized Spectral Signature of Cloud Streets

In the realm of quantum statistical mechanics, the notion of quantum temperature emerges, introducing us to a fascinating domain where particles' behaviors unfold at the quantum scale. Traditionally, temperature in classical physics gauges the average kinetic energy of particles within a system. Yet, in the quantum landscape, particles eschew predictable trajectories and embrace the uncertainties inherent in wave functions and probabilities. Let us consider,a hypothetical expression that combines quantum principles with the idea of brightness temperature:

$$\Psi|T_B\rangle = \frac{\Psi|E\rangle_i}{k - \ln\left(\frac{\Psi|N\rangle_i}{\Psi|N\rangle_0}\right)} \tag{7.15}$$

Here $\Psi|T_B\rangle$ is the quantum brightness temperature (hypothetical); $\Psi|E\rangle_i$ is the energy associated with a quantum state; k is the Boltzmann constant; $\Psi|N\rangle_i$ is the number of particles in the quantum state; and $\Psi|N\rangle_0$ is the number of particles in the ground state. Equation 7.15 combines quantum principles with the idea of brightness temperature, which is more commonly associated with classical physics and remote sensing. Therefore, Equation 7.15 expresses the Bose-Einstein distribution function that describes how particles are distributed among different energy states in a quantum system, and it accounts for quantum effects such as the tendency of bosons (particles with integer spin) to condense into the lowest energy state at very low temperatures [22–24].

Consistent with the above perspective, the main challenge question is how to determine $\Psi|N\rangle_i$? In this view, $\Psi|N\rangle_i$ can be mathematically expressed as:

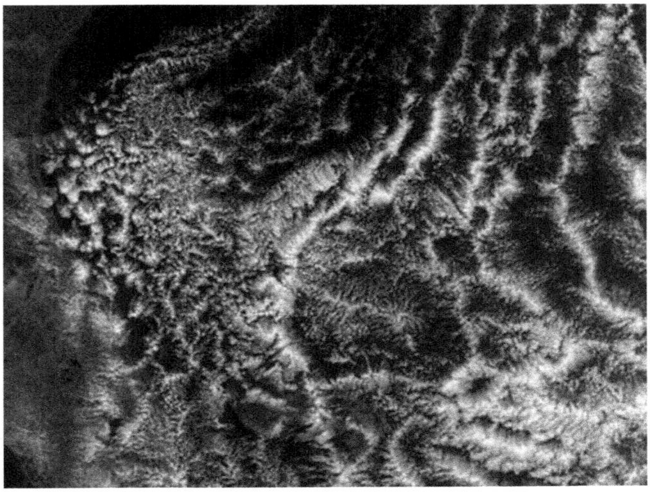

Figure 7.7: Open-cell clouds from MODIS data.

$$\Psi|N\rangle_i = \Psi\left[e^{\frac{\Psi|E\rangle_i - \mu_i}{kT_i}} - 1\right]^{-1} \tag{7.15.1}$$

where μ is the physical potential, which involves temperature, pressure, humidity, wind patterns, and more. While there isn't a single equation that encompasses all aspects of cloud behavior, here are some equations that relate to specific physical parameters: (i) Clausius-Clapeyron Equation; (ii) Relative Humidity (RH) Calculation; (iii) Lapse rate equation; and (iv) hydrostatic equation. Hence, *i* represents the count of spectral energy levels corresponding to the various physical potential factors engaged in the computation of the spectral energy signature of cloud streets [20, 22, 24], elucidated further in the subsequent description. Clausius-Clapeyron Equation (Relating Temperature and Saturation Vapor Pressure) [23–25] can be given by:

$$\ln\left(\frac{S_2}{S_1}\right) = L[R]^{-1}\left(\frac{1}{T_1} - \frac{1}{T_2}\right) \tag{7.16}$$

In equation 7.16, S_1 and S_2 are the vapor pressures at temperatures T_1 and T_2 respectively. Additionally, L is the latent heat of vaporization (or sublimation) of the substance and R is the gas constant. Suppose the vapor pressure of water at 20 °C is 2.34 kPa and at 25 °C is 3.17 kPa. Using the Clausius-Clapeyron Equation, we can calculate the latent heat of vaporization for water:

$$\ln\left(\frac{3.17 \text{ kPa}}{2.34 \text{ kPa}}\right) = \frac{2.5\times10^{6}\,\text{Jkg}^{-1}}{R}\left(\frac{1}{298\text{K}} - \frac{1}{298+5\text{K}}\right) \tag{7.17}$$

Given that the gas constant R is approximately 8.314 J/(mol·K), we can solve for L:

$$L = \frac{R\times\ln\left(\frac{3.17 \text{ kPa}}{2.34 \text{ kPa}}\right)}{\frac{1}{298} - \frac{1}{303}} = 2.46\times10^{6}\,\text{Jkg}^{-1} \tag{7.18}$$

In this example, the equation allows us to determine the latent heat of vaporization for water based on the change in vapor pressure with temperature [22–25]. The equation applies to other substances undergoing phase transitions as well.

Accordingly, Relative Humidity (RH) Calculation is casted as:

$$RH = \varepsilon\left[e_s\right]^{-1} \times 100 \tag{7.19}$$

here ε is the actual vapor pressure; and e_s is the saturation vapor pressure at the given temperature. For example, let's say the air temperature is 25 °C, and the actual vapor pressure is 15 hPa. Using meteorological tables or equations, you determine that the saturation vapor pressure at 25 °C is 25 hPa.

$$RH = \frac{15\text{ hPa}}{25\text{ hPa}} \times 100\% = 60\% \tag{7.20}$$

This means that the air is holding 60% of the maximum moisture it could hold at 25 °C. If the relative humidity were 100%, it would indicate that the air is fully saturated and cannot hold any more moisture. Conversely, if the relative humidity were 0%, the air would be completely dry. Relative humidity is crucial in weather forecasting because it influences cloud formation, precipitation, and the overall comfort level for humans. High relative humidity can make the air feel more humid and uncomfortable, while low relative humidity can lead to dry conditions. It's important to note that relative humidity is temperature-dependent; the same absolute amount of water vapor can result in different relative humidities at different temperatures [1, 22, 24].

Consequently, the Lapse rate Γ can be given by:

$$\Gamma = -\frac{dT}{dz} \tag{7.21}$$

The lapse rate is a measure of how temperature changes with altitude in the Earth's atmosphere. It indicates the rate at which temperature decreases as you move upward through the atmosphere. The lapse rate is an essential concept in meteorology and climatology, as it has a significant impact on weather patterns and atmospheric stability. For instance, the standard lapse rate is often referred to as the "dry adiabatic lapse rate" and is approximately –9.8 °C per kilometer (–3.6 °F per 1000 feet) in the troposphere, which is the lowest layer of the atmosphere.

For example, let's consider a scenario where you're climbing a mountain, and you measure the temperature change as you ascend. If you start at sea level with a temperature of 20 °C and you climb to an altitude of 2 kilometers where the temperature is now 5 °C, you can calculate the lapse rate as follows:

$$\Gamma = \frac{5\text{ °C} - 20\text{ °C}}{2\text{ km}} = -7.5\text{ °C km}^{-1} \tag{7.22}$$

In this example, the temperature decreases by 15 °C over a vertical distance of 2 kilometers, resulting in a lapse rate of –7.5 °C per kilometer. This value indicates that the atmosphere is cooling as you move higher.

Last of all, the hydrostatic equation can be expressed as:

$$\frac{dp}{dz} = -\rho g \tag{7.23}$$

Here $\frac{dp}{dz}$ is the rate of change of pressure concerning altitude; ρ is the air density; and g is the acceleration due to gravity. Suppose we measure a pressure of 1000 hPa at sea level (altitude $z = 0$) and 800 hPa at an altitude of $z = 5000$ m. The pressure difference is $\Delta P = 1000$ hPa − 800 hPa = 200 hPa. The change in altitude Δz is 5000 m. Using the hydrostatic equation (7.23):

$$\frac{dp}{dz} = -\rho g = -\left(1.2\text{ kgm}^{-3}\right) \times (9.8\text{ ms}^2) = -11.76\text{ Nm}^2 \tag{7.24}$$

Thus, one can calculate the expected change in pressure $\Delta P'$ over Δz using this value:

$$\Delta P' = \frac{dP}{dz} \times \Delta z = -11.76 \text{ Nm}^{-2} \times 5000 = -58800 \text{ Nm}^{-2} \tag{7.25}$$

Converting this to hPa:

$$\Delta P' = -58800 \text{ Nm}^{-2} \times \frac{1 \text{ hPa}}{100 \text{ Nm}^{-2}} = -588 \text{ hPa} \tag{7.26}$$

In this example, the hydrostatic equation allows us to understand how pressure changes with altitude and predict the pressure difference at a given change in altitude.

Let us hypothesize a connection between quantum energy signature spectra and brightness temperature, it might involve the idea that the thermal radiation emitted by an object could be influenced by the quantum energy transitions of its constituent particles. This could result in specific spectral features or patterns in the thermal emission, which could then be observed and quantified through brightness temperature measurements in thermal bands. To correlate between quantum energy signature spectra and brightness temperature let's denote $\Psi|E|\lambda\rangle$ as the quantum spectral energy at a specific wavelength λ related to quantum transitions; and $\Psi|T_B\rangle$ as the actual temperature of the cloud streets.

$$\Psi|T_B\rangle = \sigma_t \int_{\lambda_0}^{\lambda_1} \Psi|E|\lambda\rangle.\lambda^{-1}.\frac{1}{e^{\frac{hc}{\lambda kT}} - 1} d\lambda \tag{7.27}$$

In this equation, h is the Planck constant ($6.62607015 \times 10^{-34}$ Joule·seconds); c is the speed of light; σ_t is constant with an approximate value of 0.003 m.K and k is the Boltzmann constant (1.380649×10^{-23} Joules/Kelvin). Therefore, the simplification of quantum spectral energy signature as a function of brightness temperature can be expressed as follows:

$$\int \Psi|E\rangle = \frac{\sigma_t.h.c}{\sigma_t} \frac{\int \Psi|T_B\rangle dT_B}{dE} \tag{7.28}$$

Thus, by integrating both sides of the equation for $\Psi|E\rangle$ and $\Psi|T_B\rangle$ one can obtain:

$$\Psi|E\rangle = \sqrt{\frac{h.c}{0.003}}.T_B.|q\rangle + |C\rangle \tag{7.29}$$

In equation 7.29, $|C\rangle$ represents the quantum state associated with the constant of integration C.

Additionally, $\Psi|E\rangle$ represents the quantum state associated with spectral energy signature; and $|q\rangle$ is a quantum state factor corresponding to the brightness temperature [24-28]. The quantum state factor can $|q\rangle$ be estimated using the formula:

$$|q\rangle = \Psi|T_B\rangle[\sigma_t]^{-1} \tag{7.29.1}$$

The normalization of quantum spectral energy; hence, refers to the process of adjusting the spectral energy values so that they are scaled within a specific range or relative to a reference value. Normalization is commonly performed to make the spectral energy values more interpretable, comparable, or suitable for specific analyses [22–26]. In this understanding, mathematically, the normalized quantum spectral energy $\Psi|E_{norm}\rangle$ can be defined as:

$$\Psi|E_{norm}\rangle = \frac{\Psi|E\rangle - \Psi|E_{\min}\rangle}{\Psi|E_{\max}\rangle - \Psi|E_{\min}\rangle} \tag{7.30}$$

Here $\Psi|E_{\min}\rangle$ and $\Psi|E_{\max}\rangle$ are the minimum and maximum quantum spectral energy signatures, respectively of equation 7.29. In this normalization, $\Psi|E_{norm}\rangle$ will have a value between 0 and 1, indicating the relative position of $\Psi|E\rangle$ within the range defined by $\Psi|E_{\min}\rangle$ and $\Psi|E_{\max}\rangle$ (Figure 7.8). This normalization is known as "min-max normalization".

```
Function estimate_quantum_spectral energy spectra (BrightnessTemperature Temperature, Constant, SmallEnergy):
    # Calculate the quantum spectral energy signature based on
      BrightnessTemperature
      quantum spectral energy signature = Constant * Brightness Temperature
    # Combine the quantum spectral energy signature with the small energy term
      quantum_state = quantum_energy_state + SmallEnergy
# Normalize the quantum state to ensure its components sum up to 1
      normalized_quantum_state = normalize(quantum_state)
      return normalized_quantum_state
# Constants
      Planck_Constant = 2.89777107e3
      Small_Energy = 1.0e-20
# User input: Temperature in Kelvin
      temperature_input = get_user_input("Enter temperature in Kelvin: ")
# Estimate the quantum state
      estimated_quantum_state =
      estimate_quantum_state(temperature_input, Planck_Constant,
      Small_Energy)
# Display the normalized quantum state
      display_output("Estimated normalized quantum state:",
      estimated_quantum_state)
```

Figure 7.8: Pesudo-code of estimation min-max normalization of quantum spectral energy signatures.

7.7 Quantum Edge Detection Algorithm for Cloud Streets

From the aforementioned standpoint, the utilization of quantum spectral energy signatures presents an avenue for the formulation of an innovative edge detection algorithm tailored for linear arrays of cloud streets. In this context, the essence of edge detection lies in the identification of boundaries or transitions between distinct zones within an image or dataset, frequently employing mathematical methodologies from the realm of image processing. Consequently, if there exists an interest in venturing into a theoretical realm where quantum principles are harnessed for edge detection rooted in variations of brightness temperature, the construction of a new theoretical framework becomes imperative [29–31].

Creating a quantum representation of a histogram involves encoding the frequency distribution of pixel intensities into quantum states. In this regard, let us denote $\Psi|E_0\rangle$, $\Psi|E_1\rangle$, ..., $\Psi|E_n\rangle$ as quantum states that represent different pixel energy values (quantum basis states). Additionally, consider $f|E_i\rangle$ as the frequency or probability distribution associated with each pixel spectral energy signature value. Hence, the quantum representation of the spectra energy variation could be a quantum superposition $\Psi|S\rangle$ of these states, weighted by their corresponding probabilities:

$$\Psi|S\rangle = \sum_{i=0}^{n} \sqrt{f|E_i\rangle}\Psi|E_i\rangle \tag{7.31}$$

In this representation, the probabilities $f|E_i\rangle$ are encoded as the square of the amplitudes of the quantum states. Measuring this quantum state would reveal information about the pixel intensity distribution, analogous to how measuring a quantum state yields information about a classical observable. In other words, quantum computing can potentially represent the histogram of pixel intensities in a quantum state. Each quantum state could encode a specific pixel intensity value along with its corresponding frequency in the image. In other words, each state represents different classes (foreground and background in the context of the quantum spectra energy signature algorithm). The variance calculation involves computing the variance of the states' measurements around their mean value. In this understanding, the quantum variance $\Psi|\sigma^2\rangle$ calculation equation could be formulated as follows:

$$\Psi|\sigma^2\rangle = \sum_{i=0}^{n} \left|\Psi|E_i\rangle\right| \left(\hat{M} - \langle\hat{M}\rangle\right)^2 \left|\Psi|E_i\rangle\right| \tag{7.32}$$

where $\hat{M}$ represents the quantum observable (operator) corresponding to the measurement of a property (e.g., the quantum spectral energy signature of pixels) in the quantum states. In practice, implementing this calculation on a quantum computer involves encoding the states, performing quantum measurements, and applying quantum operators [29, 30]. Therefore, let's denote $\hat{O}$ as the quantum operator for edge detection for quantum spectral energy signature, for instance, $\Psi|E_1\rangle$, and $\Psi|E_2\rangle$. Thus, considering, $P_{|E_1\rangle}$, and $P_{|E_2\rangle}$ as projectors onto the $\Psi|E_1\rangle$, and $\Psi|E_2\rangle$, respectively. A simplified, speculative form of the quantum operator might look like:

$$\hat{O} = P_{|E_1\rangle} - P_{|E_2\rangle} \tag{7.33}$$

This hypothetical operator subtracts the projection onto $\Psi|E_2\rangle$ from the projection onto $\Psi|E_1\rangle$, potentially highlighting regions of transition or edges between different quantum spectral energy signature levels. Therefore, apply quantum optimization techniques to find the optimal threshold value that maximizes the quantum version of the spectral energy signature criterion. This optimization could be done using quantum annealing or other quantum optimization algorithms. In other words, quantum optimization involves using quantum computing to find the optimal solution to a given optimization problem. One widely known quantum optimization algorithm is the Quantum Approximate Optimization Algorithm (QAOA). Let's consider an optimization problem with a cost function $C|E\rangle$, where $|E\rangle$ is a set of variables that need to be optimized. The goal is to find the values of $|E\rangle$ that minimize the cost function $C|E\rangle$. The equation for QAOA can be expressed as:

$$\text{Minimize } \langle\Psi(\gamma, \beta)|\hat{C}|\Psi(\gamma, \beta)\rangle \tag{7.34}$$

According to equation 7.34, γ and β are vectors of variational parameters that determine the behavior of the quantum state of remote sensing data [29, 31]. Therefore, $|\Psi(\gamma,\beta)\rangle$ represents the quantum state obtained by applying quantum gates to an initial state and $\hat{C}$ is the operator corresponding to the cost function (Figure 7.9). The goal of quantum optimization is to find the values of γ and β that minimize the expectation value of $\hat{C}$. This process involves iteratively adjusting the parameters and evaluating the cost function on a quantum computer.

Utilize the quantum-computed optimal threshold within the context of the quantum histogram representation. This phase entails adapting the quantum states to classify pixels into foreground or background categories, determined by the quantum threshold value. The quantum thresholding operator $\hat{T}$ can be defined as a diagonal operator based on quantum states:

$$\hat{T} = \sum_{i=0}^{n} |P_{E_i}\rangle\langle P_{E_i}| . \Theta\left(|P_{E_i}\rangle - \theta\right) \tag{7.35}$$

here $\Theta(E)$ is the Heaviside step function, which is 0 for $E < 0$ and 1 for $E \geq 0$. Hence, $|P_{E_i}\rangle\langle P_{E_i}|$ is a projection operator onto the *i*th quantum state (Figure 7.10). Applying the quantum thresholding operator to a quantum state $|E\rangle$ results in a new quantum state that represents the thresholded version of the original state:

$$\Psi'|E\rangle = \hat{T}|P_E\rangle \tag{7.36}$$

In this view, $\Psi'|E\rangle$ quantum states corresponding to pixel quantum spectral energy signature values below the threshold θ would have their coefficients reduced to zero, effectively categorizing them as background, while states above the threshold will be retained, representing the foreground.

```
QAOA(p, H)
Input: p (number of mixing and optimization steps), H (problem Hamiltonian)
Output: Optimal angles theta
theta = RandomInitialAngles() // Initialize angles randomly
best_theta = theta
best_cost = ExpectationValue(theta, H)

for k = 1 to MaxIterations:
    for j = 1 to p:
        beta_j = UpdateBeta(j)
        gamma_j = UpdateGamma(j)
        theta[j] = [beta_j, gamma_j]

    cost = ExpectationValue(theta, H)
    if cost < best_cost:
        best_cost = cost
        best_theta = theta

return best_theta
```

Figure 7.9: Pseudo-code of QAOA.

```
function QuantumThreshold Operator(T, |E⟩):
    # T is the quantum threshold value
    # |E⟩ is the input quantum state
    # Initialize a new quantum state |E⟩
    |E⟩ = |0⟩ # Initialize to the zero state
    # Apply the quantum thresholding operator
    if measurement(|E⟩) >= T:
        |E⟩ = |1⟩ # Set to the one state if the measurement is greater than or equal to T
        return |E⟩ # Return the resulting quantum state
```

Figure 7.10: Pseudo-code of quantum thresholding of quantum spectral energy signature

Please note that this pseudocode is a simplified representation and may need to be adapted based on the specific quantum computing framework and language one is using. The 'measurement($|E\rangle$)' function represents the measurement outcome of the quantum state $|E\rangle$, and the threshold value T is a parameter that one would provide based on your requirements [29–31].

7.8 Automatic Detection Quantum Edge Algorithm of Cloud Street in MODIS Data

According to the above-mentioned perspective, the composite satellite MODIS data with visibile mid-wave Infrared (MWIR) and long-wave infrared (LWIR) data are encoded into a quantum energy state. This technique is used to represent energy distributions in a quantum energy image. A quantum state is created with the necessary number of qubits to represent the energy levels. The number of qubits needed depends on the precision required and the number of distinct energy levels. In this sense, for each energy level, its energy value is mapped to the amplitude of a quantum state (Figure 7.11). The amplitude can be scaled and normalized based on the maximum energy value.

Let's delve into the Controlled-NOT (CNOT) gate, a core two-qubit operation in quantum computing. It applies a Pauli-X gate (a bit-flip operation) to the target qubit, but only if the control

Figure 7.11: Encoding MODIS data into quantum energy state image.

qubit is in the $|1\rangle$ state. When the control qubit is $|0\rangle$, the target qubit remains unchanged. In this understanding, the control qubit comes into play, deciding whether the energy is deemed "high" or "low." In this example, if the energy value surpasses a predetermined threshold (let's say 0.5), one can set the control qubit to $|1\rangle$; otherwise, it's set to $|0\rangle$. Once determined, we unleash the power of the CNOT gate to encode the energy value into the target qubit.

```
# Quantum circuit with two qubits
qc = QuantumCircuit(2)
# Energy value (normalized) assumption: 0.7, above the 0.5 threshold
energy_value = 0.7
# Control qubit set based on energy value (above or below threshold)
if energy_value > 0.5:
    qc.x(0) # Control qubit becomes |1⟩
# CNOT gate unleashed based on control qubit
qc.cx(0, 1) # CNOT gate from qubit 0 (control) to qubit 1 (target)
# Opting for a simulator backend
simulator = Aer.get_backend('statevector_simulator')
# Executing the quantum circuit on the simulator
result = execute(qc, simulator).result()
# Extracting the state vector of the final quantum state
State vector = result.get_statevector()
print("Final state vector:", state vector)
```

Figure 7.12: Pesudo-code of CNOT gate for encoding MODIS data into thresholding high quantum energy data.

In this scenario, if the energy value crosses the threshold (0.7 > 0.5) (Figure 7.13), the control qubit becomes $|1\rangle$. Consequently, the CNOT gate toggles the state of the target qubit. Conversely, if the energy value falls below the threshold, the control qubit remains $|0\rangle$, and the target qubit remains unaltered. Therefore, Figure 4.14 shows the threshold of the quantum energy value of 1. This leads to automatic vector segmentation of the cloud streets cover zones with scattered small cumulus clouds.

Figure 7.15 demonstrates the spectral of cloud street brightness temperature obtained from quantum spectral energy signature. The top cloud streets have a lower brightness temperature of 266 K than the lower base of cloud streets. The space lines of convection are dominated by a higher brightness temperature of 291 K. This is excellent proof of convection occurrence in developing street clouds [1, 5, 8].

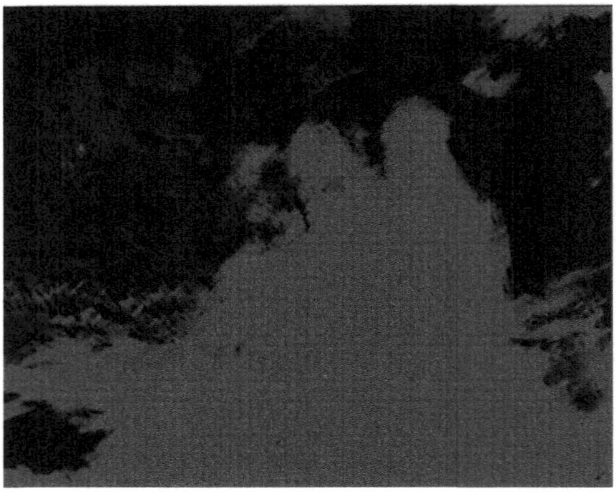

Figure 7.13: Thresholding quantum energy image above 0.5.

Figure 7.14: Quantum energy image of street cloud covers with the threshold value of 1.

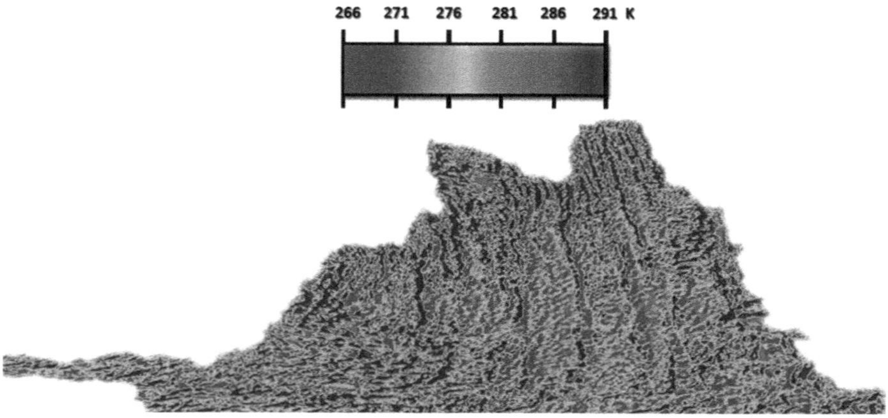

Figure 7.15: Brightness spectra temperature of cloud streets.

Why the spaces between cloud streets have a high temperature at this point is a crucial question? Understanding the reason behind the high temperature in the spaces between cloud streets is essential for several reasons. Firstly, it can provide insights into the dynamics of cloud formation and atmospheric processes. Additionally, this knowledge can help improve weather forecasting models and enhance our understanding of climate patterns on a larger scale. In this perspective, the temperature is greater in the areas between cloud streets for the reasons mentioned below. Cloud streets emerge as thermals or rising columns of warm air, transfer heat away from the ocean's surface, and rise through the sky. The elevated temperature seen in the gaps between cloud streets is a result of these rising thermals. When cold air blows over warmer water, it creates cloud streets, which are then covered by a layer of warmer air known as a temperature inversion. As a lid, the warmer air layer traps heat and moisture from the relatively warm water below. The rising temperature seen in the gaps between cloud streets is the consequence of this retained heat.

Convection rolls, in which warm air rises on one side, cools, and water vapor condenses to create a cloud, causing the formation of cloud streets. The warmer air that is ascending causes the temperature in the gaps between cloud streets to be higher. Therefore, quantum spectral signature energy demonstrates an inverse relationship with brightness temperature. In this sense, the highest brightness temperature of 291 K located between space lines of cloud streets has a lower quantum spectra energy signature of 0.35 than the top cloud streets. On the contrary, the highest quantum spectral energy signature of 0.8 is associated with the top cold surface of cloud streets of 265 K (Figure 7.16).

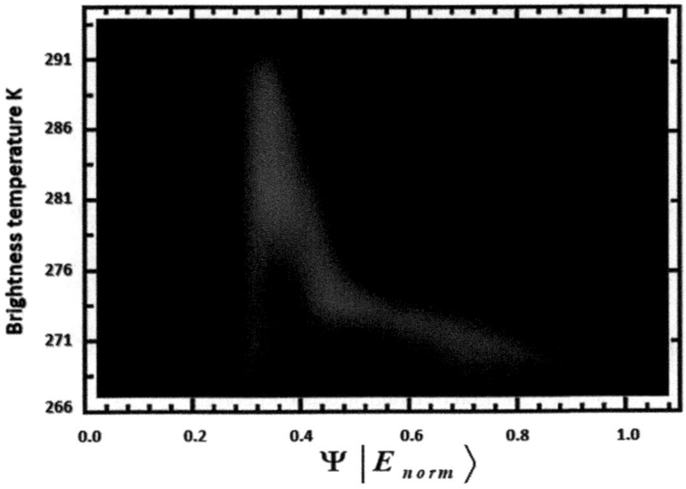

Figure 7.16: Quantum spectral energy signature of cloud street.

Could the automated identification of features from remote sensing data benefit from the thermal gradient found across cloud streets? The variation in the quantum spectral energy signature gradient can be discerned between distinct structural configurations of cloud streets, including the spacing between lines, the upper cloud surface, and the lower base of the cloud surface, as highlighted in the earlier representations seen in Figure 7.15 and 7.16. From this standpoint, leveraging a quantum edge detection algorithm can yield distinguishable outlines of cloud street formations. The dispersion pattern of space lines within cloud streets is accurately delineated, as are the discernible characteristics of cumulus and stratus clouds. Cumulus clouds exhibit an association with the topological layout of cloud streets, whereas stratus clouds are dispersed away from the core structural morphology of the cloud streets (Figure 7.17). Similar outcomes could manifest on February 26, 2023, over the expanse of the East China Sea (Figure 7.18).

Figure 7.17: Automatic detection of cloud streets using quantum edge algorithm-based quantum energy states.

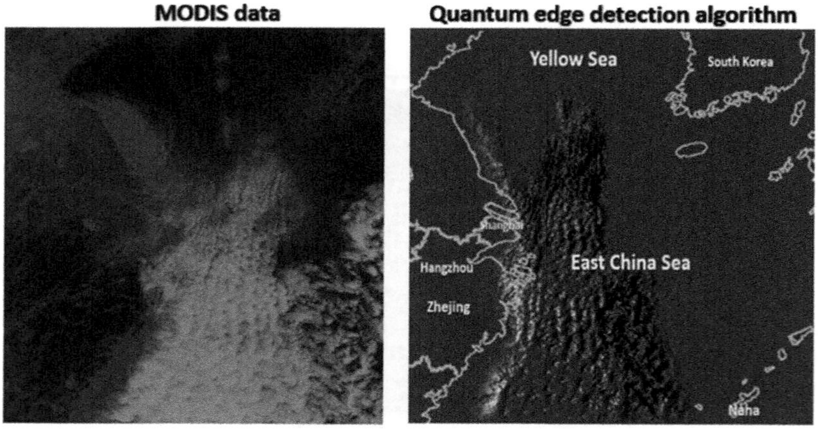

Figure 7.18: Automatic detection of cloud streets on February 26, 2023, above the East China Sea.

In a parallel scenario, IKONOS satellite data skillfully captured mesmerizing cloud streets over Johor Bharu, Malaysia, on March 23rd, 2011. Leveraging the prowess of the quantum edge detection algorithm, these ethereal formations can be automatically identified in high-resolution data, such as that obtained from IKONOS. Notably, the Panchromatic data boasts a resolution range spanning from 0.82 to 1 meter [32] (Figure 7.19).

March, in the chronological tapestry, signifies the northeast monsoon period, a phase predominantly characterized by the prevalence of high cloud covers over the east coast of peninsular Malaysia. However, it's noteworthy that the quantum edge detection algorithm, while adept at discerning cloud streets in IKONOS images, encounters limitations in automatically detecting such phenomena in Meteosat data [33–35] (Figure 7.20).

In fact, the edge boundaries of cloud street covers are distinctly delineated in Meteosat data, a testament to the precision achieved by the quantum edge detection algorithm. The forthcoming section will meticulously address this anomaly, delving into strategies and methodologies aimed at refining the algorithm's capability to identify cloud street covers in Meteosat data imagery.

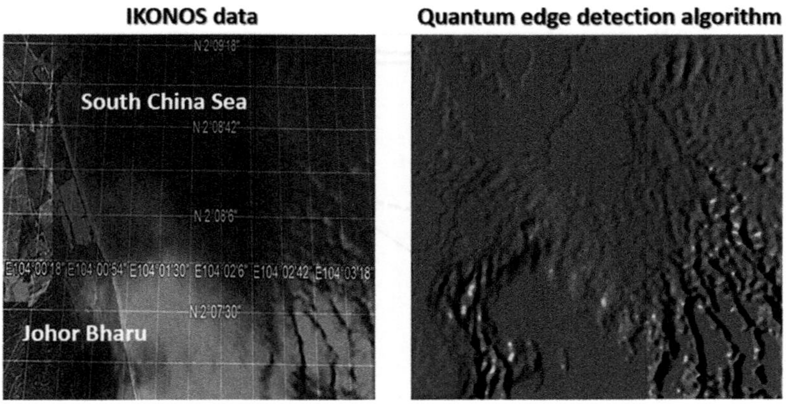

Figure 7.19: Automatic detection of cloud streets on March 23, 2011, above Johor Bharu, east coast of Malaysia in IKONOS panchromatic data.

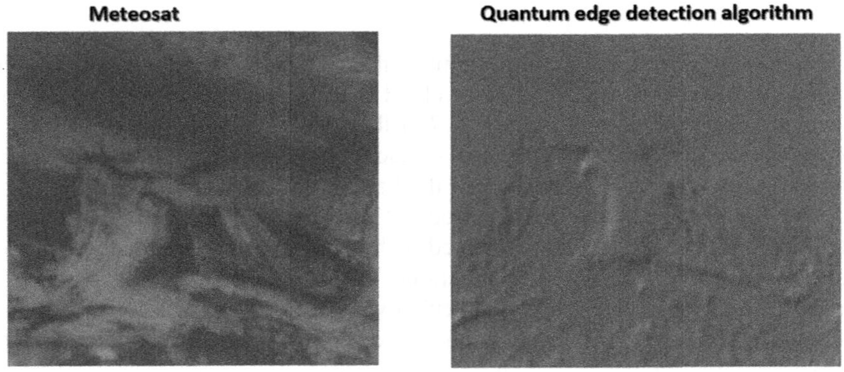

Figure 7.20: Limitations of quantum edge detection algorithm in automatic detection of cloud streets in Meteosat data.

7.9 Why Quantum Edge Detection Algorithm Accurately Detect Cloud Streets?

The Receiver Operating Characteristic (ROC) curve effectively illustrates the precision of findings when comparing data obtained from MODIS and Meteosat sources in the context of cloud street automatic detection. This assessment is conducted through the utilization of a quantum edge detection algorithm, which relies on the quantum spectral energy signature. The ROC curve provides a comprehensive visualization of the algorithm's performance, showcasing its ability to distinguish between cloud street features derived from these two distinct datasets. In this regard, the quantum edge detection algorithm accurately identified 95% of the cloud streets in both MODIS and Meteosat data (true positive rate) and also correctly identified 95% of the zones without occurrence of cloud streets (true negative rate) (Figure 7.21). The ROC curve generated for quantum edge detection of cloud streets would show a curve that rises sharply towards the upper-left corner of the graph, indicating the high true positive and true negative rates. The AUC, in this case, would be close to 1, which is the highest value possible, reflecting the excellent performance of the quantum edge detection algorithm.

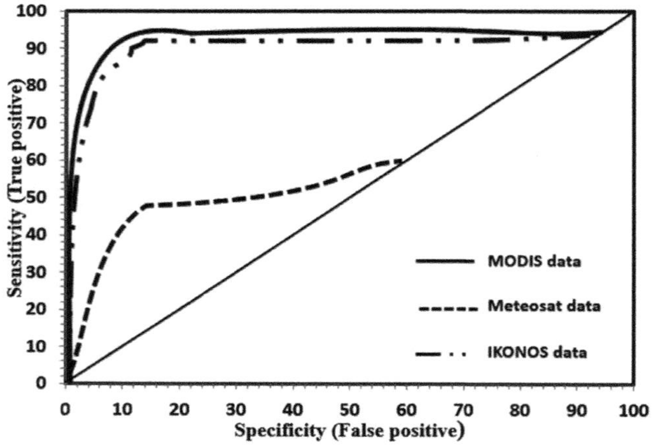

Figure 7.21: Accuracy of quantum edge detection algorithm based on quantum spectral energy signature.

In this scenario, the high AUC value signifies that the quantum edge detection algorithm has a high accuracy in the automatic detection of cloud streets, making it a reliable diagnostic tool.

Cloud streets can be detected in Meteosat 7 visible (VIS) imagery, but the detailed structure is not easily discernible due to the relatively coarse resolution of this type of imagery. Visible imagery captures the reflected sunlight from clouds and the Earth's surface, allowing for cloud detection, but fine details may be lost. In Meteosat 7 infrared (IR) imagery, cloud streets are not easily visible. This is likely because the convection associated with cloud streets is limited to the lower levels of the troposphere. IR imagery detects the temperature of the cloud tops and surfaces, and it may not capture lower-level atmospheric features as effectively. However, in Meteosat 8 high-resolution visible (HiresVIS) imagery, cloud streets cannot be detected clearly as parallel white cloud bands. The improved resolution of this imagery type allows for better visualization of the detailed structure of cloud streets, making them more visible and distinguishable [33–35].

The effectiveness of the quantum edge detection algorithm in accurately identifying cloud streets stems from various inherent factors in its design and methodology. This algorithm's success can be attributed to the following key aspects [29–31].

Firstly, the algorithm exhibits a high sensitivity to the quantum spectral energy signature. Leveraging variations in quantum spectral energy across different regions, it discerns shifts in the signature caused by distinct thermal properties of cloud street components like line spacing, cloud surfaces, and bases. This sensitivity enables precise identification of transitions between these components, effectively outlining the structural features of cloud streets. Secondly, the quantum spectral energy gradient analysis plays a pivotal role. Cloud streets display variations in thermal properties due to dynamic factors like warm air currents, cloud formation, and atmospheric conditions. The algorithm's capability to analyze the gradient of the quantum spectral energy signature provides a valuable tool to identify transitions and boundaries, contributing to the accurate detection of cloud street features.

Additionally, the algorithm excels in recognizing distinct morphological characteristics of cloud streets, encompassing line spacing, cloud type, and structural arrangement. These unique features translate into specific quantum spectral energy signatures. The algorithm's proficiency in distinguishing between these signatures enables accurate differentiation of cloud street components, capturing their distinctive structural morphology.

Moreover, the quantum edge detection algorithm capitalizes on thermal gradients and temperature differences within cloud streets. Variations in thermal gradients and temperature differences between components serve as cues for the algorithm to identify and delineate boundaries between line spacing, cloud surfaces, and base areas. This thermal information proves crucial for detecting intricate features within cloud streets.

Lastly, the algorithm showcases adaptive feature extraction capabilities. Its ability to adaptively extract features from the quantum spectral energy signature allows the accommodation of variations in cloud street patterns and atmospheric conditions. This adaptability significantly enhances the algorithm's accuracy in detecting cloud street structures, even in complex and variable scenarios.

In conclusion, this chapter has unveiled an innovative approach for the automated detection of cloud streets using optical satellite data, exemplified by MODIS. The newly developed algorithm hinges on the concept of quantum spectral energy, gleaned from the fluctuations in brightness temperature spanning cloud street formations and their adjacent surroundings. The quantum edge detection algorithm, a pivotal component, exhibits a remarkable potential for the automated identification of cloud streets within optical remote sensing datasets, including MODIS and Meteosat data. Its precision and promise in this context highlight its significance as an effective tool for cloud street detection and analysis.

References

[1] Etling, D. and Brown, R.A. (1993). Roll vortices in the planetary boundary layer: A review. *Boundary-Layer Meteorology*, 65, 215–248.

[2] Garratt, J.R. (1994). The atmospheric boundary layer. *Earth-Science Reviews*, 37(1-2), 89-134.

[3] Alpers, W. and Brümmer, B. (1994). Atmospheric boundary layer rolls observed by the synthetic aperture radar aboard the ERS-1 satellite. *Journal of Geophysical Research: Oceans*, 99(C6), 12613–12621.

[4] Alpers, W. and Brummer, B. (1993). Imaging of atmospheric boundary layer rolls by the synthetic aperture radar aboard the European ERS-1 satellite. *In:* Proceedings of IGARSS'93-IEEE International Geoscience and Remote Sensing Symposium (pp. 540–542). IEEE.

[5] Feng, X., Wu, B. and Yan, N. (2015). A method for deriving the boundary layer mixing height from modis atmospheric profile data. *Atmosphere*, 6(9), 1346–1361.

[6] Onyango, S., Anguma, S.K., Andima, G. and Parks, B. (2020). Validation of the atmospheric boundary layer height estimated from the MODIS atmospheric profile data at an equatorial site. *Atmosphere*, 11(9), 908.

[7] Hutchison, K.D., Pekker, T. and Smith, S. (2006). Improved retrievals of cloud boundaries from MODIS for use in air quality modeling. *Atmospheric Environment*, 40(30), 5798–5806.

[8] Frey, R.A., Ackerman, S.A., Liu, Y., Strabala, K.I., Zhang, H., Key, J.R. and Wang, X. (2008). Cloud detection with MODIS. Part I: Improvements in the MODIS cloud mask for collection 5. *Journal of Atmospheric and Oceanic Technology*, 25(7), 1057–1072.

[9] Wu, C.K. and Chen, J.P. (2021). Simulation of aerosol indirect effects on cloud streets over the Northwestern Pacific Ocean. *Journal of Geophysical Research: Atmospheres*, 126(11), e2020JD034325.

[10] Dror, T., Koren, I., Liu, H. and Altaratz, O. (2023). Convective steady state in shallow cloud fields. *Physical Review Letters*, 131(13), 134201.

[11] Lai, H.W., Zhang, F., Stauffer, D., Gaudet, B.J. and Clothiaux, E.E. (2018). Sensitivity of arctic boundary layer cloud streets to modeling resolution. *In:* AGU Fall Meeting Abstracts (Vol. 2018, A31J–3008).

[12] Wille, R. (1960). Karman vortex streets. *Advances in Applied Mechanics*, 6, 273–287.

[13] Lemone, M.A. (1973). The structure and dynamics of horizontal roll vortices in the planetary boundary layer. *Journal of Atmospheric Sciences*, 30(6), 1077–1091.

[14] Brown, R.A. (1980). Longitudinal instabilities and secondary flows in the planetary boundary layer: A review. *Reviews of Geophysics*, 18(3), 683–697.

[15] Weckwerth, T.M., Wilson, J.W., Wakimoto, R.M. and Crook, N.A. (1997). Horizontal convective rolls: Determining the environmental conditions supporting their existence and characteristics. *Monthly Weather Review*, 125(4), 505–526.

[16] Rebollo, T.C. and Lewandowski, R. (2014). *Mathematical and Numerical Foundations of Turbulence Models and Applications* (pp. 7–44). Birkhäuser.

[17] Kuettner, J.P. (1971). Cloud bands in the earth's atmosphere: Observations and theory. *Tellus*, 23(4-5), 404–426.
[18] HouzeJr, R.A. (2014). *Cloud Dynamics*. Academic Press.
[19] Organisationscientifique et technique internationale du vol à voile. (1993). *Handbook of Meteorological Forecasting for Soaring Flight* (Vol. 158). Secretariat of the World Meteorological Organization.
[20] Müller, D., Etling, D., Kottmeier, C. and Roth, R. (1985). On the occurrence of cloud streets over northern Germany. *Quarterly Journal of the Royal Meteorological Society*, 111(469), 761–772.
[21] Atkinson, B.W. and Wu Zhang, J. (1996). Mesoscale shallow convection in the atmosphere. *Reviews of Geophysics*, 34(4), 403–431.
[22] Sancho, P. (2018). Entanglement in absorption processes. *European Journal of Physics*, 40(1), 015404.
[23] Lü, X.Y., Si, L.G., Hao, X.Y. and Yang, X. (2009). Achieving multipartite entanglement of distant atoms through selective photon emission and absorption processes. *Physical Review A*, 79(5), 052330.
[24] Lee, D.I. and Goodson, T. (2006). Entangled photon absorption in an organic porphyrin dendrimer. *The Journal of Physical Chemistry B*, 110(51), 25582–25585.
[25] Kurpiers, P., Magnard, P., Walter, T., Royer, B., Pechal, M., Heinsoo, J. and Wallraff, A. (2018). Deterministic quantum state transfer and remote entanglement using microwave photons. *Nature*, 558(7709), 264–267.
[26] Oka, H. (2017). Generation of broadband ultraviolet frequency-entangled photons using cavity quantum plasmonics. *Scientific Reports*, 7(1), 8047.
[27] Laurel, C.O., Dong, S.H. and Cruz-Irisson, M. (2015). Equivalence of a bit pixel image to a quantum pixel image. *Communications in Theoretical Physics*, 64(5), 501.
[28] Saleh, B.E., Jost, B.M., Fei, H.B. and Teich, M.C. (1998). Entangled-photon virtual-state spectroscopy. *Physical Review Letters*, 80(16), 3483.
[29] Sundani, D., Widiyanto, S., Karyanti, Y. and Wardani, D.T. (2019). Identification of image edge using quantum canny edge detection algorithm. *Journal of ICT Research and Applications*, 13(2), 133–144.
[30] Cavalieri, G. and Maio, D. (2020). A quantum edge detection algorithm. arXiv preprint arXiv:2012.11036.
[31] Yao, X.W., Wang, H., Liao, Z., Chen, M.C., Pan, J., Li, J., ... and Suter, D. (2017). Quantum image processing and its application to edge detection: Theory and experiment. *Physical Review X*, 7(3), 031041.
[32] Karpouzli, E. and Malthus, T. (2003). The empirical line method for the atmospheric correction of IKONOS imagery. *International Journal of Remote Sensing*, 24(5), 1143–1150.
[33] Meinel, G., Neubert, M. and Reder, J. (2001). The potential use of very high resolution satellite data for urban areas – First experiences with IKONOS data, their classification and application in urban planning and environmental monitoring. *Regensburger Geographische Schriften*, 35, 196–205.
[34] Feijt, A., De Valk, P. and Van Der Veen, S. (2000). Cloud detection using Meteosat imagery and numerical weather prediction model data. *Journal of Applied Meteorology and Climatology*, 39(7), 1017–1030.
[35] Ottenbacher, A., Tomassini, M., Holmlund, K. and Schmetz, J. (1997). Low-level cloud motion winds from Meteosat high-resolution visible imagery. *Weather and Forecasting*, 12(1), 175–184.

CHAPTER
8

Development of the Superposition Vortex Distance Decomposition Algorithm for Detecting von Kármán Vortex Cloud Streets in Multi-Satellite Imagery

In the preceding chapter, cloud streets emerged as a notable feature of atmospheric turbulence, showcasing the efficacy of modern quantum computing algorithms outlined in earlier chapters. This upcoming chapter takes a novel approach by applying a distinct quantum computing algorithm to track spinning vorticity, particularly von Kármán Vortex Streets—a subject briefly touched upon in Chapter 7, Section 7.2.2. The primary focus of this chapter is to investigate the detection of von Kármán Vortex Streets, exploring whether the algorithm operates with optimal efficiency automatically or semi-automatically across multisatellite sensors. The central question to be addressed is whether the physical properties of sensors can impact the algorithm's effectiveness.

8.1 What is the Magic of von Kármán Vortex Street

The captivating swirl of vortices known as the Kármán vortex street, or von Kármán vortex street, is a recurring sight in the complex field of fluid dynamics. The interesting process known as vortex shedding, in which the fluid flow periodically separates around blunt objects, is exhibited in this mesmerizing occurrence. In this perspective, when a fluid, such as air or water, comes into contact with a solid object or an impediment, vortex shedding occurs. The fluid's route around the structure is not uniform; instead, it creates zones of high and low pressure that fluctuate. These swapping zones of pressure cause the air to gradually flow around the object, ultimately causing vortices to emerge on both sides and creating the beautiful Kármán vortex street (Figure 8.1).

Figure 8.1: Captivating Kármán vortex cloud streets is a testament to the fluid's poetic response to obstacles.

Within the atmospheric boundary layer, Kármán Vortex Streets, well-known for the flow around towering obstructions, can also take shape. Termed as atmospheric vortex streets, they consist of mesoscale eddies trailing large islands [1]. Figure 8.1 illustrates a vortex street with paired opposing revolving vortices [2]. The formation of Kármán Vortex Streets requires specific meteorological conditions [1]. There must be robust, persistent winds with a generally constant direction, and the boundary layer should be well-mixed with a strong capping inversion under the island top. In many instances, the mixed layer is capped by stratocumulus clouds, serving as a tracer. These conditions are often met during cold-air outbreaks and in trade-wind regions [3].

8.2 Mathematical Description of Kármán Vortex Streets

The mathematical description of a Kármán Vortex Streets involves the representation of vortices formed in the wake of an obstacle. One common model used to describe vortex streets is based on the Kármán Vortex Streets backside of a cylinder. In such a model, the vorticity (circulation per unit length) shed from the cylinder alternates its sign, creating a pattern of vortices in the wake. In this understanding, let's denote V as the velocity of the fluid, D as the diameter of the obstacle, and f as the shedding frequency; respectively (Figure 8.2). The vortices generated in a Kármán vortex street impart a sinusoidal force onto a body placed within the fluid flow [1–4]. This phenomenon is a result of the alternating patterns of vortices shedding from the body, creating fluctuations in the force experienced by the object. As the vortices are periodically shed on either side of the object, they induce a rhythmic force variation perpendicular to the direction of the fluid flow. Mathematically, this sinusoidal force (F) can be expressed as:

$$F(t) = F_0 \sin(\omega t) \tag{8.1}$$

Here F_0 denotes the force's amplitude, which shows the force variations at their greatest magnitude. The rate of oscillation of the force with time (t) is represented by the angular frequency, or ω [4].

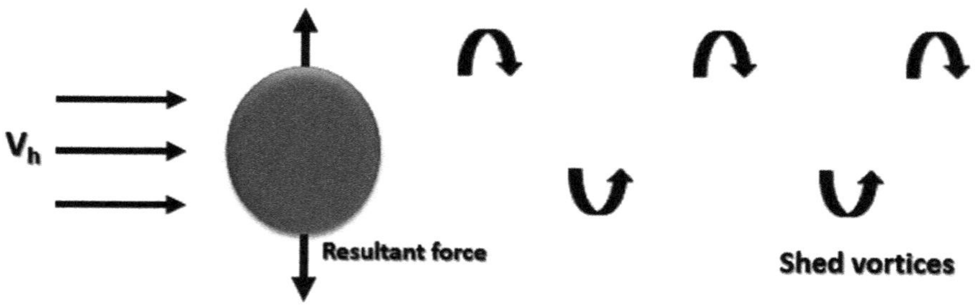

Figure 8.2: Mechanism of Kármán Vortex Streets backside of a cylinder.

Therefore, the frequency shedding is often interrelated to the velocity and obstacle size, and for Kármán Vortex Streets, it can be given by the Strouhal number:

$$St = Vf \cdot D \tag{8.2}$$

The vorticity (ω) in the vortex street, thus, is narrated to the flaking frequency and the distance (x) from the obstacle:

$$\omega = 2\pi f = 2\pi \cdot St \cdot DV \tag{8.3}$$

In this view, The Kármán Vortex Streets is depicted by a regular pattern of vortices, and its mathematical description helps understand the dynamics of these vortices in the wake of obstacles, such as cylinders. In this sense, let's consider a numerical example of the shedding frequency and vorticity in Kármán Vortex Streets. Therefore, let's assume that fluid velocity (V) = 10 m/s; the diameter of the obstacle (D) = 1 m; and Strouhal number (St) = 0.2 (a common value for Kármán Vortex Streets). Consequently, the shedding frequency (f) is based on equation 8.3 i.e., $f = St \cdot VD^{-1} = f = 0.2 \cdot \frac{10}{1} = 2$ Hz. Subsequently, the vorticity (ω) can be calculated as $\omega = 2\pi \cdot 2\text{Hz} = 4\pi$ rad s^{-1}. Accordingly, in this example, for a fluid flowing at 10 m/s past a cylinder with a diameter of 1 m and a Strouhal number of 0.2, the shedding frequency is 2 Hz, and the vorticity is 4π rads^{-1}. In this understanding, Chopra and Hubert [2] were pioneers in determining the fundamental properties of a vortex street observed in satellite imagery. They approached the atmospheric vortex street as analogous to the classical Kármán Vortex Streets that typically form in the wake of a cylinder. Their work involved recognizing and analyzing the vortex patterns in the atmospheric boundary layer, drawing parallels to the well-known Kármán Vortex Streets phenomenon. This approach allowed them to apply principles from fluid dynamics, particularly those related to vortex streets, to the atmospheric context captured in satellite images [1, 4, 6].

These characteristics comprise the aspect ratio, the eddies' speed of propagation, and the time interval between the shedding of two sequential vortices with the same rotational sense. The aspect ratio, which gives information on the shape and elongation of the vortices, is the ratio of the vortices' length to width. The eddies' propagation speed controls how fast they travel through a fluid medium and affects how they interact with nearby objects. Furthermore, the period interval between the production and detachment of two subsequent vortices with the same illusion of rotation is described by the shedding period of these vortices, providing important insights into their dynamics and stability [3–6].

The formation of lee vortices and their separation from the obstacle is influenced by factors such as obstacle height, oncoming flow speed, and the strength of stable stratification. These properties are often quantified using a dimensionless Froude number, which takes various forms in the previous works. In cases with a simple basic flow, where flow speed and density/temperature gradient are constant, the classical form of the Froude number [4]:

$$Fr = \frac{U_0}{NH} \tag{8.4}$$

Equation 8.4 represents the Froude number in the context of atmospheric flows where the flow speed (U_0) is influenced by stratification characterized by the Brunt–Väisälä frequency (N) and the depth of the flow (H). In this view, The Brunt–Väisälä frequency (N) is a measure of the stability of a stratified fluid, particularly in the atmosphere. It is defined as the square root of the buoyancy frequency and is expressed as:

$$N = \left(\frac{g}{\theta_0} \frac{\partial \bar{\theta}}{\partial z} \right)^{1/2} \tag{8.4.1}$$

The interpretation of the Froude number (Fr) in the context of atmospheric flows is as follows: $Fr=U_0N^{-1}$, where U_0 is the horizontal velocity of the fluid parcel, and N is the Brunt–Väisälä frequency. In this regard, If $Fr>1$, the horizontal velocity tends to be larger than the ratio N, indicating that the fluid parcel can achieve a maximum vertical displacement larger than the obstacle height (H). In this case, all fluid parcels, even those near the surface, can flow over the obstacle. On the contrary, If $Fr<1$, the horizontal velocity becomes smaller than the ratio N, suggesting that the maximum vertical displacement (U_0N^{-1}) is smaller than the obstacle height (H). In this situation, some part of the flow has to go around the obstacle. This condition holds for all fluid below a critical height i.e., $z=H - U_0N^{-1}$, known as the height of the dividing streamline. In this perspective, the critical height represents the boundary below which fluid flow is diverted around the obstacle. This concept is essential in understanding the interaction between atmospheric flows and obstacles [2, 4, 7].

In atmospheric fluid dynamics, therefore, the critical height, also referred to as the height of the dividing streamline, plays a crucial role in determining the behavior of fluid parcels around obstacles. The critical height (z) is represented by the formula:

$$Zc = H(1 - Fr) \tag{8.5}$$

This equation indicates that the critical height is influenced by the Froude number, a dimensionless parameter that compares inertial forces to buoyancy forces. The Froude number related to the dividing streamline height (Fr_z) is expressed as:

$$Fr_z = 1 - Z_c H^{-1} \tag{8.6}$$

In simpler terms, these formulas illustrate the interplay between obstacle characteristics, fluid velocity, and atmospheric stratification in determining the height at which the dividing streamline occurs. The critical height and its association with the Froude number are fundamental in understanding the flow dynamics around obstacles in the atmosphere.

In the context of the previous discussion about the dividing streamline, it's important to highlight that $U_0 N^{-1}$ represents the maximum vertical displacement of fluid parcels in stably stratified flows. However, in practical scenarios, not all horizontal kinetic energy can be fully converted into upward motion. Instead, it may be the vertical wind component induced by the slope of the obstacle. Therefore, in practice, a value around $U_0 H L^{-1}$, where HL^{-1} is the typical obstacle slope (typically 0.1–0.3 for isolated mountains), becomes relevant for estimating the vertical displacement [4, 7, 9].

Consequently, the dividing streamline height (Zc), signifies the lowest possible displacement. However, in reality, this height will be found at higher levels due to the considerations mentioned above. Despite this, the Froude numbers defined in equations (8.4) and (8.6); respectively can still be effectively used to characterize flow regimes, especially as observed in wake flows in stably stratified fluids. These parameters offer valuable insights into the dynamics of airflow around obstacles, even when considering the practical limitations of vertical displacement [2, 4, 7].

According to the above perspective, this chapter emphasizes a key question: What is the critical height at which the streamline of the atmospheric vorticity such as von Kármán Vortex Streets splits up, as well as how can it be recognized and determined? For several reasons, it is important to know the critical height at which the turbulence von Kármán Vortex Streets split. This knowledge may be used to predict and lessen the impact of vortex routes on major infrastructure, like as bridges, skyscrapers, and aircraft. Additionally, by determining this critical height, we may advance atmospheric research and deepen our understanding of fluid dynamics in intricate flow patterns.

8.3 Mathematical Model for Dividing Streamline of Atmospheric von Kármán Vortex Streets

The equations (8.4) and (8.6); respectively hold for straightforward flows characterized by constant flow speed (U_0) and Brunt–Väisälä frequency (N). However, in the atmospheric context, wind speed and stratification typically vary with height, represented as $U_0 = U(z)$ and $N = N(z)$. In such instances, a simplistic definition of the Froude number, as in equation 8.4, becomes inadequate. For these scenarios, a more comprehensive concept for determining the dividing streamline height, as proposed by Snyder et al. [11], is employed. The height of the dividing streamline (Zc) can be implicitly calculated by solving the following equation:

$$\frac{1}{2}\overline{v}_h^2(Z_c) = \int_{Z_c}^{H} dz N^2(z)(H - z). \tag{8.7}$$

Here, v_h represents the horizontal velocity profile, computed as the horizontal and temporal mean of the horizontal velocity components u and v. This formulation for determining the dividing streamline height has been previously applied in studies related to air pollution in complex terrain [12] and atmospheric vortex streets [1]. It's important to note that equation 8.7 assumes the conversion of all horizontal kinetic energy into potential energy. However, as discussed post equation 8.6, in reality, only portions of the energy from the horizontal flow can be converted into vertical displacements by the obstacle. Hence, similar to equation 8.4, and equation 8.7 provides the highest conceivable dividing streamline.

How does the application of equation 8.7 to the nocturnal boundary layer, characterized by continuous stable stratification, impact the determination of the dividing streamline height of von Kármán Vortex Streets?

In delving into the application of equation 8.7 to the nocturnal boundary layer, a realm defined by a persistent stable stratification ($N > 0$) between the surface and the obstacle apex (H) unfolds. The scenarios under daytime consideration, symbolic of the boundary layer over land or a marine boundary layer, exhibit a well-mixed lower portion capped by a robust inversion, where a less stably stratified layer extends aloft. Notably, in this context, N^2 is zero or possibly negative beneath the inversion height (z_i), rendering the calculation of the integral $dzN^2(z)(H-z)$ in equation 8.6 unfeasible below z_i. Consequently, for this boundary layer typology, the dividing streamline must reside above the frontier layer height (z_i). Failure to position the dividing streamline higher than the limit layer height could permit the entire fluid to traverse the obstacle, precluding flow division and, by extension, impeding vortex street development. In the specific flow configuration under scrutiny, the dividing streamline Froude number (*Fr*), as defined in equations 8.4, 8.6, and 8.7; respectively, is not tethered to the characteristics of the well-mixed layer beneath the inversion height (z_i). Instead, it emerges as a characteristic feature of the stably stratified layer situated above z_i [1, 4, 10–12].

In practical terms, solving equation 8.7 was achieved through an iterative process to determine *Zc*. The integral on the right-hand side was computed using the trapezoidal rule. In this view, the trapezoidal rule approximates the area under a curve by dividing the interval into small trapezoids and summing up their areas (Figure 8.3). The formula for the trapezoidal rule is:

$$T_n = \frac{\Delta z}{2}(f(z_0)+2f(z_1)+2f(z_2)+\cdots+2f(z_{n-1})+f(z_n)). \tag{8.8}$$

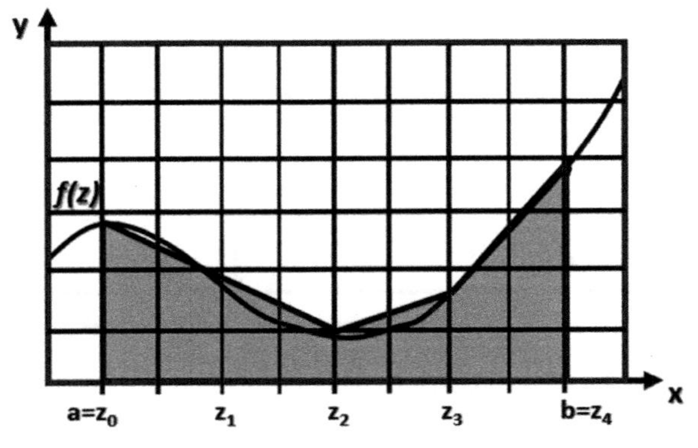

Figure 8.3: Estimation of the area under the integral curve using the trapezoidal rule.

Through an iterative loop, the lower integration limit was gradually decreased from H to z_i, representing the average height of the boundary layer determined by the height of the minimum sensible heat flux. At each step, the difference between the right and left-hand sides was calculated. The height at which this difference was minimal determined *Zc*. Consequently, the dividing streamline Froude number can be obtained using equation 8.7. To derive an average dividing streamline Froude number for the flow, profiles of u, v, and $\overline{\theta}$ can be stored every interval of specific minutes i.e. 10 minutes. The Froude number can be calculated, and the arithmetic mean can be determined for all available time steps greater than 1.5 hours to exclude the model's spin-up time [1, 4, 12].

8.4 What is the Correlation between Froude Number and Temperature Fluctuations?

The relationship between the Froude number (*Fr*) and temperature differences is crucial in understanding atmospheric phenomena, especially in the context of flows over terrain like mountains. The Froude number is a dimensionless parameter that characterizes the relative importance of inertial forces to buoyancy forces in a fluid flow. Therefore, The Froude number for this case is defined as follows:

$$Frs = \frac{\sqrt{U_0}}{g\sqrt{\Delta\theta}\left(\frac{\Delta\theta_0}{zi}\right)} \tag{8.9}$$

In equation 8.9, $\Delta\theta$ represents the temperature difference. In this sense, it corresponds to a temperature increase of 5 K over a vertical distance of 100 m in the capping inversion layer. It quantifies the vertical temperature gradient or lapse rate in the atmosphere. Therefore, $\Delta\theta_0$ is the initial or reference temperature difference, often mentioned in comparison to specific atmospheric conditions. Consequently, U_0 denotes the mean wind speed in the boundary layer. It is calculated by averaging the horizontal velocity components over all grid points that belong to the boundary layer and across all available time steps. In this perspective, Figure 8.4 demonstrates the temperature profile or lapse rate in the capping inversion layer [2, 4, 13]. In this view, the mentioned temperature increase over a specific vertical distance may be visually depicted in Figure 8.4.

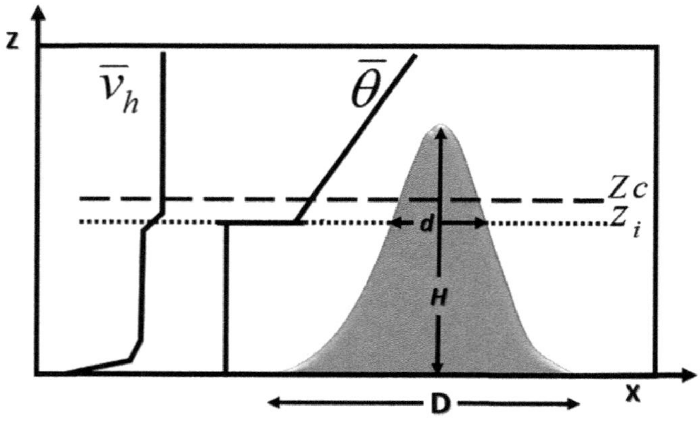

Figure 8.4: Addressing potential temperature fluctuations through the Froude number.

Additionally, the fraction $\frac{\Delta\theta_0}{zi}$ incorporates the vertical potential temperature gradient across the boundary layer. It indicates how the temperature changes with height, providing context for atmospheric stability. Consequently, gravity g is in the denominator, serving as a normalization factor. It ensures that the Froude number is dimensionless and provides a clear comparison between buoyancy and inertial forces [4, 13].

According to the above-mentioned perspective, to account for the influence of the mountain height, a dimensionless mountain height parameter M is introduced:

$$M = \frac{H}{zi} \tag{8.10}$$

Equation 8.10, therefore,encapsulates the relationship between the elevation of the terrain and the height of the atmospheric inversion layer. It serves as a quantitative measure of how the mountain's height compares to the altitude at which temperature changes significantly. This parameter plays a crucial role in understanding the dynamic interplay between topography and atmospheric conditions [13, 15, 16]. In this understanding, vortex streets can be significantly influenced by train and topography. The presence of obstacles, such as mountains or islands, can disrupt the flow of air, leading to the formation of vortex streets. The height and shape of these obstacles play a key role. Therefore, the speed and direction of the prevailing wind are crucial factors. Vortex streets often form in areas with consistent and strong winds. Therefore, daytime heating can lead to convective processes that influence the atmosphere's stability, potentially contributing to the formation of vortex streets. In this view, variations in temperature within the atmosphere contribute to the development of vortex streets [14–18]. Temperature differences can create regions of instability and influence flow patterns. Consequently, the stability of the atmosphere, characterized by factors like temperature inversions, influences the formation and persistence of vortex streets. Stable atmospheric conditions can enhance the likelihood of vortex street development [3, 8, 13].

Subsequently, the presence of clouds and moisture in the atmosphere can affect the development of vortex streets. These factors may provide visual cues and influence the overall atmospheric stability.

8.5 Formulation of Attributes: Vortex Streets

Vortex streets display various essential properties that play a crucial role in their characterization and comprehension. These properties encompass (i) the distance d_v of two vortices in the same row, and (ii) the spacing S of the two rows giving the aspect ratio S/d_v. In this view, the distance between two vortices within the same row, denoted as d_v, describes the spacing between individual vortices in a given row. It offers insights into the arrangement and regularity of the vortices. Therefore, spacing between rows is represented as S, in which the spacing between two successive rows of vortices determines the distance between vortices in different rows, crucial in shaping the overall structure and pattern of the vortex street. The aspect ratio S/d_v is often considered for a more comprehensive characterization. This ratio; consequently, provides a quantitative measure of the elongation or compression of the vortex street pattern. Different aspect ratios result in varied dynamical behaviors. Additionally, the propagation speed u_e of an individual vortex signifies the velocity at which a vortex moves downstream in the wake. This property is crucial for understanding the dynamics of the vortex street and how individual vortices evolve. In this sense, the shedding period T is the time interval between the formation of two successive vortices with the same sense of rotation. It characterizes the frequency of vortex shedding and provides information about the temporal behavior of the vortex street [11, 13, 15].

Understanding these properties is vital for unraveling the dynamic behavior of vortex streets. The distance between vortices, the spacing between rows, and the aspect ratio collectively influence the overall pattern and regularity of the vortex street. Additionally, parameters like propagation speed and shedding period contribute to the temporal aspects of the phenomenon. In this regard, the mean shedding period T can be calculated as:

$$\tilde{T} = \frac{\tilde{d}}{\tilde{u}_e}. \tag{8.11}$$

Therefore, Vortex streets can also be characterized by another property: the Strouhal number (St). The Strouhal number corresponds to the dimensionless shedding frequency of the vortices, as proposed by Etling [1] as:

$$St = \frac{d_s}{\tilde{T}\tilde{U}_0}, \tag{8.12}$$

In the context of vortex streets, the characteristic diameter d_s is considered to be the diameter of the island at the mean inversion height. The mean wind speed of the flow is denoted by $\tilde{U}_0$. As per Thomson et al. [15], this choice aligns with the selection of a characteristic value for d_s.

The Strouhal number is believed to be a constant for the same flow configuration and obstacle type. For example, in a flow behind a very long cylinder in a homogeneous fluid—the classical scenario for Kármán vortex streets—the Strouhal number has been determined as $St = 0.2$. Consequently, for a given obstacle diameter d_s, the product of the shedding period ($\tilde{T}$ and velocity $\tilde{U}_0$ remains constant. This implies that the shedding period is shorter in a fast flow than in a slow flow [4, 15, 18].

In addition to mean properties, the analysis can be delved into individual vortex characteristics. This includes the average temperature difference ($\Delta\theta_v$), average vertical vorticity (ω_v), and the radius of each vortex (r_v). These vortex-specific attributes can be computed on the original data, without coarsening or smoothing, providing detailed insights into the fine-scale features of the vortex street [17–19].

Therefore, the vertical vorticity component ξ is typically can be calculated as the vertical component of the vorticity vector, represented as:

$$\xi = \frac{\partial v}{\partial x} - \frac{\partial u}{\partial y} \tag{8.13}$$

In equation 8.13, v, and u are the northward and eastward wind flow components. This equation expresses the tendency of air parcels to rotate vertically in the atmosphere, providing valuable information about the rotational characteristics of the flow.

8.6 Quantum Vorticity States in Remote Sensing: A Novel Theoretical Framework

The incorporation of quantum mechanics into remote sensing presents a novel theory that expands our knowledge of fluid dynamics vorticity states to the domain of observational technologies. With the use of quantum mechanics, vortices—which are important in a variety of environmental phenomena—can be conceived of and examined in a new way thanks to Quantum Vorticity States in Remote Sensing (QVSRS).

In the QVSRS framework, vortex patterns observed through remote sensing techniques are considered carriers of quantum information. Each vortex within the observed patterns represents a quantum state, encoding information about the fluid dynamics and environmental conditions. In this perspective, let us $\Psi|\xi|\omega\rangle$ represent the set of vortices observed through remote sensing techniques [20–23]. Each vortex ξ_i within this set is considered a quantum state. The quantum state of a vortex ξ_i can be denoted as $\Psi|\xi_i|\omega_i\rangle$, where Ψ_i encapsulates information about the fluid dynamics and environmental conditions associated with $|\xi|\omega\rangle$. Mathematically, this can be represented as:

$$\Psi|\xi_i|\omega_i\rangle = \begin{bmatrix} \Psi|\xi_{i1}|\omega_{i1}\rangle \\ \Psi|\xi_{i2}|\omega_{i2}\rangle \\ \vdots \\ \Psi|\xi_{im}|\omega_{im}\rangle \end{bmatrix} \tag{8.14}$$

Here, $\Psi\left|\xi_{ij}\left|\omega_{ij}\right.\right\rangle$ represents the *j*-th component of the quantum state $\Psi\left|\xi_i\left|\omega_i\right.\right\rangle$, in the context of quantum vorticity chains, can be expressed as follows:

$$\Psi\left|\xi_i\left|\omega_i\right.\right\rangle = \widehat{o}_j\left|\xi_i\left|\omega_i\right.\right\rangle \tag{8.15}$$

Here $\widehat{o}_j$ is an operator corresponding to the observable related to the *j*-th component of information. The use of operators allows for a quantum mechanical description of the information associated with each vortex in the chain. In quantum mechanics, the physical properties of vortex in satellite images are represented by operators. These operators act on quantum states to extract specific information such as spectral signature and variation of visibility with different remote sensing band frequencies., In the given context, $\widehat{o}_j$ is an observable operator corresponding to a specific aspect or characteristic (such as velocity, temperature, etc.) related to the *j*-th component of the quantum state [21, 23, 25]. Needless to say, equation 8.15 signifies the application of the observable operator $\widehat{o}_j$ on the quantum state $\Psi\left|\xi_i\left|\omega_i\right.\right\rangle$ to obtain a new quantum state $\Psi\left|\xi_{ij}\left|\omega_{ij}\right.\right\rangle$. Therefore, the quantum vorticity chain C can be represented as a superposition of quantum states of its constituent vortices:

$$C = \sum_{i=1}^{n} \Psi\left|\xi_i\left|\omega_i\right.\right\rangle \tag{8.16}$$

Consequently, the evolution of the quantum state $\Psi\left|\xi_i\left|\omega_i\right.\right\rangle$ over time in remote sensing data can be described by the Schrödinger equation:

$$i\hbar\frac{\partial}{\partial t} = \widehat{H}\left|\xi_i\left|\omega_i\right.\right\rangle \tag{8.17}$$

where $\hbar$ is the reduced Planck constant, t is time, $\widehat{H}$ is the Hamiltonian operator, and $\Psi\left|\xi_i\left|\omega_i\right.\right\rangle$ is the different quantum state of the vortex as based on ξ_i, and ω_i.

8.7 Entanglement Between Vortices and Electromagnetic Waves in Remote Sensing Data

The statement "Quantum entanglement between vortices in the chain can be described using an entanglement operator $\widehat{E}(|\xi|\lambda\rangle)$" implies that the interdependence or correlation between the quantum states of different vortices in the chain and their quantum energy spectra signature $\Psi|E|\lambda\rangle$ is modeled by an entanglement operator. This operator, denoted as $\widehat{E}(|\xi|\lambda\rangle)$, captures the entangled relationship between the quantum states of distinct vortices and its quantum spectral signature $\Psi|E|\lambda\rangle$.

Mathematically, the entanglement operator acts on the tensor product of the quantum states of two distinct vortices, $\Psi\left|\xi_{ij}\left|\omega_{ij}\right.\right\rangle$, as well as the tensor product of their respective quantum energy spectra signatures, $\Psi\left|E_i\left|\lambda_i\right.\right\rangle$ and $\Psi\left|E_j\left|\lambda_j\right.\right\rangle$. The resulting entangled state can be expressed as:

$$\hat{E}\left(\left\|E_i\left|\lambda_i\right.\right\rangle\otimes\left\|E_j\left|\lambda_j\right.\right\rangle\otimes\Psi\left|E_i\left|\lambda_i\right.\right\rangle\otimes\Psi\left|E_j\left|\lambda_j\right.\right\rangle\right) = \left|\ddot{O}_{ij}\right\rangle\otimes\hat{S}_{ij} \tag{8.18}$$

here, $\left|\Phi_{ij}\right\rangle$ denotes the entangled state of vortices i and j, and $\hat{S}_{ij}$ represents the entangled quantum energy spectra signature associated with this entangled state.In simpler terms, when this operator acts on the tensor product of quantum states and energy spectra of two distinct vortices in the chain, it creates an entangled state $\left|\Phi_{ij}\right\rangle$ and an entangled quantum energy spectra signature $\hat{S}_{ij}$. This mathematical representation elegantly captures the quantum correlations and interdependencies that exist between different elements of the vortex chain [4, 19, 22].

Additionally, the concept of the entanglement operator $\widehat{E}(|\xi|\lambda\rangle)$ serves as a powerful mathematical tool to precisely characterize and model the intricate relationships within a chain of vortices in the quantum domain. This operator effectively encapsulates the entanglement between the quantum states of individual vortices and their corresponding quantum energy spectra signatures.

In the realm of remote sensing sensors, think of this entangled state as a quantum collaboration between the vortices and the electromagnetic waves. The entanglement extends its influence to the very fabric of the electromagnetic field, creating a connection that transcends classical boundaries. It's as if the vortices and the electromagnetic waves communicate in a language only the quantum world understands. Therefore, this entanglement doesn't just link the vortices; it intertwines their quantum fates with the electromagnetic waves sensed by remote sensors. The information encoded in the entangled state becomes a bridge between the fluid dynamics within the vortices and the electromagnetic signals detected by the remote sensing sensors [26].

In essence, the entanglement between vortices and electromagnetic waves adds a layer of quantum richness to our understanding of remote sensing. It's a captivating dance where the quantum states of vortices and the electromagnetic waves choreograph a symphony of information, revealing insights into the fluidic intricacies of our surroundings.

According to the above-mentioned perspective, the process of measurement quantum vorticity chain entanglements with remote sensing satellite data can be elegantly described through the application of a measurement operator $\widehat{M}$. This operator acts on the quantum states of the vortices within the chain C, and its action is represented mathematically as:

$$\widehat{M}\widehat{C} = \sum_{i=1}^{n} \widehat{M}\Psi\left|\xi_i\left|\omega_i\right.\right\rangle\left|\ddot{\mathrm{O}}_{ij}\right\rangle \otimes \hat{S}_{ij} \tag{8.19}$$

Here, $\widehat{C}$ signifies the entangled quantum vorticity chain, and the measurement operator $\widehat{M}$ is applied to each quantum state $\left|\xi_i\left|\omega_i\right.\right\rangle$ within the chain. And it is entangled in satellite remote sensing within different states of spectral signature $\left|\Phi_{ij}\right\rangle \otimes \hat{S}_{ij}$ [23–26].

8.8 Developing Quantum Distance Estimation Algorithm for Automatic Detection of von Kármán Vortex Street Features from Satellite Data

The primary challenge in extracting von Kármán Vortex Street features from satellite data is the need for high-resolution data and accurate detection algorithms. Von Kármán vortex streets are a repeating pattern of swirling vortices caused by the process of vortex shedding, which occurs when a fluid flows around an obstacle. These vortices can be seen in the climate, and their extraction from satellite information requires the capacity to distinguish and follow the vortices over the long run precisely. This is challenging due to the complex nature of the atmospheric flow and the need to distinguish the vortices from other features in the data. In this perspective, high-resolution satellite wind tracking and advanced image processing techniques are necessary to overcome these challenges. To address this challenge, advanced algorithms based on quantum computing are being explored to overcome these complexities. In this context, the Quantum Distance Estimation Algorithm is being leveraged as a promising solution to extract von Kármán Vortex Street features from the intricate and noisy surrounding environment in satellite images. This quantum algorithm offers the potential to efficiently process and analyze satellite data, even in the presence of such challenges are mentioned above. By utilizing quantum properties and computational advantages, the Quantum Distance Estimation Algorithm can enhance the accuracy and reliability of von Kármán Vortex Street features extraction in remote sensing data.

To execute the quantum distance estimation algorithm, we must first encode the generated data points into a multi-qubit quantum state. To achieve this, we employ two prevalent embedding techniques: (i) amplitude embedding and (ii) angle embedding. We then assess the performance of each method individually [26].

Amplitude embedding is a method to represent classical data in a quantum state by encoding information in the amplitudes of quantum states. Suppose we have a set of data points $\{x_1, x_2, ..., x_n\}$. To perform amplitude embedding, we map these data points to quantum states using the following process: (i) normalize the data; and (ii) quantum superposition. Normalize the data points, therefore, they lie within a specific range, often between $|0\rangle$ and $|1\rangle$ [26–28]. Encode the normalized data points into quantum superposition states. Each data point corresponds to a coefficient (amplitude) in the superposition. Mathematically, the spectral signature data points $S_1, S_2, ..., S_n$, can be represent the quantum state as:

$$\Psi\left|\hat{S}\right\rangle = \alpha_1|0\rangle + \alpha_2|1\rangle + ... + \alpha_n|n-1\rangle\,. \tag{8.20}$$

Equation 8.20 means that the quantum state of the entanglement spectra signature $\Psi\left|\hat{S}\right\rangle$ is described as a linear combination of the basis states $|0\rangle, |1\rangle, |2\rangle, ..., |n-1\rangle$ with the coefficients $\alpha_1, \alpha_2, ..., \alpha_n$ determining the probabilities associated with each basis state along the Von Kármán vortex streets features in remote sensing data. This is a fundamental representation in quantum computing, where complex amplitudes allow for the representation of quantum superposition, entanglement, and other quantum phenomena. Mathematically, the quantum state $\Psi\left|\hat{S}\right\rangle$ in angle embedding; therefore, can be represented as:

$$\Psi\left|\hat{S}\right\rangle = |0\rangle \otimes e^{(i\theta_1)}|0\rangle + |0\rangle \otimes e^{(i\theta_2)}|1\rangle + ... + |0\rangle \otimes e^{(i\theta_n)}|n-1\rangle \tag{8.21}$$

here $e^{(i\theta_1)}, e^{(i\theta_2)}, ..., e^{(i\theta_n)}$ are complex numbers represented using the exponential notation. Each $e^{(i\theta_1)}$ is associated with a specific basis state of Von Kármán vortex streets features in the satellite data and represents a complex probability amplitude [27–30]. This notation is often used in quantum computing to represent quantum states, and it allows for the description of quantum superposition and phase information associated with each basis state [31]. In this sense, these embeddings allow classical data to be efficiently processed on a quantum computer, opening up possibilities for quantum data analysis and machine learning.

Let's create a new dissimilarity measure inspired by the author for measuring dissimilarity or similarity between two vectors $\vec{a}$ and $\vec{b}$ of Von Kármán vortex streets features in satellite images. We shall call this measure "Vortex Dissimilarity Measure (VDM)". VDM is designed to capture the dissimilarity or similarity between two vectors representing Von Kármán vortex street features in satellite images. It takes into account both the differences in feature values and their relative importance through element-wise weighting (Figure 8.5). VDM is defined as follows:

$$VD\left(\left|a\middle|b\right\rangle, p\right) = \left(\sum_{i=1}^{n} W_i \delta_i \left(\left|\vec{a}_i - \vec{b}_i\right|\right)^{1+i}\right)^{p^{-1}} \tag{8.22}$$

here quantum states $|a\rangle$ and $|b\rangle$, where $|a\rangle = \sum_i \alpha_i|i\rangle$ and $|b\rangle = \sum_i \beta_i|i\rangle$ represent vectors $\vec{a}$ and $\vec{b}$ in a quantum superposition, respectively. Therefore, α_i and β_i are complex probability amplitudes for the $|i\rangle$ basis state. $\delta_i(|\alpha_i|, |\beta_i|)$ is the dissimilarity measure between the complex probability amplitudes α_i and β_i. Additionally, p is the order parameter controlling the sensitivity to individual elements. Therefore, in this pseudocode, we assume that one has functions or methods for creating quantum superposition states, obtaining amplitudes, calculating dissimilarity measures, and accessing element-wise weights. The **quantum_superposition_vdd** function iterates through elements in superposition states and calculates the VDD using the specified order parameter (**p**). The result is the von Kármán vortex street feature Distance in Quantum Superposition.

Here W_i represents the weight assigned to the dissimilarity measure for the i-th element in satellite data. Therefore, the von Kármán vortex street feature similarity scores as the inverse of the weighted distance, normalized to the range [0, 1]:

```
function quantum_superposition_vdd(|a⟩, |b⟩, p):
# Initialize variables to store intermediate values result = 0.0 n = number_of_elements_in(|a⟩)
 # Assuming |a⟩ and |b⟩ have the same number of elements
# Loop over each element in the superposition states for i in range(n):

 # Get the complex probability amplitudes
 αi = amplitude_of(|a⟩, |i⟩) βi = amplitude_of(|b⟩, |i⟩)
# Get the element-wise weight
 Wi = get_weight_for_element(i)
 # Calculate the dissimilarity measure for the element dissimilarity = calculate_dissimilarity(|αi|, |βi|)
# Apply the weight and order parameter
weighted_dissimilarity = Wi * (dissimilarity ** p)
 # Add the weighted dissimilarity to the result result += weighted_dissimilarity
# Apply the p-th root to the sum of weighted dissimilarities VDD_result = result ** (1/p)
 # Return
# Example usage: |a⟩ = create_quantum_superposition_vector(a)
# Create a quantum superposition state for vector a |b⟩ = create_quantum_superposition_vector(b)
# Create a quantum superposition state for vector b p = 2
# Choose the order parameter
distance = quantum_superposition_vdd(|a⟩, |b⟩, p)
 // Compute (VDM) using a loop
// Normalize VDM to the range [0, 1]
minVDM = 0 // Minimum possible VDM
maxVDM = 1 // Maximum possible VDM
normalizedVDM = (VDM - minVDM) / (maxVDM - minVDM)
// Compute the Vortex Similarity Score as the inverse of normalized VDM
  VortexSimilarityScore = 1 - normalizedVDM
  return VortexSimilarityScore
```

Figure 8.5: Pseudo-code for quantum superposition of Vortex Dissimilarity Distance (VDD).

$$S_{vdd} = \frac{1}{\left(1 + \sqrt{\left(\sum_i \left(W_i * \left(\vec{a_i} - \vec{b_i}\right)^2\right)\right)}\right)} \tag{8.23}$$

This ensures that higher values of S_{vdd} indicate greater similarity, while lower values indicate dissimilarity. This equation enables the calculation of the Marghany Distance between two vectors, $\vec{a}$, and $\vec{b}$ using different distance metrics (Euclidean, Manhattan, or Minkowski) by adjusting the value of p. It offers a unified dissimilarity measure capable of accommodating various distance metrics based on the specified order. in this pseudocode, we use the **quantum_superposition_vdd** function to calculate the VDD between two quantum superposition states representing von Kármán vortex street feature. The VDD is then normalized to the range [0, 1], and the von Kármán vortex street feature "Similarity Score (VSS)" is computed as 1 minus the normalized distance. This score quantifies the similarity between the von Kármán vortex street feature in the quantum superposition states (Figure 8.6).

Consequently, Quantum Element-wise Weighted Distance Decomposition harnesses the remarkable quantum property of superposition [27, 29, 32]. In classical computing, data elements are processed sequentially or in parallel, but quantum computing takes a different approach [33]. With Q-EWDD, we delve into the quantum realm, where quantum bits, or qubits, can exist in a multitude of states simultaneously. This unique property forms the foundation of Q-EWDD and empowers us to explore an array of weighted combinations of elements in an incredibly efficient manner (Figure 8.7).

```
function von Kármán Vortex Street Feature _similarity_score(|a⟩, |b⟩, p):
# Calculate Vortex Distance (VDD) using quantum_superposition_vdd function
VDD = quantum_superposition_vdd(|a⟩, |b⟩, p)
 # Normalize VDD to the range [0, 1] max_possible_distance = 1.0
# Maximum possible VDD (can be determined empirically)
if VDD > max_possible_distance: normalized_distance=1.0 else: normalized_distance = VDD / max_
possible_distance
# Calculate von Kármán Vortex Street Feature Similarity Score (VSS) VSS = 1.0 - normalized_distance
# Return VSS
# Example usage: |a⟩ = create_quantum_superposition_vector(a)
# Create a quantum superposition state for vector a |b⟩ = create_quantum_superposition_vector(b)
# Create a quantum superposition state for vector b p = 2
# Choose the order parameter
similarity_score = vortex_similarity_score(|a⟩, |b⟩, p) print("von Kármán Vortex Street Feature
Similarity Score:", similarity_score)
```

Figure 8.6: A pseudocode representation of von Kármán Vortex Street Feature Similarity Score (VSS) using the quantum_superposition_Vdd

```
# Define the quantum state for data vectors a and b # |ψ⟩ = α1|00⟩ + α2|01⟩ + α3|10⟩ + α4|11⟩
# Where α1, α2, α3, α4 are complex coefficients
# Initialize the Vortex Distance (VDD) to zero VDD = 0
# Define the element-wise weights for vectors a and b weights_a = [W1, W2, ..., Wn]
# Weight vector for vector a weights_b = [W1, W2, ..., Wn] # Weight vector for vector b
# Loop through the elements of vectors a and b for i in range(len(a)):
# Calculate the dissimilarity measure δi(ai, bi) between elements ai and bi dissimilarity =
   calculate_dissimilarity(a[i], b[i])
# Replace 'calculate_dissimilarity' with the desired distance metric
 # Multiply the dissimilarity measure by the element-wise weights
weighted_dissimilarity = dissimilarity * weights_a[i] * weights_b[i]
 # Add the weighted dissimilarity to the VDD
VDD += weighted_dissimilarity
# VDD now contains the Quantum Element-wise Weighted Distance Decomposition
```

Figure 8.7: Pseudocode for Quantum Element-wise Weighted Distance Decomposition (Q-EWDD).

In this pseudocode, we represent the quantum state $|\psi\rangle$, initialize the VDD to zero, define element-wise weights for vectors a and b, and loop through the elements to calculate the dissimilarity measure for each element. The dissimilarity measure is then weighted by the respective element-wise weights and added to the VDD. This process yields the Quantum Element-wise Weighted Distance Decomposition (Q-EWDD) for the given data vectors and $\vec{b}$ of Von Kármán vortex streets features in satellite images.

8.9 Quantum Amplitude Embedding for Identifying von Kármán Vortex Streets Features in Multi-Satellite Images

To articulate the process of analyzing decomposed distances to identify patterns linked with spiral features, we can formulate a pioneering mathematical equation that capitalizes on the quantum advantage to enable efficient exploration. This equation shall be referred to as the "Von Kármán vortex streets Pattern Identification Index (VPII)."

$$VPII(VDD) = \sum_{i=1}^{n} \left(1 - \frac{VDD_i}{\max(VDD)}\right) \cdot P_i \tag{8.24}$$

In this equation, *VPII*(*VDD*) represents the Von Kármán vortex streets Pattern Identification Index, *VDD* is the set of decomposed distances obtained through Q-EWDD, n is the number of data vectors, VDD_i is the decomposed distance for the i-th data vector, max (*VDD*) is the maximum decomposed distance, and P_i is the pattern detection coefficient for the i-th data vector. To formulate a quantum mechanics formula for the Von Kármán vortex streets Pattern Identification Index, we can start by defining the *VPII* as a quantum expectation value. Let's assume we have a quantum state $|\psi\rangle$ that encodes the information about decomposed distances and pattern detection coefficients. The *VPII* can be represented as:

$$VPII = \langle \Psi | \hat{O} | \Psi \rangle \tag{8.25}$$

In equation 8.25, $|\Psi\rangle$ is the quantum state that encodes decomposed distances, and $\hat{O}$ is the quantum operator corresponding to the VPII calculation. The quantum operator $\hat{O}$ should be constructed based on the specific equation or method used for *VPII* calculation. This equation often involves weighted sums or integrations over the decomposed distances. The exact form of $\hat{O}$ would depend on the mathematical representation of *VPII*. For example, if *VPII* involves a weighted sum of decomposed distances, the quantum operator could be constructed as:

$$\hat{O} = \sum_i W_i \widehat{VDD_i} \tag{8.26}$$

where W_i are the pattern detection coefficients, and $\widehat{VDD_i}$ are quantum operators corresponding to decomposed distances. This quantum mechanics formula allows you to calculate the SPII using quantum states and operators. The specific construction of $\hat{O}$ and the quantum state $|\Psi\rangle$ would depend on the mathematical expression of SPII and the details of the quantum computing system being used. In this sense, quantum computers perform operations in parallel, which dramatically speeds up the computation of distances for large datasets.SPII and Q-EWDD can handle high-dimensional data with ease. Q-EWDD employs a quantum circuit that leverages quantum parallelism to calculate the weighted distances in parallel.

Therefore, a mathematical representation of the step where quantum clustering algorithms or techniques are utilized to group similar data points based on dissimilarity values from the distance matrix is:

$$\text{Minimize: } J = i = \sum_{i=1}^{n} \sum_{j=1}^{k} X_i VDD_{ij}^2 \tag{8.27}$$

Subject to the constraint that each data point i is assigned to exactly one cluster:

$$\sum_{i=1}^{k} X_i = 1, \forall_i \in 1, 2, \ldots, n \tag{8.28}$$

Possibly using a quantum optimization algorithm, will yield the assignment of data points to clusters, effectively grouping similar data points based on the dissimilarity values in the distance matrix. Consequently, the Quantum Approximate Optimization Algorithm (QAOA)utilizes the quantum nature of qubits to explore multiple potential solutions simultaneously. This quantum parallelism allows them to search for the optimal configuration of cluster assignments more efficiently than classical optimization methods [34–36]. The quantum optimization algorithm iteratively adjusts the values of the binary variables X_i to minimize the cost function J. These adjustments aim to group data points into clusters such that the overall dissimilarity, as represented by the cost function, is minimized. The result is an assignment of data points to clusters that best capture the inherent patterns in the data.

QAOA starts by preparing a quantum state $|\Psi(\beta,\gamma)\rangle$, which depends on two sets of parameters β and γ. This state is created by applying quantum operators to an initial state, typically the uniform superposition state $|+\rangle = N^{-1}\sum_x |x\rangle$. The expectation value of the objective function is calculated concerning the quantum state $|\Psi(\beta,\gamma)\rangle$. This is done by measuring the Hamiltonian $\hat{H}$ in this state:

$$\langle \Psi(\beta,\gamma) | \hat{H} | \Psi(\beta,\gamma) \rangle \tag{8.29}$$

Computing this expectation value requires running quantum circuits and measuring the results. QAOA is designed to optimize the parameters β and γ to find the values that minimize the expectation value of the objective function [33, 36].

8.10 Tested Remote Sensing Satellite Data

This captivating image captured by SeaWiFS (Sea-viewing Wide Field-of-view Sensor) showcases the mesmerizing von Kármán cloud vortex streets gracefully being shed by the picturesque Canary Islands, which was acquired on June 4, 2000 (Figure 8.8). The intricate interplay of atmospheric dynamics and island topography is vividly depicted, offering a stunning view of the natural beauty and complexity inherent in the von Kármán vortex phenomena. In this view, SeaWiFS features an eight-band across-track scanner that operates across a spectral range from 0.402 to 0.885 mm. This wide spectral coverage allowed SeaWiFS to capture diverse information about the Earth's surface and atmosphere, enabling the observation and analysis of various environmental phenomena, including the intricate von Kármán cloud vortex streets showcased in the imagery [37].

Figure 8.8: Occurrence of von Kármán cloud vortex streets in SeaWiFS data.

SeaWiFS had a nominal spatial resolution of approximately 1.1 kilometers at nadir. This means that each pixel in the SeaWiFS image represented an area on the Earth's surface that was approximately 1.1 kilometers by 1.1 kilometers. Consequently, Figure 8.9 demonstrates the sharp edges of von Kármán cloud vortex streets. Although the surrounding area is dominated by heavy cloud covers, the newly developed quantum algorithm based on quantum superposition and Vortex Distance Decomposition (quantum_superposition_VDD) delivers sharper patterns of von Kármán cloud vortex streets.

On June 12, 2022, the Visible Infrared Imaging Radiometer Suite (VIIRS) instrument documented the presence of von Kármán vortex streets near the Canary Islands in the Atlantic Ocean. These distinctive vortex patterns were observed to undulate along the coastal waters of Morocco and Mauritania. VIIRS, renowned for its ability to capture both visible and infrared imagery, facilitates comprehensive global observations encompassing land, atmosphere, cryosphere, and oceans [38, 40]. The captured imagery on this date vividly portrays the fascinating atmospheric phenomenon of von Kármán vortex streets in the specified geographical region (Figure 8.10).

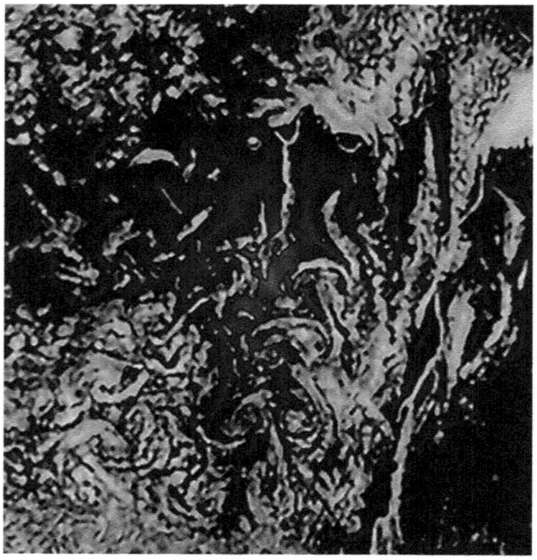

Figure 8.9: Detection of von Kármán cloud vortex streets pattern in SeaWiFS data using quantum superposition vortex distance decomposition (VDD).

Figure 8.10: Occurrence of von Kármán cloud vortex streets in VIIRS data.

In this regard,the VIIRS imagery bands, characterized by a moderate resolution of 750 meters for I-Bands and a dual-resolution of 750 meters (visible) and 375 meters (infrared) for the Day-Night Band (DNB), provide detailed and versatile visual and infrared information. In parallel, the Environmental Data Records (EDRs) offer a broader perspective with moderate resolutions of 1.5 kilometers for bands M1-M15, with some bands aggregated at 6 kilometers.

Within the I-Bands, spanning from Red (I1) to Near-Infrared (I5), VIIRS excels in capturing both daytime and nighttime imagery. The unique capability of the Day-Night Band to operate in low-light conditions, gathering visible and infrared data during the night, enhances its observational prowess. Meanwhile, the Environmental Data Records (EDRs) encompass a range of spectral bands (M1-M15), serving various purposes in environmental monitoring, including visible, near-infrared, and thermal infrared observations [38–40]. According to this mentioned perspective, the quantum_superposition_VDD algorithm delivers the sharpest edges of von Kármán vortex streets than SeaWiFS (Figure 8.11).

In a recent Terra MODIS image acquired on October 1, 2023, a visually striking von Kármán vortex street unfolds slightly southwest of the Canary Islands in the vast expanse of the Atlantic Ocean (refer to Figure 8.12). Notably, the quantum_superposition_VDD algorithm, akin to its VIIRS counterpart, exhibits a remarkable capability to render the sharpest edges and intricate details of these vortex streets when compared to SeaWiFS (see Figure 8.13).

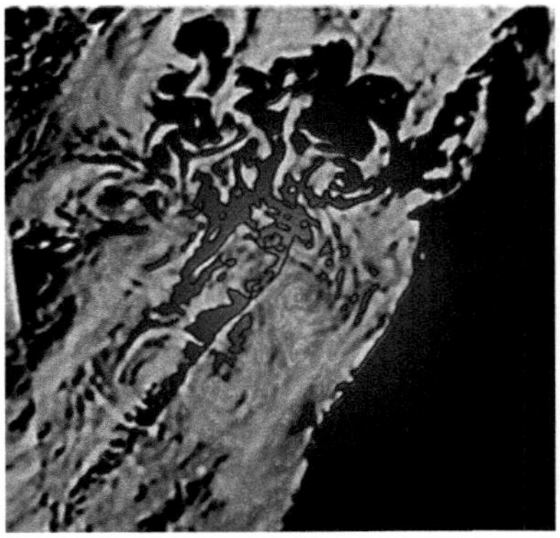

Figure 8.11: Detection of von Kármán cloud vortex streets pattern in VIIRS data using quantum superposition vortex distance decomposition (VDD).

Figure 8.12: Occurrence of von Kármán vortex street southwest of Canary Island in Terra MODIS data.

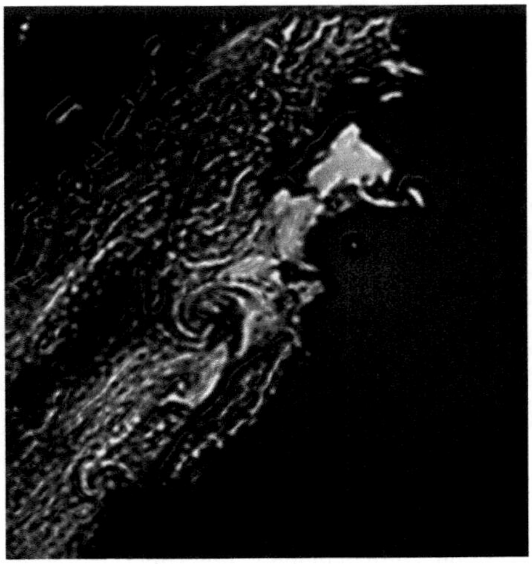

Figure 8.13: Detection of von Kármán cloud vortex streets pattern in *v* using quantum superposition vortex distance decomposition (VDD).

A fascinating visualization of atmospheric phenomena has been made possible by the recent observation of von Kármán cloud vortex streets throughout the Atlantic Ocean zone from Los Lianos to Villa de Valverde, which includes the La Frontera and La Restinga Islands (Figure 8.14). Sentinatel-A satellite data, more precisely in the C-VV polarized band, was used to record the event (Figure 8.15).

Notably, the imagery depicts the dynamic movement of swirling vortices originating from Los Lianos and progressing towards La Restinga Islands, exhibiting well-defined edges despite the inherent challenge of high levels of speckles dominating the radar data (Figure 8.15).

From the aforementioned perspective, the application of the quantum superposition vortex distance decomposition (VDD) algorithm in Sentinel-1A reveals a more distinct boundary edge

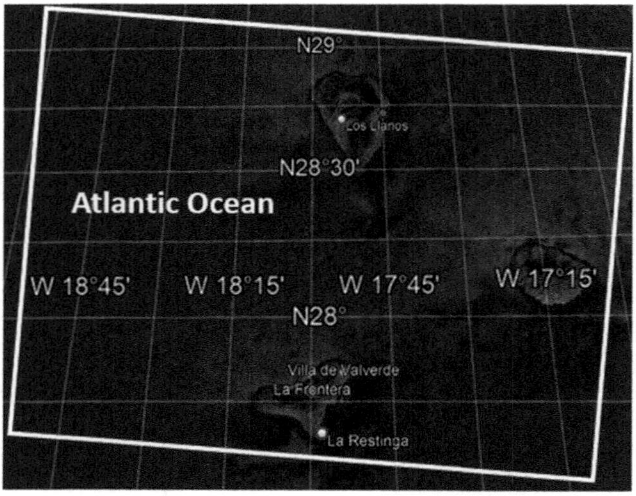

Figure 8.14: Location of Sentinel-1A satellite overpass over Los Lianos Island.

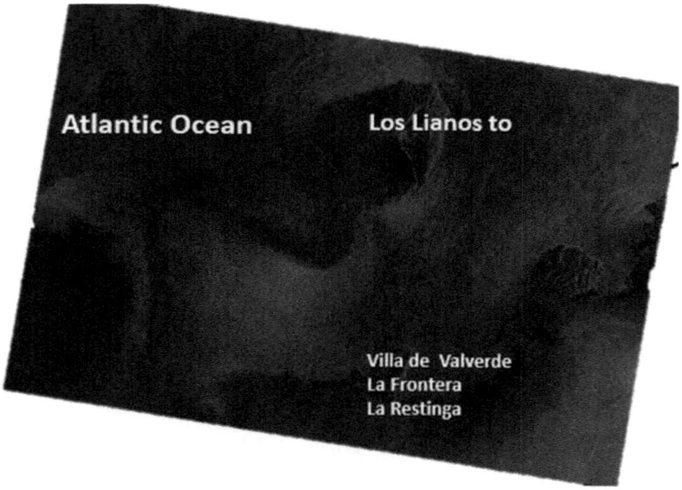

Figure 8.15: Occurrence of von Kármán cloud vortex streets in Sentinel-1 A on August 31, 2023.

for von Kármán cloud vortex streets compared to optical satellite imagery. This heightened clarity is attributed to the heightened sensitivity of radar sensors, specifically synthetic aperture radar, to sea surface roughness. The movement of this vortex type amplifies surface roughness, resulting in a robust backscatter to radar images. Consequently, the VDD algorithm exhibits greater efficacy in discerning the sharpest boundary edges in radar images, a task that proves more challenging when relying on spectral signature variations in optical satellite images (Figure 8.16).

The phenomenon of a vortex street materializes when oceanic clouds undergo disturbance from winds interacting with landmasses or other obstacles above the sea surface, as is the case here with Ilha da Madeira. The southeastward drift of low-level winds induces the alignment of clouds in a uniform direction, forming what is known as a "street." The interaction of the winds with the islands generates mesmerizing swirls termed vortices [1–4].

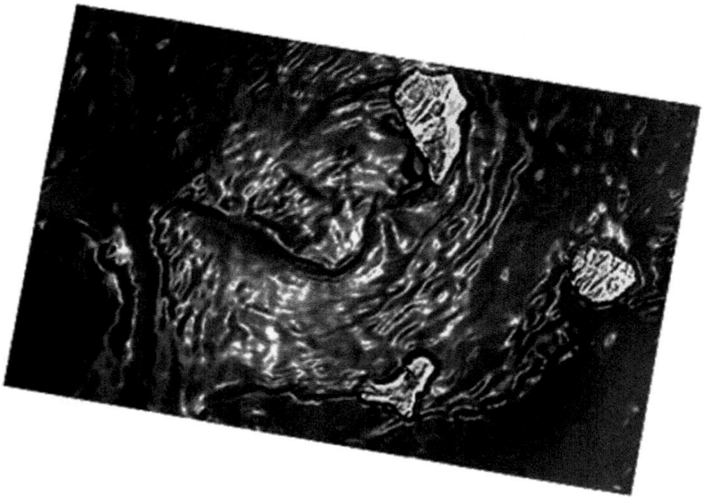

Figure 8.16: Automatic detection of von Kármán cloud vortex streets in Sentinel-1A using quantum superposition vortex distance decomposition (VDD) algorithm.

The clouds contributing to this vortex street are of a specific type known as "closed cell." These closed cell clouds often organize themselves into roughly hexagonal arrays within a layer of air exhibiting fluid-like behavior, a common occurrence in the atmosphere. The initiation of convection, triggered by either heating at the base or cooling at the top, leads to the formation of these closed-cell clouds. Within these clouds, warm air ascends at their centers, while cooler air descends around the edges, resulting in the creation of a distinctive honeycomb-like pattern. This image captures the intricate beauty of atmospheric dynamics and the interplay between wind patterns and geographical features [2, 4, 14].

8.11 What makes the Quantum Superposition VDD Algorithm a Valuable Tool for Automated von Kármán Cloud Vortex Streets Detection?

The ROC curve provides a visual representation of the algorithm's precision in distinguishing cloud street features between these diverse datasets. Remarkably, the VDD algorithm achieved a 98% accurate identification of von Kármán cloud vortex streets in radar Sentinel-1A satellite data. In contrast, both VIIRS and Terra MODIS data exhibited a high true positive rate (94%) for identifying von Kármán cloud vortex streets and an equally impressive true negative rate for discerning surrounding cloud patterns (Figure 8.17).

When visualizing the ROC curve for the VDD algorithm applied to von Kármán cloud vortex streets, it would manifest as a curve sharply rising towards the upper-left corner of the graph, indicating the algorithm's high true positive and true negative rates within Sentinel-1A data. The Area Under the Curve (AUC) in this scenario would approach 1, signifying exceptional performance in quantum edge detection.

Certainly, the Vortex Distance Decomposition (VDD) algorithm serves as an automatic detection tool when applied to radar images and functions as a semi-automatic tool when utilized with optical satellite data like VIIRS and MODIS. In this view, the algorithm's sensitivity to object roughness precedes its reliance on spectral signatures for effective detection. This means that VDD excels at capturing features based on the surface roughness of objects, making it particularly adept at detecting

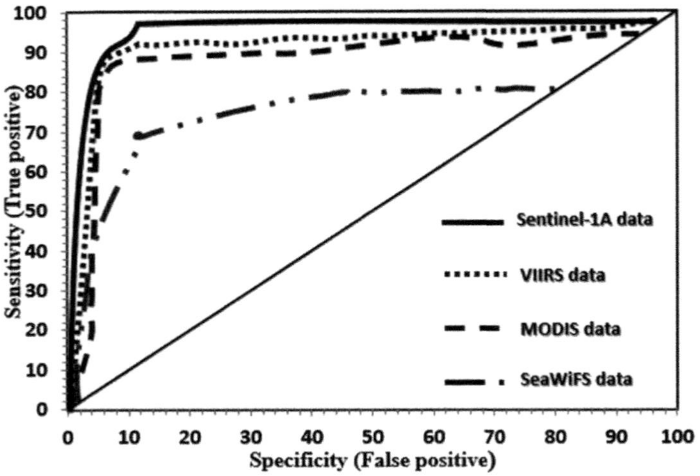

Figure 8.17: Accuracy comparison of the Quantum Superposition Vortex Distance Decomposition (VDD) algorithm across different satellite sensors.

patterns in radar imagery and providing valuable insights in the analysis of optical satellite data. In addition, the algorithm's ability to prioritize object roughness allows it to differentiate between various materials and textures, accurately identifying different objects in radar and optical data. This unique capability makes VDD a versatile tool for various applications, including urban planning, environmental monitoring, and disaster response.

The Quantum_Superposition_VDD algorithm stands out as a groundbreaking solution for automated von Kármán cloud vortex streets detection. The pronounced sharpness observed in patterns generated by this newly devised quantum algorithm can be attributed to a synergy of key factors.

Firstly, the utilization of quantum parallelism empowers the algorithm to process data concurrently across multiple quantum states. This unique capability enables Quantum_Superposition_VDD to explore a multitude of possibilities simultaneously, resulting in a more intricate and precise pattern recognition compared to its classical counterpart.

The concept of quantum superposition, a defining feature of quantum algorithms, further enhances the algorithm's pattern recognition prowess. By allowing quantum qubits to exist in multiple states simultaneously, the algorithm comprehensively examines various combinations of data elements, providing a more nuanced understanding of complex patterns. These are agreed with studies of Etling [1], Tomesh and Martonosi [33], and Crooks [35].

The inclusion of Vortex Distance Decomposition (VDD) introduces a unique approach to evaluating dissimilarity or similarity between data points. By assigning specific weights to elements and considering individual contributions, VDD excels at highlighting subtle variations that might be overlooked by conventional clustering techniques.

Quantum_Superposition_VDD's adept handling of high-dimensional data is another key advantage. In scenarios where data points exhibit numerous dimensions or features, the quantum algorithm efficiently reduces computational complexity, delivering clearer insights into intricate patterns. Therefore, optimization techniques, such as the use of Quantum Approximate Optimization Algorithm (QAOA), further refine patterns and contribute to the algorithm's ability to deliver sharper results.

In summary, the synergy of quantum computing principles, Vortex Distance Decomposition, and specific quantum capabilities positions Quantum_Superposition_VDD as a frontrunner in pattern recognition. Its quantum advantage proves particularly promising for applications requiring critical pattern recognition and data analysis. This innovative algorithm not only revolutionizes pattern recognition in multispectral imagery but also contributes to a deeper understanding of complex atmospheric turbulence dynamics such as von Kármán cloud vortex streets.

This chapter introduces an innovative tutorial on the creation of an automatic detection algorithm for identifying von Kármán cloud vortex streets across multiple satellite images, leveraging the capabilities of quantum computing. The algorithm, named Vortex Distance Decomposition (VDD), was developed by the author and further optimized using the Quantum Approximate Optimization Algorithm (QAOA). A notable observation is the algorithm's efficiency, which exhibits variations depending on the sensor employed. This suggests that the physical properties of sensors, such as spectral signatures in optical sensors and sea roughness backscatter in radar sensors, significantly influence the accuracy of the algorithm. The chapter delves into an exploration of how the VDD algorithm functions as a semi-automatic tool when applied to optical images but transforms into an automatic tool when dealing with SAR images.

In the upcoming chapter, there will be a focused exploration of radar satellite images, delving into the tracking of atmospheric turbulence features like gravity waves, typhoons, and hurricanes.

References

[1] Etling, D. (1989). On atmospheric vortex streets in the wake of large islands. *Meteorology and Atmospheric Physics*, 41(3), 157–164.

[2] Chopra, K.P. and Hubert, L.F. (1965). Karman vortex streets in wakes of islands. *AIAA Journal*, 3(10), 1941–1943.

[3] Young, G.S. and Zawislak, J. (2006). An observational study of vortex spacing in island wake vortex streets. *Monthly Weather Review*, 134(8), 2285–2294.

[4] Heinze, R., Raasch, S. and Etling, D. (2012). The structure of Kármán vortex streets in the atmospheric boundary layer derived from large eddy simulation. *Meteorologische Zeitschrift*, 21(3), 221.

[5] Letzel, M.O., Krane, M. and Raasch, S. (2008). High resolution urban large-eddy simulation studies from street canyon to neighbourhood scale. *Atmospheric Environment*, 42(38), 8770–8784.

[6] Letzel, M.O. and Raasch, S. (2002). Large-eddy simulation of thermally induced oscillations in the convective boundary layer. *Proceedings of Hydraulic Engineering*, 46, 67–72.

[7] Ahlborn, B., Seto, M.L. and Noack, B.R. (2002). On drag, Strouhal number and vortex-street structure. *Fluid Dynamics Research*, 30(6), 379.

[8] Wang, A.B., Trávníček, Z. and Chia, K.C. (2000). On the relationship of effective Reynolds number and Strouhal number for the laminar vortex shedding of a heated circular cylinder. *Physics of Fluids*, 12(6), 1401–1410.

[9] Igarashi, T. (1999). Flow resistance and Strouhal number of a vortex shedder in a circular pipe. *JSME International Journal Series B Fluids and Thermal Engineering*, 42(4), 586–595.

[10] Griffin, O.M. (1978). A universal Strouhal number for the 'locking-on' of vortex shedding to the vibrations of bluff cylinders. *Journal of Fluid Mechanics*, 85(3), 591–606.

[11] Snyder, W.H., Thompson, R.S., Eskridge, R.E., Lawson, R.E., Castro, I.P., Lee, J.T., ... and Ogawa, Y. (1985). The structure of strongly stratified flow over hills: Dividing-streamline concept. *Journal of Fluid Mechanics*, 152, 249–288.

[12] Ryan, W. and Lamb, B. (1984). Determination of dividing streamline heights and Froude numbers for predicting plume transport in complex terrain. *Journal of the Air Pollution Control Association*, 34(2), 152–155.

[13] Raasch, S. and Franke, T. (2011). Structure and formation of dust devil–like vortices in the atmospheric boundary layer: A high-resolution numerical study. *Journal of Geophysical Research: Atmospheres*, 116(D16).

[14] Taubin, G. (1995). Curve and surface smoothing without shrinkage. *In:* Proceedings of IEEE International Conference on Computer Vision (pp. 852–857). IEEE.

[15] Thomson, R.E., Gower, J.F.R. and Bowker, N.W. (1977). Vortex streets in the wake of the Aleutian Islands. *Mon. Wea. Rev.*, 105, 873–884.
[16] Tsuchiya, K. (1969). The clouds with the shape of Kármán vortex street in the wake of Cheju Island, Korea. *J. Meteor. Soc. Japan*, 47, 457–465.
[17] Vosper, S.B. (2000). Three-dimensional numerical simulations of strongly stratified flow past conical orography. *J. Atmos. Sci.*, 57, 3716–3739.
[18] Vosper, S.B., Castro, I.P., Snyder, W.H. and Mobbs, S.D.(1999). Experimental studies of strongly stratified flow past three-dimensional orography. *J. Fluid Mech.*, 390, 223–249.
[19] Young, G.S. and Zawislak, J.(2006). An observational study of vortex spacing in island wake vortex streets. *Mon. Wea. Rev.*, 8, 2285–2294.
[20] Billam, T.P., Reeves, M.T. and Bradley, A.S. (2015). Spectral energy transport in two-dimensional quantum vortex dynamics. *Physical Review A*, 91(2), 023615.
[21] Dominici, L., Dagvadorj, G., Fellows, J.M., Ballarini, D., De Giorgi, M., Marchetti, F.M., ... and Sanvitto, D. (2015). Vortex and half-vortex dynamics in a nonlinear spinor quantum fluid. *Science Advances*, 1(11), e1500807.
[22] Wallraff, A., Lukashenko, A., Lisenfeld, J., Kemp, A., Fistul, M.V., Koval, Y. and Ustinov, A.V. (2003). Quantum dynamics of a single vortex. *Nature*, 425(6954), 155–158.
[23] Wisniacki, D.A., Pujals, E.R. and Borondo, F. (2007). Vortex dynamics and their interactions in quantum trajectories. *Journal of Physics A: Mathematical and Theoretical*, 40(48), 14353.
[24] Golosovsky, M., Tsindlekht, M. and Davidov, D. (1996). High-frequency vortex dynamics in YBa2Cu3O7. *Superconductor Science and Technology*, 9(1), 1.
[25] Wang, C.C.J., Duine, R.A. and MacDonald, A.H. (2010). Quantum vortex dynamics in two-dimensional neutral superfluids. *Physical Review A*, 81(1), 013609.
[26] Marghany, M. (2022). *Remote Sensing and Image Processing in Mineralogy*. CRC Press.
[27] Gu, M., Weedbrook, C., Menicucci, N.C., Ralph, T.C. and van Loock, P. (2009). Quantum computing with continuous-variable clusters. *Physical Review A*, 79(6), 062318.
[28] Khan, T.M. and Robles-Kelly, A. (2020). Machine learning: Quantum vs classical. *IEEE Access*, 8, 219275-219294.
[29] Khan, R.A. (2019). An improved flexible representation of quantum images. *Quantum Information Processing*, 18, 1–19.
[30] Khan, M., Hussain, I., Jamal, S.S. and Amin, M. (2019). A privacy scheme for digital images based on quantum particles. *International Journal of Theoretical Physics*, 58, 4293–4310.
[31] Anand, A., Lyu, M., Baweja, P.S. and Patil, V. (2022). Quantum image processing. arXiv preprint arXiv:2203.01831.
[32] Wereszczyński, K., Michalczuk, A., Paszkuta, M. and Gumiela, J. (2022). High-precision voltage measurement for optical Quantum computation. *Energies*, 15(12), 4205.
[33] Tomesh, T. and Martonosi, M. (2021). Quantum codesign. *IEEE Micro*, 41(5), 33–40.
[34] Choi, J. and Kim, J. (2019). A tutorial on quantum approximate optimization algorithm (QAOA): Fundamentals and applications. *In:* 2019 International Conference on Information and Communication Technology Convergence (ICTC) (pp. 138–142). IEEE.
[35] Crooks, G.E. (2018). Performance of the quantum approximate optimization algorithm on the maximum cut problem. arXiv preprint arXiv:1811.08419.
[36] Shaydulin, R. and Alexeev, Y. (2019). Evaluating quantum approximate optimization algorithm: A case study. *In:* 2019 Tenth International Green and Sustainable Computing Conference (IGSC) (pp. 1–6). IEEE.
[37] Lillesand, T., Kiefer, R.W. and Chipman, J. (2015). *Remote Sensing and Image Interpretation*. John Wiley & Sons.
[38] Schueler, C.F., Lee, T.F. and Miller, S.D. (2013). VIIRS constant spatial-resolution advantages. *International Journal of Remote Sensing*, 34(16), 5761–5777.
[39] Hillger, D., Seaman, C., Liang, C., Miller, S., Lindsey, D. and Kopp, T. (2014). Suomi NPP VIIRS imagery evaluation. *Journal of Geophysical Research: Atmospheres*, 119(11), 6440–6455.
[40] Liang, C.K., Mills, S., Hauss, B.I. and Miller, S.D. (2014). Improved VIIRS day/night band imagery with near-constant contrast. *IEEE Transactions on Geoscience and Remote Sensing*, 52(11), 6964–6971.

CHAPTER

9

Exploring the Principles of Synthetic Aperture Radar

In the previous chapter, challenges arose in the automatic detection of cloud streets owing to the complexity of the surrounding heavy cloud cover, particularly in optical data. In contrast, radar images have proven to be a successful tool for monitoring turbulence wakes. Readers need to grasp the theoretical mechanisms behind Synthetic Aperture Radar (SAR) images and understand their role in capturing specific atmospheric turbulence phenomena, such as gravity waves. Although optical satellite data can image these phenomena, automatic detection in the surrounding environment poses significant difficulties.

Radar, an acronym for Radio Detection and Ranging, serves as a meteorological powerhouse by leveraging its capability to remotely sense and measure precipitation and atmospheric phenomena. In meteorology, radar systems are used to investigate the spatial distribution, intensity, and movement of precipitation particles, and offer invaluable insights into weather patterns and severe weather events.

By emitting radio waves and analyzing signals reflected from atmospheric elements, radar meteorology empowers scientists to observe and analyze precipitation types, rainfall rates, and storm structures. This information is crucial for turbulence monitoring and contributes to our comprehensive understanding of atmospheric turbulence processes detectable by optical satellite images.

Radar meteorology plays a pivotal role in advancing our understanding of the Earth's atmosphere by detecting severe weather events, such as thunderstorms and hurricanes, to monitor cloud dynamics and study precipitation behavior. This introduction lays the foundation for the upcoming chapters' exploration of atmospheric turbulence features, such as gravity waves and hurricanes. This chapter aims to provide a quick background concerning the fundamentals of SAR images for capturing atmospheric turbulence features, subsequently setting the stage for a deeper dive into the intricacies of these phenomena in subsequent discussion.

9.1 Principles of Microwave Bands

The microwave segment, nestled within the electromagnetic spectrum, spans approximately from a wavelength λ of 1 mm to 1 m. This translates to signal frequencies f ranging between 300 GHz and 300 MHz, respectively, with the speed of light ($\lambda f = c$) playing a key role. Unlike its showy cousin, the visible spectrum, microwave wavelengths are significantly larger, mingling with targets in a manner distinct from passive spectra.

In this intricate dance of signals, microwaves exhibit considerably lower energy compared to passive spectra, making them less likely to induce molecular resonance. Take, for example, the passive spectra of visible and near-infrared, which dive into the chemical structures of the atmosphere and soil with their high energy. Microwave signals, on the flip side, maintain just enough energy to play with the resonant spin of specific dipole molecules, syncing up with the frequency changes in the electric field.

In this cosmic tango, objects may appear dark because they are tinier than the radar wavelength, lacking the gusto to reflect abundant energy. Conversely, radars with shorter wavelengths can discern subtler variations in irregularity compared to their long-wavelength counterparts. In simple terms, the roughness of an object decides to shimmy with the wavelength.

Enter SAR sensors, the keen observers of an object's physical traits – permittivity ε, surface roughness, morphology, and geometry. These active microwave wavelengths elegantly divide themselves into distinct regions or bands, each with its unique charm and ability to unveil the secrets of the world (Table 9.1).

Table 9.1: Physical characteristics of microwave bands

Microwave Band	Wavelength Range	Frequency Range
L-Band	1 m–2 m	1 GHz–2 GHz
S-Band	15 cm–30 cm	2 GHz–4 GHz
C-Band	3.8 cm–7.5 cm	4 GHz–8 GHz
X-Band	2.4 cm–3.8 cm	8 GHz–12 GHz
Ku-Band	1.7 cm– 2.4 cm	12 GHz–18 GHz
Ka-Band	0.75 cm–1.7 cm	26.5 GHz–40 GHz

In the realm of microwave sensors, frequency in Hertz serves as an alternative descriptor for defining a band range. It's important to note that satellite active microwave sensors, in contrast to multispectral counterparts, don't capture images across multiple microwave bands. Rather, they emphasize acquiring data for a single band that corresponds to a particular wavelength or frequency. A comprehensive review of surface properties and ambient factors, this feature expands the range and versatility of applications for microwave detection.

However, advancements in technology are evident, with recent airborne sensors like NASA's Jet Propulsion Laboratory AIRSAR system capable of capturing data across multiple frequency bands, such as C, L, and P-bands. By considering a more in-depth analysis of surface characteristics and environmental conditions, this capability enhances the adaptability and scope of microwave-detecting applications.

9.2 Radio Detecting and Ranging

In the quest to unravel the mysteries concealed within synthetic aperture radar (SAR) data capturing the beforemath of atmospheric turbulence such as hurricanes, a nuanced exploration of SAR's fundamental characteristics becomes paramount. Active microwave Earth observation, often denoted as RADAR (RAdio Detection And Ranging), emerges as the key player in this captivating narrative. The nomenclature "radio" harks back to the origins of radar, where long wavelengths (1 to 10 m) are nestled within the radio band of the electromagnetic spectrum [1–3].

A symphony of technological components orchestrates the radar configuration, with an antenna, transmitter, and receiver taking center stage. The transmitter, akin to a maestro, generates the energy that propels the radar beam towards its target. The transmitted signal, a staccato burst rapidly repeated, forms the pulsating essence of radar's communication dance. As these signals traverse the vast expanse at a speed of 300,000 km/s, they encounter targets that absorb, reflect, refract, or backscatter the energy, creating a dynamic interplay [2].

In this intricate spinning of microwave radiation within the targets, the receiver assumes a dual role, seamlessly integrated with the transmitter. It not only captures the pulsed return signal but also engages in a delicate dance of signal strength determination, drawing correlations between the received and transmitted signals. These data, once processed, metamorphose into a visual masterpiece—a grayscale tone painting the canvas of the SAR image [4–5].

The term "active" in microwave imaging becomes the protagonist, signifying the system's ability to transmit its microwave energy—pulses—across a specific wavelength for a defined duration. This pulsating symphony, encapsulated in the concept of pulsed coherent radar, paints a vivid picture of the radar system's functionality, as depicted in Figure 9.1.

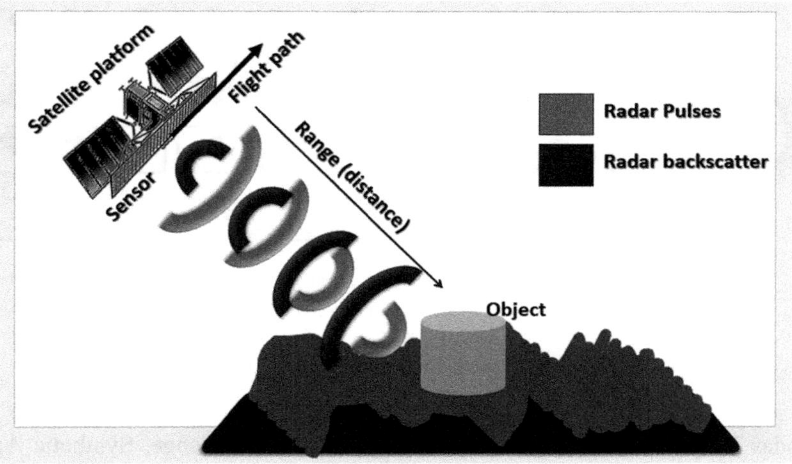

Figure 9.1: Transmitted pulse and its backscatter return in a radar system.

With the stage set by these fundamental concepts, the subsequent sections weave through the intricate tapestry of SAR image data, beckoning us to delve deeper into the realms where microwaves and imagery converge [1, 3, 6].

9.3 What is meant by Synthetic Aperture Radar?

The term "synthetic aperture" refers to the ability of SAR to simulate a much larger antenna or aperture by combining the signals from multiple positions along the flight path of the radar sensor. In traditional radar systems, the antenna size is limited by the physical constraints of the platform (aircraft or satellite). However, SAR overcomes this limitation by effectively creating a virtual aperture through the movement of the radar sensor. As the sensor moves, it collects radar echoes from different positions, and by processing these echoes together, SAR achieves the resolution equivalent to a much larger antenna.

To comprehend how SAR works for imaging atmospheric turbulence and other environmental phenomena, one must have a solid understanding of the concept of synthetic aperture. In this regard, the term "aperture" in photography refers to the diameter of the lens opening, playing a crucial role in determining the amount of light collected [4, 7]. Aperture is measured in focal length (*f*) stops, with a smaller aperture allowing less reflected light to enter the camera (see Figure 9.2). To clarify, a wide-open aperture can lead to an unclear picture. Conversely, a smaller aperture, associated with a greater depth of field, results in a sharp, dark picture (Figure 9.2).

In a radar system, obtaining high-quality images requires an understanding of these aperture dynamics. In this sense, aperture dynamics refer to the changes in the size and shape of the radar antenna's aperture during operation. These progressions can fundamentally affect the goal and precision of the radar framework's imagery. By accomplishing it and overseeing aperture elements, designers can enhance the radar framework's presentation and guarantee excellent and high-resolution imagery obtaining.

Correspondingly, the antenna length of the radar partly specifies the area through which it collects radar signals. This length, known as the aperture, plays a crucial role in determining the amount of unique information obtainable about a viewed object. A larger antenna generally provides more information, resulting in improved image resolution. However, placing very large

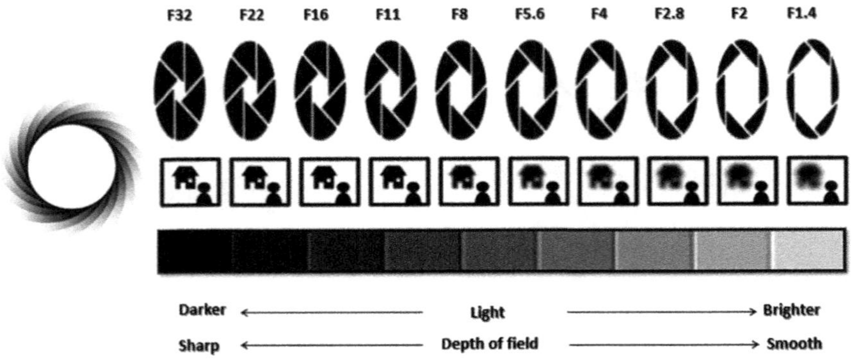

Figure 9.2: Impact of aperture on picture quality.

radar antennas in space is cost-prohibitive. To address this challenge, Synthetic Aperture Radar (SAR) utilizes a synthetic aperture antenna. This approach allows short antennas with wide beams to behave as though they are much longer [1, 8, 10].

In practical terms, SAR can transmit several hundred pulses as its parent spacecraft passes over a specific object, effectively creating a synthetic aperture with increased capabilities [5, 8, 10]. Compared to real-aperture radar, SAR employs synthetic aperture processing to enhance azimuth resolution. This sophisticated data processing involves receiving signals and phases from moving targets using a small antenna [3–6].

The complexities of a radar system's image geometry differ greatly from the framing and scanning techniques often used in optical remote sensing—a discussion covered in Chapter 3. Similar to optical frames, the platform makes a progressive traverse in the direction of flight indicated by (A), with the nadir (B) being immediately below the trajectory of the platform. The microwave beam deviates from the optical norm by assuming a transmission trajectory that is obliquely slanted at right angles to the flight direction. This causes the beam to illuminate a swath (C) that shows a clear offset from the nadir. "Azimuth" (E) describes the dimension that runs down the track, whereas "Range" (D) refers to the dimension that crosses the track and is perpendicular to the flight path positioned in line with the flight path (Figure 9.3). Whether in an airborne or spaceborne configuration, imaging radar systems are distinguished by their unique side-looking viewing geometry.

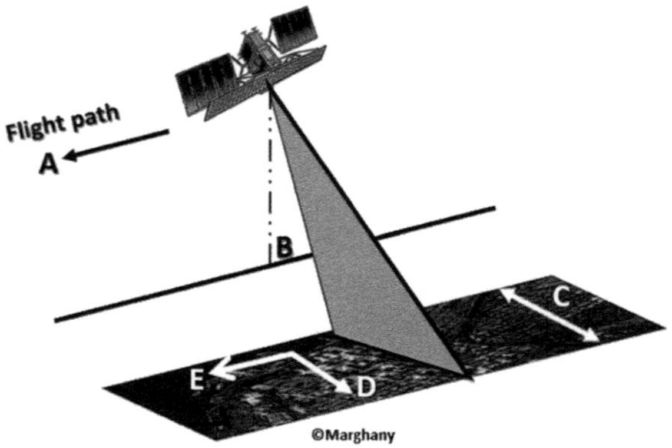

Figure 9.3: Geometry of real aperture radar.

Consequently, the segment of the image swath that lies in proximity to the nadir track of the radar platform is termed the near range (N), whereas the portion situated farthest from the nadir is denoted as the far range (F) (Figure 9.4).

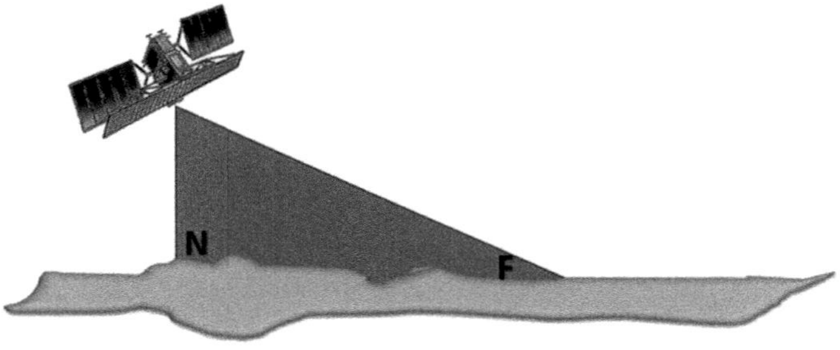

Figure 9.4: Near and far ranges of SAR geometry.

9.4 Radar Resolution

Radar resolution refers to the ability of a radar system to distinguish between closely spaced objects or reflectors in its field of view. It is a critical aspect of radar performance, impacting the system's ability to provide detailed and accurate information about targets. There are two primary types of resolution in radar: range resolution and azimuth resolution. However, another crucial aspect of resolution that demands exploration in radar is spatial resolution.

9.4.1 Spatial Resolution

Spatial resolution serves as the cornerstone in discerning the physical attributes captured by radar sensors. It denotes the ability of a radar sensor to distinguish the proximity between two objects as distinct points. For example, a radar system that can differentiate between two closely spaced internal wave crests is considered to have higher resolution, while a lower resolution system might only perceive them as a single internal wave crest. The clarity of internal wave imagery is notably enhanced with finer spatial resolution, such as 25 meters, compared to resolutions of 30 meters and 100 meters, respectively.

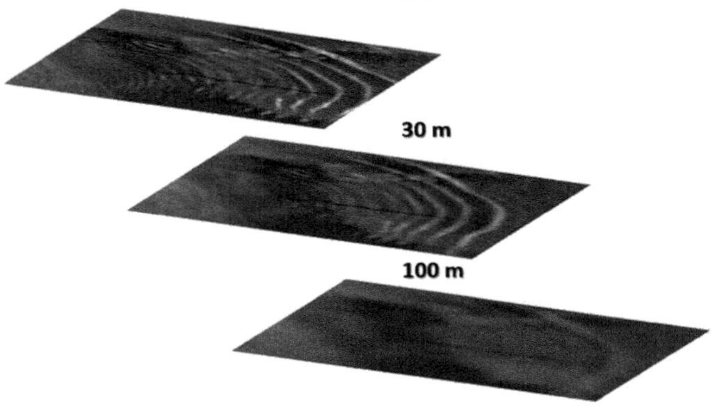

Figure 9.5: Different SAR image resolutions.

The distinctive characteristics of the radar system and sensor describe the spatial resolution of SAR, which is obtained in the azimuth and range directions.In principle, the azimuth resolution is equal to twofold the radar antenna's length. Significantly, azimuth resolution is independent of range. On the other hand, range resolution relies on the circulated pulse's frequency bandwidth and, as a consequence, on the range-focused pulse's temporal length (width). It is noteworthy that lower-focussed pulse widths are produced by wider bandwidths [5, 7, 9].

Conversely, the angular resolution represents the smallest angular difference needed to discern between two identical targets at a comparable distance (Figure 9.6). The distance that exists between two objects, known as the angular resolution, depends on the slant range and may be written arithmetically as [2–6]:

$$S_A \leq 2R.\sin\frac{\theta}{2} \text{ [m]} \tag{9.1}$$

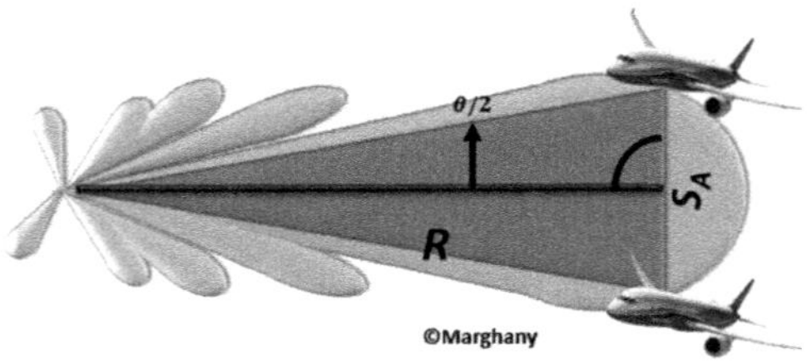

Figure 9.6: Angular resolution.

Equation 9.1 demonstrated the intricacies of radar angular resolution, expressed through parameters like the antenna beam width (θ). Therefore, the space between two objects is known as the angular resolution S_A, which stands as a pivotal characteristic. These features are meticulously determined by the antenna's –3 dB angle, defined by its half-power points, highlighting the importance of this aspect in radar technology [8]. The –3 dB beam width serves as a crucial specification for angular resolution, signifying that when two identical targets at an equal distance surpass this beam width separation, they are effectively resolved in angle. Consequently, a narrower beam width correlates with the sharp directivity of the radar antenna, ultimately contributing to superior bearing resolution [10].

9.4.2 Slant and Ground Range Resolution

In the realm of imaging SARs, the genuine distance between the leading and trailing edges of the pulse takes precedence, as range resolution hinges on the signal pulse length. The visualization of these pulses can be accomplished by observing the signal wavefronts emitted from the SAR sensor. It is important to note that the resolution distance in the ground range invariably surpasses the slant range resolution when these wavefront arcs are projected onto a "flat" Earth's surface, as depicted in Figure 9.7. Notably, at lower incidence angles, the ground range resolution experiences a substantial increase [9].

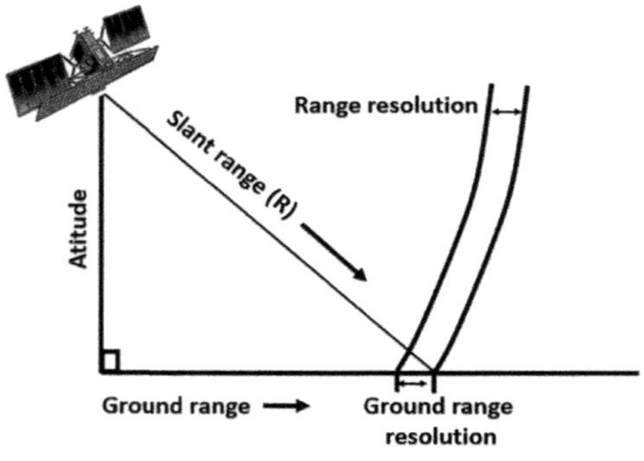

Figure 9.7: Slant and ground range resolution.

Consequently, the speculative range resolution of a radar system is appraised through:

$$S_r = \frac{c_0 . \tau}{2} \text{ [m]} \quad (9.2)$$

Equation 9.2 illuminates that c_0 represents the speed of light, τ denotes the transmitter pulse width, and S_r signifies the range resolution, delineated as the distance between two targets. In the realm of pulse compression systems, it's crucial to note that the radar's range resolution is dictated by the bandwidth of the transmitted pulse (B_{tx}), not its pulse width, as expounded in reference [8–11]. In this circumstance,

$$S_r \geq \frac{c_0}{2B_{tx}} \text{ [m]} \quad (9.3)$$

In this equation, S_r stands for the range resolution, representing the distance between two targets, and B_{tx} denotes the bandwidth of the transmitted pulse. This formulation enables the attainment of exceptionally high resolution with a prolonged pulse, consequently achieving a higher average power, as elucidated in references [7, 11].

9.4.3 Resolution Cell

In accordance with Moreira [5], the interplay of range and angular resolutions gives rise to what is known as the resolution cell. The significance of this cell is lucid: the ability to distinguish two targets within the same resolution cell is contingent on potential differences in Doppler shift. A smaller resolution cell is achieved with a shorter pulse duration (τ) or a broader spectrum of the transmitted pulse, coupled with a narrower aperture angle. This results in heightened interference immunity for the radar station.

Notably, ground-range resolution exhibits its weakest performance in the near-range segment of the SAR image, reaching its peak effectiveness in the far-range sector. The depression angle (β) illuminates the near-range portion of the swath, with smaller depression angles accentuating the far-range sector of the beam. The correlation between ground resolution and depression angle is formulated as:

$$R_{GR} = \frac{\tau c}{2\cos\beta} \quad (9.4)$$

Equation 9.4 illustrates that R_{GR}, representing ground resolution, is influenced by c (the speed of light) and β (the depression angle). This equation signifies that pulse duration, ground range, and beam width collectively govern the dimensions of the ground resolution cell. Pulse duration and ground range, as a result, dictate the spatial resolution along the path of energy transmission, denoting the range resolution. The spatial resolution in the direction of flight is referred to as azimuth resolution and is contingent upon the beamwidth.

9.4.4 Ambiguous Range

The primary challenge in range measurement with pulsed radars lies in unambiguously resolving the range of the object. Ambiguous range consequences from the diffusion of a sequence of pulses. In this context, a pulse repetition interval (PRI) is established, also known as the spacing between spread pulses, and alternatively referred to as pulse repetition frequency (PRF). Accordingly, the delay time is generated due to the spacing between transmit and return pulses. This delay time introduces ambiguity along the range direction or uncertainty. The unambiguous range R_A, as a function of delay time τ_{PRI}, is well-defined as:

$$R_{amb} = \frac{c\tau_{PRI}}{2} \quad (9.5)$$

In the context of equation 9.5, a radar can reliably regulate its range unambiguously when the R_{amb} is larger than the target range. Conversely, unambiguous range determination becomes impossible when the object range exceeds R_{amb}. In such scenarios, the radar must ensure that the pulse repetition interval (PRI) is greater than the range delay associated with the longest target ranges to circumvent range ambiguities. An alternative approach to mitigate the ambiguous range problem involves employing multiple PRIs with different waveforms. In this scenario, the waveforms play a crucial role in altering the spacing between transmit pulses and identifying instances where the target range is ambiguous, allowing for the easy disregard of the return pulse. An advanced method for resolving range delays relies on a range resolution algorithm, which calculates the precise target range [3, 5].

Under the circumstance of equation 9.5, radar can determine its range unambiguously when the object range is smaller than R_{amb}. On the contrary, an impossible range unambiguously accounts for when the object range is larger than R_{amb}. Under these circumstances, the radar must have PRI greater than the range delay frequented with the longest target ranges to circumvent range ambiguities. An alternative method to avoid the ambiguous range problem is based on multiple PRIs with waveforms. In this understanding, the waveforms assist in changing the positioning between radiated pulses and detect that the object range is ambiguous. Then it can be easy to ignore the return pulse. The advanced method to resolve range delays is based on a range resolve algorithm to calculate the precise target range [3, 5].

9.4.5 Range-Rate Measurement (Doppler)

In delving into the intricacies of range rate measurement, a pivotal avenue to explore is the utilization of Doppler frequency. This intricate frequency phenomenon manifests as a consequence of the nuanced distinctions in frequency arising from the spacing intricacies between the signals in their transmission and reception phases. An enlightening scenario to consider unfolds when contemplating the linear velocity of an aircraft, elegantly symbolized as v with different times dt. This contemplation unfolds across discrete temporal intervals, elegantly encapsulated by the variables denoted as and the dynamic variation in ranges, symbolized by dR. This dynamic interplay corresponds to the dynamic motion of the aircraft, gracefully transitioning from the spatial coordinates of point A to the destination marked as point B, an unfolding visual depicted in the illustrative Figure 9.8. Within this nuanced framework, the range rate, a pivotal metric indicative of the rate of change in range, crystallizes into a precise formulation articulated by the following mathematical expression:

$$\dot{R} = \frac{dR}{dt} \tag{9.6}$$

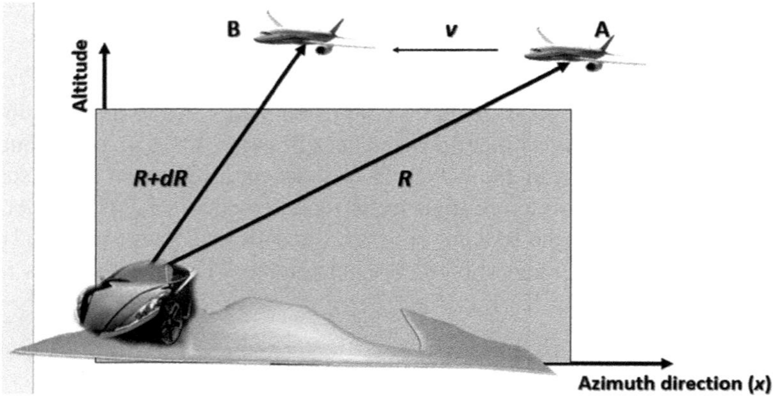

Figure 9.8: Range rate measurement by Doppler.

Mathematically, the transmitted pulse is expressed as:

$$v_T(t) = \text{rect}\left[\frac{t}{\tau_p}\right]\cos(2\pi f_c t) \tag{9.7}$$

where

$$\text{rect}\left[\frac{t}{\tau_p}\right] = \begin{cases} 1 & 0 \le t < \tau_p \\ 0 & \text{elsewhere} \end{cases}. \tag{9.8}$$

Within the intricate fabric of radar dynamics, the carrier frequency, encapsulated by the symbol f_c, assumes a pivotal role, intricately interwoven with the pulse width, denoted by $\tau_p = 1\ \mu\text{s}$, and the temporal nuances embodied in the variable *t*, elegantly representing the duration of the transmitted pulse. In stark contrast, the received pulse, a multifaceted entity, unfolds its essence through the nuanced lens of delay time or the dynamic parameter of range rate, symbolized by v_R. This intricate dance of frequencies and temporal dimensions unveils a complex symphony, where each element plays a distinctive role, orchestrating a narrative that transcends the boundaries of conventional radar discourse.

$$v_R(t) = \xi v_T(t - \tau_R) \tag{9.9}$$

In the intricate realm of radar intricacies, the amplitude scaling factor, artfully represented by , takes center stage, coalescing with the ethereal concept of range delay denoted by the symbiotic τ_R. This harmonious interplay of mathematical elements crafts a narrative of precision, where the amplitude factor scales the symphony of signals, and the temporal intricacies of range delay unfold, creating a mathematical tapestry that resonates with the cadence of radar intricacies. In this sense, the range delay is given by

$$\tau_R(t) = \frac{2R(t)}{c} \tag{9.10}$$

As a function of a range delay, the Doppler frequency is then given mathematically by

$$v_R(t) = \xi\,\text{rect}\left[\frac{t - \tau_R}{\tau_p}\right]\cos\left(2\pi(f_c + f_d)t + \phi_R\right) \tag{9.11}$$

The intricate function $\phi_R = -2\pi f_c(2R/c)$ in this equation represents the phase shift caused by the range delay, and the incomprehensible Doppler frequency shift $f_d = -f_c(2\dot{R}/c)$, tells with its narrative in an attractive manner as a harmonic function of the radar wavelength represented by the λ. The equation reveals the subtle interactions between satellite and object mobilities under the impact of the Doppler frequency shift. Understanding how radar signals behave in many situations, such as monitoring moving objects or examining satellite data, depends on these interactions. With the inclusion of the phase shift and Doppler frequency shift, this formula offers a thorough foundation for analyzing the intricate dynamics of radar systems.

$$f_d = -2\dot{R}/\lambda\,. \tag{9.11.1}$$

Equation 9.11 shows how the frequency of the reflected signal, represented by $f_c + f_d$, is not just a straightforward number but rather a celestial symbol f_c. The Doppler frequency, denoted by f_d, is like the spinning of these frequencies – comparing how the transmitted and received signals move together. Once we figure it out, we can use it to calculate the range rate using equation 9.11.1.

Theoretically, it is more difficult to compute Doppler frequency than equation 9.11 suggests. The challenge arises from the relative magnitudes of f_d and f_c. Even though the Doppler frequency quantity is challenging, it is possible. The transmit signal needs to be very lengthy (ms or longer, instead of μs) or the Doppler frequency needs to be calculated by processing many signals.

9.5 Radar Range Equation

In the realm of radar principles, the radar range equation stands as a straightforward mathematical expression. Despite its simplicity, or perhaps because of it, this equation becomes a formidable challenge for many radar analysts to fully grasp and is frequently mishandled. The complexity does not lie within the equation itself but rather in the multitude of terms that constitute it. Understanding the radar range equation in depth provides a robust foundation for comprehending radar principles. It's worth noting that the radar equation is the culmination of several more basic formulas.

Let's assume that E_T is the transmitted energy density, which represents the energy per unit area within a specific range R. The scientific representation of the power density at the distance of the emitted signal can be articulated mathematically as:

$$E_T = \frac{P_T.\tau.G_T}{4\pi R^2} \tag{9.12}$$

In the mathematical expression presented as Equation 9.12, the term $4\pi R^2$ establishes a connection between the power transmitted by the radar and an isotropic sphere, indicating that electromagnetic energy is uniformly radiated in all directions. Additionally, G represents the antenna gain, and the focus of the antenna signal is influenced by the ratio of $\frac{G_T}{4\pi R^2}$. Consequently, P_T is referred to as the peak transmit power in watts, and denoting the maximum power when the radar transmits a signal. This quantity can be specified at the output of the transmitter, with a duration time τ, or at another point, such as the output of the antenna radiation.

From this standpoint, Equation 9.12 can be further elaborated by introducing the concept of radar cross-section (RCS), representing the backscatter size of the target upon which the radar signal is concentrated. In this context, the estimated reradiated energy of an object is expressed as:

$$E_\sigma = \frac{P_T.\tau.G_T.\sigma}{4\pi R^2} \tag{9.13}$$

Equation 13.3 articulates the backscatter power density in terms of radar cross-section (RCS). Thus, RCS is contingent upon the object's distinctive scattering characteristics. Additionally, the target's radar cross-section, denoted as RCS, is measured in square meters (m^2).

According to the above-mentioned perspective, the energy of the receiving antenna signal S, denoted as E_r, is contingent upon the backscatter and is positioned similarly to the transmit antenna, characterizing it as a monostatic radar system. In this configuration, the term $4\pi R^2$ in the previous equation transforms into $(4\pi)^2 R^4$ with the inclusion of the parameter A_R in the numerator. The term A_R, on the contrary, represents the effective area of the receiving antenna and serves as the numerator in a ratio involving the second $4\pi R^2$. This accounts for the isotropic radiation attributable to a target's radar cross-section. Thus, under these conditions, equation 9.13 can be adapted as follows:

$$S = \frac{P_T.\tau.G_T.\sigma.A_r}{(4\pi)^2 R^4} \tag{9.14}$$

Equation 9.14 elucidates the quantity of signal reaching the receiving antenna, giving rise to the potential generation of noise within the radar system's receiver. This noise is quantified by the signal-to-noise ratio (SNR), expressed in units of watts per watt (W/W). Moreover, the radar system is subject to additional influential factors, including thermal noise temperature (T_0), Boltzmann's constant (K), and their relationship given by $T_0 = K/B$, where B represents the receiver bandwidth. The losses within the system are denoted by L. To complete the formulation of equation 9.14, a connection between antenna gain, effective area, and signal wavelength (λ) needs to be established as:

$$A_R = \frac{G_R \lambda^2}{4\pi} \tag{9.14.1}$$

In this regard, Equation 9.14.1 delineates the relationship between antenna area and gain, specifically portraying the correlation between gain and effective aperture. The ultimate radar formula is derived by substituting A_r and the noise constants into the previously developed formulas. In this context, the scientific representation of radar range can be expressed mathematically as:

$$SNR = \frac{P_S}{P_N} = \frac{P_T G_T G_R \lambda^2 \sigma}{(4\pi)^3 R^4 k T_0 B F_n L} \tag{9.15}$$

Hence, equation 9.15 illustrates F as the radar noise figure, a dimensionless quantity, or has the units of w/w, and L is a term encompassing all losses that must be taken into account when using the radar range equation, with units of W/W. Equation 9.15 represents the final iteration of the radar equation, and additional terms can be incorporated or substituted for different scenarios. For example, at times, the transmission interval may indicate the amalgamation of numerous signals occurring over the total time t. The pivotal concept to grasp from the radar equation is the intricate interplay of various aspects within the radar.

9.6 Exploring Radar Backscattering

The synthetic aperture radar (SAR) backscattering cross section essentially represents the area that scatters signals uniformly in all directions, creating an echo at the radar comparable to that from the target [5, 8]. In simpler terms, it quantifies how much of the radar signal is reflected. The radar cross-section is a crucial metric, denoting the equivalent surface area of the target observed by the radar. If this area intercepted and scattered the incident radar power uniformly, it would generate a return at the radar receiver equivalent to that from the target [4, 9]. This parameter is commonly known as an effective echo area or simply an echo area. Mathematically, radar cross-section (σ) is defined as 4π power unit solid angle scattered back to the receiver per power unit area incident on the target [2–5].

As the radar beam reflects from the surface, three key properties come into play: (i) dielectric constant (or permittivity), (ii) roughness (height relative to a smooth surface), and (iii) local slope [6, 11]. The interaction of electromagnetic waves with a surface is termed scattering, which can be categorized into two types: surface scattering and volume scattering. Surface scattering takes place at the interface between two different homogeneous media, such as the atmosphere and the Earth's surface. On the other hand, volume scattering results from the interaction with particles within a non-homogeneous medium [1–6]. According to this perspective, the reflection of electromagnetic waves from smooth surfaces, like mirrors or a calm body of water, results in specular reflection (Figure 9.9). In contrast, reflection from rough surfaces, such as clothing or paper, leads to diffuse reflection (Figure 9.10).

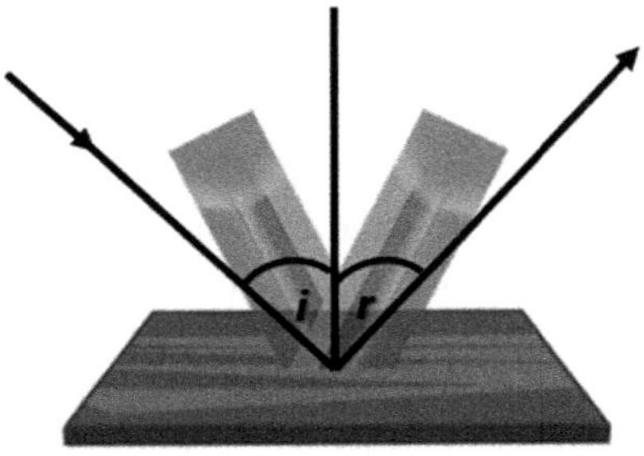

Figure 9.9: Specular scattering.

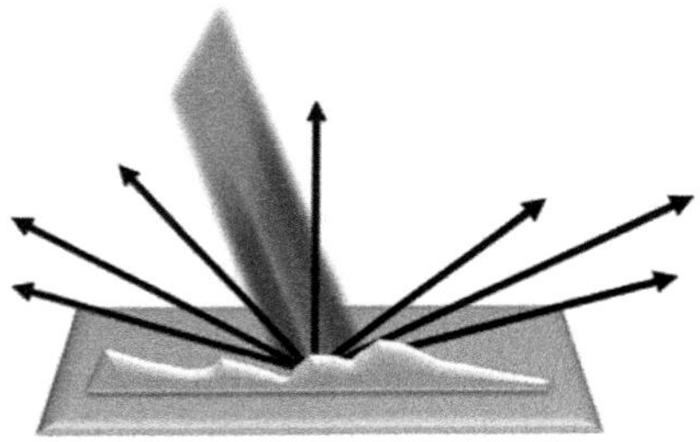

Figure 9.10: Diffuse reflection.

In other words, rough surfaces scatter a beam of light because a beam of electromagnetic wave consists of individual photon rays that are parallel to each other. When these individual rays encounter a rough surface, each point of incidence on the surface has a different orientation from the normal line. As a result, when the individual rays reflect off the rough surface following the law of reflection, they scatter in different directions due to the varying orientations at different points of incidence. This scattering phenomenon is responsible for the diffuse reflection observed from rough surfaces.

Additionally,volume scattering refers to the scattering that occurs within a medium when electromagnetic radiation transitions from one medium to another. The schematic model in Figure 9.11 illustrates volume scattering widely due to distributed particles like raindrops Examples of volume scattering include scattering by trees or branches (Figure 9.12), subsurface or soil layers, and snow layers.

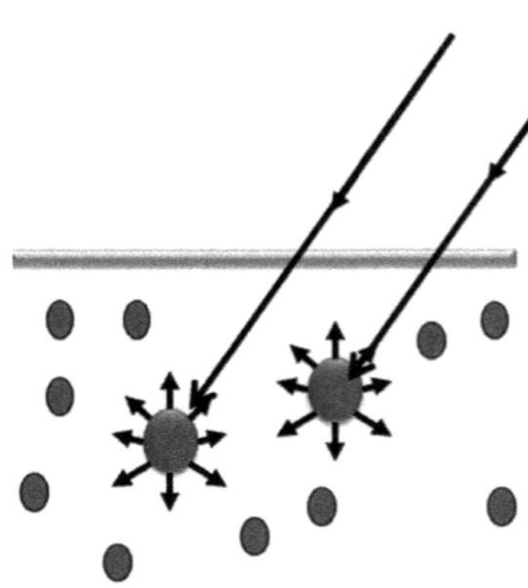

Figure 9.11: Volume scattering.

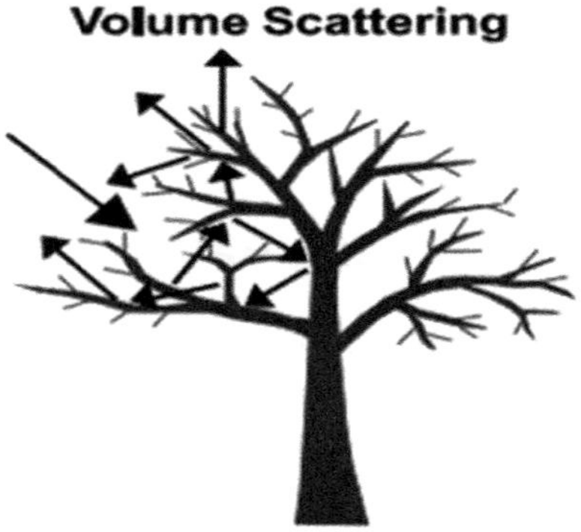

Figure 9.12: Volume scattering by tree branches.

Microwave radiation penetrating a medium allows the observation of volume scattering, with the penetration depth defined as the distance at which the incident power attenuates to 1/e (exponential coefficient). Therefore, the intensity of volume scattering is proportionate to the discontinuity inductivity in a medium and the density of the heterogeneous medium. The scattering angle is influenced by surface roughness, average relative permittivity, and wavelength.

Based on the perspective mentioned above, the radar backscatter is significantly influenced by an electric property of the reflected substances known as the complex permittivity (ε_c). This property, also referred to as the dielectric constant, can be mathematically obtained by:

$$\varepsilon_c = \varepsilon' + i\varepsilon'' \tag{9.16}$$

In equation 9.16, ε' is the dielectric constant of the substance, ε'' is the "lossy" part of the dielectric constant, and i is the imaginary part (square root of -1). ε' represents the material's response to an electrical field. When an electrical field interacts with the medium, molecules attempt to align themselves with the field's polarity to the lowest power state. However, due to their crystalline structure, molecules cannot fully align with an electrical field. As a result, the lag time between the electric field and the molecules' response is described by the dielectric loss factor (ε'').

Needless to say, radar backscatter is directly proportional to the dielectric properties of the material. Higher backscatter is observed with higher dielectric materials. For instance, the sea surface, with its high dielectric, prevents the penetration of microwave spectra. Conversely, dry materials with lower dielectric properties allow radar signals to penetrate, resulting in volume reflection and generating a significantly brighter backscatter at the antenna receiver.

9.7 Mechanism of Surface Backscattering

In accordance with Kingsley [9], the radiance captured in a radar image is directly proportional to the localized radar backscatter. This backscatter, influenced by surface roughness, behaves distinctively on different surfaces. A smooth surface, acting like a mirror, redirects radiation away from the radar. Conversely, a rough surface scatters more power back into the radar. Significant backscatter is observed when the surface height standard deviation coincides with the radar wavelength [12]. The change from an extremely smooth surface to one that is somewhat rougher causes energy to scatter at different angles, with just a little portion—referred to as diffuse reflectance—reflecting the SAR sensor, which was confirmed by Lillesand et al. [11]. Surfaces like sparse vegetation, bare agriculture fields, and other rough terrains exhibit diffuse reflectance, resulting in intermediate tones on SAR imagery compared to specular reflectors (Figure 9.9). Coarse river gravel and water surfaces disturbed by wind, with stone dimensions comparable to the radar wavelength, exemplify this type of reflectance, appearing notably bright on SAR images [13].

The numerous dispersion of a radar signal within a medium, like the plant canopy of a cornfield or forest (Figure 9.12), or beneath layers of extremely dry soil, sand, or ice, is characterized by volume scattering. A SAR sensor may record backscatter from the target's internal volume as well as its surface when volume scattering is used [13]. Ahern [12] observes that sparse vegetation or crops exhibit an intermediate degree of volume dispersion. Three major processes of scattering are involved in these circumstances (Figure 9.13); they are (i) diffuse scattering from the ground (1), (ii) direct (single-bounce) scattering from different plant components (2 and 3), and (iii) double-bounce vegetation-ground interaction (4).

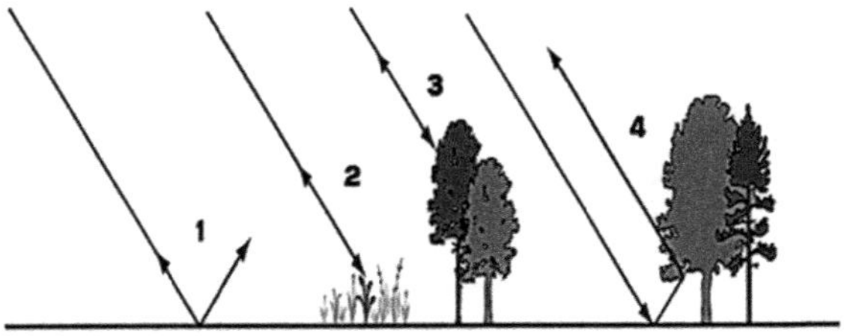

Figure 9.13: Intermediate degree of volume scattering.

9.8 What is the Backscatter Coefficient?

Targets disperse the energy transmitted by the radar in various directions, with the radar capturing the energy scattered backward. The intensity of each pixel in a radar image correlates with the ratio between the density of scattered energy and the density of energy transmitted from Earth's land surface targets [10–12]. The backscatter coefficient (σ°) is defined as:

$$\sigma^{\circ} = 10 \log \sigma^{\circ} \tag{9.17}$$

The backscattered energy is associated with a variable known as radar cross-section, representing the amount of transmitted power absorbed and reflected by the target. The backscatter coefficient (σ°) quantifies the radar cross-section per unit area on the ground. σ° serves as a characteristic of the scattering behavior exhibited by all targets within a pixel. Due to its wide-ranging values, σ° is expressed in logarithmic decibel units [11]. The large range of values that σ° might take is more easily represented thanks to the logarithmic decibel units. This makes it simple for radar analysts to evaluate and compare how various targets scatter inside a pixel. In radar remote sensing applications, the backscatter coefficient is also a crucial metric as it offers essential details about the target's composition and physical characteristics.

The backscatter is measured as a complex number, which contains information about the amplitude easily converted to σ° by specific equations and the phase of the backscatter [10].Contain speckle, an interference phenomenon produced between backscatter coming from many random targets within a pixel. The speckle represents a true electromagnetic scattering and influences the interpretation of SAR images [15]. Backscattering types in marine environments include volume scattering, subsurface (volume) scattering, surface scattering, and corner reflector-like scattering.

9.9 Principles of SAR Bragg Scattering

In light of the above perspectives, backscatter is primarily modeled in two ways: specular reflection and Bragg scattering. Specular reflection occurs when the water surface is tilted, creating a small mirror pointing to the radar. For a perfect specular reflector, radar returns (backscatter) exist only near vertical incidence, typically at a 90°-depression angle or the slope of the surface. The reflected energy is concentrated in small angular regions around the angle of reflection. Even for non-vertical incidence, backscatter can occur for a rough subsurface if the radar can penetrate deep enough.

Bragg or resonant scattering involves scattering from a regular surface pattern. Resonant backscattering occurs when phase differences between rays backscattered from a subsurface pattern interfere constructively. The resonance condition is given by $2\lambda \sin\theta = \lambda'$, where λ and λ' are the water wavelength and radar wavelength, respectively, and θ is the local angle of incidence (refer to Figure 9.14). In this context, short Bragg-scale waves form in response to wind stress, particularly if the sea surface is rippled by a light breeze and no longer waves are present. The radar backscatter is a result of the component of the wave spectrum resonating with the radar wavelength.

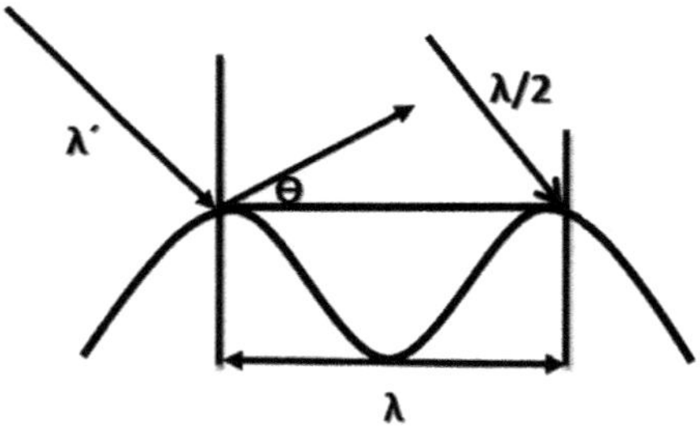

Figure 9.14: Concept of Bragg scattering.

Thus, the swell waves being captured have far longer wavelengths than the Bragg resonance-inducing short gravity waves. It is also possible to think of an oceanic Bragg resonance as coming from facets. A facet is a relatively flat section of the long wave structure that behaves like a specular point and is covered with ripples that include Bragg-resonant facets. Each facet's length in the proper direction determines its beam-width and scattering gain.

The normalized cross-section as a function of the conventional Bragg/composite surface scattering theory may be represented mathematically as follows:

$$\sigma_o = \sigma_c + \sigma_b \tag{9.18}$$

Where σ_c is the composite surface cross-section, and the assured wave cross-section σ_b are expressed respectively, by:

$$\sigma_c = \iint \sigma_B(\theta_o+\gamma, \alpha)\, P_f\, P(\gamma, \alpha \mid f)\, d\gamma d\alpha \tag{9.19}$$

$$\sigma_b = \iint \sigma_B(\theta_o+\gamma, \alpha)\, P_b P(\gamma, \alpha \mid b)\, d\gamma d\alpha \tag{9.20}$$

Standard Bragg scattering cross section σ_B serves as the basis for both equations, which are associated with three types of wave probabilities: (i) the probability of finding free waves (P_f); (ii) the probability of finding bound waves (P_b); and (iii) the probability distribution of a wave sort, whichever free or bound ($P(\gamma, \alpha \mid x)$). Thus, θ_o is the nominal incidence angle, while γ, α are the long wave slopes in and perpendicular to the plane of incidence. Therefore, σ_B on a typical Bragg scattering cross-section may be attained by:

$$\sigma_B = 16\, \pi\, k_o^4 |\, F\, (\theta_o+\gamma,\, \alpha)|^2\, \mu(2k_o \sin\, (\theta_o+\gamma),\, 0) \tag{9.21}$$

In equation 9.21, the microwave number is k_o, and the wave height variance spectrum is μ which is a function of ($2k_o\sin\, (\theta_o+\gamma)$, 0). Finally, F is a function of incidence angle θ and dielectric ε.

In essence, the intensity of the radar cross-section is intricately linked to the incidence angle θ. When the radar signal meets the surface head-on or at a high incidence angle, its reflection is notably stronger compared to a scenario where it strikes at a low, grazing angle. In the latter case, a significant portion of the radar energy veers away from the radar receiver, resulting in a subdued or dark response on the image. Hence, the local incidence angle emerges as a pivotal factor influencing the Bragg scattering cross-section's intensity [1, 11, 15].

In this sense, the incident angle emerges as a pivotal factor shaping radar backscatter and the visual representation of targets in images. This angle, present at any given point within the range, is the divergence between the direction of the radar beam (line of look) and a line perpendicular (normal) to the surface. Notably, this line can exhibit inclination at varying angles, dictated by the slope orientation in non-flat topography. As one progresses from near to far range, the depression angle, which is complementary to the incident angle, decreases.

Within the context of a flat surface, the incident angle essentially serves as the complement of the depression angle, as illustrated in Figure 9.15 [7, 11, 15]. Every pixel in radar data can be associated with a local incident angle, contributing to the nuanced variations in pixel brightness. As a general trend, reflectivity from distributed scatter tends to diminish with escalating incident angles. A smaller incident angle correlates with heightened backscatter, although it is noteworthy that for very rough surfaces, backscatter remains independent of θ [6, 11, 15].

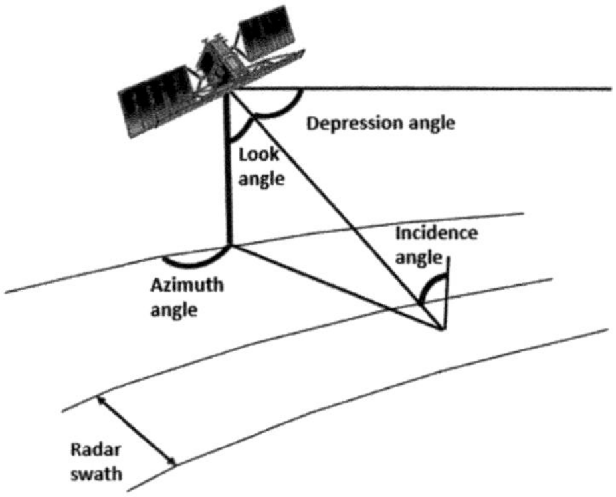

Figure 9.15: Incident and depression angles.

The peak radar backscatter typically emanates from the slope of the wave oriented towards the radar. Consequently, an image modulated solely by tilt would depict a plane parallel swell-wave field as a sequence of parallel light and dark lines, each corresponding to slopes facing either toward or away from the radar—effectively offset by a 90° phase difference from the lines representing troughs and crests.

In the broader context, synthetic aperture radar (SAR) ocean surface imagery relies heavily on the intricate interplay between incident angles and radar backscatter signals. This interplay is intricately linked with variations in both local geometry and the spectral density distribution of short gravity and gravity-capillary waves. Resonant surface waves, notably shorter at more oblique incidence angles, become a key player. Simply put, as incidence angles increase, the ocean backscatter returns decrease, primarily because larger oblique angles capture smaller amplitude Bragg waves, resulting in diminished backscatter [5, 9, 13, 15].

9.10 SAR Polarization

The polarization of an electromagnetic wave signifies the orientation of the electric field intensity vector. In the realm of Synthetic Aperture Radar (SAR), the standard transmission involves a horizontally polarized wave. While the majority of the received energy retains its horizontal polarization, a fraction undergoes depolarization through interactions with the terrain, spawning diverse components at varied polarization angles [2–5].

To isolate the desired polarization—either horizontal or vertical—a filter at the antenna can screen out all other polarizations. The configuration with both transmission and reception in the horizontal plane (HH) is termed the like-polarized return, whereas the configuration with horizontal transmission and vertical reception (HV) is termed the cross-polarized return. Due to its considerably lower energy, the HV return necessitates a much higher antenna gain than the HH return. Consequently, images derived from these two returns may exhibit disparities owing to the distinct scattering processes involved [7, 11, 13].

Depolarization, often attributed to volume scatter or multiple reflections, can result in nearly identical like- and cross-polarized images at short wavelengths when the terrain is considered very rough. However, noticeable differences emerge at longer wavelengths when the terrain is relatively smooth, leading to limited geological applications for identifying rock types [9–11].

Recent advancements in polarimetric SARs enable the simultaneous transmission and reception of both horizontal and vertical polarizations—HH, HV, VH, and VV returns. Mathematical analysis of these returns provides a geometric basis, allowing the synthesis of images for any conceivable transmit/receive polarization across the entire 360° spectrum. Certain objects, particularly man-made ones, exhibit enhanced visibility at specific transmit/receive polarizations, facilitating their detection. Over the ocean, where multiple reflections are scarce, ocean-viewing sensors typically employ HH or VV polarizations, with the polarization ratio being larger, as vertically polarized radar reflects more strongly than horizontally polarized radar [1, 7, 13, 15].

9.11 Speckles

Speckle, essentially a form of noise, introduces degradation to the quality of an image, complicating its visual or digital interpretation. A speckle pattern emerges as a random intensity pattern resulting from the mutual interference of wavefronts with different phases. In this scenario, these wavefronts combine, forming a resultant wave with a randomly varying amplitude and, consequently, intensity. If each wave is conceptualized as a vector, adding numerous vectors with random angles creates a 2-dimensional random walk, colloquially known as a "drunkard's walk."

When a surface is illuminated by microwaves, following diffraction theory, each point on the illuminated surface serves as a source of secondary spherical waves. The microwave spectra at any point in the scattered microwave field comprise waves scattered from every point on the illuminated surface. If the surface exhibits sufficient roughness to induce path-length differences surpassing one wavelength, leading to phase changes exceeding 2π, the amplitude, and thus the intensity, of the resultant backscattered microwave exhibits random variations [11, 13].

In the realm of radar imagery, a common phenomenon is the presence of what we term a radar speckle, observable to some extent in all radar images. This speckle manifests as a grainy "salt and pepper" texture, as depicted in Figure 9.16. The origin of this texture lies in the realm of random constructive and destructive interference arising from multiple scattering returns, a phenomenon occurring within each resolution cell, as illustrated in Figure 9.17 [2, 10, 13].

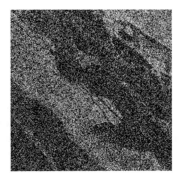

Figure 9.16: SAR data with speckles.

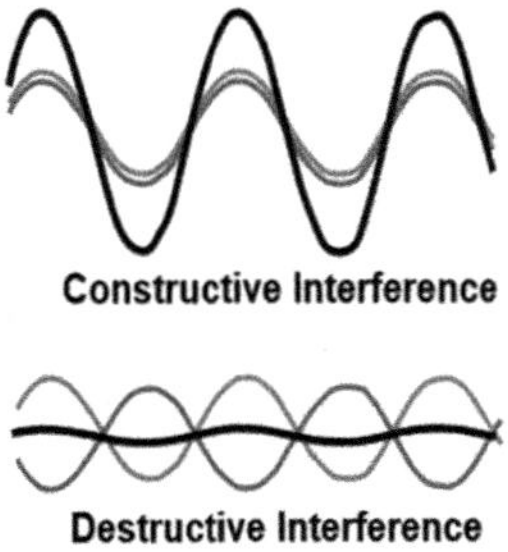

Figure 9.17: Constructive and destructive interferences.

The dynamics of this interference are noteworthy. Constructive interference leads to an augmentation from the mean intensity, resulting in the manifestation of bright pixels. Conversely, destructive interference entails a reduction from the mean intensity, giving rise to the appearance of dark pixels, as depicted in Figure 9.18. This interplay of constructive and destructive interference contributes to the distinctive speckle patterns encountered in radar imagery.

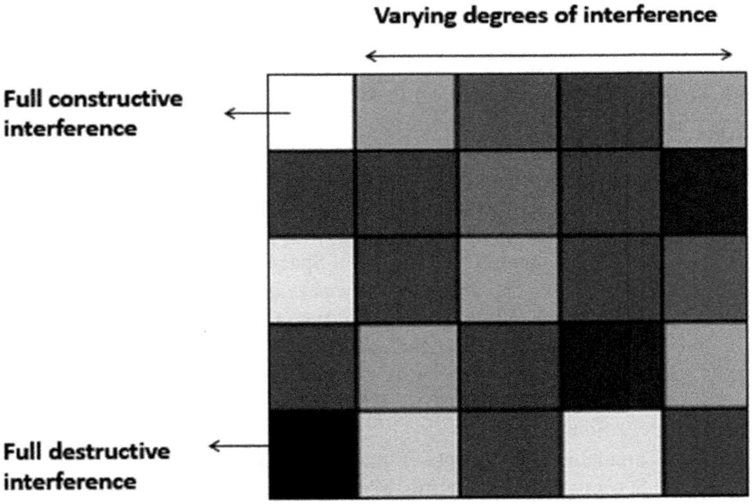

Figure 9.18: Impact of constructive and destructive interference on pixel's brightness.

Consider a homogeneous target, like an expansive grass-covered field, which, in the absence of speckle, would typically yield light-toned pixel values in an image. However, the reflections from individual blades of grass within each resolution cell introduce variability in pixel values, some appearing brighter and others darker than the average tone. This intricate interplay creates a speckled appearance across the field, showcasing the impact of speckles in radar imagery.

The prevalence of high speckle noise in SAR images has presented substantial challenges in the accurate inversion of these images for mapping morphological features. Speckle, a product of coherent interference effects among scatterers randomly distributed within each resolution cell, introduces a level of complexity in feature detection [2, 3, 8]. The size of the speckle becomes contingent on the spatial resolution, thereby introducing errors in the detection of morphological feature signatures. To mitigate these speckle effects, various filters, such as Lee, Gaussian, and others [11, 15], can be applied during the preprocessing stage. It is crucial to note, however, that the effectiveness of these speckle-reducing filters is subject to local factors and specific applications. Furthermore, every speckle present in SAR images is intricately linked to local changes in Earth's surface roughness.

This chapter has delved into the fundamentals of radar, with a particular focus on grasping the concept of radar resolution. The crux of understanding radar lies in comprehending the radar equation and the speculation surrounding Bragg scattering. By understanding the radar equation, which relates the power received by the radar to various parameters such as transmitted power, antenna gain, and target range, one can effectively analyze and interpret radar data. Additionally, the speculation surrounding Bragg scattering sheds light on how electromagnetic waves interact with small particles or irregularities in the atmosphere, providing valuable insights into radar signal propagation. This knowledge serves as a foundational framework for a deeper understanding of the application of SAR data in studying air turbulence phenomena, notably gravity waves. The upcoming chapter will delve into these applications with more comprehensive details.

References

[1] Ahern, F.J. (1995). *Fundamental Concepts of Imaging Radar: Basic Level*. Unpublished Manual, Canada Centre for Remote Sensing, Ottawa, Ontario, 87 p.
[2] Chan, Y.K. and Koo, V.C. (2008). An introduction to synthetic aperture radar (SAR). *Progress in Electromagnetics Research B*, 2, 27–60.
[3] Bamler, R. (2000). Principles of synthetic aperture radar. *Surveys in Geophysics*, 21(2), 147-157.
[4] CCRS. (1993). Radar basics—Introduction to Synthetic Aperture Radar. Unpublished Manual, Canada Centre for Remote Sensing, Ottawa, Ontario, 75 p.
[5] Moreira, A. (1992). Real-time synthetic aperture radar (SAR) processing with a new subaperture approach. *IEEE Transactions on Geoscience and Remote Sensing*, 30(4), 714–722.
[6] Ramirez, A.B., Rivera, I.J. and Rodriguez, D. (2005). SAR image processing algorithms based on the ambiguity function. *In:* Circuits and Systems, 2005. 48th Midwest Symposium (pp. 1430–1433). IEEE.
[7] Zhang, S., Long, T., Zeng, T. and Ding, Z. (2008). Space-borne synthetic aperture radar received data simulation based on airborne SAR image data. *Advances in Space Research*, 41(11), 1818–1821.
[8] Hovanessian, A. (1984). *Radar System Design and Analysis*. Artech.
[9] Kingsley, S. and Quegan, S. (1999). *Understanding Radar Systems*. Scitech Publishing, Inc, New York.
[10] Marghany, M. (2021). *Nonlinear Ocean Dynamics: Synthetic Aperture Radar*. Elsevier.
[11] Lillesand, T., Kiefer, R.W. and Chipman, J. (2014). *Remote Sensing and Image Interpretation*. Sixth Edition. John Wiley & Sons.
[12] Ahern, F.J. (1995). Fundamental Concepts of Imaging Radar: Basic Level. Unpublished Manual, Canada Centre for Remote Sensing, Ottawa, Ontario, 87 p.
[13] Marghany, M. and van Genderen, J.L. (2021). Sea surface current velocity retrieving from TanDAM-X satellite data. *International Journal of Geoinformatics*, 17(4).
[14] López-Martínez, C. and Pottier, E. (2021). Basic principles of SAR polarimetry. *Polarimetric Synthetic Aperture Radar: Principles and Application*, 1-58.
[15] Franceschetti, G. and Lanari, R. (2018). *Synthetic Aperture Radar Processing*. CRC Press.

CHAPTER

10

Developing Quantum Soliton-Inspired Particle Swarm Optimization Algorithm for Automatic Detection of Atmospheric Gravity Waves in Multisar Satellite Data

The phenomenon of atmospheric turbulence includes the presence of gravity waves. While many studies have explored the occurrence of these waves from a meteorological perspective, utilizing advanced microwave remote sensing synthetic aperture radar (SAR) technology, a comprehensive connection between SAR technology and the physics of atmospheric gravity waves remains lacking. This chapter—aims to bridge this gap by introducing a novel theory grounded in quantum mechanics. The primary objective is to develop an innovative automatic detection algorithm designed to identify atmospheric gravity waves in multiSAR satellite data.

Furthermore, this chapter focuses on extending the tracking of gravity waves to regions that have not been explored before, specifically the coastal zones of Malaysia, Pattani, and Brunei. Remarkably, research in these areas has predominantly centered on fundamental studies for several decades, without even little emphasis on delivering novel insights. This chapter endeavors to fill this void by introducing an inventive perspective and innovative studies in these coastal zones.

10.1 Atmospheric Waves

The existence of waves in water is a well-established phenomenon, but what about waves in the atmosphere and their connection to turbulence? In the preceding two chapters, we delved into various aspects of atmospheric turbulence, laying the groundwork for a deeper exploration of the intricate relationship between atmospheric waves and turbulence. In this view, atmospheric waves represent periodic disturbances in various atmospheric variables, including surface pressure, geopotential height, temperature, and wind velocity. These waves exhibit a diverse range of characteristics, either propagating as traveling waves or remaining stationary as standing waves [1–3].

From large-scale planetary waves, such as Rossby waves, to the minutiae of sound waves, atmospheric waves span a spectrum of spatial and temporal scales. Notably, atmospheric tides refer to waves with periods harmonically related to one solar day, encompassing intervals like 24 hours, 12 hours, and 8 hours, among others. This intricate interplay of atmospheric phenomena forms a compelling subject for exploration in understanding the complexities of our atmosphere [1, 4, 6].

10.2 Atmospheric Tide

Atmospheric tides, akin to their oceanic counterparts, manifest as global-scale periodic oscillations within the Earth's atmosphere. These tides find their origin in various factors, including the routine day-night cycle influenced by the Sun's heating of the atmosphere (insolation) (Figure 10.1).

Additionally, the gravitational pull of the Moon contributes (Figure 10.2) to the generation of atmospheric tides, reflecting the intricate interplay between celestial bodies and our atmospheric dynamics. The complexity; therefore, deepens as non-linear interactions emerge between these atmospheric tides and planetary waves, adding layers to the multifaceted nature of atmospheric phenomena. Furthermore, large-scale latent heat release, a consequence of profound convection in tropical regions, serves as another impetus for the activation of atmospheric tides [3–6].

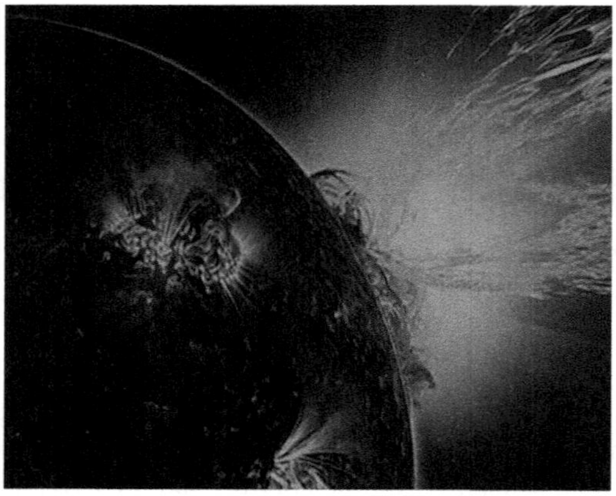

Figure 10.1: Sun thermal heating.

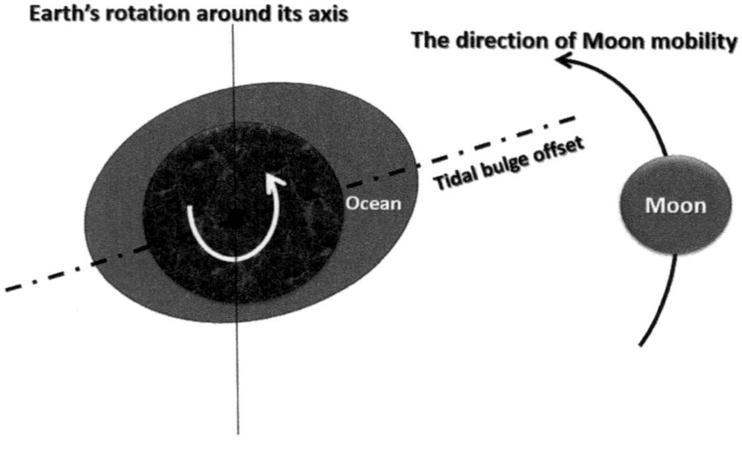

Figure 10.2: The gravitational tide is caused by the gravity force between the Earth and the Moon.

In delving into the realms of atmospheric tides, we unveil a symphony of influences, combining the rhythmic mobility of celestial bodies, the nuanced interplay of planetary waves, and the dynamic energy exchanges occurring in the tropics. This orchestration of forces paints a vivid canvas of atmospheric dynamics, inviting exploration and understanding at the intersection of celestial mechanics and meteorology. These atmospheric tides, originating predominantly in the troposphere and stratosphere, experience a transformative expedition. Propagating into the mesosphere and thermosphere, they leave an indelible mark on the atmospheric dynamics, demonstrating a unique interplay with Earth's atmospheric layers. While sharing commonalities with ocean tides, atmospheric tides exhibit distinct characteristics that distinguish them from their marine counterparts [3, 5, 8]. Primarily driven by the Sun's thermal influence, atmospheric tides synchronize their oscillations with the solar day, attaining a rhythmic cadence attuned to the

24-hour cycle. This stands in contrast to ocean tides, influenced by both lunar and solar gravitational forces, navigating a more intricate temporal choreography linked to the solar and lunar days.

The atmospheric stage, therefore, is characterized by its variable density with altitude, which imparts an exponential amplification to the amplitude of atmospheric tides during their ascent. This remarkable crescendo, governed by the delicate dynamic fluctuations of atmospheric density with elevation, elevates atmospheric tides to prominence in the mesosphere. Ascending to heights of 50–100 km, these tides attain amplitudes surpassing 50 m/s, establishing their dominance in shaping atmospheric motion. In this view, the underlying mechanism driving this amplitude amplification lies in the conservation of kinetic energy density. As atmospheric tides ascend through regions of diminishing atmospheric density, their amplitudes burgeon to uphold energy conservation principles. This intricate conveyor belt of energy preservation transforms subtle ground-level fluctuations into awe-inspiring oscillations that govern the celestial heights [9].

The grand orchestration of atmospheric tides unfolds as a consequence of the Sun's rhythmic influence, conducting a celestial symphony within Earth's atmospheric domains. This symphony is composed of thermal tides, emanating from the periodic diurnal cycle of the Sun's heating – a radiant performance where the atmosphere basks in daytime warmth and rests in nocturnal coolness [2–5]. For instance, the arrows in Figure 10.3 show the movement of the particles of the atmosphere pushed from the hot spot (sub-stellar) toward the cold spots.

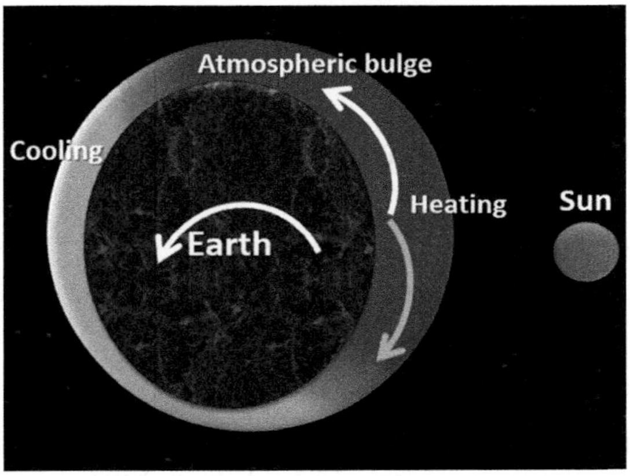

Figure 10.3: Resonance of atmospheric tides.

Intriguingly, the diurnal heating, expected to yield tides within 24 hours, unveils a more complex rhythm. Observations reveal the emergence of tides with periods not only of 24 hours but also 12, 8, and 6 hours. In this sense,these thermal solar tides find resonance in atmospheric constituents, with water vapor, ozone, molecular oxygen, and molecular nitrogen becoming key players in this cosmic performance. Solar energy's absorption, sculpted by the altitude and density of these atmospheric species, modulates the amplitude of the solar tides, creating a dynamic interplay of celestial forces and atmospheric composition [7, 9].

10.3 Harmony in the Skies: Unveiling the Cosmic Symphony of Solar Atmospheric Tides

As the solar tides traverse the atmospheric stage, their journey is not without influence from the environment they traverse. The global distribution and density of atmospheric constituents act as dynamic companions, shaping the amplitude of the solar tides in their celestial voyage.

Within the intricate choreography of solar tides, two distinct components emerge migrating and non-migrating. Each component adds its unique cadence to the cosmic symphony, with migrating tides embarking on a journey across latitudes, guided by the Sun's gravitational influence. In contrast, non-migrating tides carve their patterns closer to home, resonating within specific regions of the atmosphere. The solar tides, driven by the Sun's radiant energy, transcend mere thermal fluctuations; they become cosmic melodies, intricately woven into the fabric of Earth's atmospheric tapestry. As the Sun's influence conducts this symphony, the atmospheric tides harmonize with the rhythms of celestial forces, creating a celestial masterpiece that unfolds across the vast expanse of the atmosphere [2, 5].

10.4 Mathematical Descriptions of Atmospheric Tide

Embarking on a study into the intricacies of atmospheric tides, the classical tidal theory unveils fundamental characteristics that shape our understanding of these celestial phenomena. In this theoretical realm, the classical tidal theory gracefully sidesteps the influence of mechanical forcing and dissipation, casting atmospheric wave motions as linear perturbations. Within this conceptual framework, the atmosphere takes on the guise of an initially quiescent zonal mean state, characterized by horizontal stratification and isothermal conditions [5–7].

$$\frac{\partial u'}{\partial t} - 2\Omega \sin \varphi v' + \frac{1}{r \cos \varphi} \frac{\partial \Phi'}{\partial \lambda} = 0 \tag{10.1}$$

$$\frac{\partial v'}{\partial t} - 2\Omega \sin \varphi u' + \frac{1}{r} \frac{\partial \Phi'}{\partial \varphi} = 0 \tag{10.2}$$

Therefore, the energy can be revealed by:

$$\frac{\partial^2}{\partial t \partial z} \Phi' + N^2 w' = \frac{\kappa J'}{H} \tag{10.3}$$

Consequently, the atmospheric tide can be expressed using the continuity formula;

$$\frac{1}{a \cos \varphi} \left(\frac{\partial u'}{\partial \lambda} + \frac{\partial}{\partial \varphi} (v' \cos \varphi) \right) + \frac{1}{Q_o} \frac{\partial}{\partial z} (Q_o w') = 0 \tag{10.4}$$

In the domain of atmospheric dynamics, several key parameters weave a complex tapestry, each playing a distinct role in shaping the intricate dance of atmospheric phenomena. The eastward zonal wind denoted as u, portrays the component of wind moving parallel to the equator. Concurrently, the northward meridional wind, symbolized as v, illustrates the component of wind directed toward the north pole. Adding another layer to this dynamic symphony, the upward vertical wind, represented by w, provides a glimpse into the vertical motion within the atmosphere, unraveling the secrets of its ascent [1, 4, 7].

The geopotential, therefore, denoted by the symbol Φ, serves as a measure of the Earth's gravitational potential, offering insights into the distribution of mass across the globe. Delving into the fabric of atmospheric stratification, the square of Brunt-Vaisala (buoyancy) frequency, symbolized as N^2 captures the essence of buoyancy-induced oscillations within the atmosphere.

As we pivot towards the celestial stage, the angular velocity of the Earth, portrayed by the symbol Ω, dictates the rate of Earth's rotation, an omnipresent force influencing atmospheric dynamics. Consequently, the altitude, denoted as z, takes center stage, acting as a variable parameter that unfurls the layers of the atmosphere with each ascent. Geographic longitude, λ, and latitude, φ, add a spatial dimension to the narrative, guiding us through the Earth's coordinates as we navigate the atmospheric landscape. In this view,the heating rate per unit mass, symbolized by J, emerges as a dynamic force, influencing temperature variations and setting the stage for atmospheric transformations [1, 3, 6].

In this intricate tidal atmospheric dynamics, the radius of the Earth, r, gravity acceleration, g, and the constant scale height, H, come together to define the foundational elements that govern the atmospheric realm. As time, t, becomes the silent orchestrator, orchestrating the temporal evolution of atmospheric phenomena, the density, Q_o, undergoes dynamic fluctuations, responding to the ever-changing forces in this scenario. Amidst this symphony of atmospheric parameters, an exponential relation unfolds, with density Q_o showing a proportional decrease with altitude z following the expression $\exp(-z/H)$. In this intricate interplay of atmospheric parameters, each element contributes its unique signature to the grand tapestry of atmospheric dynamics, unveiling the captivating symphony that unfolds in the celestial expanse [3–6].

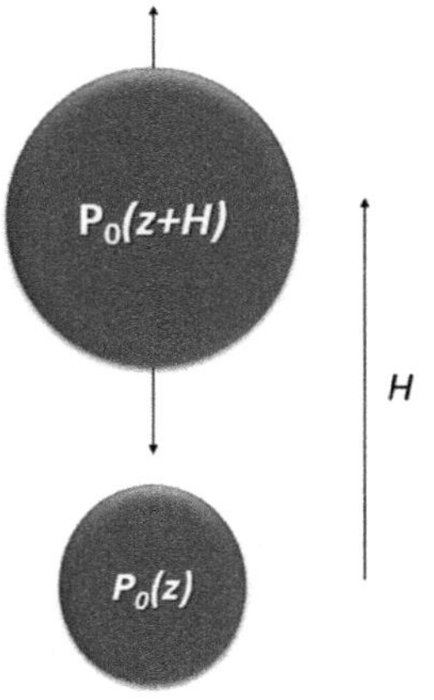

Figure 10.4: Oscillation of an atmospheric parcel.

The system of equations unravels the secrets of atmospheric tides, revealing them as longitudinally propagating waves characterized by the zonal wavenumber s and frequency σ. In this context, zonal wavenumber s takes on positive integer values, where positive σ signifies eastward propagating tides, while negative values denote westward propagating tides [1, 5, 7]. A pivotal technique in unraveling this atmospheric tapestry involves a separation approach, depicted by the equation:

$$\Phi'(\varphi, \lambda, z, t) = \hat{\Phi}(\varphi, z)e^{i(s\lambda-\sigma t)} \tag{10.5}$$

$$\Phi'(\varphi, z) = \sum_n \Theta_n(\varphi)G_n(z) \tag{10.6}$$

The intricate mobility of atmospheric tides unfolds through the lens of the horizontal structure equation, a venerable mathematical expression often referred to as Laplace's tidal equation:

$$L = \frac{\partial}{\partial\mu}\left[\frac{(1-\mu^2)}{(\eta^2-\mu^2)}\frac{\partial}{\partial\mu}\right] - \frac{1}{\eta^2-\mu^2}\left[-\frac{s\,(\eta^2+\mu^2)}{\eta\,(\eta^2-\mu^2)} + \frac{s^2}{1-\mu^2}\right][\theta_n] + \frac{(2\Omega a)^2}{gh_n}[\theta_n] \tag{10.7}$$

here $\mu = \sin\varphi$, and $\eta = \dfrac{\sigma}{2\Omega}$. This elegant equation encapsulates the nuanced interplay between latitude and the zonal wavenumber, providing a mathematical framework for understanding the latitudinal intricacies of atmospheric tides [4–7].

The attenuation of atmospheric tides finds its stage predominantly in the lower thermosphere, where various factors contribute to the damping of these celestial oscillations. Turbulence induced by the rupture of gravity waves plays a pivotal role in this intriguing process. Analogous to the rhythmic breaking of ocean waves upon a shore, the energy carried by atmospheric tides disperses into the ambient atmosphere, harmonizing with the intricate movements of atmospheric dynamics [6–8].

Furthermore, as we ascend to higher altitudes within the lower thermosphere, molecular diffusion emerges as a significant player in the damping mechanism. In this ethereal realm, where the atmosphere becomes increasingly rarefied, the mean free path of molecules lengthens, fostering the prominence of molecular diffusion. The intricate study of atmospheric tides as they travel through the planetary realms is vividly depicted by this ethereal interaction of turbulence and molecular diffusion [3, 7, 10].

10.5 Generation Atmospheric Waves

From the perspective of air parcels in the domain of wave motion, the complex choreography of atmospheric waves is interpreted as a symphony created by the interaction of forces through the air. This dynamic performance, developed by the forces shaping the air's mobility, unveils an ironic tapestry of wave types, each weaving its unique narrative across latitudinal landscapes.

The influence of the Coriolis effect, akin to a celestial conductor guiding the orchestra, imparts distinct characteristics to the propagation of these waves. This influence attains its zenith at the poles, where horizontal flow encounters the maximal Coriolis effect, while at the equator, this effect recedes to zero as discussed earlier in Chapter 2.

Within this atmospheric stage, four distinct waves take center stage, each with its own unique composition and propagation characteristics. Sound waves, though often excluded from atmospheric equations due to their high frequency, resonate as longitudinal or compression waves, creating a rhythmic dance of compressions and expansions parallel to their journey through the atmosphere.

In the ethereal realm where stable stratification prevails, internal gravity waves make their presence known, their undulating movements a testament to the delicate balance within the atmospheric layers. Inertio-gravity waves, bearing the imprint of a significant Coriolis effect, stand as a harmonious fusion of inertia and gravity.

The atmospheric stage further unfolds with the entrancing performance of Rossby waves, gracefully tracing the contours of 500 hPa geopotential in response to the ballet of midlatitude cyclones and anticyclones. At the equator, the ensemble takes a captivating turn, featuring mixed Rossby-gravity and Kelvin waves in a celestial display of atmospheric artistry. Each wave type, a distinctive instrument in the orchestra of atmospheric dynamics, contributes to the ever-evolving masterpiece of Earth's atmospheric mobility [1–4].

Dynamic effects and heating are two of the many mechanisms involved in the creation of atmospheric waves. Small-scale processes like convection, which produces gravity waves, or large-scale events like temperature differences between continents and seas, which, particularly during the winter months in the Northern Hemisphere, produce Rossby waves, may both trigger these waves. Furthermore, barriers like as mountain ranges – such as the Alps in Europe or the Rocky Mountains in the United States – can impede flow and add to dynamic impacts.

As atmospheric waves propagate, they transport momentum, influencing the background flow as the waves dissipate. This wave-induced forcing is particularly significant in the stratosphere, where planetary-scale Rossby waves contribute to sudden stratospheric warmings, and gravity waves contribute to the quasi-biennial oscillation.

Certainly, let's represent the atmospheric waves mathematically using spherical harmonics. The spherical harmonics, denoted as $Y_l^m(\theta,\phi)$, are solutions to Laplace's equation on a sphere and can be expressed as follows:

$$Y_l^m(\theta,\phi) = A_l^m P_l^m(\cos\theta)e^{im\phi} \tag{10.8}$$

In the mathematical representation of atmospheric waves, spherical harmonics play a crucial role. These harmonics, denoted as $Y_l^m(\theta,\phi)$, embody solutions to Laplace's equation on a sphere and are characterized by specific parameters. The degree l delineates the number of nodal lines on the sphere, while the order m signifies the quantity of full oscillations around the axis. The polar angle θ and azimuthal angle ϕ describe the orientation of the harmonic in spherical coordinates. The associated Legendre polynomial $P_l^m(\cos\theta)$ encapsulates the angular dependence and A_l^m serves as a normalization constant. The azimuthal dependence is introduced by $e^{im\phi}$. This systematic breakdown of parameters elucidates the intricate components embedded in spherical harmonics, forming the basis for understanding atmospheric wave dynamics. In this understanding, let us express the atmospheric wave as the sum of these spherical harmonics by thinking of the wave as represented along a latitude circle:

$$\Phi'(\theta,\phi,\lambda,t)=\sum_{l=0}^{\infty}\sum_{m=-l}^{l}\Theta_l^m(\lambda)Y_l^m(\theta,\phi)e^{i\sigma t} \tag{10.9}$$

In the realm of atmospheric wave representation, the perturbation of the geopotential, denoted as $\Phi'(\theta,\phi,\lambda,t)$, takes center stage. This perturbation encapsulates the dynamic alterations in the Earth's gravitational field. Accompanying this, the amplitude of the spherical harmonics in the longitudinal direction, represented by $\Theta_l^m(\lambda)$, signifies the strength and spatial distribution of the wave's components along the meridional lines. The frequency of the wave, denoted as σ, quantifies the rate of oscillation or repetition of the atmospheric disturbance. These interrelated components contribute to the comprehensive understanding of atmospheric waves and their behavior over time and space. Needless to say, this mathematical representation allows us to decompose the atmospheric waves into different spherical harmonic modes, providing insights into their structure and behavior [3–7].

10.6 Deriving the Linearity of Gravity Waves

$$\frac{\partial u}{\partial t}+u\frac{\partial u}{\partial x}+w\frac{\partial u}{\partial z}=-\frac{1}{\rho}\frac{\partial p}{\partial x} \tag{10.10}$$

$$\frac{\partial w}{\partial t}+u\frac{\partial w}{\partial x}+w\frac{\partial w}{\partial z}=-\frac{1}{\rho}\frac{\partial p}{\partial z}-g \tag{10.11}$$

$$\frac{\partial u}{\partial x}+\frac{\partial w}{\partial z}=0 \tag{10.12}$$

$$\frac{\partial \rho}{\partial t}+u\frac{\partial \rho}{\partial x}+w\frac{\partial \rho}{\partial z}=0 \tag{10.13}$$

The set of equations presented describes the dynamics of a two-dimensional, irrotational, and inviscid flow under the Boussinesq approximation, with consideration for the x and z-directions of momentum, mass continuity, and conservation of thermal energy. Equation (10.10) represents the momentum equation in the x-direction, while Equation (10.11) addresses the momentum equation in the z-direction. Equation (10.12) accounts for mass continuity, ensuring the balance between inflow and outflow. Furthermore, Equation (10.13), derived from $\frac{DP}{Dt}=\left[\left(\frac{c_p}{c_v}\right)\left(\frac{p}{\rho}\right)\right]^2\frac{D\rho}{Dt}=0$, signifies the conservation of thermal energy. To facilitate analysis and solution, a linearization process is applied to these equations, allowing for a more tractable representation that captures the essential characteristics of the atmospheric system under study [13–15].

$$q(x,\ z,\ t)\ =\ q_0(z)\ +\ q_1(x,\ z,\ t), \tag{10.14}$$

Incorporating the steady, horizontally uniform background value denoted as $q_0(z)$ and the first-order perturbation $q_1(x, z, t)$, the equations (10.10) to (10.13) can be expressed under the assumption of a background flow in hydrostatic balance, as indicated by:

$$\frac{\partial p}{\partial z}=-\rho g \tag{10.15}$$

This set of equations governs the dynamics of the atmospheric system, where $q_1(x, z, t)$ represents the first-order perturbation that captures the deviations from the background state. The hydrostatic balance ensures a stable equilibrium between vertical pressure gradients and gravitational forces, contributing to a comprehensive understanding of atmospheric oscillations under buoyancy [5, 8, 11, 13, 15].

$$\frac{\partial u_1}{\partial t} + u_0 \frac{\partial u_1}{\partial x} + w_1 \frac{\partial u_0}{\partial z} = -\frac{1}{\rho_0} \frac{\partial p_1}{\partial x} \tag{10.16}$$

$$\frac{\partial w_1}{\partial t} + u_0 \frac{\partial w_1}{\partial x} = -\frac{1}{\rho_0} \frac{\partial p_1}{\partial z} - \frac{\rho_1}{\rho_0} g \tag{10.17}$$

$$\frac{\partial u_1}{\partial x} + \frac{\partial w_1}{\partial z} = 0 \tag{10.18}$$

$$\frac{\partial \rho_1}{\partial t} + u_0 \frac{\partial \rho_1}{\partial x} + w_1 \frac{\partial \rho_0}{\partial z} = 0 \tag{10.19}$$

Incorporating the background atmospheric density denoted as ρ_0, the equations (10.16) to (10.19) can be further analyzed. One approach involves employing a double Fourier transform in both the spatial variable x and the temporal variable t. Alternatively, we can explore wave-like solutions, assuming a specific form that captures the oscillatory behavior of the atmospheric system. These mathematical techniques provide valuable insights into the characteristics of atmospheric oscillations under buoyancy (Figure 10.5), allowing for a more detailed examination of the underlying dynamics.

$$u_1(x, z, t) = \tilde{u}(z) e^{i(kx-\omega t)} \tag{10.20}$$

$$\rho_1(x, z, t) = \tilde{\rho}(z) e^{i(kx-\omega t)} \tag{10.21}$$

$$p_1(x, z, t) = \tilde{p}(z) e^{i(kx-\omega t)} \tag{10.22}$$

$$w_1(x, z, t) = \tilde{w}(z) e^{i(kx-\omega t)} \tag{10.23}$$

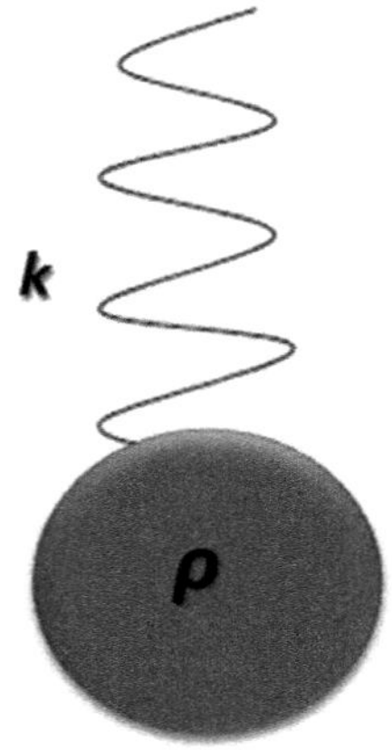

Figure 10.5: Atmospheric oscillations under buoyancy.

The functions $\tilde{\rho}_1, \tilde{w}_1$, etc., being solely dependent on z, allow us to express the derivatives as total rather than partial. Introducing the intrinsic frequency, denoted as Ω, becomes pertinent in this context. The intrinsic frequency represents the frequency of a wave relative to the flow, specifically the frequency measured by an observer moving with the fluid at the velocity u_0. This consideration provides a valuable perspective for understanding the dynamics of waves in the given fluid system [9, 13, 15].

Gravity wave propagation can indeed be described in terms of polarization propagation. In the context of gravity waves, polarization refers to the oscillation pattern perpendicular to the direction of wave travel. This phenomenon arises from the variable strengths of gravitational forces acting on different segments of the wave. In this context, equation 10.23 can be used to express the gravity wave propagation from the point of view of polarization propagation. Mathematically, polarization in the context of gravity waves can be expressed using complex amplitudes [12–16].

$$w_1(x,\ z,\ t) \ = \ \tilde{w}(z)\left(\frac{\rho_s}{\rho_0(z)}\right)e^{i(kx-\omega t)}. \tag{10.24}$$

In equation 2.24, it can be observed that as the wave propagates upward, there is a corresponding decrease in the background air density denoted by ρ_0. This alteration in density implies an elevation in the wave amplitude, represented by w_1, as indicated in the same equation. This intriguing relationship between wave propagation, air density, and amplitude sets the stage for further exploration. Therefore, the interconnected dynamics of wave behavior and atmospheric parameters unfold with each layer of insight, promising a comprehensive understanding of the underlying physics.

$$\tau(z) = -\rho_0\overline{u}_1\overline{w}_1, \tag{10.25}$$

The concept of a horizontal average, denoted by the overline symbol, is conventionally applied over a wavelength, especially when dealing with nearly monochromatic waves. In the scenario where the background air density, ρ_0, diminishes with height, a consequential effect on the upward transport of momentum becomes apparent. To maintain constant momentum, the product $\overline{u}_1\overline{w}_1$ must increase, serving as a compensatory mechanism to counterbalance the decline in density. This increase is precisely accounted for by the factor $\left(\frac{\rho_s}{\rho_0(z)}\right)^{1/2}$.

Drawing an analogy to the mechanics of a bullwhip, a similar principle is at play. The bullwhip's thickness decreases gradually from approximately 4 cm at the handle to about 0.1 cm at the end. This reduction in linear density prompts the amplification of the wave's amplitude as it traverses from the handle towards the end of the whip. This phenomenon finds resonance in the audible "crack" of the bullwhip, where the perturbation velocity at the whip's end surpasses the speed of sound, resulting in a distinctive and loud sound. Therefore, through these comparisons and insights, a more thorough comprehension of the intricate interplay between density variations and momentum transport emerges.

10.7 The Influence of Wind on the Propagation of Atmospheric Gravity Waves

Embarking on an exploration of solutions to the Taylor–Goldstein equation (2.34), we commence with the straightforward scenario of constant background stratification, N, and a wavelength significantly shorter than λ_s. In other words, when surface tension is disregarded, the interface between two fluids in parallel motion, each with different velocities and densities, exhibits instability to short-wavelength perturbations across all speeds. However, the introduction of surface tension can stabilize this short-wavelength instability, imposing a threshold velocity beyond which stability is compromised.

In the scenario where density and velocity exhibit continuous spatial variations—with lighter layers positioned uppermost, ensuring Rayleigh–Taylor stability—the dynamics of the gravity wave instability can be elegantly described by the Taylor–Goldstein equation:

$$(U-c)^2\left(\frac{d^2\overline{\phi}}{dz^2}-k^2\overline{\phi}\right)+\left[N^2-(U-c)\frac{d^2U}{dz^2}\right]\overline{\phi}=0, \tag{10.26}$$

In the absence of any background wind speed, the Taylor–Goldstein equation (10.26) simplifies to:

$$\frac{d^2\overline{\phi}}{dz^2}+k^2N^2\left[\overline{\phi}\right]^{-2}k^2\overline{\phi}=0 \tag{10.27}$$

The general solution to this differential equation is:

$$\overline{\phi}(z)=Ae^{imz}+Be^{-imz} \tag{10.28}$$

Here, the vertical wavenumber mm is determined by:

$$m=k^2\overline{\phi}^{-2}N^2-1 \tag{10.29}$$

Solving (10.29) for ω yields the dispersion relation:

$$\omega=\pm kN\sqrt{\frac{k^2+m^2}{2}} \tag{10.30}$$

The dispersion relation holds paramount importance in linear wave theory, serving as a crucial link between the angular frequency of the wave and the atmospheric characteristics governing its structure. This relation uniquely determines all variables of the wave field—*k*, *m*, and ω – making them eigenvalues for a given atmospheric condition. Therefore, examining the negative branch of (10.29) gives:

$$\omega=ck=\pm kN\sqrt{\frac{k^2+m^2}{2}} \tag{10.31}$$

This indicates that the horizontal phase velocity (c) is negative, implying that the wave travels in the negative x-direction. Adhering to this convention, where $k < 0$, ensures $\omega > 0$ for all conditions. Thus, the dispersion relation and its implications shed light on the directionality and characteristics of the wave as it propagates through the atmosphere. When $\omega < N$, the vertical wavenumber mm is real, indicating that the wave is propagating or internal. Conversely, when $\omega > N$, m becomes imaginary, signifying that the wave is evanescent or external. In the evanescent case ($\omega > N$), we can express m as:

$$m=\pm ik\sqrt{\frac{1-N^2}{\omega^2}}=\pm iq \tag{10.32}$$

The general solution for the evanescent wave in the case of $\omega > N$ is given by:

$$w_1=Ae^{-q(z-zL)}e^{i(kx-\omega t)} \quad \text{for} \quad z>zL \tag{10.33}$$

$$w_1=Be^{q(z-zL)}e^{i(kx-\omega t)} \quad \text{for} \quad z<zL \tag{10.34}$$

Here, A and B are the amplitude of the wave, q is the imaginary part of the vertical wavenumber, zL is the height where the evanescent wave is initiated (Figure 10.6), k is the horizontal wavenumber, ω is the angular frequency, x is the horizontal coordinate, and t is time. This expression describes how the evanescent wave varies with height (z), horizontal position (x), and time (t), emphasizing the exponential decrease in amplitude with increasing height [11, 14].

Taking into account the influence of a constant background wind on wave propagation, where u_0 represents the component of the background wind velocity aligned with the direction of wave propagation, we introduce the phase angle:

$$\phi=kx+mz-\omega t \tag{10.35}$$

In this view, the vertical wavenumber m is expressed as:

$$m=\pm\sqrt{\frac{k^2N^2}{(\omega-u_0k)^2}-k^2} \tag{10.36}$$

Figure 10.6: Generation of evanescent waves at altitude *zL*.

This formulation, which considers the background wind speed u_0, necessitates a careful approach, as assigning a branch to (10.36) in advance could determine the sign of the vertical group velocity. It is crucial not to let the mathematics dictate the physics of the problem. Instead, the approach involves selecting the physics and letting the mathematics follow.

Solving equation 10.36 for ω yields:

$$\omega = u_0 k \pm \sqrt{\frac{k^2 N^2}{k^2 + m^2}} \tag{10.37}$$

Therefore, the phase velocities can be obtained by:

$$c = u_0 + k^{-1} N \cos\beta \tag{10.38}$$

$$cz = u_0 mk^{-1} + m^{-1} N \cos\beta \tag{10.39}$$

In this understanding, the group velocity components can then be given by:

$$u_g = u_0 + m^2 N^2 (c - u_0)^3 \tag{10.40}$$

$$w_g = -kmN^{-2}(c - u_0)^3 \tag{10.41}$$

To illustrate the selection of appropriate wavenumbers, consider a scenario where both u_g and w_g are positive, and $c > u_0$. Accordingly, in the situation where $c < u_0$, the choice of m becomes crucial, particularly when w_g is positive. In this case, it is necessary to select $m > 0$ to ensure the positivity of w_g.

Now, let's examine the scenario where $u_{g>0}$ and $w_{g<0}$, signifying downward-propagating energy. In this instance, the choice of m is contingent on the sign of c: $m > 0$ when $c > 0$ and $m < 0$ when $c < 0$. This careful selection of m aligns with the specific conditions of downward energy propagation in the atmosphere. The impact of a constant background wind extends to the polarization equations [14, 16]. Take, for instance, equation (10.42), which links the horizontal wind perturbation (u_1) (Figure 10.7) to the pressure perturbation (p_1) as:

$$i\Omega\tilde{u} - \tilde{w}\frac{du}{dz} = —k\tilde{p} \tag{10.42}$$

This term $i\Omega\tilde{u} - \tilde{w}\frac{du_0}{dz}$ involves the perturbation in the horizontal wind ($\tilde{u}$) and the vertical wind ($\tilde{w}$) scaled by the vertical gradient of the background wind $\left(\frac{du_0}{dz}\right)$. The i term typically

Figure 10.7: Wind perturbation.

represents the imaginary unit in mathematical notation. Therefore, $\frac{i}{\rho_0}k\tilde{p}$ involves the background air density (ρ_0), the horizontal wavenumber (k), and the perturbation in the vertical displacement ($\tilde{p}$). For a constant background wind, this equation takes the form:

$$\tilde{u}_1 = \tilde{p}_1\left(\rho_0\left(c - u_0\right)\right)^{-1} \tag{10.43}$$

Here, $\tilde{u}_1$ represents the horizontal wind perturbation, $\tilde{p}_1$ is the pressure perturbation, ρ_0 is the background air density, c is the phase speed, and u_0 is the background wind speed. As per this formulation, if $c > u_0$, the pressure and wind speed perturbations are in phase. Conversely, if $c < u_0$, the perturbations are 180° out of phase. This relationship underscores the influence of the background wind on the synchronization or phase opposition of pressure and wind speed perturbations [11–16].

10.8 Mechanisms of SAR Imaging of Atmospheric Gravity Waves

Deriving exact mathematical formulas for gravity wave imaging by normalized radar cross-section (NRCS) under strong wind involves a complex interplay of atmospheric and surface parameters. The process requires a detailed understanding of the atmospheric dynamics, wave characteristics, and radar sensing principles. However, we can outline a simplified approach by incorporating key elements.

As a starting point, consider a modified Kirchhoff model (Chapter 9):

$$\sigma(\theta_i,\theta_r) = \sigma_0\left|\frac{\sin(\theta_i - \theta_r)}{\cos(\theta_i) + \cos(\theta_r)}\right|^2 \tag{10.44}$$

here $\sigma(\theta_i,\theta_r)$ is the radar cross-section; σ_0 is the backscattering coefficient for a smooth surface; are the incidence and scattering angles, respectively. Introducing non-Gaussian height distributions, anisotropic scattering, and wavelength-dependent effects into the radar cross-section (RCS) equation can significantly enhance the model's realism. Wind can influence surface roughness, affecting the scattering behavior [17–19]. Increased wind speed may lead to more significant wave-induced roughness, altering this scattering term. Below is a modified radar cross-section equation incorporating these elements:

$$\sigma(\theta_i,\theta_r) = \sigma_0\left|\frac{\sin(\theta_i - \theta_r)}{\cos(\theta_i) + \cos(\theta_r)}\right|^2 S(h)G(\theta_i,\theta_r)W(\lambda) \tag{10.45}$$

In this equation, $S(h)$ represents the non-Gaussian height distribution; $G(\theta_i, \theta_r)$ accounts for anisotropic scattering effects; and $W(\lambda)$ incorporates wavelength-dependent effects. Let's break down each component:

$$S(h) = e^{\left(-\frac{h^2}{2\sigma_h^2}\right)} \tag{10.45.1}$$

Equation 10.45.1 represents the non-Gaussian height distribution. In this equation, h is the height of the surface, and σ_h is the standard deviation of the height distribution. This Gaussian distribution can be modified based on the specific non-Gaussian characteristics of surface features in SAR data [17, 20, 21]. In this sense, higher wind speeds may lead to larger wave amplitudes, impacting the surface roughness modeled by this term. Therefore, anisotropic scattering effects $G(\theta_i, \theta_r)$ can be given by:

$$G(\theta_i, \theta_r) = e^{\left(-\frac{(\theta_i-\theta_r)^2}{2\sigma_\alpha^2}\right)} \tag{10.45.2}$$

This factor accounts for anisotropic scattering effects, where σ_α represents the standard deviation of the scattering angle distribution. Adjusting σ_α will control the anisotropy of the scattering behavior. Therefore, wind direction can influence the anisotropy of surface roughness. For example, wind blowing in a specific direction may align surface waves, impacting the scattering pattern. Consequently, $W(\lambda)$ wavelength-dependent effects $W(\lambda)$ can be computed as:

$$W(\lambda) = \left(\frac{\lambda}{\lambda_0}\right)^\beta \tag{10.45.3}$$

This factor introduces wavelength-dependent effects, where λ is the wavelength of the radar signal, λ_0 is a reference wavelength, and β is a parameter controlling the wavelength dependence. Indeed, Wind-induced changes in surface roughness can affect the radar response at different wavelengths. For instance, larger waves may become more prominent with increasing wind speeds, altering the wavelength dependence [18, 22, 23]. As a result, the given equation represents a modified radar cross-section $\sigma(\theta_i, \theta_r)$ for the imaging of atmospheric gravity waves using Synthetic Aperture Radar (SAR):

$$\sigma(\theta_i, \theta_r) = \sigma_0 \left|\frac{\sin(\theta_i - \theta_r)}{\cos(\theta_i) + \cos(\theta_r)}\right|^2 e^{\left(-\frac{h^2}{2\sigma_h^2}\right)} \left(\frac{\lambda}{\lambda_0}\right)^\beta e^{\left(-\frac{(\theta_i-\theta_r)^2}{2\sigma_\alpha^2}\right)} \tag{10.46}$$

In summary, this equation combines geometric scattering effects, non-Gaussian height distributions, anisotropic scattering, and wavelength-dependent effects to model the radar cross-section for SAR imaging of atmospheric gravity waves [23–25]. Adjusting the parameters (σ_h, σ_α, β) allows customization based on specific characteristics of the rough surface and atmospheric conditions. Needless to say, while the specific wind speed is not explicitly present in the equation, the influence of wind on the surface roughness components (height distribution, anisotropy, and wavelength dependence) indirectly affects the radar cross-section. The relationship between wind speed and the radar cross-section involves the dynamic interplay of wind-induced surface variations, impacting how gravity waves are imaged by SAR.*

10.9 Quantized SAR Imaging Mechanism of Atmospheric Gravity Waves

In the realm of atmospheric physics, a theoretical abstraction is considered, drawing inspiration from the principles of quantum mechanics to represent atmospheric gravity waves. Within this speculative framework, a wave function $\Psi(z,t)$ is introduced to describe the state of atmospheric gravity waves at a given altitude z and time t. The dynamics of these waves are envisioned through a hypothetical gravity wave Hamiltonian, denoted as $\hat{H}_{GW}$, operating on the wave function [26–28]. This speculative Hamiltonian incorporates terms analogous to energy, potential, and other features characterizing atmospheric gravity waves. Therefore, expressed in a Schrödinger-like equation, the quantum-inspired wave equation becomes:

$$i\hbar\frac{\partial}{\partial t}\Psi(z,t) = -\frac{\hbar^2}{2m}\frac{\partial^2}{\partial z}\Psi(z,t) + V_{GW}(z,t)\Psi(z,t) \tag{10.47}$$

here is $V_{GW}(z,t)$ a potential term for atmospheric gravity wave propagation. Consequently, ensuring that the probability density of finding the atmospheric gravity waves at any position is normalized as:

$$\int_{-\infty}^{\infty} i\hbar\frac{\partial}{\partial t}|\Psi(z,t)|^2 = 1 \tag{10.48}$$

Therefore, a modified radar cross-section $\sigma(\theta_i,\theta_r)$ (equation 10.46) for the imaging of atmospheric gravity waves using SAR can be given from the point of view of quantum mechanics:

$$i\hbar\frac{\partial\Psi}{\partial t} = \hat{H}\Psi + f\left(\sigma(\theta_i,\theta_r)\right)\Psi \tag{10.49}$$

Let's represent the nonlinear relationship between $\hat{H}$ and atmospheric wave gravity characteristics $f(\sigma(\theta_i,\theta_r,\lambda,c_g))$ more explicitly:

$$\frac{\partial\Psi}{\partial t} = \frac{1}{i\hbar}\Psi\left[f(\sigma(\theta_i,\theta_r,\lambda,c_g)) + \hat{H}\right] \tag{10.50}$$

To delve into the specifics, we can express $\hat{H}$ and $f(\sigma(\theta_i,\theta_r,\lambda,c_g))$ as follows:

$$\hat{H} = -\frac{\hbar^2}{2m}\nabla^2 + V(r) \tag{10.51}$$

here m is the mass of the particle associated with the wave function; $V(r)$ is the potential energy function; and ∇^2 is the Laplace operator [27, 29, 31]. Therefore, $f(\sigma(\theta_i,\theta_r,\lambda,c_g))$, we can consider a hypothetical function that captures the nonlinear dependence:

$$\Psi\left|(\sigma(\theta_i,\theta_r,\lambda,c_g))\right\rangle = \Psi\left|\left((\frac{\lambda^2\omega^2}{4\pi k^2})\right)\right\rangle + \Psi\left|\frac{\lambda^2}{4\pi}|K(\theta,\phi)|^2\right\rangle \tag{10.52}$$

Equation 10.52, expressed the quantum mechanism of SAR imagine AGW. In this view, the normalized radar cross-section quantum state involves two states. First is $\Psi\left|\left(\frac{\lambda^2\omega^2}{4\pi k^2}\right)\right\rangle$ presents the physical characteristics of AGW i.e., wavelength frequency; wave number; angular frequency i.e. group velocity $c_g^2 = \frac{\omega^2}{k^2}$. Second is $\Psi\left|\frac{\lambda^2}{4\pi}|K(\theta,\phi)|^2\right\rangle$ which presents the radar backscatter (σ) from an AGW rough surface which is proportional to the product of the incident electric field, the reflected electric field, and a scattering coefficient [24, 26, 29].

In the realm of quantum mechanics, Therefore, the entanglement Hamiltonian $(H^{\text{entangled}})$ is a crucial concept that encapsulates the intricate relationship between the Synthetic Aperture Radar (SAR) and Atmospheric Gravity Waves (AGW). The entanglement Hamiltonian is elegantly expressed as a sum of individual Hamiltonians for SAR and AGW:

$$H^{\text{entangled}} = H^{\text{SAR}} \otimes I^{\text{AGW}} + I^{\text{SAR}} \otimes H^{\text{AGW}} + H^{\text{interaction}} \tag{10.53}$$

here, H^{SAR} and H^{AGW} represent the Hamiltonians governing the dynamics of SAR and AGW, respectively. The operators I^{SAR} and I^{AGW} are identity operators associated with SAR and AGW, ensuring a seamless integration. Additionally, the term $H^{\text{interaction}}$ captures the essence of their entanglement. It symbolizes the intricate interplay and mutual influence between SAR and AGW. This interaction term is at the heart of understanding how the quantum states of SAR and AGW become entangled, leading to the fascinating phenomena observed in their joint dynamics. In this understanding, The entanglement process can be mathematically described by the interaction Hamiltonian $\text{H}^{\text{interaction}}$ acting on the entangled state:

$$H^{\text{interaction}} \left| \Psi_{\text{entangled}} \right\rangle = (A \otimes B)\Psi \left| (\sigma(\theta_i, \theta_r, \lambda, c_g)) \right\rangle \otimes \Psi \left| \left(\frac{\lambda^2 \omega^2}{4\pi k^2} \right) \right\rangle \tag{10.54}$$

Here, A and B are operators associated with SAR and AGW, respectively. In summary, the interaction term $H^{\text{interaction}}$ encapsulates the intricate interplay between SAR and AGW, and its form is derived based on the underlying physics of their entanglement [28–31].

10.10 Automatic Detection of Atmospheric Gravity Waves Using Quantum Particle Swarm Optimization Algorithm

The understanding of quantized atmospheric gravity waves (Section 10.9) can be approached through the lenses of wave function, superposition, and entanglement, according to the author. These fundamental quantum mechanisms gain clarity when wave–particle duality [29] is incorporated. The application of an algorithm capable of handling atmospheric gravity wave imaging in SAR images can be beneficial from this perspective.

Although quantum effects are more evident at the microscale, the principles of quantum mechanics fundamentally govern nature. The concept of wave-particle duality, which portrays a particle's information through a wave function $\Psi(r, t)$, introduces the idea of normalized quantum eigenstates [27–30]. This wave function calculates the probability of finding a particle at a specific location at a given time. Additionally, superposition allows a particle to exist simultaneously in multiple states until it is observed or measured.

AGW packets, which are superpositions of waves, theoretically represent a localized particle in space when subjected to a physical potential. However, the uncertainty principle broadens this representation. The Heisenberg uncertainty principle dictates the impossibility of precisely determining both the position and velocity of a quantum particle simultaneously, leading to the broadening of the energy levels in the physical potential.

Inspired by the probabilistic nature of quantum mechanics, the quantum particle swarm optimization (QPSO) algorithm seeks to identify the correct wave function corresponding to a quantum particle in a potential field. Utilizing a quantum probability density function, QPSO directs particles to the most likely positions to find optimal answers and simultaneously explores multiple solutions through the concept of superposition.

The solution, guided by the probabilistic nature of quantum mechanics and the correlation between quantum particles, is expected to be identified in the SAR area of the search extent with enhanced efficiency. Quantum computing, which allows the exploration of multiple solutions simultaneously, is expected to lead to faster convergence rates [29]. Quantum entanglement further enhances the efficiency of quantum search algorithms, facilitating faster communication and information processing.

By shifting the focus from quantum particles, let M represent the quantity of particle-like 2-D quantum soliton wave packets in the SAR image along pixels i and j, denoted as $i = 1, 2, \ldots, M$, and $j = 1, 2, \ldots, G$. These AGW exhibited nonlinear interactions and could traverse significant distances without dispersion. Considering this perspective, the best local and global positions are combined into a linear convex sensor, resembling the local harmonic oscillator in the SAR image, defined as follows:

$$\psi\left|L_{i,j}(t)\right\rangle = \frac{\psi\left|l_1 r_1\right\rangle}{\psi\left|(l_1 r_1 + l_2 r_2)\right\rangle}\psi\left|P_{Beast_{i,j}}(t)\right\rangle + 1 - \left[\frac{\psi\left|l_1 r_1\right\rangle}{\psi\left|(l_1 r_1 + l_2 r_2)\right\rangle}\right]\psi\left|\mathfrak{g}_{Beast_j}(t)\right\rangle \tag{10.55}$$

Consequently, the analysis of the trajectory determines the optimal local $\psi\left|P_{Beast_{i,j}}(t)\right\rangle$ and global positions $\psi\left|\mathfrak{g}_{Beast_j}(t)\right\rangle$ for the entire swarm, leading to a novel movement approach. The adjusted position of this model can be expressed as follows:

$$\psi\left|L_i(t)\right\rangle = \psi\left|x_i(t+1)\right\rangle - \psi\left|\lambda_{d_s}\left(X_i(t), \cos h^{-1}\left(\sqrt[-1]{\frac{\left|\psi\left(\lambda_{d_s}\right)\right|^2}{max\left|\psi\left(\lambda_{d_s}\right)\right|^2}}\right)\right)\right\rangle \tag{10.56}$$

Hence, in alignment with the perspective, the updated position for exploring the quantum probability density function of AGW in the SAR image can be derived from the distribution function $F\left(\lambda_{d_s}\right) = \cos h^{-1}\left(\sqrt[-1]{\frac{\left|\psi\left(\lambda_{d_s}\right)\right|^2}{max\left|\psi\left(\lambda_{d_s}\right)\right|^2}}\right)$:

$$\psi\left|L_i(t)\right\rangle = \psi\left|x_i(t+1)\right\rangle + \cos h^{-1}\left(\sqrt[-1]{\frac{\left|\psi\left(\lambda_{d_s}\right)\right|^2}{max\left|\psi\left(\lambda_{d_s}\right)\right|^2}}\right) \tag{10.57}$$

Quantum mechanics posits that the quantum state (position) remains unknown until a measurement is conducted. This implies that equation 10.57 can undergo modification $\psi\left|x_i(t+1)\right\rangle$. It essentially exists in a superposition of the two states in the absence of observation [32–35]. The random function serves a similar role to the experimenter in the computational process (Figure 10.8).

Therefore, let's consider the following criteria for implementing QSPSO:

$$\psi\left|x_i(t+1)\right\rangle = \begin{cases} \text{Equation 5.85 if } \frac{1}{2} \le \psi|0|1> \le 1 \\ \text{Equation 5.86 if } 0 \le \psi|0|1> < \frac{1}{2} \end{cases} \tag{10.58}$$

$$\psi\left|\lambda_{d_s}\right\rangle = \begin{cases} \left|\psi\left|X_i(t)\right\rangle - w\otimes\psi\left|M^{-1}\sum_{i=1}^{M} P_{Best_i}(t)\right\rangle\right| \cos h^{-1}\left(\sqrt[-1]{\frac{\left|\psi\left(\lambda_{d_s}\right)\right|^2}{max\left|\psi\left(\lambda_{d_s}\right)\right|^2}}\right) & \text{if } \psi|0|1> < \frac{1}{2} \\ \left|\psi\left|X_i(t)\right\rangle + w\otimes\psi\left|M^{-1}\sum_{i=1}^{M} P_{Best_i}(t)\right\rangle\right| \cos h^{-1}\left(\sqrt[-1]{\frac{\left|\psi\left(\lambda_{d_s}\right)\right|^2}{max\left|\psi\left(\lambda_{d_s}\right)\right|^2}}\right) & \text{if } \psi|0|1> \ge \frac{1}{2} \end{cases}$$

(10.59)

When the search radius is multiplied by the following factor, w, the search radius decreases linearly:

$$w = w_1 + \frac{(w_0 - w_1)\times(maxI_{it} - t)}{maxI_{it}} \tag{10.60}$$

Input: I (σ_0) → SAR data
M → population size
G → Dimensions
w_0, w_1 → expansion
$maxI_{it}$ → *maximum number of iterations*
Being
Initialize the initial location of the particle position $\Psi|X>$
Compute $\psi\left|M^{-1}\sum_{i=1}^{M} P_{Best_i}(t)-M_{g_{Best}}(t)\right\rangle$
and
$\psi\left|g_{Beast_j}(t)\right\rangle$

While $t\neq maxI_{it}$ & $\psi\left|M^{-1}\sum_{i=1}^{M} P_{Best_i}(t)-M_{g_{Best}}(t)\right\rangle \leq \propto$
$\propto$ → *accuracy*
Determine w → Equation 10.60
for i=1 to M do
Compute $\psi\left|L_{i,j}(t)\right\rangle$ → *Equation 10.55*
for j=1 to G do
if $\psi|0|1>\geq\frac{1}{2}$ then

$$\psi\left|\lambda_{d_s}\right\rangle=\left|\psi\left|X_i(t)\right\rangle+w\otimes\psi\left|M^{-1}\sum_{i=1}^{M} P_{Best_i}(t)\right\rangle\right|\cos h^{-1}\left(\sqrt[-1]{\frac{\left|\psi\left(\lambda_{d_s}\right)\right|^2}{max\left|\psi\left(\lambda_{d_s}\right)\right|^2}}\right)$$

Else

$$\psi\left|\lambda_{d_s}\right\rangle=\left|\psi\left|X_i(t)\right\rangle-w\otimes\psi\left|M^{-1}\sum_{i=1}^{M} P_{Best_i}(t)\right\rangle\right|\cos h^{-1}\left(\sqrt[-1]{\frac{\left|\psi\left(\lambda_{d_s}\right)\right|^2}{max\left|\psi\left(\lambda_{d_s}\right)\right|^2}}\right)$$

Update $\psi\left|P_{Beast_{i,j}}(t)\right\rangle$ & $\psi\left|g_{Beast_j}(t)\right\rangle$

Return
End For
End if
While maximum iterations or minimum error criteria are not attained
End While

Figure 10.8: Pseudo-code of the quantum soliton-inspired particle swarm optimization (QSPSO) algorithm.

here t is the current search iteration numbers, and $maxI_{it}$ is the maximum number of iterations. Consequently, w_0 and w_1 represent w's initial and final values, respectively.

QSPSO algorithm performance can be optimized by modifying the number of particles, fitness, and clamping velocity. For SAR data, the $\psi\left|P_{Beast_{i,j}}(t)\right\rangle$ to $\psi\left|g_{Beast_j}(t)\right\rangle$ QSPSO is implemented in this demonstration. To prevent from becoming trapped in local optima, this algorithm commences with a zero-radius neighborhood and progresses to $\psi\left|g_{Beast_j}(t)\right\rangle$ as the radius of the neighborhood is gradually raised [36–38]. Using the $\psi\left|g_{Beast_j}(t)\right\rangle$ technique, the algorithm converges on the best option.

According to the aforementioned perspective, RADARSAT-1 standard-2 data acquired on March 31, 2005, were analyzed using QSPSO. As a result, the range of the normalized radar cross-section was observed to be between −25 dB and −12 dB (Figure 10.9). A challenge arises owing to the correlation between the occurrence of atmospheric gravity waves and internal waves. In this context, variations in backscatter are primarily influenced by the incidence angle, wavelength, and sea surface conditions, with wind speeds ranging from to 2–6 m/s. The wavelength of the RADARSAT-1 SAR Standard-2 mode was 5.3 cm, which is approximately equal to its Bragg scattering wavelength. Notably, the impact of wind stress on the imaging of AGW was significant. In the SAR imaging of atmospheric gravity waves, wind stress plays a crucial role by influencing the roughness of the ocean surface. Wind-induced surface waves generate surface roughness, which affects the backscattered radar signal received by the SAR sensor. This interaction results in scattering, causing distortions and speckles in the SAR image, making accurate detection and analysis of atmospheric gravity waves challenging. The presence of surface wave-induced roughness and low wind stress patterns in SAR images can obscure AGW signals, leading to reduced visibility and potentially affecting the interpretation of the wave patterns. The dark zone in the SAR images suggests low wind stress and the lowest backscatter of –25 dB contributes to the reduced visibility of atmospheric gravity waves. Additionally, modulation in short waves is attributed to current shear, wind speed, long gravity waves associated with eddies, and internal waves. These factors contribute to a high degree of backscatter variability, as is evident in the Standard-2 mode data. The large incidence angle of Standard-2 mode data, ranging from 20° to 49°, and the high radar backscatter contrast enabled significant discrimination between water and land boundaries. These findings confirm the work done by Li et al. [20]; Li [21]; and Vesecky and Stewart [22].

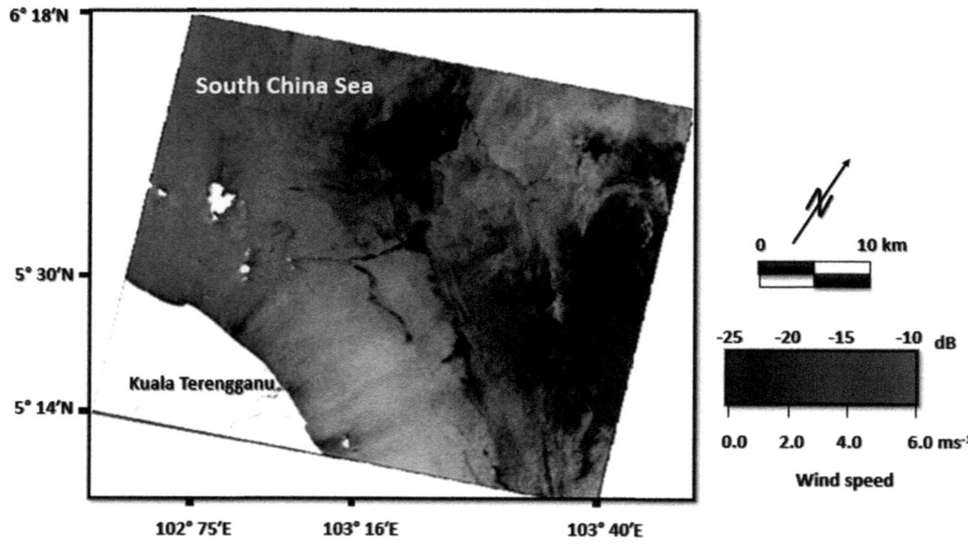

Figure 10.9: Backscatter of RADARSAT-1 standard-2, which was acquired on March 31st, 2005.

The initial stage of QSPSO operation in the SAR image is depicted in (Figure 10.10) with a maximum particle size of more than 2000 of 70 iterations and a root mean square error (RMSE) of ±45. The preliminary QSPSO result exposes some details regarding the pixel brightness and dark variation patterns. As a result, the essential details in the identification of the shoreline in the SAR image are revealed.

The subsequent phase delineates the process of converting a SAR image into qubits. In line with this approach, the exploration of internal waves' topology is translated into qubits, represented as |0> and |1>. This transformation is achieved with a maximum particle size of 3000 for 100 iterations, yielding an RMSE of ±39 (Figure 10.11). This methodology ensures the creation of an optimally fitting land mask.

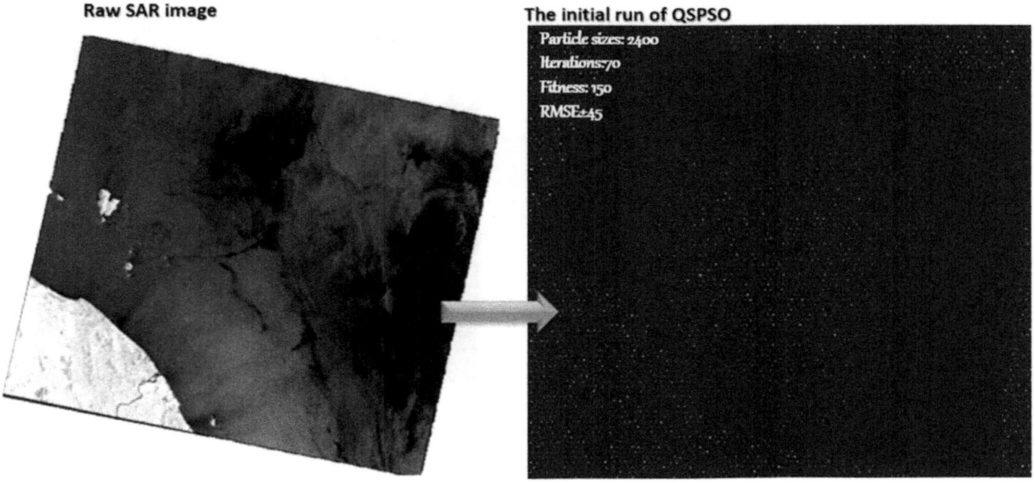

Figure 10.10: The initial run of QSPSO algorithm.

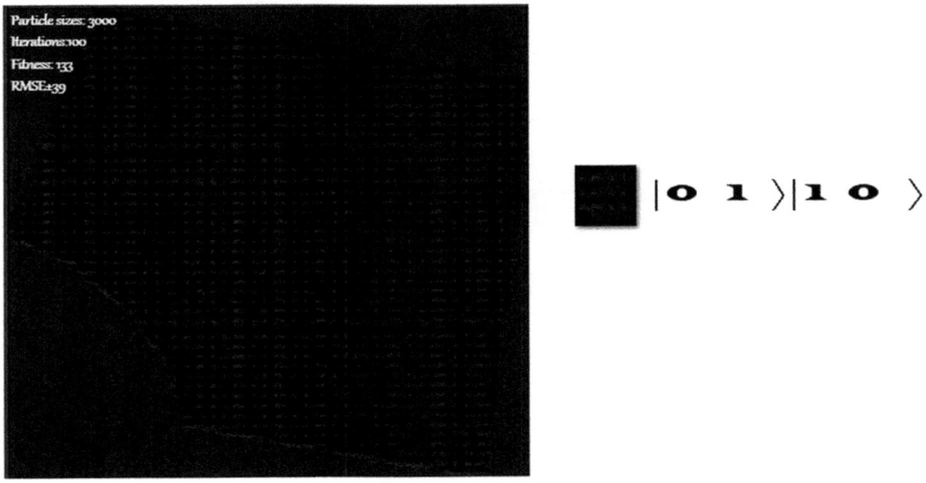

Figure 10.11: QSPSO turns the SAR image into the qubit topology.

The QSPSO algorithm efficiently identifies and improves the representation of atmospheric gravity wave topologies in SAR images. A well-organized depiction of atmospheric gravity wave morphology structures is precisely achieved in SAR images by increasing the number of particle swarms of 2650 particle sizes. This accomplishment is supported by the best fitness function of 100 and an RMSE of ±0.22. Consequently, utilizing 3280 particle sizes, along with the best fitness function of 120 and an RMSE of ±0.14, ensures the successful automatic segmentation of atmospheric gravity waves (Figures 10.12).

Similarly, ALOSPALSAR data in L-band HH polarization were collected along the coastal waters of Sarawak, Malaysia, in December 2010, investigating the occurrences of atmospheric gravity waves approximately 70 km away from the coast (Figure 10.13). In contrast, HV polarization is unable to capture the presence of AGW as effectively compared to HH polarization (Figure 10.14).

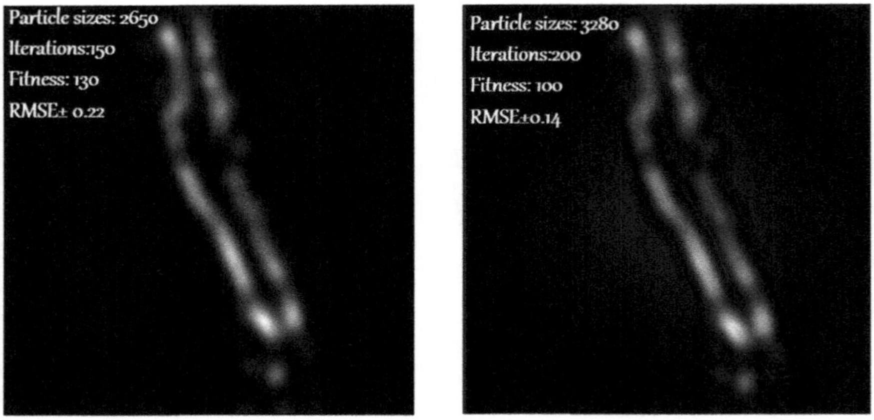

Figure 10.12: Precise automatic detection of atmospheric gravity waves using QSPSO algorithm..

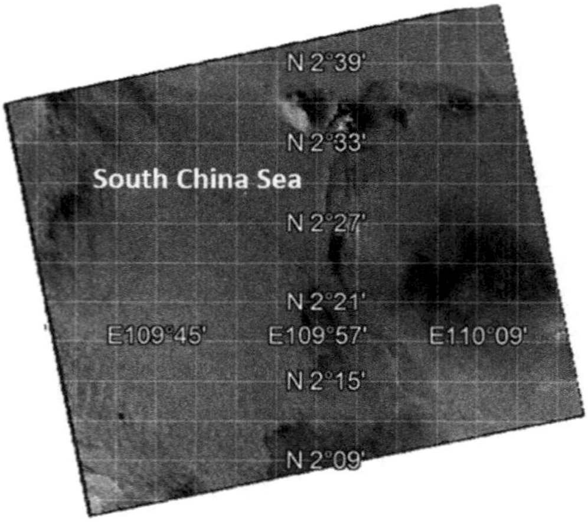

Figure 10.13: A clear occurrence of atmospheric gravity waves in AlosPALSAR LHH-band.

Figure.10.14: Absent of atmospheric gravity waves footprint signatures in AlosPALSAR LHV-band.

The difference in polarization effectiveness in capturing the presence of Atmospheric Gravity Waves (AGW) between HV (Horizontal-Vertical) and HH (Horizontal-Horizontal) polarizations can be attributed to the sensitivity of radar signals to the orientation and characteristics of surface features. In this view, HH polarization involves transmitting and receiving signals with the same horizontal orientation. This configuration is often more sensitive to surface roughness and can provide better imaging of certain types of features, including AGW. In contrast, HV polarization transmits horizontally and receives vertically, which may not be as effective in capturing the specific characteristics associated with AGW [19–22].

Therefore, AGW can induce changes in the surface roughness of the ocean, affecting the backscattered radar signal. HH polarization might be more adept at capturing these changes and revealing the presence of AGW in SAR images due to its sensitivity to variations in surface roughness. HV polarization, on the other hand, may not be as responsive to the specific characteristics of AGW, leading to reduced effectiveness in detecting and imaging these atmospheric phenomena.

The QSPSO (Quantum Particle Swarm Optimization) algorithm plays a crucial role in automatically identifying Atmospheric Gravity Waves (AGW) within a given environment. In the context of the study, the algorithm utilizes 3000 particle sizes, with a best fitness function of 100 and an RMSE (Root Mean Square Error) of ±0.132, as depicted in Figure 10.15.

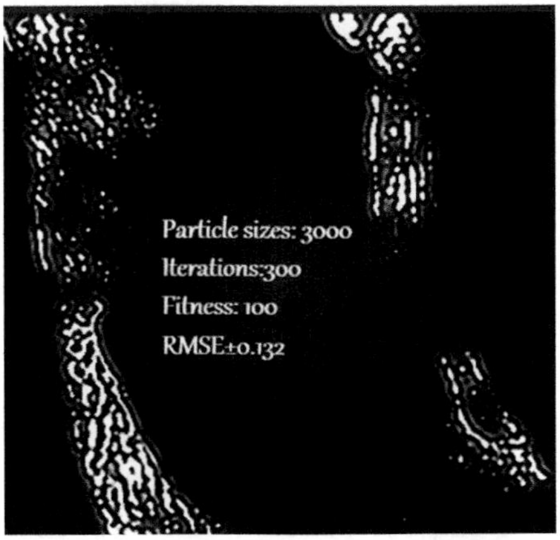

Figure 10.15: Automatic detection of atmospheric gravity waves in ALOSPALSAR L-HH band using QSPSO algorithm.

The effectiveness and performance of the QSPSO algorithm are influenced by the physical parameters and errors associated with the complexity of the surrounding environment where AGW is present. The algorithm is adaptive, and its ability to accurately detect AGW is contingent upon the characteristics of the environmental conditions.In the first scenario, where AGW is associated with other complicated features such as internal wave occurrences, the QSPSO algorithm must navigate through a more intricate and diverse environment. The presence of additional features can introduce complexities and challenges for the algorithm in distinguishing AGW from other phenomena [21, 31, 35, 38].

Conversely, in the second scenario where AGW is surrounded by a homogeneous sea environment, the task becomes relatively more straightforward. The algorithm operates in a more uniform and predictable setting, making it easier to isolate and identify the distinctive characteristics of AGW without the interference of diverse features. Therefore, the QSPSO algorithm's performance is context-dependent, with its efficacy varying based on the environmental complexity. The algorithm showcases its adaptability by successfully identifying AGW in different scenarios, demonstrating its potential for automatic detection and characterization of atmospheric phenomena in diverse environments.

Recently, Sentinel-1A satellite data in the C-VV band, acquired on November 23, 2023, along the coastal waters of Pattani, Thailand, reveals the presence of atmospheric gravity waves (Figure 10.16). Similarly, the Quantum Particle Swarm Optimization (QSPSO) algorithm demonstrates its ability to automatically detect the edges and boundaries of atmospheric gravity waves in Sentinel-1A data. This accomplishment is achieved using 3400 particle sizes, an RMSE of 0.133, 100 fitness, and 200 iterations (Figure 10.17). In this scenario, the sequences of the sharpest atmospheric gravity wave boundaries are successfully identified. These results align with the previously mentioned outcomes.

Figure 10.16: A clear occurrence of atmospheric gravity waves in C-VV band polarization of Sentinel-1A satellite data.

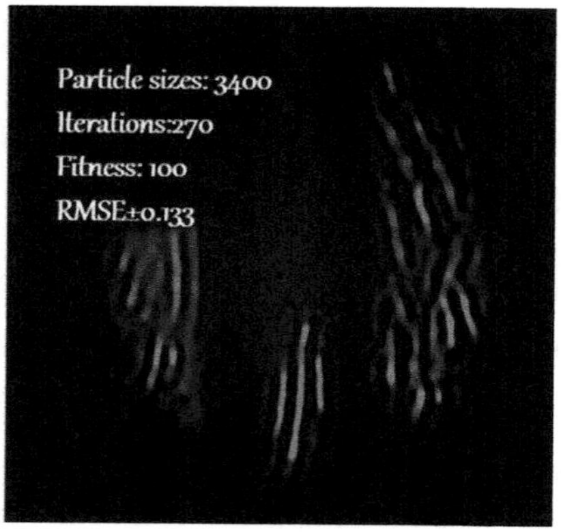

Figure 10.17: Automatic detection of atmospheric gravity waves in Sentinel-1A data C-VV band using QSPSO algorithm.

The exploration of atmospheric gravity waves in Malaysian coastal waters, particularly in the context of nonlinear features, remains an untouched area in scientific research. The author is at the forefront, pioneering the investigation of these unique atmospheric phenomena within both Malaysian coastal waters and Brunei coastal waters. The approach involves leveraging SAR (Synthetic Aperture Radar) technology and employing advanced quantum image processing algorithms. This pioneering initiative aims to fill the existing gap in understanding nonlinear atmospheric features, paving the way for new insights and discoveries in the field.

Moreover, in tropical regions characterized by high moisture intensity, the occurrence of atmospheric gravity waves is expected to be frequent. This is attributed to the rapid generation of dense cloud covers, as illustrated in Chapter 7, Figure 7.19, depicting the occurrence of street clouds observed in the coastal waters of Malaysia.

According to the above-mentioned consequences,and in discriminating between radar signatures of solitary Oceanic Internal Waves (OIWs) and Atmospheric Gravity Waves (AGWs), multiple criteria can be applied. Firstly, if a wave-like pattern on the radar image starts with a bright band in front, followed by a narrow dark line, it may suggest a nonlinear OIW packet. Conversely, when the front of a wave packet comprises a bright band bordered by two narrow dark bands, it is likely the radar signature of the leading solitary wave in an AGW packet (Figures 10.15 and 10.17). Additionally, the wave pattern being quasi-sinusoidal or having broader bright bands compared to the dark bands may indicate a weakly nonlinear AGW (Figure 10.14).

Additionally, the consideration of polarization diversity is crucial in this discrimination process. Specifically, if the modulation depth is more prominent in SAR images acquired at HH polarization compared to VV polarization, the pattern may lean towards indicating Oceanic Internal Waves (OIWs) [24]. Nevertheless, this finding diverges from Alpers et al.'s perspective [24], as Figures 10.14 to 10.17 demonstrate that HH polarization provides accurate detection of AGW occurrences. However, the applicability of this criterion is constrained by the limited availability of multi-polarization SAR data.

10.11 Why QSPSO Cable for Automatic Detection of Atmospheric Gravity Waves in SAR data?

The foundation of Quantum Particle Swarm Optimization (QSPSO) lies in the quantum Schrödinger equation, a common tool for explaining the behavior of particles in limited potential environments. Solitons emerge as prevalent solutions to the quantum nonlinear Schrödinger equation, showcasing localization, stability, reproduction, and reorganization properties even without a trapping potential. These characteristics make solitons promising candidates for applications in quantum information processing and communication, serving as robust qubits for quantum computing due to their resistance to decoherence and interaction capabilities without information loss as was investigated by Ding et al. [36]; Flori et al. [37]; and Zhou et al. [38].

The algorithm's mobility scenario incorporates the probability density function of quantum solitons, enhancing its global search capabilities. This motion scenario allows particles to explore a larger potential space, reducing the likelihood of becoming trapped in a global optimum. The QSPSO algorithm demonstrates overall performance in terms of high accuracy of 95% and reliability, surpassing other optimization algorithms in convergence speed and computational efficiency. These results position QSPSO as a promising solution for complex optimization problems across various domains. The multiSAR data, employing various bands and VV and HH polarizations, exhibits high accuracy, achieving a low error standard deviation of less than 0.20. This accuracy is realized under a wind speed of 6 m/s and a maximum iteration count of approximately 270. The high accuracy achieved in the multiSAR data, utilizing different bands and VV and HH polarizations, can be attributed to several factors. Firstly, the utilization of various bands and polarizations allows for a comprehensive analysis of the atmospheric conditions and their interaction with the radar signals. This diversity enables the system to capture a more detailed and nuanced representation of the studied area.

In the analysis of radar signatures, a substantial modulation depth in regions where wind speeds exceed 10 m/s is often indicative of Atmospheric Gravity Waves (AGWs) [22–24]. Interestingly, even when wind speeds approach 6 m/s, MultiSAR data can still effectively capture distinct features of AGWs (Figure 10.18). Furthermore, the specific selection of a 6 m/s wind speed is pivotal. Wind speed significantly influences the dynamics of atmospheric features, including gravity waves. Setting it at 6 m/s likely corresponds to conditions that optimize the interaction of radar signals, leading to more accurate measurements and detection of AGW.

Lastly, the high iteration count of approximately 270 indicates a thorough exploration of the optimization process, allowing the algorithm to fine-tune its parameters and converge toward an

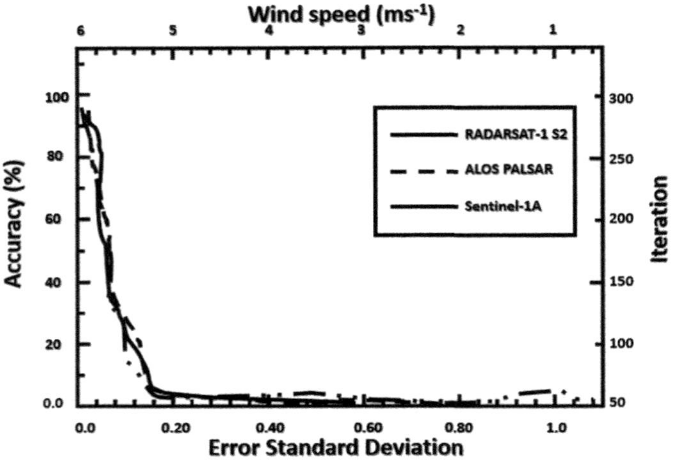

Figure 10.18: Accuracy of QSPSO under wind speed impacts.

optimal solution. This extensive iteration contributes to minimizing errors and enhancing the overall precision of the results by approximately 270.

In practice, QSPSO synchronizes the sequence sides of the atmospheric gravity wave edges in SAR data. The algorithm initiates by assigning particles random locations within an initialization pixel of the atmospheric gravity waves. To prevent premature particle departure during initial iterations, particle speeds, and positions can be adjusted. The iterative rationalization of particle positions continues until an end condition is met. The QSPSO algorithm efficiently constructs the discontinuity of atmospheric gravity wave edge pixels in a quality-ordered manner, crucial for handling speckle noises of the SAR data. Its core loop ensures the accurate identification of edge elements, contributing to the algorithm's effectiveness in processing SAR images [35–38].

This chapter introduces a ground breaking method utilizing Quantum Particle Swarm Optimization (QSPSO) for the automatic detection of atmospheric gravity waves in multiSAR data across different wavelength and polarization bands. Notably, it reveals the first documented occurrences of atmospheric gravity waves in the coastal waters of Malaysia, Pattani (Thailand), and Kelantan, situated in the South China Sea. The study also highlights the significant impact of the highest recorded wind speed at 6 m/s, showcasing how it enhances QSPSO's capability to automatically detect clear features of atmospheric gravity waves in multiSAR data. Overall, these findings suggest that QSPSO holds great promise as an automatic detection algorithm for atmospheric gravity waves in SAR data.

References

[1] Longuet-Higgins, M.S. (1964). Planetary waves on a rotating sphere. *Proceedings of the Royal Society A*, 279, 446–473.

[2] Toorn, Ramses van der (2019). Elementary properties of non-Linear Rossby-Haurwitz planetary waves revisited in terms of the underlying spherical symmetry. *AIMS Mathematics*, 4(2), 279–298.

[3] Hagan, M.E. and Forbes, J.M. (2003). Migrating and nonmigrating semidiurnal tides in the upper atmosphere excited by tropospheric latent heat release. *Journal of Geophysical Research*, 108(A2), 1062.

[4] Chapman, S. and Lindzen, R.S. (1970). *Atmospheric Tides*. Norwell, Massachusetts: D. Reidel.

[5] Holton, J. (1975). The dynamic meteorology of the stratosphere and mesosphere. *Meteorological Monographs*, 15(37). Massachusetts: American Meteorological Society.

[6] Oberheide, J. (2007). On large-scale wave coupling across the stratopause. Archived July 22, 2011, at the Wayback Machine. Appendix A2, pp. 113–117. University of Wuppertal.
[7] Longuet-Higgins, M.S. (1968). The eigenfunctions of Laplace's equations over a sphere. *Philosophical Transactions of the Royal Society*, London, A262, 511.
[8] Volland, H. (1988). *Atmospheric Tidal and Planetary Waves*. Dordrecht: Kluwer.
[9] Mitchell, R.N. and Kirscher, U. (2023). Mid-proterozoicday length stalled by tidal resonance. *Nature Geoscience*, 1–3.
[10] Revol, A., Bolmont, É., Tobie, G., Dumoulin, C., Musseau, Y., Mathis, S., ... and Brun, A.S. (2023). Spin evolution of Venus-like planets subjected to gravitational and thermal tides. arXiv preprint arXiv:2303.00084.
[11] Hines, C. (2002). Nonlinearities and linearities in internal gravity waves of the atmosphere and oceans. *Geophysical & Astrophysical Fluid Dynamics*, 96(1), 1–30.
[12] Madsen, P.A. and Schäffer, H.A. (1998). Higher–order Boussinesq–type equations for surface gravity waves: Derivation and analysis. Philosophical Transactions of the Royal Society of London. *Series A: Mathematical, Physical and Engineering Sciences*, 356(1749), 3123–3181.
[13] Elgar, S. and Guza, R.T. (1985). Shoaling gravity waves: Comparisons between field observations, linear theory, and a nonlinear model. *Journal of Fluid Mechanics*, 158, 47–70.
[14] Pierson Jr, W.J. (1955). Wind generated gravity waves. *Advances in Geophysics*,2, 93–178. Elsevier.
[15] Bona, Chen and Saut. (2002). Boussinesq equations and other systems for small-amplitude long waves in nonlinear dispersive media. I: Derivation and linear theory. *Journal of Nonlinear Science*, 12, 283–318.
[16] Hasselmann, K. (1962). On the non-linear energy transfer in a gravity-wave spectrum Part 1. General theory. *Journal of Fluid Mechanics*, 12(4), 481–500.
[17] Chunchuzov, I., Vachon, P.W. and Li, Xiaofeng. (2000).Analysis and modeling of atmospheric gravity waves observed in RADARSAT SAR images. *Remote Sensing of Environment*, 74.3, 343–361.
[18] Liu, S., Li, Z., Yang, X., Pichel William, G., Yu, Y., Zheng, Q. and Li, X. (2010). Atmospheric frontal gravity waves observed in satellite SAR images of the Bohai Sea and Huanghai Sea. *Acta Oceanologica Sinica*, 29, 35–43.
[19] Vachon, P.W., Johannessen, J.A. and Browne, D.P. (1995). ERS-1 SAR images of atmospheric gravity waves. *IEEE Transactions on Geoscience and Remote Sensing*, 33(4), 1014–1025.
[20] Li, X., Zheng, W., Yang, X., Zhang, J.A., Pichel, W.G. and Li, Z. (2013). Coexistence of atmospheric gravity waves and boundary layer rolls observed by SAR. *Journal of the Atmospheric Sciences*, 70(11), 3448–3459.
[21] Li, X. (2017). SAR imaging of internal gravity waves: From atmosphere to ocean. *In:* 2017 IEEE International Geoscience and Remote Sensing Symposium (IGARSS) (pp. 1499–1501). IEEE.
[22] Vesecky, J.F. and Stewart, R.H. (1982). The observation of ocean surface phenomena using imagery from the SEASAT synthetic aperture radar: An assessment. *Journal of Geophysical Research: Oceans*, 87(C5), 3397–3430.
[23] Cheng, C.M. and Alpers, W. (2010). Investigation of trapped atmospheric gravity waves over the South China Sea using Envisat synthetic aperture radar images. *International Journal of Remote Sensing*, 31(17–18), 4725–4742.
[24] Alpers, W. and Huang, W. (2010). On the discrimination of radar signatures of atmospheric gravity waves and oceanic internal waves on synthetic aperture radar images of the sea surface. *IEEE Transactions on Geoscience and Remote Sensing*, 49(3), 1114–1126.
[25] Li, X., Jackson, C.R. and Apel, J.R. (2004). Atmospheric vortex streets and gravity waves. *Synthetic Aperture Radar Marine User's Manual*, 341, 354.
[26] Midgley, J.E. and Liemohn, H.B. (1966). Gravity waves in a realistic atmosphere. *Journal of Geophysical Research*, 71(15), 3729–3748.
[27] Bresson, A., Bidel, Y., Bouyer, P., Leone, B., Murphy, E. and Silvestrin, P. (2006). Quantum mechanics for space applications. *Applied Physics B*, 84, 545–550.
[28] Selvam, A.M. (2005). A general systems theory for chaos, quantum mechanics and gravity for dynamical systems of all space-time scales. arXiv preprint physics/0503028.
[29] Greiner, W. (2000). *Quantum Mechanics: An Introduction* (Vol. 1). Springer Science & Business Media.
[30] Valentini, A. (1992). On the pilot-wave theory of classical, quantum and subquantum physics. Ph.D. Theses. International School for Advanced Studies (SISSA),Trieste,Italy.
[31] Duval, C., Gibbons, G. and Horváthy, P. (1991). Celestial mechanics, conformal structures, and gravitational waves. *Physical Review D*, 43(12), 3907.

[32] Fallahi, S. and Taghadosi, M. (2022). Quantum-behaved particle swarm optimization based on solitons. *Scientific Reports*, 12(1), 13977.

[33] Mulumba, D.M., Liu, J., Hao, J., Zheng, Y. and Liu, H. (2023). Application of an optimized PSO-BP neural network to the assessment and prediction of underground coal mine safety risk factors. *Applied Sciences*, 13(9), 5317.

[34] Yu, L., Ren, J. and Zhang, J. (2023). A quantum-based beetle swarm optimization algorithm for numerical optimization. *Applied Sciences*, 13(5), 3179.

[35] Fang, W., Sun, J., Ding, Y., Wu, X. and Xu, W. (2010). A review of quantum-behaved particle swarm optimization. *IETE Technical Review*, 27(4), 336–348.

[36] Ding, S., Zhang, Z., Sun, Y. and Shi, S. (2022). Multiple birth support vector machine based on dynamic quantum particle swarm optimization algorithm. *Neurocomputing*, 480, 146–156.

[37] Flori, A., Oulhadj, H. and Siarry, P. (2022). Quantum particle swarm optimization: An auto-adaptive PSO for local and global optimization. *Computational Optimization and Applications*, 82(2), 525–559.

[38] Zhou, N.R., Xia, S.H., Ma, Y. and Zhang, Y. (2022). Quantum particle swarm optimization algorithm with the truncated mean stabilization strategy. *Quantum Information Processing*, 21(2). https://doi.org/10.1007/s11128-021-03380-xferences

CHAPTER

11

Quantum Multiobjective Algorithm for Automatic Detection of Tropical Cyclone in Synthetic Aperture Radar Satellite Data

One of the main features of atmospheric turbulence is tropical cyclones. In this sense, high turbulence energy input by a tropical cyclone refers to the substantial amount of turbulent kinetic energy introduced into the atmosphere due to the cyclonic activity. Tropical cyclones, also known as hurricanes or typhoons depending on the region, are powerful storm systems characterized by low-pressure centers, organized thunderstorms, and strong winds. According to this perspective, an understanding of tropical cyclonic structures is required to understand the mechanism of how they input high energy turbulence which can cause the massive destruction of everything in the environment.

The novel approach; therefore, is to determine the physical characteristics of the tropical cyclone from SAR images and simulate the wind speed associated with the tropical cyclone. This chapter aims at delivering a novel algorithm for the automatic detection of tropical cyclone structure involving an accurate algorithm such as a multiobjective algorithm.

11.1 What are the Distinctions between Cyclones, Hurricanes, and Typhoons?

Before delving into discussions about tropical cyclones (Figure 11.1), it is essential to understand the distinctions between cyclones, hurricanes, and typhoons. The intriguing inquiry into the distinctions between cyclones, hurricanes, and typhoons reveals these weather phenomena as distinct entities, each marked by specific regional occurrences and rotation patterns. Cyclones, prevalent in the Indian Ocean and South Pacific, exhibit a counterclockwise rotation in the Southern Hemisphere. In the realm of tropical cyclones, the wind direction follows a distinct pattern—anti-clockwise or cyclonic in the northern hemisphere and clockwise or anti-cyclonic in the southern hemisphere (Figure 11.2). This behavior is attributed to the Coriolis Effect, a phenomenon resulting from the Earth's rotation. The Coriolis Effect causes the wind to deflect to the right side of its actual movement in the northern hemisphere and the left side in the southern hemisphere. It plays a pivotal role in shaping the movement of winds around low-pressure systems, inducing an anti-clockwise motion in the northern hemisphere and a clockwise motion in the southern hemisphere. Conversely, this directionality reverses around high-pressure systems, contributing to the intricate dynamics of atmospheric circulation [1–3].

Conversely, hurricanes (Figure 11.3) predominantly manifest in the North Atlantic, central and east North Pacific, and South Pacific, displaying a counterclockwise rotation in the Northern Hemisphere. Typhoons (Figure 11.4) found in the Northwest Pacific, share the counterclockwise rotation characteristic with hurricanes, both occurrences transpiring in the Northern Hemisphere. Therefore, a nuanced understanding of these meteorological phenomena unveils their unique

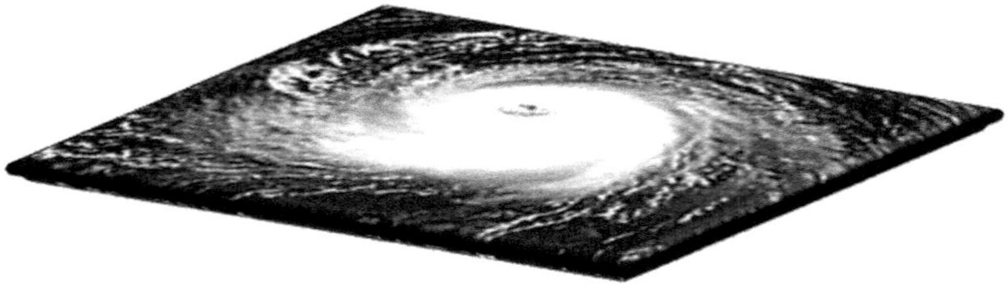

Figure 11.1: Tropical cyclonic.

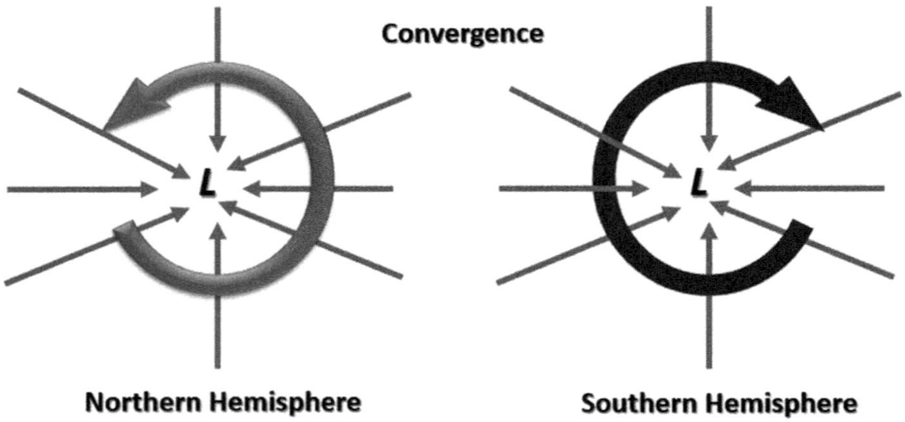

Figure 11.2: Cyclonic and anti-cyclonic in the northern hemisphere and southern hemisphere; respectively.

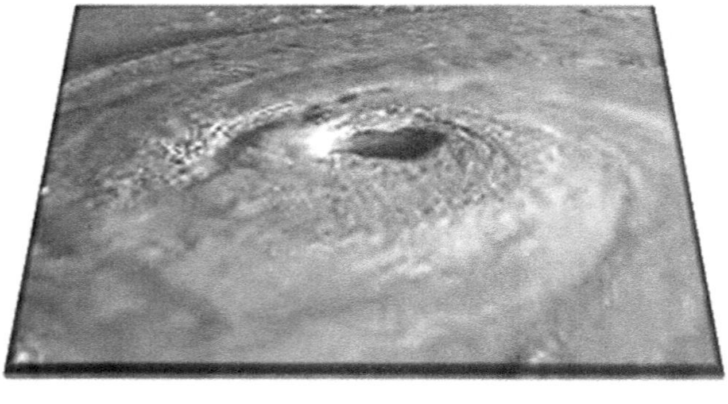

Figure 11.3: Hurricane occurrence.

Figure 11.4: Typhones mobility.

geographical manifestations and rotational dynamics. Consequently, the exploration of cyclones, hurricanes, and typhoons enriches our comprehension of the intricate dynamics shaping global weather patterns. In this perspective, Table 11.1 summarizes the differences between cyclones, hurricanes, and typhoons.

Table 11.1: Contrasts in physical characteristics across tropical cyclone types

Aspect	Cyclones	Hurricanes	Typhoons
Region	Indian Ocean and South Pacific	North Atlantic, central and east North Pacific, South Pacific	Northwest Pacific

11.2 What are Tropical Cyclones?

Is there a single terminology that can encompass cyclones, hurricanes, and typhoons? Indeed, there is a unified terminology that encompasses cyclones, hurricanes, and typhoons—they are collectively referred to as tropical cyclones. The term "tropical cyclone" serves as a comprehensive descriptor for these meteorological phenomena, irrespective of their specific regional occurrences or rotation patterns. Tropical cyclones represent a category of intense circular storms fueled by warm ocean waters, characterized by low atmospheric pressure and distinct wind patterns. Therefore, adopting the term "tropical cyclone" provides a cohesive and inclusive designation for these weather events, facilitating a clearer and more encompassing communication in the field of meteorology.

According to the above-mentioned perspective, the acronym "tropical" refers to the region from which these weather systems originate, primarily over tropical oceans. The term "cyclone" describes the winds' swirling close to a distinct core eye. Winds from the surface follow a counterclockwise path in the North Hemisphere and a clockwise path in the South Hemisphere; the Coriolis effect determines the direction of this circulation. Tropical cyclones usually form over large areas of warm water, using the energy released by water vapor evaporating off the ocean's surface.

Saturation-level cooling of wet air transforms this energy into clouds and precipitation. In this sense, the temperature fluctuations can cause cyclonic storms across the mid-latitude, such as nor'easters and European windstorms. Therefore, tropical storms usually have great diameters of 2000 km. Every year, these atmospheric events affect a variety of locations throughout the world, including Bangladesh, Australia, India, and the North American Gulf Coast [1, 4, 6].

11.3 Formation Mechanism

Tropical cyclones typically emerge during the summer months, although instances have been observed nearly every month across most tropical cyclone basins. The genesis of tropical cyclones on both sides of the Equator is generally associated with the Intertropical Convergence Zone, characterized by prevailing winds from the northeast or southeast. Within this expansive area of low pressure, atmospheric heating occurs over warm tropical oceans, prompting the ascent of air in distinct parcels and the formation of thundery showers. While these showers disperse relatively swiftly, they can coalesce into extensive clusters of thunderstorms. This amalgamation generates a flow of warm, moist air rapidly ascending, which initiates cyclonic rotation as it interacts with the Earth's rotational forces [1, 7].

Several pivotal factors have come into play in the intricate dance of tropical cyclone development. Foremost among these is the requirement for sea surface temperatures hovering around 27 °C (81 °F). This thermal backdrop sets the stage for atmospheric performance, working in tandem with the minimal vertical wind shear surrounding the system, which is an essential ingredient for the orchestration of thunderstorm development. In addition, this meteorological symphony is an element of atmospheric instability, which is a key player in the evolving composition. Tropical cyclones are dynamic and intricately structured, partly because of the rise of warm, humid air in the lower to middle troposphere, which produces thunderous showers [1, 8].

Moreover, the stage is further set by the involvement of a sufficient Coriolis force, a celestial choreographer that orchestrates the formation of a low-pressure center. This, coupled with the presence of a pre-existing low-level focus or disturbance, completes the ensemble, bringing the elements into a harmonious convergence. The intensity of these tropical cyclones unfolds as a nuanced narrative intricately woven into the tapestry of their journey. Water temperatures along their path and upper-level divergence emerged as pivotal plot points, shaping the intensity trajectory [1, 4, 8, 11]. However, constraints exist in all narratives. A limitation on tropical cyclone intensity was unveiled, and a narrative arc was strongly influenced by the factors previously elucidated.

From a global perspective, the meteorological stage has an annual performance, with an average of 86 tropical cyclones attaining tropical storm intensity. However, within this meteorological ensemble, 47 cyclones ascended to strengths exceeding 119 km/h (74 mph), which is a testament to the dynamic interplay between atmospheric elements [5–9]. Further elevating this climatic drama, 20 cyclones ascended to the grandeur of intense tropical cyclones, achieving at least Category 3 intensity on the Saffir–Simpson scale—a climax in the meteorological storyline [6, 9, 11]. Thus, within the nuanced choreography of tropical cyclones, each factor and its interplay contribute to the captivating tale of their development and intensity [1, 8, 10]. From this standpoint, the pivotal question becomes: What is the primary technique for comprehending the intensity development of tropical cyclones?

11.4 What is Rapid intensification?

Rapid intensification refers to a meteorological phenomenon in which a tropical cyclone undergoes a substantial increase in its maximum sustained winds over a relatively short period. Specifically, for a tropical cyclone to be considered to have undergone rapid intensification, its maximum sustained winds must increase by at least 30 knots (about 35 miles per hour) within a 24-hour period. This phenomenon is a crucial aspect of tropical cyclone behavior and can have significant implications for forecasting and disaster preparedness. Rapid intensification events are often associated with specific environmental conditions, such as warm sea surface temperatures, low vertical wind shear, and high levels of atmospheric moisture. These conditions provide the necessary fuel for the storm to strengthen rapidly [11, 13].

The ability to accurately predict and understand rapid intensification events is of great importance for coastal communities and emergency management, as it can result in a sudden and significant escalation of the storm's destructive potential. Forecasters use advanced models and satellite observations to monitor and predict these intensification events, enhancing our ability to respond effectively to the impacts of tropical cyclones [12–14].

Therefore, rapid intensification in tropical cyclones is a complex process influenced by various atmospheric and oceanic factors. While the exact mechanisms can vary, a general understanding involves the interaction of several key elements [13–15]:

1. **Warm Sea Surface Temperatures (SST):** Rapid intensification often occurs over warm ocean waters (typically above 26.5°C or 80°F). Warm sea surface temperatures provide the energy needed for the storm to strengthen. The process involves the transfer of heat from the ocean surface to the atmosphere through evaporation. In this understanding, the process of rapid intensification often involves a feedback loop, where warm sea surface temperatures lead to increased evaporation and atmospheric moisture. This, in turn, enhances convection and latent heat release, further intensifying the cyclone.
2. **Deep Warm Ocean Layer:** Rapid intensification is more likely when warm sea surface temperatures extend to a considerable depth beneath the ocean surface. A deep layer of warm water allows for sustained energy transfer to the atmosphere, supporting the storm's development.
3. **Low Vertical Wind Shear:** Vertical wind shear refers to the change in wind speed and direction with altitude. Low vertical wind shear is conducive to rapid intensification because it

allows the storm to remain vertically aligned. Strong wind shear can disrupt the organization of the storm and hinder intensification.

4. **High Atmospheric Moisture:** A high level of moisture in the lower to middle troposphere is essential for the development of deep convection and thunderstorms within the cyclone. This moisture, when lifted and condensed, releases latent heat, further fueling the storm's intensification.
5. **Coriolis Effect:** A tropical cyclone requires the Coriolis effect to develop and intensify. The Coriolis effect, caused by the rotation of the Earth, contributes to the cyclonic rotation of the storm. Rapid intensification is facilitated when the storm's core aligns with the Coriolis force, promoting a well-defined center.
6. **Pre-existing Disturbance:** Often, rapid intensification is associated with a pre-existing disturbance or organized weather system. These disturbances can provide a starting point for cyclone development, especially when they move into favorable environmental conditions.

The interaction of these factors in a favorable environment creates a feedback loop: warm sea surface temperatures lead to increased evaporation and moisture in the atmosphere, which, in turn, enhances convection and latent heat release, further intensifying the cyclone. Forecasting rapid intensification remains a challenge, but advances in satellite technology and numerical weather prediction models have improved our ability to monitor and predict these events.

According to the above perspective, the primary hallmark of rapid intensification is a significant surge in the storm's maximum sustained winds. To meet the criteria, these winds must increase by at least 30 knots (about 35 miles per hour) within a 24-hour timeframe. In this view, rapid intensification occurs over a relatively brief period, typically within a 24-hour window. This rapid strengthening distinguishes it from the more gradual intensification observed in standard cyclone development [12, 15].

11.5 Tropical Cyclone Anatomy: Unveiling the Inner Workings of These Powerful Storms

The lifecycle stages of a tropical cyclone involve (i) the formative phase; (ii) the developing or original phase (Figure 11.5); (iii) the mature phase (Figure 11.6); and (iv) the dissipating phase (Figure 11.7). In the initial stage of the formative phase, within the proximity of the Intertropical Convergence Zone (ITCZ), there exists preliminary cyclonic vorticity. The creation process involves the assistance of an easterly wave, fostering a feeble 'cold core' low. This low exhibits relatively cooler air at lower altitudes and warmer air aloft. The induction of cyclonic vorticity results in horizontal inflow, prompting the ascent of air through dry adiabatic processes. Visualized through satellite imagery, this phase unveils spiraling clouds with intermittent open spaces (Figure 11.8).

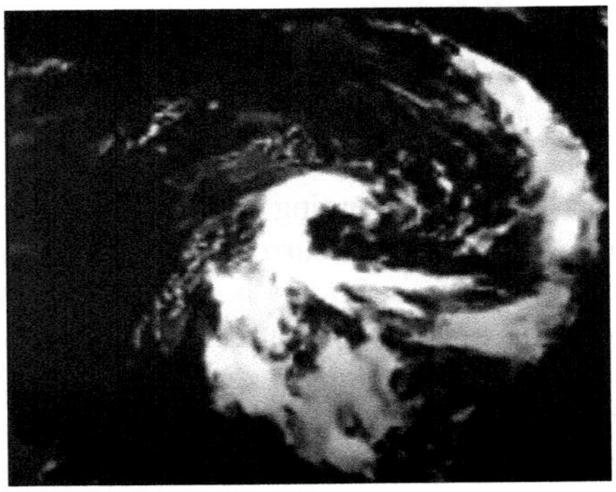

Figure 11.5: The developing or original phase of tropical cyclonic formation.

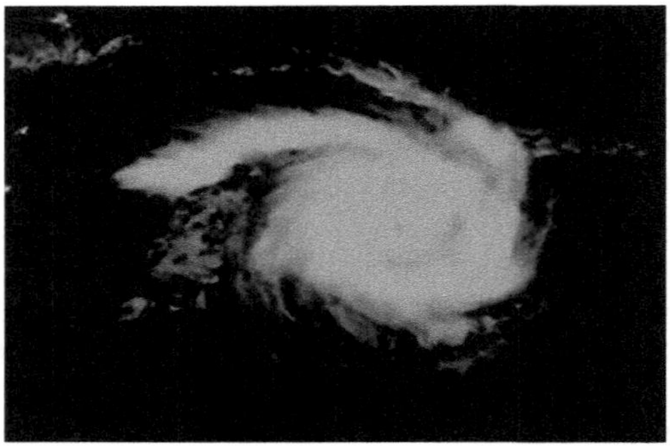

Figure 11.6: Mature formation phase of tropical cyclonic.

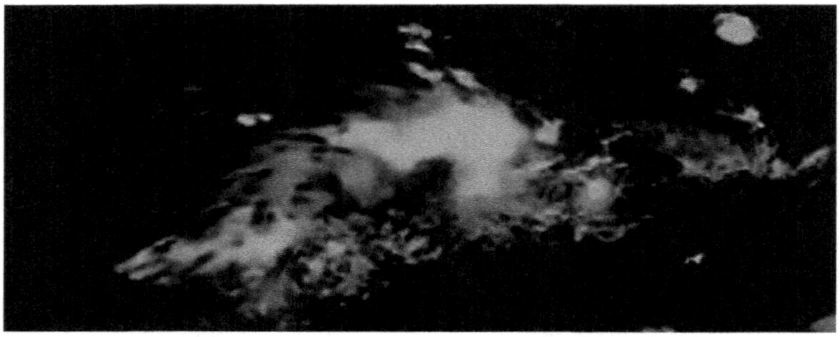

Figure 11.7: The dissipating phase of tropical cyclonic.

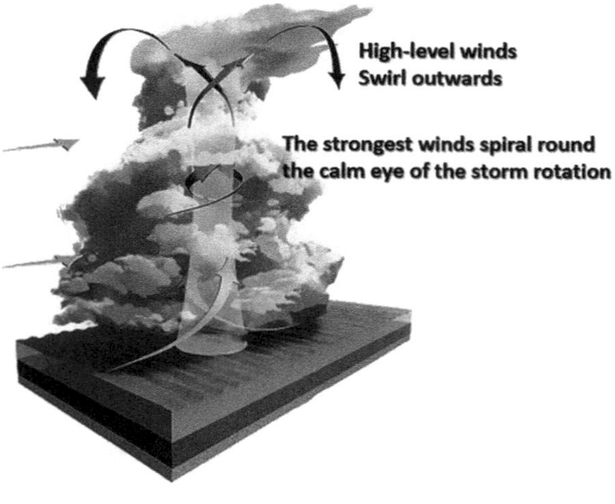

Figure 11.8: Spiraling cloud mobilities.

Within the developing phase, as the ascending air surpasses the condensation level, it reaches a state of saturation, becoming warmer than the surrounding environment. The significant release of latent heat during condensation becomes a primary heat source, setting forth convective instability within the lower and middle troposphere. The subsequent vertical acceleration induces inward spiraling of air, with the warm core descending to sea level. This transformative process converts the entire vortex into a warm core system [16–18].

Consequently, the developing phase turns into the mature phase. In this sense, the winds intensify, atmospheric pressure drops and substantial rainfall occurs at the surface. Initiating at the top of the warm core, a descending motion gradually penetrates downward, ultimately leading to the formation of the eye. The cyclonic vortex drifts within the prevailing air current, fostering interaction and potential modification.

Lastly, the dissipating phase takes place in which landfall and the deprivation of warm moist air, coupled with encounters with rough terrain, the cyclone reverts to a cold core system. It undergoes a weakening process, transforming into a depression that gradually dissipates. In certain instances, a cyclone that has dissipated over land may re-emerge into the ocean. For example, a Bay of Bengal storm might weaken after crossing the Indian peninsula but persist westward, eventually re-intensifying over the Arabian Sea. Alternatively, some storms may shift to higher latitudes, adopting characteristics of an extra-tropical cyclone. The overall lifespan of a tropical cyclone can vary, ranging from 5 to 15 days [1, 16, 19].

Consistent with the above-mentioned perspective, the key components of a tropical cyclone include (Figure 11.9) (i) eye and eyewall; (ii) spiral bands;(iii) central dense overcast (CDO); and (iv) moat. In the former stage, the eye is the center of the cyclone and is characterized by calm and clear weather. It is a region of low pressure and often has a circular or elliptical shape. The eye is typically surrounded by the eyewall. In this view, the eyewall is a ring of intense thunderstorms surrounding the eye. It is where the strongest winds and heaviest rainfall occur in a tropical cyclone. The eyewall is a region of high convective activity and is responsible for the cyclone's most severe weather [17–19].

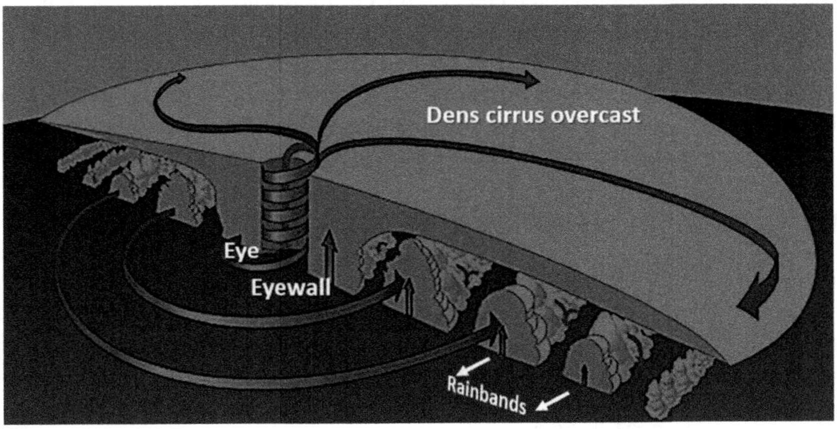

Figure 11.9: Structure components of tropical cyclones.

Consequently, in tropical cyclones, convection is intricately organized into extended, slender rain bands that align with the horizontal wind direction. These bands, often referred to as spiral bands, exhibit a distinctive spiraling pattern toward the center of the cyclone. Along these bands, there is a notable emphasis on low-level convergence and upper-level divergence. The width of these rain bands measures approximately 100 km near the outer periphery of the storm but gradually diminishes towards the center.

Severe convective rain rates, exceeding 3 cm/hr or more, are commonly associated with these spiral bands. While these bands move in tandem with the cyclone system, they also exhibit a slower circular motion around the cyclone's center, following the direction of the tangential wind. The persistence of spiral bands spans over a few days, yet the individual cells within these bands have relatively short-lived lifespans. These dynamic features contribute significantly to the overall structure and behavior of tropical cyclones, influencing precipitation patterns and the distribution of convective activity within the storm system [17,19].

Consequently, CDO is situated between the eyewall and the outer bands of a tropical cyclone. In this view, CDO manifests as a protective shield of cirrus clouds. This expansive cirrus cloud shield originates from the intense thunderstorms within the cyclone's eyewall and its associated rain bands. In the initial stages, particularly before the storm achieves hurricane strength of 64 knots, the CDO appears uniform. However, as the cyclone intensifies, an eye may become discernible in infrared or visible channel images. Circular formations within the CDO are indicative of favorable conditions, particularly in environments characterized by low vertical shear [16, 20].

The presence and characteristics of the CDO play a crucial role in understanding the development and strength of tropical cyclones. Observing changes in the CDO provides valuable insights into the storm's evolving structure and the potential for further intensification.

The term "moat"designates a distinct structure within tropical cyclones. Specifically, the moat refers to the region positioned between the eyewall and an outer rain band. In meteorological terms, the moat signifies an area characterized by relatively light rainfall. This meteorological feature, the moat, plays a role in delineating the inner and outer regions of a tropical cyclone. While the eyewall and outer rain bands are associated with more intense and concentrated precipitation, the moat serves as a transitional zone marked by lighter rainfall. Understanding the dynamics of the moat contributes to a comprehensive comprehension of the varied precipitation patterns within the cyclonic system, shedding light on the complexity of tropical cyclone structures.

According to the above-mentioned perspective, tropical cyclones are characterized by a distinct spiral pattern in their cloud structure. The spiral pattern is a result of the rotation of the storm around its center, known as the eye. The structure of a tropical cyclone includes features such as spiral rain bands extending outward from the center, converging toward the eye. These bands contribute to the overall spiral appearance of the cyclone. In this view, the spiral pattern is a fundamental aspect of the organization of tropical cyclones and is influenced by the Earth's rotation (Coriolis effect) and the dynamics of the storm's development. The rotation of the cyclone is evident in both satellite imagery and radar observations, highlighting the characteristic spiraling cloud structure associated with these powerful weather systems [17–20].

11.6 Mathematical Description of Tropical Cyclones

Mathematically describing tropical cyclones entails the formulation of a set of intricate equations that aim to encapsulate the dynamic and intricate interactions among a multitude of atmospheric and oceanic factors. These equations serve as a mathematical framework, allowing scientists and meteorologists to model and understand the complex behaviors exhibited by tropical cyclones.

11.6.1 Conservation of Mass

The conservation of mass equation is a fundamental component of atmospheric dynamics. It expresses that the change in air density over time, plus the divergence of the product of density and wind vector, is zero. This equation highlights the essential concept that mass is conserved within the atmospheric system.

$$\frac{\partial \rho}{\partial t} + \nabla \cdot (\rho V) = 0 \tag{11.1}$$

This equation represents the conservation of mass, where ρ is the air density, t is time, V is the wind vector, and ∇ is the del operator. At its core, the equation reflects the principle of mass conservation, asserting that the total mass within a given volume of air remains constant over time. Mathematically, this is expressed as the divergence of the mass flux being equal to the rate of change of density concerning time. Therefore, the term $\nabla \cdot (\rho V)$ represents the divergence of the mass flux. In the context of tropical cyclones, this divergence captures the changes in airflow patterns associated with the cyclone's formation. As air converges toward the low-pressure center of the developing cyclone, the divergence term accounts for the increase in air density. In the specific context of tropical cyclones, the equation implies that there is a converging flow of air towards the center of low pressure, signifying inflow. Simultaneously, there is an outflow of air at higher altitudes, ensuring that the total mass within the system remains constant [20–22].

As tropical cyclones form, the conservation of mass is integral to understanding how the system evolves. Warm air over the ocean surface rises, creating a low-pressure area. The equation encapsulates the inflow of surrounding air to fill the low-pressure void, contributing to the formation and intensification of the cyclone. Needless to say, the conservation of mass equation encapsulates the core principle that the mass of air within a defined region remains constant over time. In the context of tropical cyclones, it helps elucidate the intricate dynamics associated with the inflow and outflow of air during the formation and evolution of these powerful weather systems [20, 22].

11.6.2 Momentum Equation

The momentum equation accounts for the conservation of momentum in the atmosphere. It incorporates factors such as pressure gradients, the Coriolis effect, gravitational effects, and external forces. This equation is crucial for understanding how the winds within a tropical cyclone evolve.

$$\frac{\partial \vec{\mathbf{V}}}{\partial t} + (\vec{\mathbf{V}} \cdot \nabla)\vec{\mathbf{V}} = -\frac{1}{\rho}\nabla p - 2\vec{\Omega} \times \vec{\mathbf{V}} - \nabla\Phi + \vec{\mathbf{F}} \tag{11.2}$$

This term $\frac{\partial \vec{\mathbf{V}}}{\partial t}$ represents the rate of change of wind velocity (V) for time (t). In the context of tropical cyclones, it accounts for how the wind speed and direction evolve. Therefore, $(\vec{\mathbf{V}} \cdot \nabla)\vec{\mathbf{V}}$ known as the advection term, describes the advection or transport of momentum by the wind. It signifies how the wind carries its momentum as it moves through space. Consequently, The negative gradient of pressure (∇p) divided by air density (ρ) represents the pressure gradient force. In tropical cyclones, the low-pressure center is associated with a pressure gradient that influences the wind flow.The convergence of air at the surface results in the formation of a low-pressure center. The conservation of mass implies that air converging at the surface must ascend, and this ascending air further lowers the pressure at the center of the disturbance.

In this regard, $2\vec{\Omega} \times \vec{\mathbf{V}}$ accounts for the Coriolis effect, where Ω is the Earth's angular velocity. The cross product $\Omega \times V$ represents the Coriolis force, which deflects the wind direction due to the Earth's rotation. In other words, the Earth's rotation (Coriolis effect) influences the direction of the ascending air. In the Northern Hemisphere, the Coriolis effect causes the air to deflect to the right, leading to the formation of a cyclonic (counterclockwise) circulation around the low-pressure center. In the Southern Hemisphere, the deflection is to the left, resulting in an anticyclonic (clockwise) circulation [23–26]. Thus, the geopotential gradient term ($-\nabla\Phi$) accounts for the influence of gravity on vertical motion. This contributes to the development of the eyewall and the vertical structure of the cyclone. Lastly, the external force term (F) represents factors like friction and latent heat release, which can influence the wind dynamics within the developing cyclone. In this understanding, the momentum equation provides a comprehensive understanding of the dynamic forces influencing wind patterns in tropical cyclones. It considers the intricate interplay of pressure gradients, the Coriolis effect, gravitational effects, and external forces, offering insights into the complex mechanisms driving the formation and intensification of these powerful weather systems [20–23].

11.6.3 Conservation of Energy

The conservation of energy equation focuses on the temperature aspect of the atmosphere. It describes how temperature changes over time due to advection by winds. The equation also includes a term related to the heating rate, providing insight into the energy dynamics within the tropical cyclone.

$$\frac{\partial T}{\partial t} + \vec{\mathbf{V}} \cdot \nabla T = \frac{Q}{C_p} \tag{11.3}$$

Let's break down and expand the components of this equation to understand its implications for the conservation of energy in the context of atmospheric processes, such as those occurring in tropical cyclones. In this view, $\frac{\partial T}{\partial t}$ represents the rate of change of temperature (T) concerning time (t). It accounts for how the temperature at a particular location evolves periodically. The advection term ($V \cdot \nabla T$) describes the advection or transport of temperature by the wind (V). It signifies how the wind carries temperature with it as it moves through space. This term represents the heating rate (Q) divided by the specific heat at constant pressure (Cp). The specific heat at constant pressure is a measure of how much heat is needed to raise the temperature of a unit mass of air by one degree Celsius at constant pressure. Therefore, equation 11.3, in essence, captures the energy exchange processes within the atmosphere. Heating processes (positive Q) contribute to temperature increases while cooling processes (negative Q) lead to temperature decreases [21, 24, 27].

Understanding this conservation of energy equation is fundamental to grasping the thermodynamic processes that govern temperature changes within the atmosphere, contributing to the broader understanding of meteorological phenomena like tropical cyclones.

11.6.4 Thermodynamic Equation

The thermodynamic equation relates the rate of change of pressure to various atmospheric parameters, including temperature and the gas constant. This equation is vital for understanding the thermodynamic processes occurring within the tropical cyclone, such as how changes in pressure are linked to temperature variations.

$$\frac{Dp}{Dt} = -\rho \nabla \cdot \vec{\mathbf{V}} - \frac{\rho R T}{p} \frac{Dp}{Dt} \tag{11.4}$$

In equation 11.4, the $\frac{Dp}{Dt}$ term captures the overall changes in pressure over time, reflecting the combined impact of various atmospheric processes, including vertical motion, adiabatic heating or cooling, and temperature changes. The term $-\rho \nabla \cdot \vec{\mathbf{V}}$; therefore, accounts for changes in pressure due to adiabatic processes, where air parcels are compressed or expanded without heat exchange with the surroundings [25-27]. Divergence of velocity ($\nabla \cdot V$) indicates the spatial variations in the velocity field. Thus, the term $\frac{\rho R T}{p}$ captures the effects of temperature changes on pressure. As temperature increases, pressure tends to decrease, and vice versa. This term considers the relationship between pressure, temperature, and density in the ideal gas law. In this view, equation 11.4 reflects the dynamic processes occurring in the atmosphere, such as vertical motion, adiabatic cooling or heating during ascent or descent, and the impact of temperature changes on pressure. Needless to say, the thermodynamic equation provides valuable insights into how pressure changes over time, considering the interplay of vertical motion, adiabatic processes, and temperature variations within the atmosphere [20, 22, 25, 27].

These equations collectively form part of the Navier-Stokes equations, which govern the fluid dynamics of the Earth's atmosphere. The complexity arises from the dynamic interactions and feedback loops inherent in tropical cyclones. These storms involve the exchange of energy between the ocean and the atmosphere, convective processes, and the influence of factors like the Earth's rotation.

11.7 How Does SAR Imagine Tropical Cyclone Pattern?

To convey the advantages of SAR over optical satellite sensors in observing the 2-D sea surface wind field, the following statement can be made: Compared with optical satellite sensors, SAR holds distinct advantages in capturing the two-dimensional sea surface wind field with high resolution and expansive spatial coverage, regardless of weather conditions. Hurricanes have been regularly observed since the first SAR satellite image was made available in 1978, however until recently, the coverage of SAR images over the entire hurricane system has been limited. Notably, the acquisition of large numbers of hurricane images by SARs such as RADARSAT, ENVISAT, and Sentinel-1 has significantly expanded our ability to study and monitor hurricanes [28–30].

In synthetic aperture radar (SAR) imaging, complex streak patterns, including those observed in tropical cyclones, are attributed to the interaction between radar waves and the Earth's surface characteristics. The surface elevation profile is denoted as $Z(x, y)$, where x and y are spatial coordinates. The deviation of this surface profile from a smooth surface is characterized by $H(x, y)$, representing the surface roughness.

In this sense, the SAR system transmits radar waves toward the Earth's surface, and the interaction with surface features, such as ocean waves in the case of tropical cyclones, results in variations in the backscattered radar signal. These variations are then captured by the SAR instrument, producing complex streak patterns in the imagery. The deviation $H(x, y)$ accounts for irregularities in the surface elevation, which may include features like ocean waves, wind patterns, and other surface structures. These deviations influence how radar waves interact with the surface, leading to the observed streak patterns in SAR images. One can represent this deviation as $H(x, y)$:

$$H(x, y) = Z(x, y) - Z_{\text{smooth}}(x, y) \tag{11.5}$$

The radar wave interaction with surface roughness can be characterized by the radar backscatter coefficient (σ_0), which is dependent on the specific characteristics of the surface. The radar equation establishes a relationship between the backscatter coefficient and surface roughness:

$$\sigma_0 = K \cdot \frac{\left|\int\int H(x,y)\, e^{i2\pi(f/c)\cdot R}\, dx\, dy\right|^2}{\int\int \left|H(x,y)\right|^2 dx\, dy} \tag{11.6}$$

In this equation σ_0 is the radar backscatter coefficient; $H(x, y)$ represents the surface roughness deviation; f is the radar frequency; R is the range. Therefore, K is a constant and c is the speed of light. This equation captures the intricate relationship between the spatial tropical cyclone modulation of surface roughness and the resulting variations in radar backscatter, ultimately leading to the observation of spiral patterns of tropical cyclones in SAR images [28, 30, 32]. The integration terms account for the spatial distribution and characteristics of surface properties, providing a mathematical representation of the complex interaction between radar waves and the cyclone's surface. The mathematical expression is:

$$\sigma_0(x, y) = A\cos(kx + \phi_s) = f(H_0 \cos(kx + \phi_s)) \tag{11.7}$$

here $\sigma_0(x, y)$ is the radar backscatter coefficient at coordinates (x, y); A is the amplitude; and k is the wave number. Therefore, ϕ_s is the phase shift, and H_0 represents the surface roughness deviation. This equation describes the radar backscatter coefficient as a function of the spatial modulation of surface roughness ($H_0 \cos(kx+\phi_s)$). The function f encapsulates the overall relationship, indicating that the radar backscatter is influenced by the modulated surface roughness of tropical cyclones.

In Synthetic Aperture Radar (SAR) images, therefore, the strong roughness associated with the high velocity of tropical cyclones leads to increased backscatter. The turbulent and rough surfaces within the cyclone, including its spiral bands and eyewall, result in a higher intensity of radar signals being scattered back to the sensor [30]. In this understanding, the surface wind field for an elliptical flow of tropical cyclone is:

$$V(r,\theta) = \begin{cases} V_{\max} \cdot \dfrac{r}{r_m(\theta)}, & \text{if } r \leq r_m(\theta) \\ V_{\max} \cdot \dfrac{r_m(\theta)^\alpha}{r}, & \text{if } r_m(\theta) < r \leq 150 \text{ km} \end{cases} \tag{11.8}$$

here $V_{\max}$ represents the maximum velocity; r is the radial distance; $r_m(\theta)$ is a function that likely depends on the angle θ; and α is a parameter. In this sense, the function is defined differently for two ranges of r, depending on whether r is less than or equal to $r_m(\theta)$ or falls within the range $r_m(\theta) < r \leq 150$ km. This expression appears to describe the radial wind velocity $V(r, \theta)$ in a tropical cyclone, where the velocity is a function of radial distance and angular direction. The function $r_m(\theta)$ likely represents the radius at which the maximum wind occurs, and α influences the decay of the velocity beyond that radius [28, 30]. In this regard, the radar backscatter is influenced by the surface roughness and the dielectric properties of the medium, which can be related to the wind-induced roughening of the sea surface in the case of a tropical cyclone. This can be expressed mathematically as:

$$\sigma_0(r,\theta) = a + b \cdot \log_{10}\left(V(r,\theta)\left(1-\frac{r}{R}\right)e^{\left(-\frac{r}{R}\right)}\right) + c \cdot \log_{10}^2\left(V(r,\theta)\left(1-\frac{r}{R}\right)e^{\left(-\frac{r}{R}\right)}\right) \tag{11.9}$$

here a, b, and c are empirical coefficients. However, the eye of a tropical cyclone appears dark in radar images. This darkness in the eye is primarily due to the presence of calmer and smoother conditions. The eye is characterized by relatively light winds, reduced convective activity, and a generally clear sky. These factors contribute to lower backscatter in radar images compared to the surrounding areas with higher wind speeds and more turbulent conditions.

Certainly, the complexity of radar interactions with tropical cyclones is well-captured by considering both the phase shift ($\varphi(\theta)$) and scattering amplitude ($F(\theta)$) as intricate functions. These functions are influenced by factors such as the cyclone's geometry and orientation. The interplay of these parameters is pivotal in shaping the radar wave interactions with the cyclone structure, thereby influencing the spatial distribution of radar return signals. Mathematically, the radar return signal (denoted as $S(\theta)$) can be expressed as:

$$S(\sigma_0) = \sigma_0(r,\theta)e^{i\cdot\phi(\theta)} \tag{11.10}$$

The backscatter probability spectrum, denoted as $P_S(f)$, can be obtained by combining the amplitude and phase probability spectra [30, 33]. Considering the amplitude and phase probability densities, the backscatter probability spectrum is given by:

$$P_S(f) = \int_0^\infty \int_{-\pi}^{\pi} P_{\sigma_0}(\sigma_0) \cdot P_\phi(\phi) \cdot \delta\left(f - \sigma_0(r,\theta) \cdot e^{i\cdot\phi}\right) d\sigma_0 \, d\phi \tag{11.11}$$

This equation represents the probability distribution of radar return signals as a function of frequency (f). It considers the statistical characteristics of both the amplitude and phase components, providing insights into the variability of backscatter signals from tropical cyclones. The Dirac delta function δ ensures that the integral is evaluated at the specific frequency corresponding to the radar return signal.

Therefore, there are two main approaches to understanding how rain affects the $P_S(f)$. Formal is attenuation and volumetric scattering. When microwaves are transferred through the atmosphere, raindrops cause attenuation and volumetric scattering. Raindrops add to the total radar cross-section as they scatter as they descend through the sky. Furthermore, rain modifies air conditions, which impacts microwave signal propagation.

Later is the modification of ocean surface irregularity. In this scenario, rainfall significantly affects the ocean's surface roughness. Raindrops hitting the water's surface produce splash products, such as vertical stalk creation, crater collapse, and concentric ring waves from the initial impact. Potential microwave radiation scatterers are these splash products. On the ocean's surface, rain also causes gravity waves to be attenuated. Rain-induced turbulence damping, temperature and salinity variations in the rain, nonlinear wave-wave interactions, and surface boundary condition modifications are some of the mechanisms responsible for this damping [30–33].

This study's composite model attempts to account for the impact of precipitation on the ocean's surface as well as the atmosphere. Raindrops have two effects on SAR observations that are taken into account: they weaken the $P_S(f)$ from the sea surface and cause volume backscatter in the atmosphere. This thorough method sheds light on the intricate relationships that exist between rain and SAR readings during storm conditions.

11.8 Quantum-Enhanced SAR Image Processing for Automated Tropical Cyclone Detection

In conventional image processing, data is commonly represented using bits, and binary units that can be in states of 0 or 1. Classical image processing algorithms manipulate these bits for various image operations. Quantum image processing introduces a significant difference by employing qubits instead of classical bits. Qubits, the quantum counterparts to classical bits, exist in a superposition of states. Unlike classical bits limited to states of 0 or 1, qubits can exist in a linear combination of these states, showcasing a unique property derived from quantum mechanics. This characteristic is frequently elucidated using a mathematical framework known as Hilbert space. In quantum image processing, the shift occurs in the utilization of qubits ($|\psi\rangle|\psi\rangle$) instead of classical bits. Qubits are quantum counterparts to classical bits but exist in a superposition of states. Unlike classical bits, which can only exist in a state of 00 or 11, qubits can exist in a linear combination of these states, expressed as:

$$|\psi\rangle = \alpha|0\rangle + \beta|1\rangle \tag{11.12}$$

here, α and β are complex numbers, and the probabilities of measuring $|0\rangle$ or $|1\rangle$ are given by $|\alpha|^2$ and $|\beta|^2$, respectively. This unique property is a consequence of quantum mechanics and is often described using a mathematical framework called Hilbert space. The linear superposition of quantum states in a Synthetic Aperture Radar (SAR) image can be expressed as:

$$|\Psi\rangle = \sum_i c_i |\psi_i\rangle \tag{11.13}$$

here, $|\psi_i\rangle$ represents individual quantum states, c_i are complex coefficients, and the sum extends over all relevant quantum states in the SAR image. This formulation allows for a comprehensive representation of the quantum information encoded in the SAR image through the superposition of its constituent states. In the context of radar systems, the representation of radar signals using quantum states involves encoding information about radar echoes into the quantum states of qubits. Quantum superposition, a fundamental quantum property, can enhance the sensitivity and resolution of radar systems. Unlike classical states, quantum states can exist in a linear combination of multiple states simultaneously, allowing for a more efficient and complex representation of radar signals.

The utilization of quantum superposition in radar systems has the potential to revolutionize signal processing capabilities. By incorporating coherent superposition, probability amplitudes can be manipulated coherently, introducing a degree of fuzziness to adapt to the dynamic nature of Synthetic Aperture Radar (SAR) data [34–37]. This adaptability can lead to more advanced and precise sensing technologies in radar applications.

$$\Psi|C_f\rangle = \sum_i f_i c_i |\phi_i\rangle \tag{11.14}$$

where f_i represents the fuzziness factor and C_f is coherent fuzziness due to tropical cyclone occurrences in SAR data. Consequently, the quantum interference is expressed through interference terms, introducing implicit parallelism:

$$\Psi \left| I_{\text{Interfered}} \right\rangle = \sum_{ij} I_{ij} f_i c_i \left| \phi_i \right\rangle + I_{ij} f_j c_j \left| \phi_i \right. \tag{11.15}$$

The interference factor I_{ij} between quantum phases i and j can be mathematically expressed using the inner product of the corresponding quantum states $|\psi_i\rangle$ and $|\psi_j\rangle$ [36-39]. The definition of the interference factor is given by:

$$I_{ij} = \langle \psi_i | \psi_j \rangle \tag{11.16}$$

here, $\langle \psi_i|$ represents the conjugate transpose (also known as the adjoint or Hermitian conjugate) of the quantum state $|\psi_i\rangle$. If $|\psi_i\rangle$ and $|\psi_j\rangle$ are represented as column vectors, the interference factor can be calculated using the dot product:

$$I_{ij} = \begin{bmatrix} \alpha_i^* & \beta_i^* \end{bmatrix} \begin{bmatrix} \alpha_j \\ \beta_j \end{bmatrix} \tag{11.17}$$

Equation 11.17 demonstrates that the complex coefficients α_i and β_i correspond to the quantum state $|\psi_i\rangle$, and α_j and β_j correspond to the quantum state $|\psi_j\rangle$. The asterisk (*) denotes the complex conjugate. This equation quantifies the degree of overlap or interference between the quantum states $|\psi_i\rangle$ and $|\psi_j\rangle$, providing a measure of their correlation in the context of the Quantum-Enhanced SAR Image Processing algorithm [34–36].

Therefore, the quantum state equations can distinguish between strongly scattered signals due to rough eddy and vorticity occurrences on the sea surface and weak ones that are scattered from a smooth sea surface state. This is because quantum states can provide more information about the radar signal, allowing for better differentiation between different types of scattering. Additionally, quantum sensors can be used to detect very weak radar signals that would otherwise be undetectable with traditional radar systems.

Equation 11.12 illustrates that the qubit is in a linear superposition of these base states, achieved by varying the values of α and β with rotation angles θ and ϕ. This process can be mathematically explored using the Bloch sphere, as depicted in Figure 11.10, and is expressed as follows:

$$I_{ij} = \langle \psi_i | \psi_j \rangle | \psi \rangle = \cos\left(\frac{\theta}{2}\right)|0\rangle + e^{i\phi} \quad \sin\left(\frac{\theta}{2}\right)|1\rangle \tag{11.18}$$

here, θ and ϕ are rotation angles, and the Bloch sphere provides a geometric representation of the qubit state (Figure 11.10). The superposition of base states is achieved by adjusting θ and ϕ, allowing for an adaptable representation of quantum information [35]. This mathematical formulation and geometric interpretation showcase the fundamental principles of qubits and their manipulation in quantum computing and quantum information processing [34–36]. The Bloch sphere provides a concise visual representation of the qubit's state and its evolution through quantum operations (Figure 11.11).

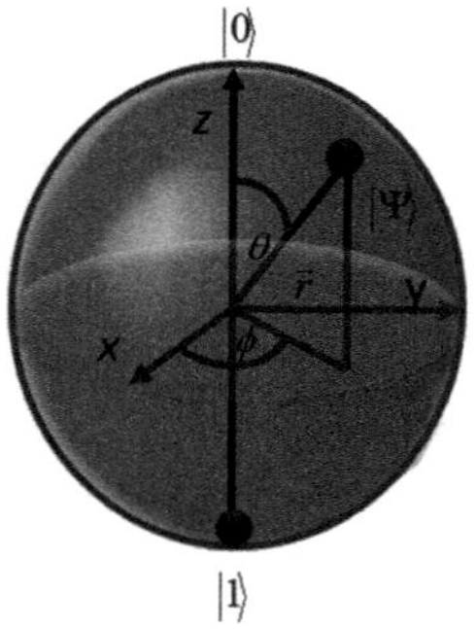

Figure 11.10: 3-D illustration of Bloch sphere.

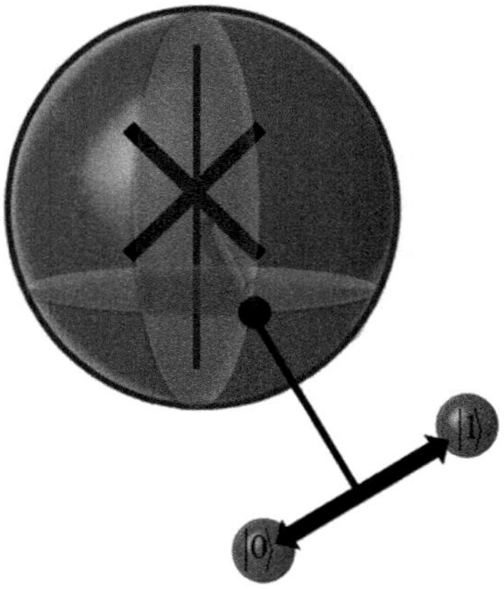

Figure 11.11: Sketch of a qubit.

In the context outlined above, the Bloch sphere plays a crucial role in distinguishing between qubits and classical bits. The qubit, characterized by spins both up and down and existing in superposition simultaneously in less than a second offers a distinct advantage over conventional bits. Unlike classical bits, which can only be either 0 or 1, the qubit's ability to exist in multiple states concurrently allows for faster computational processes [39].

The remaining question now is: How can qubits be leveraged to explore multiobjective evolution, leading to the development of a novel quantum multiobjective evolution (QME) for the automatic detection of turbulence structures based on tropical cyclone patterns in SAR images?

11.9 Quantum Multiobjective Evolutionary Algorithm (QMEA)

According to the above perspective, when the projection on a particular basis is recognized, the quantum system is considered to be collapsed. The measures or decoherence are other names for wave function $|\Psi\rangle$ collapsing. In other words, it should be $|\Psi\rangle = \alpha|0\rangle$ in the particular instance of the projection of $|\Psi\rangle$ on the $|0\rangle$ state. Consequently, on the state $|0\rangle$, the probability of the qubit collapsing would approach $|\dot{\alpha}^2|$. Accordingly, a novel qubit chromosome representation is adopted in quantum multiobjective evolution (QME) on the assumption and core principles of quantum computing, including such qubits and linear superposition [40]. The representation's distinguishing feature is its ability to represent any linear superposition of solutions. A quantum bit (qubit) is the smallest component of data that can be stored in a dual-state quantum computer. It can be in either the "1" or "0" states or any superposition of both $|0\rangle$ and $|1\rangle$. Equation 11.19 can be utilized to represent the state of a qubit. In this view, let us consider P is the turbulent flow pattern induced by tropical cyclones in SAR data, which has the wave function:

$$|\Psi(P)\rangle = \sum_{m \in P} \Psi(P)|P\rangle \tag{11.19}$$

The pattern-sorting process is depicted by m in equation 11.19, which includes the pattern P [39–43]. Accordingly, m patterns of the tropical cyclonic turbulence flow features are assumed as a training set D in the SAR data, which is described as:

$$D = \{\Psi(P)\} \tag{11.20}$$

According to this perspective, the coherence state of the SAR features $|\Psi_{SAR}\rangle$ is mathematically described as follows:

$$|\Psi_{SAR}\rangle = |x_1 x_2 \ldots\ldots, x_n, R_1 R_2, \ldots, R_{n-1}, C_1 C_2\rangle \tag{11.21}$$

where n is the total number of the SAR coherence features $x_1x_2.......,x_n$, which are stored in patterns P. Besides, the controlled register patterns are depicted by $R_1R_2,....,R_{n-1}$ accompanied by C_1C_2.

Equation 11.21 indicates the restoring P patterns as $x_1x_2.......,x_n$, with the total number n. Also, $R_1R_2,....,R_{n-1}$ as well as C_1C_2 are controlled register patterns. The Grover algorithm, therefore, involves three operations, which deliver the coherent state of the SAR features $|\psi_{SAR}\rangle$. In this understanding, exploring Grover algorithms, n qubits are vital to signify P patterns, as well as n+1 qubits as demanded in controlling qubits in QME algorithms. That delivers the entirety of 2n+1 qubits, which are imposed to differentiate m patterns. Therefore, the Grover algorithm also provides the quantity of 2n+1 encodes in embedding the QME network algorithms [44]. Consequently, the M matrix generation is the first operation, and it is cast as follows:

$$M = \begin{bmatrix} 1 & 0 & 0 & 0 \\ 0 & 1 & 0 & 0 \\ 0 & 0 & \sqrt{\dfrac{p-1}{p}} & \dfrac{\Psi(m)}{\sqrt{p}} \\ 0 & 0 & \dfrac{\Psi(m)}{\sqrt{p}} & \sqrt{\dfrac{p-1}{p}} \end{bmatrix} \tag{11.22}$$

The main question is now: how to achieve M? In the circumstance of $1 \leq P \leq m$, and 1<= P<=m, M can be accurately achieved. The flip transformation Ψ^0, which is identified by the matrix transformation, is then cast as the second giving operation:

$$\Psi^0 = \begin{bmatrix} \Psi & 0 \\ 0 & I \end{bmatrix} \tag{11.23}$$

In such cases of dual qubits, one for the tropical cyclonic turbulent flow and the other for the surrounding environment, one of them must regulate the flip transformation on the other, and equation 11.23 is exploited. In other words, let us assume that the SAR image, $\Psi^0|\text{turbulent}\rangle$ represents the state of the turbulent flow, and that $|\Psi \geq \beta|0\rangle + \alpha||1\rangle$ represents the state of the surrounding environment [40–44]. Using the M matrix, the pattern state detection of tropical cyclone turbulent flow is given by:

$$\Psi^0|\text{turbulent}\rangle\langle\text{non}-\text{turbulent}| = \begin{bmatrix} 0 & 1 & 0 & 0 & 1 & \beta \\ 1 & 0 & 0 & 0 & 1 & \alpha \\ 0 & 0 & 0 & 1 & 0 & \beta \\ 0 & 0 & 1 & 0 & 0 & \alpha \end{bmatrix} \tag{11.24}$$

Therefore, achieving the third operation requires a 3-qubit operation, which is a function of transformation Ψ^{00} [42–44]. In this perspective, a 3-qubit operation can be recognized as $|\Psi_1 \geq 1|0\rangle$, $|\Psi_2 \geq 1|1\rangle$, and $|\Psi_3 \geq \beta|0\rangle + \alpha|1\rangle$ in which the cyclonic turbulent flow levels, i.e., strong, medium, and light, are encoded, respectively, in the SAR data. Therefore, 3-qubit operation can be identified as:

$$\Psi^{00}|\Psi_1\Psi_2\Psi_3\rangle = \begin{bmatrix} \alpha \\ \beta \\ 0 \\ 0 \\ 0 \\ 0 \\ 0 \\ 0 \end{bmatrix} \tag{11.25}$$

where equation 11.25 represents the different levels of the cyclonic turbulent pattern flows $|\Psi_1\rangle$, $|\Psi_2\rangle$, and $|\Psi_3\rangle$, respectively, in the SAR image.

In line with Wang et al. [41], evolutionary computing with qubit representation has a better characteristic of population diversity than other representations since it can represent a linear superposition of state probabilities. In binary representation, at least eight strings (000), (001), (010), (011), (100), (101), (110), and (111) are required, whereas the above single qubit is enough to depict eight states (Figure 11.12). In this view, the quantum evolutionary algorithm (QMEA) and quantum genetic algorithms (QGA) operate efficiently on probability amplitudes of fundamental quantum states, whereas in the latter two (QEA), active crossovers and identified mutations are properly used to maintain population diversity [46].

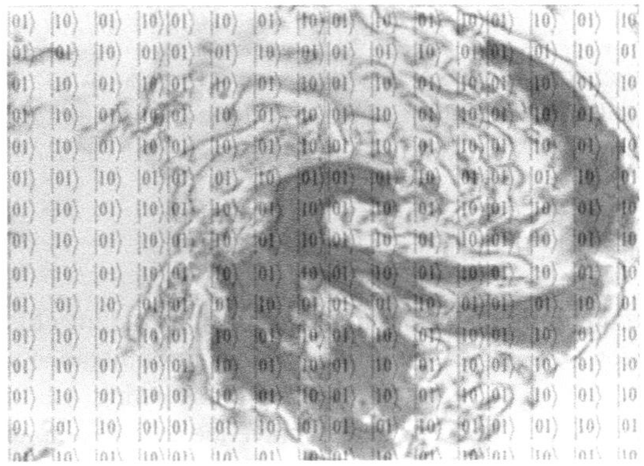

Figure 11.12: Qubits of cyclonic turbulent generation in SAR image.

The initial stage of QMEA is to generate accurate individual qubit populations of the cyclonic turbulence flow levels in an SAR image, which can be signified as $|Q(t)\rangle = \{|q_1^t\rangle, |q_2^t\rangle, \ldots, |q_n^t\rangle\}$ (Figure 11.13). In this sense, let us consider that

$$|q_1^t\rangle = \begin{Bmatrix} |\alpha_{i1}^t\rangle, |\alpha_{i2}^t\rangle, \ldots\ldots, |\alpha_{im}^t\rangle \\ |\beta_{i1}^t\rangle, |\beta_{i2}^t\rangle, \ldots\ldots\ldots, |\beta_{im}^t\rangle \end{Bmatrix}$$

where (i = 1, 2, ..., n) is a qubit individual, n is the size of the population, and m is the span of the individual.

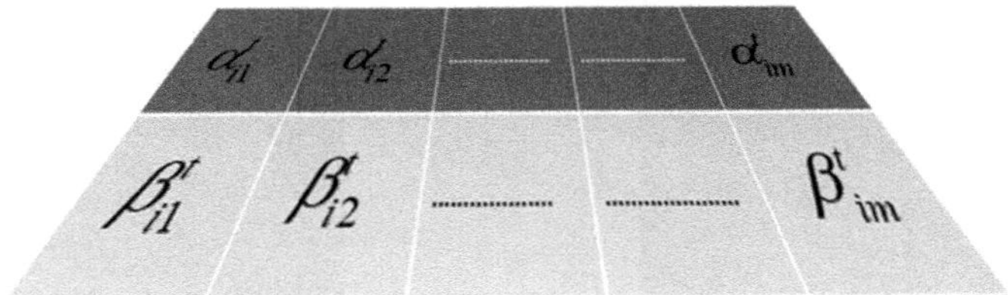

Figure 11.13: An array of one qubit q_1^t populations.

11.10 Generation of Quantum Turbulent Flow Population Pattern

Following the above, let us consider the constraint of the random qubit generation of the cyclonic turbulence flow as $Q|N_O\rangle \in |0,1\rangle$. In this, the random number $Q|N_O\rangle$ is confirmed as a turbulent flow if $Q|N_O\rangle > |\alpha_i^t|^2$ in which the equivalent qubit equals 1, and is dissimilar when q_n^t is equivalent to 0, and cyclonic turbulence does not exist in the SAR data. Consequently, the clustering centre point of the cyclonic turbulence is depicted by qubits $|1\rangle$ individually [43–46].

11.10.1 Quantum Fitness

The qubit chromosome form needs to be coded with the boundaries of the cyclonic turbulence footprint. Every qubit in the qubit chromosome $|q_m\rangle$ in this issue corresponds to a coefficient in the north-order surface fitting polynomial, which is given by:

$$f|i,j\rangle_q = |q_0\rangle + |q_1 i\rangle + |q_2 j\rangle + |q_3 i^2\rangle + |q_4 ij\rangle + |q_5 j^2\rangle + \ldots + |q_m j^n\rangle \tag{11.26}$$

The connectedness metric is then employed to assess fitness. It is conceivable to mathematically express the level of the qubit chromosomes that are signed from adjacent pixels in the SAR image and clustered as the turbulent flow as follows:

$$conn|q^t\rangle = \sum_{i=1}^{m}\left(\sum_{j=1}^{L}|q_{i,j}^t\rangle\right) \tag{11.27}$$

Equation 11.27 explains how to distinguish cyclonic turbulent clusters in the environment of several qubit chromosomes L of m size. In this particular circumstance, each 10 qubit chromosome for every pixel closest to pixel i and its adjoining jth is distinguished as the cyclonic turbulent patterns. In contrast, the absence of turbulent occurrence in SAR images and features belonging to the surrounding sea surface are recognized as well $|q_{i,j}^t\rangle = |j^{-1}\rangle$.

11.10.2 Quantum Mutation

The quantum mutation is carried out through the quantum rotation gate $G(\theta)$ with the assistance of the existing truly outstanding individual qubit, which behaves as a mutation operator in QEA (Figure 11.14):

$$G(\theta) = \begin{bmatrix} \cos\theta & -\sin(\theta) \\ \sin\theta & \cos\theta \end{bmatrix} \tag{11.28}$$

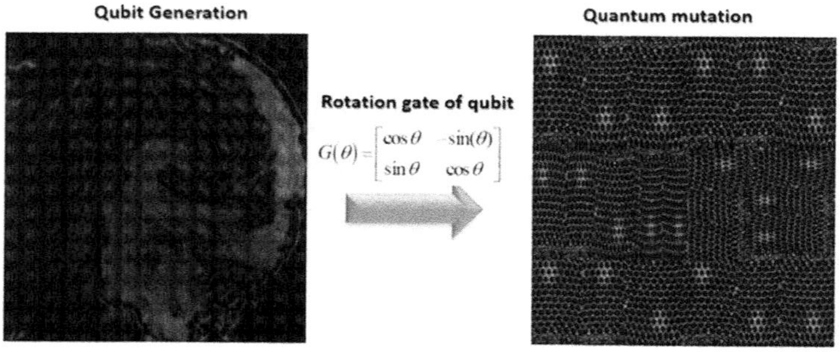

Figure 11.14: Quantum mutation of SAR image for turbulent detection.

Equation 11.28 demonstrates the quantum rotation gate for searching in the quantum pixel in a SAR image (Figure 11.15) as a function of rotation angle, which is cast as:

$$\theta = k \times f|\alpha_i, \beta_i\rangle \quad (11.29$$

In this case, the value of the coefficient k affects how quickly convergence arises. The number

k must be selected wisely. The search grid of the algorithm would be large if k is too large, which would cause the solutions to diverge or prematurely converge to a local optimum. Conversely, if k is too small, the search grid would also be small, which would induce the algorithm to potentially become stuck. Additionally, $f|\alpha_i, \beta_i\rangle$ maintains the direction of convergence of the search to a global optimum [41].

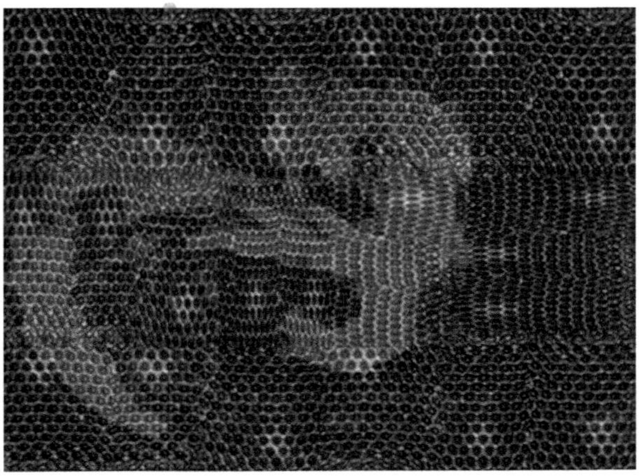

Figure 11.15: Quantum rotation gate for turbulent detection in SAR data.

The quantum gate determines whether the clustering of turbulence detection is a false alarm or a true clustering. Figure 4.15 reveals how a quantum rotation gate can be applied to exhibit the turbulent detection as bright patches for the initial moment.

The rules in the opposite direction can be provided by:

$$|q_i^{t+1}\rangle = G(t) \times |q_j^t\rangle \tag{11.30}$$

In this sense, the probability amplitude of the qubit is signified by an individual population at t-th creation as t is the evolutionary initiation at the quantum gate $G(t)$ [44–49].

11.11 Pareto Optimal Solution for Quantum Non-dominated Sort and Elitism (QNSGA-II)

In short, the second version of the "*Non-dominated Sorting Genetic Algorithm*" introduced by Prof. Kalyanmoy Deb for solving *non-convex* and *non-smooth* single and multi-objective optimization problems is then developed by Marghany [44] to deliver QNSGA-II. Let us assume that the qubit chromosome is $Q|q_1^t\rangle, \ldots\ldots\ldots\ldots\ldots, Q|q_N^t\rangle$ for population size number N. In this regard, each solution m would be compared with every other solution in the population to find if it is dominated. Therefore, $S_{Q|q_1^t\rangle}$ is considered as every quantum multiobjective solution that would be compared to other solutions to resolve the level of domination, i.e., initialize $S_{Q|q_1^t\rangle} = \Psi(Q|q_1^t\rangle)$ [50]. In this sense, the main features of QNSGA-II involve: (i) a sorting non-dominated procedure of qubit chromosome $Q|q_N^t\rangle$ where all the individuals are sorted according to the level of non-domination; (ii) it implements elitism which stores all non-dominated solutions, and hence enhancing convergence properties; (iii) it adopts a suitable automatic mechanics based on the crowding distance to guarantee diversity and spread of solutions; and (iv) constraints are implemented using a modified definition of dominance without the use of penalty functions [44,51]. To this end, the best classification technique of Deb et al. [50] can be examined in tropical cyclonic turbulence automatic detection as addressed in pseudo-code (Figure 11.16).

In main population *P*, every individual qubit $Q\left|q_N^t\right\rangle$ is validated:
Let F_1 be the first front,
Set $Q\left|q_1\right\rangle = 1$.
Individuals dominate $Q\left|q_1\right\rangle$ then $Q\left|q_1\right\rangle \in F_1$,
Set first front set $F_1\left(Q\left|q_1\right\rangle\right)$,
Update by adding $Q\left|q_1\right\rangle$ to front 1,
$F_1\left(Q\left|q_1\right\rangle\right) = F_1\left(Q\left|q_N^t\right\rangle\right) \cup Q\left|q_1\right\rangle$
Then $F_i\left(Q\left|q_N^t\right\rangle\right) \neq \Psi\left(Q\left|q_N^t\right\rangle\right)$
Hence set $Q(q_1)_{rank} = i + \left|1\right\rangle$. Update the set $Q\left|q_i^t\right\rangle$ with individual *m*, i.e., $Q\left|q_i^t\right\rangle = Q\left|q_i^t\right\rangle \cup m$.

- increment the front of one.
- For turbulence, now the set $Q\left|q_i^t\right\rangle$ is the next front,

$F_i(Q\left|q_i^t\right\rangle) = Q\left|q_i^t\right\rangle$.
Return
End

Figure 11.16: Pesudo-code for tropical cyclonic turbulence automatic detection.

Consequently, the quantum probability of the maximum sorting of the qubit chromosome is $P_{\max}(Q\left|q_i^t\right\rangle)$, which is also corresponds to quantum qubit chromosomes of probability amplitude of tropical cyclonic turbulence footprint in SAR data and should comply with:

1. The probability of turbulent flow pixels is signified as ($P_{\max}(Q\left|q_i^t\right\rangle)$): the discrepancy of maximum probability of the quantum qubit chromosomes containing the tropical cyclonic turbulence quantum $P_{\max}(Q\left|q_i^t\right\rangle) = \max\left\{P_1(Q\left|q_1^t\right\rangle), \ldots, P_k(Q\left|q_k^t\right\rangle)\right\}$, which designates the probability existence of the turbulence in Pareto Front j, $\forall j = 1, 2, \ldots, k$.
2. Each row and column in the SAR image involves the sum of the quantum probability of tropical cyclonic turbulence qubit chromosomes: $\sum P_i(Q\left|q_i^t\right\rangle)_i$.

According to the above, Pareto optimal solutions are operated to hold the perceptiveness of tropical cyclonic turbulence variety and its encompassing circumstances in SAR images. Therefore, computing the Pareto front grade of convergence would be a function of the following giving the Euclid distance:

$$P_f\left(P_i(Q\left|q_i^t\right\rangle)\right) = \sum_{\omega=1}^{\omega} \sqrt{\sum_{m=1}^{N} \min_{u=1,U}\left(P_{um}(Q(\left|q_{um}^t\right\rangle) - \left|a_{\omega m}\right\rangle\right)^2} \tag{11.31}$$

Therefore, equation 11.31 is applied to compute the Pareto front for automatic detection of the turbulence flows in SAR data based on the QMEA [44, 51]. In this understanding, *um* represents the number of Pareto front points, and *N* represents the number of optimized criteria for the automatic detection of turbulent flow in SAR data (Figure 11.17). Indeed, they are completely excluded from the possibility of finding quantum qubit chromosomes in tropical cyclonic turbulence flow.

In this comprehension, the quantity of points constituting the Pareto front, contingent upon the optimized criteria for autonomously discerning tropical cyclonic turbulence within SAR data,

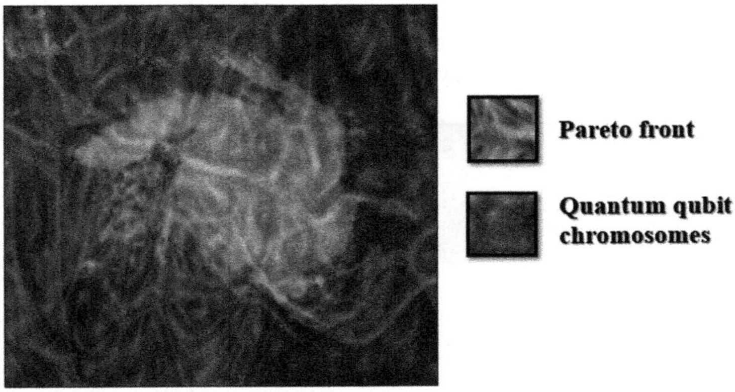

Figure 11.17: Pareto front curves engendered by quantum qubit chromosomes in SAR data.

adeptly traces the coherence of spiral pixels in SAR images. Significantly, these entities are unequivocally precluded from the prospect of encountering quantum qubit chromosomes within the cyclonic turbulence mobility. In the realm of multi-objective optimization quandaries, the Pareto front embodies a collection of optimal solutions strategically employed for delineating trade-offs among diverse objectives, emphasizing prioritization over mere accuracy assessment. Researchers can judiciously select the most optimal solution, grounded in a profound comprehension of the Pareto front.

11.12 Automated Identification of Cyclonic Turbulent Patterns in Synthetic Aperture Radar (SAR) Imagery

Satellite assessments of tropical cyclones routinely incorporate advanced technologies, encompassing high-resolution infrared and visible imagery. These images, acquired from geosynchronous orbit, exhibit remarkable detail with a pixel resolution of less than 1 kilometer. Additionally, observations from polar orbit observation satellites contribute images from low to moderate-resolution passive radiometers and active microwave scatterometers, where the former measure various forms of radiation (microwaves, visible light, and infrared) emitted or reflected by Earth, and the latter transmit microwave pulses to gauge the energy scattered back from the planet's surface.

In this understanding, conventional satellite observations utilizing visible, infrared, and microwave sources play a crucial role in tropical cyclone forecasting by offering insights into storm characteristics and the underlying atmospheric and oceanic factors influencing them. However, each of these methods has its constraints. Visible and infrared imagery primarily focuses on features at or near cloud tops. Microwave sensors, while capable of observing beneath the clouds, face limitations in horizontal resolution, preventing the capture of steep wind speed gradients around the storm's eye. To address these shortcomings, synthetic aperture radar emerges as an alternative sensor.

During the tracking of Super Typhoon Goni over the Pacific Ocean, the Radarsat-2 ScanSAR Wide mode captured critical data on October 30, 2020. This specific timeframe coincided with the storm's proximity to peak intensity, characterized by formidable wind speeds surpassing 155 knots. The utilization of ScanSAR Wide mode in this scenario underscores its efficacy in monitoring and documenting extreme weather events, providing valuable insights into the dynamics of powerful tropical cyclones. The data obtained during such instances contributes significantly to meteorological research and enhances our understanding of the behavior and impact of intense storms (Figure 11.18). Therefore, RADARSAT-2 SAR ScanSAR wide mode with HH polarization offers expansive coverage, encompassing a vast area of 500 kilometers. This wide coverage is complemented by an incidence angle range spanning from 20° to 49°, signifying the angles at which the radar beam interacts with the Earth's surface. The configuration employs a 4×4 arrangement of looks, and strategically chosen sub-images that play a crucial role in the processing of radar data.

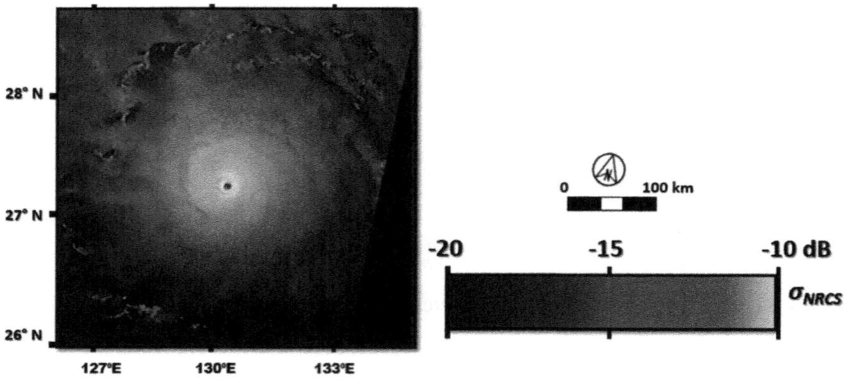

Figure 11.18: Typhoon Gani in RADARSAT-2 SAR imagery.

In the realm of radar imaging, the term "looks" refers to the subdivisions used in the creation of the final image. In this context, the 4×4 configuration indicates a structured approach to image processing. Moving to the pixel size, denoted as 100×100, this detail sheds light on the resolution of the images. Specifically, it implies that the images are constructed with pixels organized in a 100 by 100 grid.

The backscattered signals obtained through Synthetic Aperture Radar (SAR) directly capture numerous recognizable features of tropical storms, akin to those observed in visible or infrared images. These features include the eyewall, mesovortices (small circulations within the eyewall), boundary layer rolls, outflow boundaries, and rainbands (Figure 11.18).

The maximum normalized backscatter cross-section, indicated by a value of –10 dB, is observed prominently across the eyewall and mesovortices of the tropical cyclone. This measurement signifies a relatively strong intensity of backscattered signals in these regions. Notably, the eyewall, being a primary component of the cyclonic structure, exhibits the highest backscatter intensity.

In contrast, the center of the eyewall presents an interesting characteristic. Despite being a core element of the cyclonic structure, it paradoxically displays the lowest backscatter with a value of –20 dB. This lower backscatter at the eyewall's center may be attributed to specific conditions such as reduced turbulence or changes in surface roughness, leading to a decrease in the intensity of backscattered signals. The contrasting backscatter values within the eyewall and its center provide valuable insights into the complex dynamics and variations in scattering properties across different regions of the tropical cyclone.

Notably, SAR possesses a distinctive capability to directly measure the roughness of the ocean surface. This unique attribute enables the creation of high-resolution 2D maps depicting surface wind speed, significantly enhancing the accuracy of eye location estimates. It is worth mentioning that while outflow boundaries are evident in satellite imagery, they may not be as well-defined.

The veracity of this observation is further substantiated by the visible-light image captured by Japan's Himawari-8 satellite within a 500-meter resolution (Figure 11.19). In this sense, the Advanced Himawari Imager (AHI) aboard the Himawari-8 satellite is equipped with 16 channels, encompassing 3 visible bands, 3 near-infrared bands, and 10 infrared bands. These channels offer varying spatial resolutions of 0.5 km, 1 km, and 2.0 km, and observations are conducted at a frequency of 10 minutes [52]. For this investigation, the AHI visible bands covering the wavelength range from 0.47 μm to 0.64 μm were employed [53]. Notably, the reflectance percentage falls within the range of 0.8% to 1.0% (Figure 11.20). This observation is substantiated by the distinctive reflectance variations observed throughout the cyclone. It is noteworthy that the highest reflectance is observed away from the eyewall of the hurricane, while the lowest reflectance is observed away from the vortex of the cyclone. The spectral signature alone proves challenging for identifying the cyclone's structure.

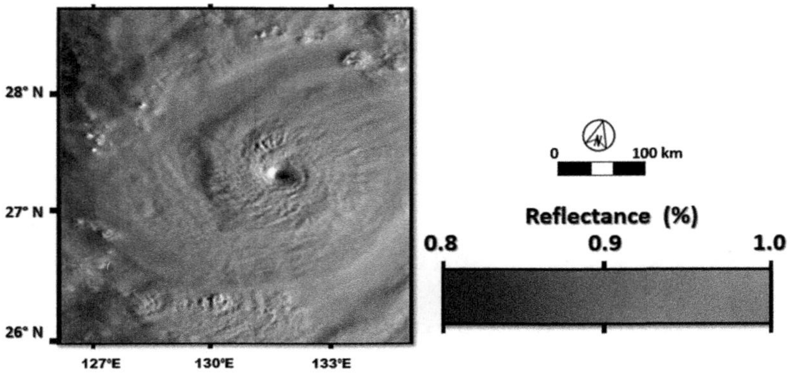

Figure 11.19: AHI tracks the Gani Typhoon.

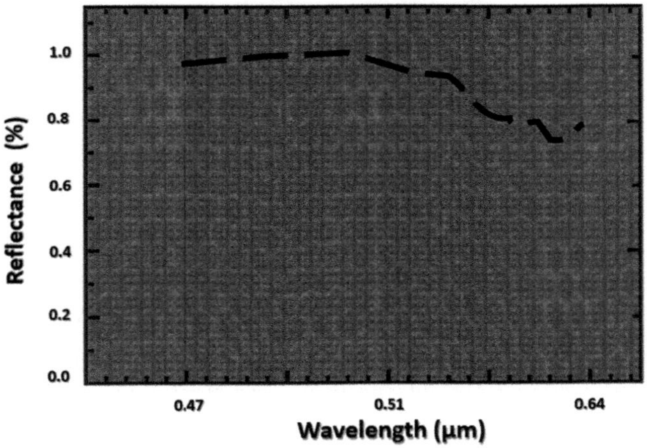

Figure 11.20: Quantum spectral reflectance curve across the Gani typhoon.

This imagery unveils intricate details, including the reflection of sunlight from the cloud tops, transverse bands within the cirrus clouds, a distinctly sloping eyewall, and spiral banding features encompassing more than 360°. The integration of data from multiple satellite sources, such as Radarsat-2 and Himawari-8, offers a comprehensive and synergistic perspective, enhancing our ability to comprehend the complexity and characteristics of significant meteorological phenomena. To address this, the QNSGA-II algorithm is employed to effectively track the cyclone's structure.

Therefore, the application of QNSGA-II facilitates the exploration of the Reynolds number variation corresponding to cyclonic wind speeds. Within this framework, the cyclonic wind speed spans a range from 10 to 50 ms^{-1}, resulting in a maximum Reynolds number reaching 1.456×10^{11} (Figure 11.21). This peak value of the Reynolds number signifies the heightened turbulence levels induced by the Gani typhoon. The typhoon manifests significant atmospheric eddy flows, extending across a span of 500 km.

Consequently,QNSGA-II proves effective in discerning the vorticity pattern within the cyclonic flow. Notably, the edges of vorticity become distinctly identifiable through the application of

QNSGA-II, particularly in RADARSAT-2 SAR HH polarization. The cyclonic vorticity is observed to spin in a clockwise direction (Figure 11.22).

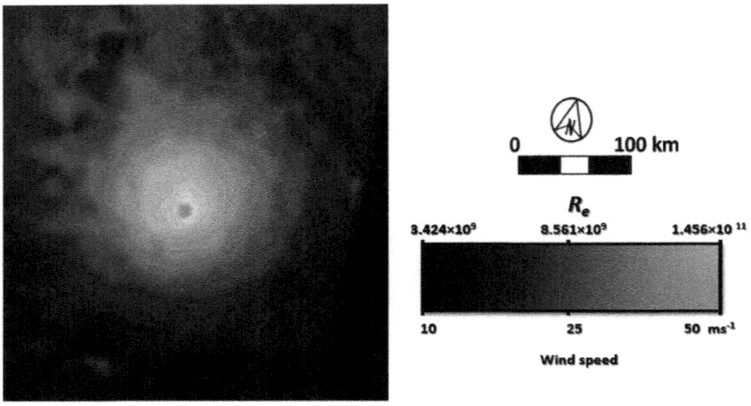

Figure 11.21: Cyclonic turbulence level simulated by QNSGA-II algorithm.

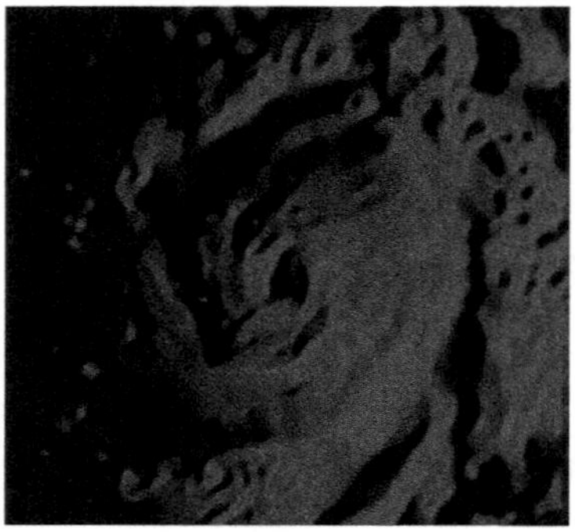

Figure 11.22: Automatic detection of the vorticity formation in RADARSAT-2 SAR using QNSGA-II.

Following this, QNSGA-II is capable of providing precise identification of cyclonic clusters, which encompass meso-vortices, bright rain bands, arc clouds, and cyclonic eye (Figure 11.23). In this regard, meso-vortices are small-scale, swirling air circulations within a larger cyclonic system. They are often associated with intense convective activity and can contribute to the overall organization and dynamics of the cyclone.

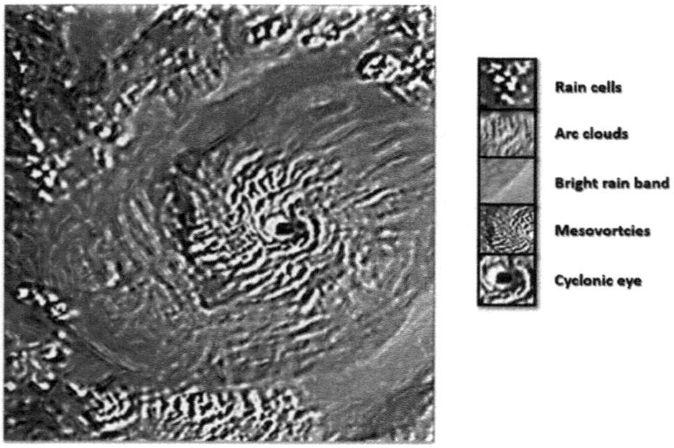

Figure 11.23: QNSGA-II automatically detects cyclonic component clusters.

Meso-vortices typically have a diameter on the order of tens to a few hundred kilometers. They are regions of increased wind speed and rotation within the broader cyclonic circulation. In this comprehension, meso-vortices are identified as regions characterized by the maximum wind speed of 50 ms^{-1} and the highest turbulence, reaching 1.456×10^{11}, in comparison to the surrounding cyclonic structural topology. Therefore, bright rain bands refer to elongated, intense bands of precipitation within a tropical cyclone that appears brighter in satellite imagery due to the presence of rain clouds. These bands are associated with heavy rainfall and can extend outward from the center of the cyclone. They often play a role in the transport of moisture and energy within the system.

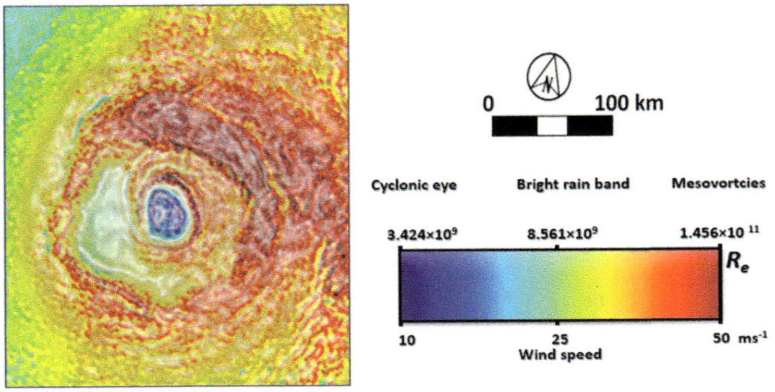

Figure 11.24: Reynolds number for each cyclonic cluster component by using QNSGA-II.

Consequently, arc clouds are formations of clouds that exhibit an arc-like shape within the cyclonic system. These clouds are often observed in the outer bands of a tropical cyclone and can indicate areas of convective activity and atmospheric instability with Reynold's number of 8.561×10^9 and wind speed of 25 ms^{-1}, which confirm the results in Figures 11.18,11.19; and 11.22.

Each cluster is distinctly characterized by its defined edge, effectively separated from every other class. This results in the creation of a quantized topology that accurately represents the cyclonic pattern flows. Conversely, the cyclonic eye is distinguished by the minimal wind speed of 10 ms^{-1} and the lowest Reynold number, measured at 3.424×10^9. In this sense,the cyclonic eye is the central and relatively calm region at the center of a tropical cyclone. The eye is characterized by clear skies, low wind speeds, and descending air. It is surrounded by the eyewall, which is a ring of intense convection and strong winds of 40 to 50 ms^{-1}. The presence of an eye is a defining feature of mature and well-organized tropical cyclones.

11.13 The Significance of Pareto Optimization in the Automatic Detection of Cyclonic Clusters Using QNSGA-II

In the realm of optimization, the QNSGA-II algorithm stands out as an exceptional blend of local scrutiny and global exploration, offering heightened precision compared to traditional evolutionary algorithms. This algorithm, proficient in both function optimization and multiuser detection within code division multiple access, gains versatility through its integration with quantum computing, thereby enhancing accuracy through a genetic algorithm. In this view, the culmination of these advancements is particularly noteworthy in the domain of quantum image processing, where QNSGA-II plays a pivotal role in advancing synthetic aperture radar (SAR) image segmentation and processing.

QNSGA-II's capabilities extend to the nuanced identification of morphological boundaries within SAR data, discerning cyclonic turbulent flow magnitudes across various segmentation layers, ranging from robust to subtle. Notably, QNSGA-II doesn't produce a singular solution but rather a set of compromised solutions known as Pareto optimal solutions. This collection of alternatives enables decision-makers to make informed choices based on specific criteria. This QNSGA-II algorithm categorized as an a posteriori approach, emphasizes that decisions are made after the completion of the search process.

Crucially, the quantum Pareto-optimization approach employed by QNSGA-II eliminates the need for a priori preference decisions amid conflicting objectives, such as mesovortex turbulent flow magnitudes and delineations between different cyclonic cluster components. Quantum Pareto optimal points converge into a Pareto front within the multiobjective function of SAR feature space, as illustrated in Figures 11.25 and 11.26, facilitating a comprehensive and nuanced understanding of the intricate SAR landscape.

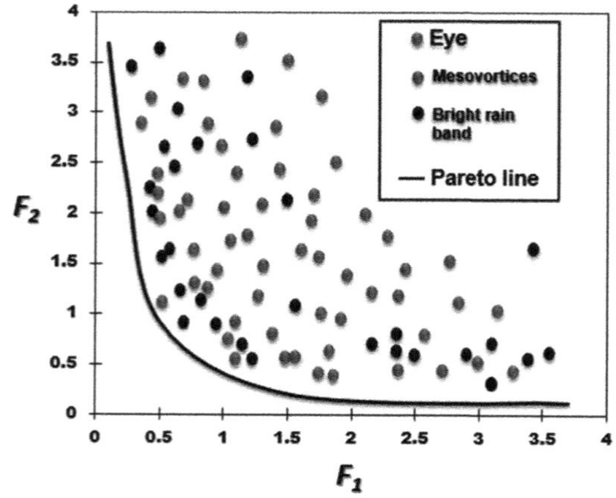

Figure 11.25: Nondominated solutions provided by Q-MEA.

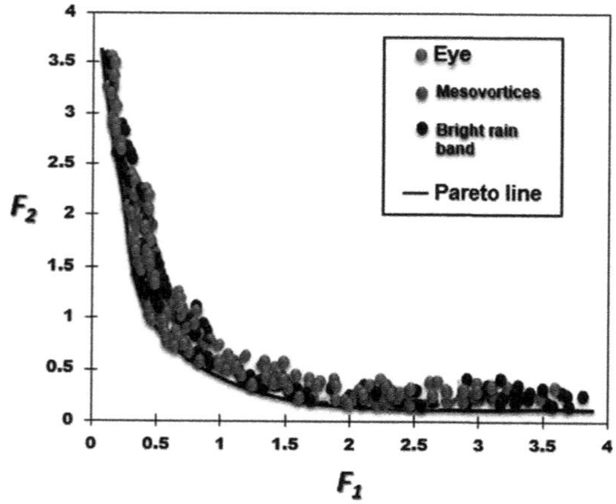

Figure 11.26: Nondominated solutions delivered by QNSGA-II.

The capability of QNSGA-II to efficiently distribute its population along the acquired Pareto front eclipses that of Q-MEA. This is emblematic of QNSGA-II's prowess in leveraging quantum computing principles, including uncertainty, superposition, and interference, to enhance the quality and diversity of the nondominated set in multiobjective problems involving polarimetric SAR features. QNSGA-II's adeptness in the automatic detection of cyclonic turbulent flow magnitudes, and surrounding cyclonic footprint boundaries is further highlighted by its accurate positioning in proximity to the Pareto optimal front (Figure 11.26).

In essence, QNSGA-II not only refines the accuracy of the Pareto optimal front but also preserves a comprehensive diversity by embodying the compensatory mechanisms intrinsic to a quantum-inspired evolutionary algorithm (QEA). This nuanced approach aims at improving the proximity to the Pareto optimal front, thereby facilitating the identification of superior solutions, deemed virtuous individuals through a fitness function.

Therefore, the QNSGA-II can also better detect the varying level cyclonic turbulence flows as a function of Reynold number, than QMEA. This accuracy is well shown by the Pareto front of value 3.7 in both dimensions of the front one (F1) and front two (F2) axes. In conclusion, QNSGA-II represents a promising tool for the automatic detection of atmospheric cyclonic turbulent flows in such tropical zones as the Pacific Ocean. However, the effectiveness of QNSGA-II may be influenced by the availability and quality of input data.

This chapter introduces a groundbreaking technique for the automated identification of tropical cyclonic vorticity flows in Synthetic Aperture Radar (SAR) data, leveraging platforms like RADARSAT-2 SAR ScanSAR wide mode in HH polarization data. The Quantum Multiobjective Evolutionary Algorithm (QMEA) is developed and evaluated using SAR imagery captured over typhoon Gani in the Pacific Ocean. The approach involves the instantaneous transformation of SAR backscatter and texture information into qubit chromosomes.

QMEA employs a novel qubit chromosomal representation based on the principles of quantum computing, incorporating qubits and nonlinear superposition. The algorithm efficiently produces a distinct classification map of cyclonic turbulent flow without well-defined boundaries and edges. To enhance its efficacy, Quantum Non-dominated Sort and Elitism (QNSGA-II) is judiciously employed in conjunction with QMEA for the automatic detection of cyclonic turbulent flow causes in RADARSA-T2 SAR data.

Yet, the application of the most advanced image processing tools was essential to fully comprehend the intricacies of atmospheric turbulence flows. The upcoming chapter will delve into the exploration of cutting-edge image processing algorithms that have never been employed in air turbulence studies within the realm of remote sensing technology.

References

[1] Spiridonov, V. and Ćurić, M. (2021). *Fundamentals of Meteorology* (pp. 219–228). Cham: Springer.

[2] Nott, J. (2011). A 6000 year tropical cyclone record from Western Australia. *Quaternary Science Reviews*, 30(5), 713–722. doi:10.1016/j.quascirev.2010.12.004

[3] World Meteorological Organization. (2018, April 17). *Global Guide to Tropical Cyclone Forecasting: 2017* (PDF). Archived from the original on July 14, 2019. Retrieved September 6, 2020.

[4] Emanuel, K. (2003). Tropical cyclones. *Annual Review of Earth and Planetary Sciences*, 31(1), 75–104.

[5] Walsh, K.J., McBride, J.L., Klotzbach, P.J., Balachandran, S., Camargo, S.J., Holland, G., ... and Sugi, M. (2016). Tropical cyclones and climate change. *Wiley Interdisciplinary Reviews: Climate Change*, 7(1), 65–89.

[6] Anthes, R. (Ed.). (2016). *Tropical Cyclones: Their Evolution, Structure and Effects* (Vol. 19). Springer.

[7] Gray, W.M. (1998). The formation of tropical cyclones. *Meteorology and Atmospheric Physics*, 67(1–4), 37–69.

[8] Yanai, M. (1964). Formation of tropical cyclones. *Reviews of Geophysics*, 2(2), 367–414.

[9] Knutson, T.R., McBride, J.L., Chan, J., Emanuel, K., Holland, G., Landsea, C., ... and Sugi, M. (2010). Tropical cyclones and climate change. *Nature Geoscience*, 3(3), 157–163.

[10] Ooyama, K. (1969). Numerical simulation of the life cycle of tropical cyclones. *Journal of the Atmospheric Sciences*, 26(1), 3–40.

[11] Lighthill, J., Holland, G., Gray, W., Landsea, C., Craig, G., Evans, J., ... and Guard, C. (1994). Global climate change and tropical cyclones. *Bulletin of the American Meteorological Society*, 2147–2157.

[12] Kaplan, J., DeMaria, M. and Knaff, J.A. (2010). A revised tropical cyclone rapid intensification index for the Atlantic and eastern North Pacific basins. *Weather and Forecasting*, 25(1), 220–241.

[13] Molinari, J. and Vollaro, D. (2010). Rapid intensification of a sheared tropical storm. *Monthly Weather Review*, 138(10), 3869–3885.

[14] Miyamoto, Y. and Takemi, T. (2015). A triggering mechanism for rapid intensification of tropical cyclones. *Journal of the Atmospheric Sciences*, 72(7), 2666–2681.

[15] Lee, C.Y., Tippett, M.K., Sobel, A.H. and Camargo, S.J. (2016). Rapid intensification and the bimodal distribution of tropical cyclone intensity. *Nature Communications*, 7(1), 10625.

[16] Wang, Y.Q. and Wu, C.C. (2004). Current understanding of tropical cyclone structure and intensity changes – A review. *Meteorology and Atmospheric Physics*, 87(4), 257–278.

[17] Kepert, J.D. (2010). Tropical cyclone structure and dynamics. *Global Perspectives on Tropical Cyclones: From Science to Mitigation*, 3–53.

[18] Wang , Y. (2012). Recent research progress on tropical cyclone structure and intensity. *Tropical Cyclone Research and Review*, 1(2), 254–275.

[19] Wang, Y. (2009). How do outer spiral rainbands affect tropical cyclone structure and intensity? *Journal of the Atmospheric Sciences*, 66(5), 1250–1273.

[20] Holland, G.J. and Merrill, R.T. (1984). On the dynamics of tropical cyclone structural changes. *Quarterly Journal of the Royal Meteorological Society*, 110(465), 723–745.

[21] Holton, J.R. (1973). An introduction to dynamic meteorology. *American Journal of Physics*, 41(5), 752–754.

[22] Panchev, S. (2012). *Dynamic Meteorology* (Vol. 4). Springer Science & Business Media.

[23] Gordon, A., Grace, W., Byron-Scott, R. and Schwerdtfeger, P. (2016). *Dynamic Meteorology*. Routledge.

[24] Aouaouda, M., Ayadi, A. and Yashima, H.F. (2019). Mathematical modeling of tropical cyclones on the basis of wind trajectories. *Computational Mathematics and Mathematical Physics*, 59, 1493–1507.

[25] Bryan, G.H. and Rotunno, R. (2009). The maximum intensity of tropical cyclones in axisymmetric numerical model simulations. *Monthly Weather Review*, 137(6), 1770–1789.

[26] Varotsos, C.A., Krapivin, V.F. and Soldatov, V.Y. (2019). Monitoring and forecasting of tropical cyclones: A new information-modeling tool to reduce the risk. *International Journal of Disaster Risk Reduction*, 36, 101088.

[27] Rumpf, J., Weindl, H., Höppe, P., Rauch, E. and Schmidt, V. (2007). Stochastic modelling of tropical cyclone tracks. *Mathematical Methods of Operations Research*, 66, 475–490.

[28] Abhyankar, A., Patwardhan, A. and Inamdar, A. (2006). Identification of completely submerged areas due to tropical cyclone using satellite data: An Indian case study. *In:* 2006 IEEE International Symposium on Geoscience and Remote Sensing (pp. 3305–3308). IEEE.

[29] Kiage, L.M., Walker, N.D., Balasubramanian, S., Babin, A. and Barras, J. (2005). Applications of Radarsat-1 synthetic aperture radar imagery to assess hurricane-related flooding of coastal Louisiana. *International Journal of Remote Sensing*, 26(24), 5359–5380.

[30] Zhang, G., Li, X. and Perrie, W. (2017). Synthetic aperture radar observations of extreme hurricane wind and rain. *Hurricane Monitoring with Spaceborne Synthetic Aperture Radar*, 299–346.

[31] Gao, Y., Sun, J., Zhang, J. and Guan, C. (2021). Extreme wind speeds retrieval using Sentinel-1 IW Mode SAR Data. *Remote Sensing*, 13(10), 1867.

[32] Zhang, G., Zhang, B., Perrie, W., Xu, Q. and He, Y. (2014). A hurricane tangential wind profile estimation method for C-band cross-polarization SAR. *IEEE Transactions on Geoscience and Remote Sensing*, 52(11), 7186–7194.

[33] Zhang, G., Li, X., Perrie, W., Hwang, P.A., Zhang, B. and Yang, X. (2017). A hurricane wind speed retrieval model for C-band RADARSAT-2 cross-polarization ScanSAR images. *IEEE Transactions on Geoscience and Remote Sensing*, 55(8), 4766–4774.

[34] Steane, A. (1998). Quantum computing. *Reports on Progress in Physics*, 61(2), 117.

[35] O'brien, J.L. (2007). Optical quantum computing. *Science*, 318(5856), 1567–1570.

[36] Hirvensalo, M. (2003). *Quantum Computing*. Springer Science & Business Media.

[37] Williams, C.P. (2010). *Explorations in Quantum Computing*. Springer Science & Business Media.

[38] Gruska, J. (1999). *Quantum Computing* (Vol. 2005). London: McGraw-Hill.

[39] Hey, T. (1999). Quantum computing: An introduction. *Computing & Control Engineering Journal*, 10(3), 105–112.

[40] Kaye, P., Laflamme, R. and Mosca, M.A. (2007). *Introduction to Quantum Computing*. Oxford University Press.

[41] Wang, L., Tang, F. and Wu, H. (2005). Hybrid genetic algorithm based on quantum computing for numerical optimization and parameter estimation. *Applied Mathematics and Computation*, 171(2), 1141–1156.

[42] Williams, C.P. and Williams, C.P. (2011). Quantum cryptography. *Explorations in Quantum Computing*, 507–563.

[43] Bardin, J.C., Sank, D., Naaman, O. and E. Jeffrey. (2020). Quantum computing: An introduction for microwave engineers. *IEEE Microwave Mag 21*, 8, 24–44.

[44] Marghany, M. 2021. *Nonlinear Ocean Dynamics: Synthetic Aperture Radar*. Elsevier.

[45] Cuomo, D., Caleffi, M. and Cacciapuoti, A.S. (2020). Towards a distributed quantum computing ecosystem. *IET Quantum Communication*, 1(1), 3–8.

[46] Kim, Y., Kim, J.H. and Han, K.H. (2006). Quantum-inspired multiobjective evolutionary algorithm for multiobjective 0/1 knapsack problems. *In:* 2006 IEEE International Conference on Evolutionary Computation (pp. 2601–2606). IEEE.

[47] Balicki, J., Balicka, H.T., Masiejczyk, J. and Zacniewski, A. (2010). Multi-criterion decision making in distributed systems by quantum evolutionary algorithms. *In:* Proceedings of the 12th European Conference on Computer Science, Puerto de la Cruz, Spain (pp. 328–333).

[48] Balicki, J. (2014). Quantum-inspired multi-objective evolutionary algorithms for decision making: Analyzing the state-of-the-art. *In:* J. Balicki (Ed.), *Advances in Applied and Pure Mathematics*. Proceedings of the 2nd International Conference on Mathematical, Computational and Statistical Sciences MCSS'14, Gdańsk, Poland, May 15-17, 2014. WSEAS Press (pp. 383–389).

[49] Barenghi, C.F., Skrbek, L. and Sreenivasan, K.R. (2014). Introduction to quantum turbulence. *Proceedings of the National Academy of Sciences*, 111(supplement 1), 4647–4652.

[50] Deb, K., Agrawal, S., Pratap, A. and Meyarivan, T. (2000). A fast elitist non-dominated sorting genetic algorithm for multi-objective optimization: NSGA-II. *In: International Conference on Parallel Problem Solving from Nature* (pp. 849–858). Berlin Heidelberg: Springer.

[51] Deb, K. (2001). Nonlinear goal programming using multi-objective genetic algorithms. *Journal of the Operational Research Society*, 52(3), 291–302.

[52] Iwabuchi, H., Saito, M., Tokoro, Y., Putri, N.S. and Sekiguchi, M. (2016). Retrieval of radiative and microphysical properties of clouds from multispectral infrared measurements. *Progress in Earth and Planetary Science*, 3, 1–18.

[53] Tana, G., Ri, X., Shi, C., Ma, R., Letu, H., Xu, J. and Shi, J. (2023). Retrieval of cloud microphysical properties from Himawari-8/AHI infrared channels and its application in surface shortwave downward radiation estimation in the sun glint region. *Remote Sensing of Environment*, 290, 113548.

CHAPTER

12

Four-Dimensional Quantum Hologram Radar Interferometry Radar for Tropical Cyclonic Tracking

In the earlier chapters, the exploration of the air turbulence phenomena primarily involved the application of two-dimensional quantum image processing techniques. For instance, Chapter 11 showcased the utilization of the two-dimensional quantum framework, specifically the Quantum Non-dominated Sorting Genetic Algorithm II (QNSGA-II), for the automated detection and classification of tropical cyclones within two-dimensional SAR data. Building on this foundation, the current chapter aims to introduce a groundbreaking algorithm tailored for the four-dimensional reconstruction of cyclonic topology within the context of two-dimensional SAR images, such as those obtained from RADARSAT-2 SAR imagery.

This represents a notable departure from the conventional focus on two-dimensional analyses, as the algorithm presented here delves into the complexities of cyclonic dynamics in a more comprehensive four-dimensional space. By extending the analysis to four dimensions, which include both spatial and temporal coordinates, the algorithm seeks to capture a richer and more nuanced representation of cyclonic behavior. The choice of RADARSAT-2 SAR imagery as the input data source underscores the practical application of the algorithm in real-world scenarios, where such satellite data is commonly utilized for monitoring and studying atmospheric phenomena.

The motivation behind this shift to four-dimensional reconstruction lies in the recognition that cyclonic events unfold dynamically over time and space. A two-dimensional perspective may not fully capture the intricate evolution and interplay of cyclonic features. The algorithm's innovative approach aims to overcome this limitation by providing a more holistic view of cyclonic topology, incorporating the temporal dimension to better understand the development stages, interactions, and behaviors of cyclones.

As such, the algorithm contributes to advancing our capabilities in cyclonic research and monitoring, presenting a valuable tool for scientists and meteorologists seeking a deeper understanding of these complex atmospheric phenomena. The discussion and application of the algorithm within the framework of RADARSAT-2 SAR imagery signal a practical and relevant extension of quantum image processing techniques to address the challenges posed by the inherently four-dimensional nature of cyclonic dynamics.

Before delving into four-dimensional hologram interferometry, it's crucial to address the fundamental question: what do four dimensions entail? Marghany's efforts to introduce 4-D visualization in remote sensing represent pioneering work, although the mathematical philosophy underpinning 4-D geometry remains unexplored in his work. This chapter seeks to provide mathematical speculations to reconstruct the 4-D concept, facilitating a deeper understanding of the dynamic mobility of air turbulence in SAR imagery.

12.1 What Characteristics Define a Space with Four Dimensions?

Visualizing a three-dimensional space requires no effort—it comes naturally to humans. Imagine the experience of living within a three-dimensional cube; this visualization is as innate as the acts of breathing or blinking. Envisioning the universe within the cube, complete with its six square walls and eight corners, is a seamless mental process. Human's ability to navigate and explore this space mentally is effortless and intuitive.

The question now arises: Can one visualize what it would be like to inhabit the four-dimensional analog of a cube, a four-dimensional cube or "tesseract"? Unlike the effortless immediacy with which one can visualize a three-dimensional space, the visualization of a tesseract is more challenging. It is doubtful that one can do it effortlessly. Yet, this difficulty is one of the very few restrictions. Therefore, one can still determine all the properties of a tesseract and understand what it would be like to live in one. Various techniques exist for accomplishing this task. Consequently, one of these methods involves progressing through the sequence of dimensions, extrapolating the natural inferences at each step up to the fourth dimension. Once readers see how it is done for the special case of a tesseract, readers should be able to apply it to other cases without difficulty [1–3].

Let's contemplate the one-dimensional analog of a cube, which can be represented as an interval. This interval is created by taking a dimensionless point and moving it along a distance. This distance can be denoted as "d," whether it's 2 meters, 3 kilometers, or any other quantity. In this context, the interval should possess a length denoted as "d," with its boundaries defined by two points, corresponding to the two ends of the interval (Figure 12.1).

Figure 12.1: Concept of one-dimension.

Consequently, a cube's two-dimensional structure equivalent is shaped like a square. The one-dimensional interval is extended over a distance "d" in the second dimension to generate this square. Four equal sides and four right angles make up the square created by extending the one-dimensional interval through a distance "d" in the second dimension (Figure 12.2). It may be seen as a level surface whose length and width are equally equal to "d."

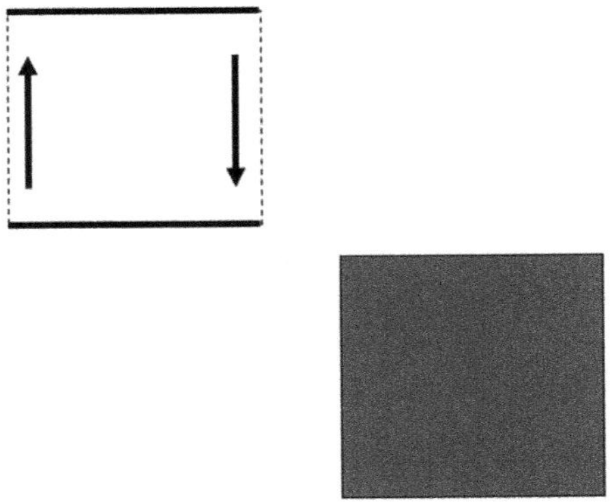

Figure 12.2: Concept of the 2-D square.

From this standpoint, the square has an area of d^2. It is enclosed by faces on all four sides, and these faces are intervals of length “d.” The count of these faces is determined by the fact that their two-dimensional axes must be terminated on both ends by additional faces. In this context, to obtain four faces, the two dimensions need to be multiplied by two faces each. The collective result is a perimeter of 4×d in dimension (Figure 12.3).

The primary question is how to construct a cube from a dimension with a perimeter of 4×d? To shape a cube, begin with a square and extend it by a distance of d in the third dimension. The resulting cube possesses a volume of d^3 and is enclosed by faces on six sides. Each face represents a square with an area of d^2, totaling six faces due to the three-dimensional axes being capped on both ends by faces [2,4,6]. This results in 3 dimensions times by 2 faces each, equating to 6 faces (Figure 12.4). The combined faces create a surface with an area of $6 \times d^2$.

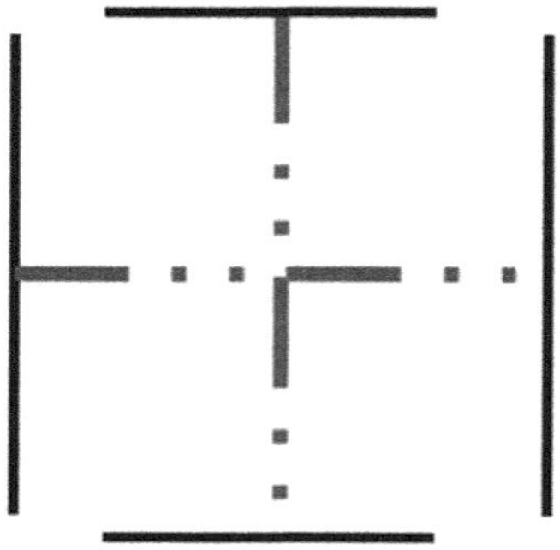

Figure 12.3: A dimension with a perimeter of 4×d.

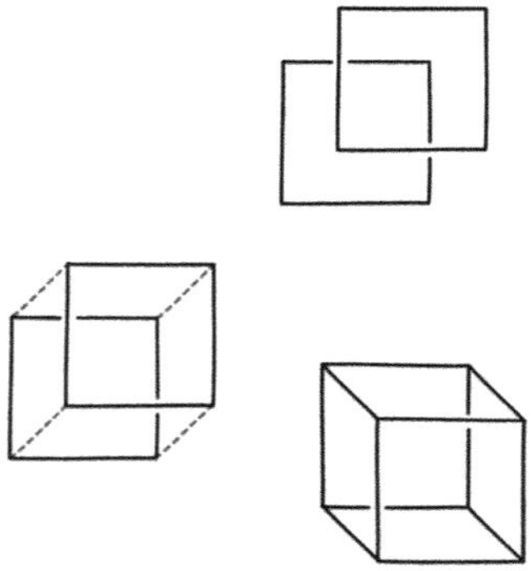

Figure 12.4: Construction of 3-D cube shape.

In light of the aforementioned perspective, the process of constructing from 1-D to 3-D appears straightforward. The encoding of 2-D within 1-D and, similarly, the encoding of 3-D within 2-D, align with this conceptual framework. Extending this understanding, the natural expectation is that 4-D should be encoded within 3-D. However, the challenge arises in deciphering how to reconstruct 4-D from the encoded 3-D representation.

Elaborating further on the establishment of 4-D, the keystone lies in the concept of a tesseract. The tesseract serves as a crucial building block in comprehending and visualizing four-dimensional space. This geometric figure, also known as a hypercube, represents an extension of the familiar cube into the fourth dimension. While our innate ability to visualize three-dimensional spaces is effortless, envisioning the tesseract requires a more intricate mental process due to its existence in the fourth dimension.

To grasp the idea of a tesseract, one can initiate the understanding by considering its lower-dimensional analogs. Beginning with a one-dimensional interval—a length denoted as “d”—and progressing to a two-dimensional square by extending the interval into a second dimension, the foundational steps are laid. The square encompasses an area of d^2 and is bounded by four faces, each an interval of dimension d. Extending this progression to three dimensions results in a cube, with a volume of d^3 and six square faces forming its boundaries.

Therefore, in the Euclidean 4-space, the standard tesseract is defined as the convex hull of the points (±1, ±1, ±1, ±1). This implies that it consists of all points (x_1, x_2, x_3, x_4) in $\mathbb{R}^4$, where each coordinate xi is constrained within the range $-1 \le x_i \le 1$. Mathematically, this can be expressed as:

$$\{(x_1, x_2, x_3, x_4) \in \mathbb{R}^4 : -1 \le x_i \le 1\} \tag{12.1}$$

In this Cartesian frame of reference, the tesseract has a radius of 2 and is enclosed by eight hyperplanes ($x_i = \pm 1$). The intersection of each non-parallel pair of hyperplanes forms 24 square faces within the tesseract. At every edge, three cubes and three squares intersect, and each vertex is a meeting point for four cubes, six squares, and four edges [1–5].

The tesseract, a four-dimensional polytope, can be analyzed by decomposing it into smaller 4-polytopes. Specifically, it is the convex hull of the compound of two demi-tesseracts, also known as 16-cells. Additionally, the tesseract can be triangulated into 4-dimensional simplices, irregularly shaped 5-cells, with shared vertices.

12.2 Topology of 4-D Reconstruction

It is noteworthy that there are a total of 92,487,256 such triangulations of the tesseract, and each of these triangulations contains a minimum of 16 4-dimensional simplices. The dissection of the tesseract into instances of its characteristic simplex represents a fundamental and direct method for constructing the tesseract. This characteristic simplex is a specific orthoscheme with a Coxeter diagram. In essence, this simplex serves as a foundational region within the tesseract, intimately connected to the tesseract's defining symmetry group known as the B4 polytopes. In geometry, therefore, a Coxeter–Dynkin diagram, also known as a Coxeter diagram or Coxeter graph, is a graphical representation featuring numerically labeled edges, referred to as branches. These branches convey the spatial relationships among a set of mirrors or reflecting hyperplanes. The diagram captures the essence of a kaleidoscopic construction, with each graph "node" representing a mirror or domain facet. The numerical label attached to a branch encodes the dihedral angle order between two mirrors on a domain ridge. This order denotes the factor by which the angle between the reflective planes can be multiplied to yield 180 degrees. For instance, Figure 12.5 demonstrates that the finite Coxeter groups can be classified into three one-parameter families of increasing rank: A_nA_n, B_nB_n, D_nD_n, one one-parameter family of dimension two, I_2(p), and six exceptional groups: E_6, E_7, E_8, F_4, H_3, and H_4. Additionally, there are three infinite one-parameter families: $I_2(p)$, I_n, and B_n. The product of finitely many Coxeter groups from this list is again a Coxeter group, and it's noteworthy that all finite Coxeter groups can be constructed in this manner [7].

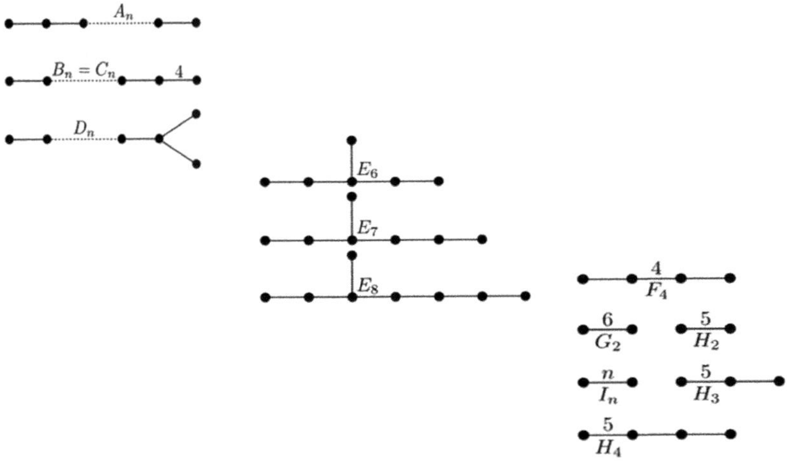

Figure 12.5: Simplification of Coxeter–Dynkin diagrams.

Accurately, a Coxeter group is demarcated by the following mathematical expression:

$$\left\langle r_1, r_2, \ldots, r_n \mid (r_i r_j)^{m_{ij}} = 1 \right\rangle \tag{12.2}$$

where $m_{ii} = 1$ and $m_{ij} \geq 2$ for $i \neq j$. The condition $m_{ij} = \infty$ means that no relation of the form $(r_i r_j)^m$ should be imposed. The pair (W, S), where W is a Coxeter group with generators $S = \{r_1,\ldots,r_n\}$, is referred to as a Coxeter system. It is important to note that, in general, the set S is not uniquely determined by the Coxeter group W. For instance, the Coxeter groups of type A_3, B_3, and $A_1 \times A_3$ are isomorphic, but the Coxeter systems are not equivalent [7–9].

According to the above perspective, many of these Coxeter groups are Weyl groups, and notably every Weyl group can be realized as a Coxeter group. The Weyl groups encompass the families A_n, B_n, D_n, and $I_2(p)$, and the exceptional groups E_6, E_7, E_8, F_4, and G_2. The non-Weyl groups include the exceptions H_3 and H_4, and the family $I_2(p)$, except in cases where it coincides with one of the Weyl groups [7, 9].

This classification is established by comparing the restrictions on (undirected) Dynkin diagrams with the constraints on Coxeter diagrams of finite groups. Formally, the Coxeter graph is derived from the Dynkin diagram by discarding the direction of edges, replacing double edges with edges labeled 4, and triple edges with edges labeled 6. Every finitely generated Coxeter group is also proven to be an automatic group. Geometrically, this corresponds to the crystallographic restriction theorem, indicating that excluded polytopes do not fill space or tile the plane. For specific cases like H_3 and H_4, certain polyhedra do not fill space. Additionally, the directed Dynkin diagrams B_n and C_n give rise to the same Weyl group (hence Coxeter group), illustrating that direction matters for root systems but not for the Weyl group. This phenomenon is analogous to the hypercube and cross-polytope being distinct regular polytopes but sharing the same symmetry group [8, 10].

In the context of 4-D reconstruction, Coxeter groups play a crucial role in understanding and representing the symmetries and structures of objects in four-dimensional space. The significance of Coxeter groups lies in their ability to provide a formal mathematical framework for describing the relationships and transformations between mirrors or hyperplanes in higher-dimensional spaces.

12.3 Is the Existence of N-dimensional Space a Reality?

An essential inquiry arises: do dimensions beyond the conventional four—namely the 5th, 6th, 7th, 8th, 9th, 10th, and 11th, including the hyperdimensional manifolds proposed by string theory—exist? According to this perspective, these additional dimensions are believed to manifest within the fabric of space-time, potentially constituting the innermost building blocks of particles, forming intricate inner manifolds that compose the entirety of the universe. This concept has evolved into what is currently recognized as an 11-dimensional framework. Notably, Calabi-Yau manifolds emerge as mathematical hyperdimensional structural entities, encompassing the extra dimensions theorized by superstring theory or the theory of everything. In the realm of algebraic geometry, a Calabi-Yau manifold, or Calabi-Yau space, stands as a distinctive type of manifold characterized by properties such as Ricci flatness, yielding functions of significance in theoretical physics [11].

In this understanding, Calabi-Yau spaces play a crucial role in string theory, where a prevalent model posits the universe's geometry as a ten-dimensional space represented by $M \times V$. Here, M denotes a four-dimensional manifold (space-time), and V signifies a six-dimensional compact Calabi-Yau space. Calabi-Yau spaces, alternatively referred to as Calabi-Yau manifolds or Calabi-Yau varieties, are not only significant in theoretical physics but also fascinating from a purely mathematical perspective [12, 14].

Despite the generalizability of the definition to any dimension, Calabi-Yau spaces are typically considered to have three complex dimensions. Viewing them as possessing six real dimensions with a fixed smooth structure proves convenient due to the variability in their complex structure. Examining the local scenario using coordinates, let us consider $\mathbb{R}^6$ with coordinates x_1, x_2, x_3, and y_1, y_2, y_3, chosen such that:

$$z_j = x_j + iy_j \tag{12.3}$$

This choice imparts the structure of C^3. Subsequently, the expression:

$$\phi_z = dz_1 \wedge dz_2 \wedge dz_3 \tag{12.4}$$

Equation 12.4 represents a local section of the canonical bundle. Introducing a unitary change of coordinates, $w = A_z$, where A is a unitary matrix, transforms ϕ by the determinant of A, denoted as det$A$. In mathematical terms:

$$\phi_w = \det A \cdot \phi_z \tag{12.5}$$

It is noteworthy that if the linear transformation A possesses a determinant of 1, indicating a special unitary transformation, then ϕ is consistently defined as either ϕ_w or ϕ_z. Therefore, an n-dimensional sphere can be characterized by examining the (real) solutions of the equation:

$$x_1^2 + \ldots + x_{2n+1}^2 = r^2. \tag{12.6}$$

Similarly, an instance of a complex three-dimensional Calabi-Yau manifold is defined by the (complex) solutions to the equation:

$$x_5^1 + x_5^2 + x_5^3 + x_5^4 + \phi x_1 x_2 x_3 x_4 = 1, \tag{12.7}$$

where ϕ is a parameter. This renowned Calabi-Yau manifold is commonly known as the 'quintic' (Figure 12.6).

Specifically within superstring theory, the higher dimensions of space-time are occasionally conjectured to adopt the structure of a 6-dimensional Calabi-Yau manifold, introducing the concept of replicate symmetry (Figure 12.7). Importantly, these dimensions are presumed to exist not only abstractly but also within the fabric of space and time. The 11 dimensions are posited to coexist—the four familiar ones and the remaining seven, which are exceptionally intricate [11, 13].

Figure 12.6: Calabi-Yau manifold 'quintic'.

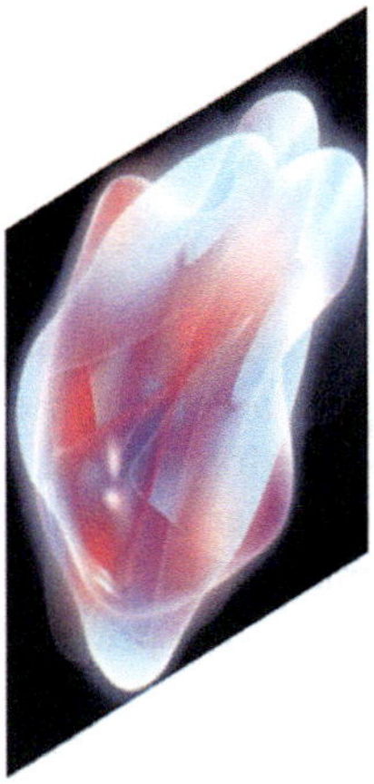

Figure 12.7: 6-dimensional Calabi-Yau manifold.

Viewed through the lens of Hilbert space and the Hilbert-Einstein collaborations, an infinite number of vectors can exist within the Hilbert space. Extending this concept to dimensions suggests the theoretical generation of an infinite number of dimensions. This gives rise to mathematical abstractions for which we currently lack explanatory or descriptive mathematical frameworks [11–14].

As a result, theoretically and abstractly, the existence of interdimensional entities appears conceivable. Moreover, these entities would likely exhibit a complexity that interweaves with the current structure of the universe. Advancements in nanotechnology and Femto-technology, integral to quantum computing, may potentially surmount the challenge posed by the absence of specific mathematical algorithms to reconstruct complex interdimensional forms.

12.4 What is the Role of Calabi-Yau Manifolds in Hologram Construction?

The connection between holograms and Calabi-Yau manifolds lies in the theoretical framework of string theory and the exploration of extra dimensions beyond the familiar four dimensions of space and time.

In string theory, which is a theoretical framework attempting to reconcile quantum mechanics and general relativity, the fundamental building blocks of the universe are not point particles but tiny, vibrating strings. To make the mathematics of string theory work requires more than the usual four dimensions of space and time. In many formulations of string theory, the additional dimensions are assumed to be compactified or curled up in a microscopic space. Calabi-Yau manifolds, being special six-dimensional spaces with specific geometric properties, are often proposed as candidates for these compactified dimensions [11, 14].

Now, how does this relate to holograms? Holography, in a simplified sense, is a method of capturing and reconstructing three-dimensional images. The concept of holography is also applied in theoretical physics through the Holographic Principle, which suggests that a higher-dimensional space can be fully described by information on its lower-dimensional boundary. This principle has its roots in string theory.

Theoretical physicists, like Juan Maldacena, have proposed a particular realization of the Holographic Principle known as the AdS/CFT correspondence. According to this conjecture, a certain type of space called Anti-de Sitter space (AdS), which has negative curvature, is holographically equivalent to a conformal field theory (CFT) defined on its boundary. In the AdS/CFT correspondence, the extra dimensions of AdS space play a role similar to the compactified dimensions in string theory. The geometry of AdS space is described by equations, including the metric tensor [15–17]. In five dimensions, AdS5 is often used in the AdS/CFT correspondence.

$$ds^2 = \frac{R^2}{z^2}\left(-dt^2 + d\vec{x}^2 + dz^2\right) \tag{12.8}$$

here, R is the AdS radius, and z is the extra dimension. In a conformal field theory, the correlation functions play a central role. The basic equations for correlation functions in a 2D CFT involve conformal transformations, primary fields, and conformal Ward identities.

Let's consider a 2D Euclidean CFT with coordinates z and $\overline{z}$, where z represents a complex plane. Primary fields in CFT are operators that transform in a specific way under conformal transformations. The primary field $\phi_i(z,\overline{z})$ has conformal weights $(h_i,\overline{h}_i)$. In this sense, Conformal transformations in 2D CFT are given by holomorphic and antiholomorphic functions $f(z)$ and $\overline{f}(\overline{z})$, respectively. The transformation of a primary field under a conformal transformation is given by:

$$\phi_i'(z',\ddot{z}') = \left(\frac{\partial f}{\partial z}\right)^{h_i}\left(\frac{\partial \ddot{f}}{\partial \ddot{z}}\right)^{\ddot{h}_i}\phi_i(z,\ddot{z}) \tag{12.9}$$

In equation 12.9, the correlation function $\left\langle \phi_1(z_1,\ddot{z}_1)\ldots\phi_n(z_n,\ddot{z}_n)\right\rangle$ represents the statistical correlation between n primary fields. In a CFT, these correlation functions are constrained by conformal symmetry and satisfy Ward identities. Therefore, Ward identities express the invariance of correlation functions under infinitesimal conformal transformations [16–19]. For example, the holomorphic part of the Ward identity for a two-point function is given by:

$$\left(z_1 \frac{\partial}{\partial z_1} + h_1\right)\langle \phi_1(z_1)\phi_2(z_2)\rangle = 0 \tag{12.10}$$

Similar identities exist for higher-point functions. It is important to note that the actual mathematical formulation of CFT involves more sophisticated techniques, including the use of operators, OPE (Operator Product Expansion), and the Virasoro algebra.

While the direct connection between everyday holograms and Calabi-Yau manifolds might not be immediately apparent, their linkage becomes evident in the theoretical explorations of string theory, extra dimensions, and the holographic nature of certain spacetime geometries. It is a fascinating interplay between abstract mathematical structures and our attempts to understand the fundamental nature of the universe.

Holographic mapping; therefore, involves correlating observables in AdS with those in the CFT. Mathematically, this can be expressed through correlation functions. For instance, the expectation value of a CFT operator $\mathcal{O}$ is related to the bulk field Φ in AdS:

$$\langle \mathcal{O}(x)\rangle_{\mathrm{CFT}} = \lim_{z\to 0} z^{\Delta}\Phi(x,z) \tag{12.11}$$

here, Δ is the scaling dimension of the operator. The correlation functions in the CFT are related to the behavior of fields in AdS. In the context of the AdS/CFT correspondence, the holographic mapping involves satisfying Ward identities and other constraints. While the precise mathematical formulation can be quite involved, we can express the general idea using some mathematical notation. Let's denote a CFT operator as $\mathcal{O}(x)$, where x represents spacetime coordinates. The holographic mapping involves relating the CFT correlation functions, denoted by $\langle \mathcal{O}_1(x_1)\mathcal{O}_2(x_2)\ldots\rangle_{\mathrm{CFT}}$ to the behavior of fields in AdS. In this context, holographic mapping imposes constraints on the behavior of fields in AdS. These constraints ensure the equivalence between CFT correlation functions and the behavior of fields in AdS. Mathematically, this can be expressed as a set of conditions on the bulk fields $\Phi(x, z)$. Some constraints on $\Phi(x, z)$ ensure equivalence.

12.5 Hologram Quantum Interferometry

Hologram quantum interferometry refers to the use of holographic techniques in quantum interferometry experiments. Quantum interferometry involves manipulating the quantum states of particles and using their interference patterns to make highly precise measurements. The integration of holography into these experiments can enhance the precision and capabilities of quantum interferometry setups. The mathematics behind holography involves the principles of wave optics and interference. In quantum interferometry, holographic techniques can be applied to manipulate the wavefronts of quantum particles [17–19]. The wave equation describes the behavior of light. For a monochromatic wave of angular frequency ω, the equation is:

$$E(x,t) = E_0 \cos(kx - \omega t + \phi) \tag{12.12}$$

where $E(x,t)$ is the electric field; E_0 is the amplitude; k is the wave number; ω is the angular frequency across the position x within time t and phase ϕ. Therefore, the interference pattern is created by the superposition of two waves:

$$I(x) = I_1 + I_2 + 2\sqrt{I_1 I_2}\cos(\Delta\phi) \tag{12.13}$$

Here $I(x)$ is the intensity pattern; $\Delta\phi$ is the phase difference; I_1 and I_2 are the intensities of the two interfering waves.

Consequently, in quantum interferometry, quantum particles are described by wavefunctions. The wavefunction (ψ) for a quantum particle can be manipulated using holographic techniques:

$$\Psi(x) = \Psi_1 + \Psi_2 + 2\sqrt{\Psi_1\Psi_2}\cos(\Delta\Phi) \tag{12.14}$$

where $\Psi(x)$ is the quantum state; Ψ_1 and Ψ_2 are the states of the two interfering quantum waves. Thus $\Delta\Phi$ is the phase difference in the quantum context [19–22]. The holographic principles in quantum interferometry involve encoding information about quantum states in interference patterns, enabling complex manipulations, and shaping quantum wavefunctions.

The interference pattern in holography is directly influenced by the holographic phase:

$$I(x) \propto |E(x,t)|^2 \tag{12.15}$$

Similarly, in quantum interferometry, the interference pattern is influenced by the quantum wavefunction:

$$P(x) \propto |\psi(x,t)|^2 \tag{12.16}$$

where $P(x)$ is the probability density. The modulation of the holographic phase corresponds to the modulation of the quantum phase:

$$\frac{d\phi_h}{dt} = \frac{d\phi_q}{dt} \tag{12.17}$$

This equation ensures a consistent modulation effect between holography and quantum interferometry. These equations capture the modulation of phases in both holography and quantum interferometry and how it directly influences the respective interference patterns. The correspondence equation emphasizes the parallelism in phase modulation between the two phenomena [19, 22].

12.6 Quantized Radar Hologram Interferometry

The Mach-Zender interferometer, depicted with dual mirrors and dual beam splitters in Figure 12.8, operates on the principle of interference. This optical device introduces a delay phase, denoted as φ, through one of its lengths. The purpose of this delay is to manipulate the interference of light waves within the interferometer. The delay phase φ is a crucial parameter that influences the interference pattern within the interferometer. To determine the value of φ, an approach involves gauging the intensity of photons in the dual output beams. This measurement allows for the computation of φ, a quantity essential for understanding the interference phenomena occurring in the device. In this context, the estimation of φ is subject to statistical considerations [20–22].

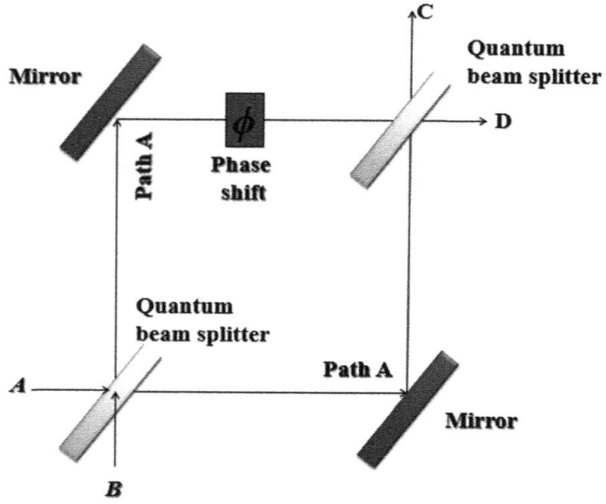

Figure 12.8: Mach-Zender Interferometer utilizing dual mirrors and dual beam splitters.

The value of φ can be estimated with a statistical error that is proportional to $\frac{1}{\sqrt{N}}$ where N represents the number of non-entangled photons involved in the measurement process. However, it is noteworthy that this interferometric quantity is constrained by standard quantum principles, imposing limitations on the precision achievable in the measurement. In this sense, the principles employed in the Mach-Zender interferometer find application in interferometric synthetic aperture radar (IFSAR). IFSAR leverages similar techniques to determine the range of a target.

Therefore, the circumstances surrounding the straightforward computation of the mean propagation time entail an error formulation. This error, denoted as a function of N, represents the statistical uncertainty associated with the traditional method, implying that the precision of the range valuation may be compromised. In this circumstance, the error is formulated as:

$$\partial R \approx O\left[\Delta\omega\sqrt{N}\right]^{-1} \tag{12.18}$$

Equation 12.18 introduces a relationship that validates the impact of the inverse of the bandwidth $\Delta\omega$ on the error incurred due to the fluctuation of the range ∂R. In other words, it establishes a connection between the precision of the measurement (related to the error) and the characteristics of the system, specifically the bandwidth.

Conventionally, quantum constraints, such as those described by the Heisenberg limit, pose limitations on the precision of certain measurements. However, the passage suggests a groundbreaking approach—leveraging entangled photons. Entangled photons, in this context, serve as a means to transcend or break down the conventional quantum constraint. To explore the quantum mechanism of hologram interferometry, the passage directs attention to a specific quantum state—the two-mode, path-entangled, photon-number state, commonly known as the N00N state. This state is a type of Schrödinger Cat state, a concept in quantum mechanics that involves the superposition of distinct states. Consequently, the introduction of the NOON state suggests a novel avenue for quantum hologram interferometry. Entangled states, like the NOON state, can potentially enhance precision beyond the limits imposed by conventional quantum constraints [20, 23]. This quantum enhancement could lead to advancements in holographic interferometry, offering unprecedented precision in measurements related to range and other parameters. This mathematically can be expressed as:

$$|\psi_{NOON}\rangle = \left[\sqrt{2}\right]^{-1}\left(|NO\rangle + |ON\rangle\right) \tag{12.19}$$

In this equation, $|\Psi\rangle$ represents the quantum state created by the entangled-photon source. It is a superposition of states where all photons are either in the upper mode A or all in the lower mode B. The coefficient $\left[\sqrt{2}\right]^{-1}$ signifies the equal probability amplitude associated with each state, reflecting the entangled nature of the photons. This expression captures the essence of entanglement, where the entangled-signal source generates superpositions of states without specifying which mode the photons are demonstrated in Figure 12.9, emphasizing the fundamental principles of quantum superposition and entanglement in hologram interferometry, which can be given by:

$$|NOON\rangle = |up\rangle + |down\rangle = |N\rangle_A |O\rangle_B + |O\rangle_A |N\rangle_B \tag{12.20}$$

To elaborate on this, let's consider the entangled state $|\Psi\rangle$ generated by the entangled-signal source in the context of the Mach-Zender interferometer. In this scenario, each half of the entangled state traverses a distinct length of the interferometer. Mathematically, this can be expressed as follows:

$$|\Psi\rangle = \frac{1}{\sqrt{2}}\left(e^{i\phi_1}|N,0\rangle + e^{i\phi_2}|0,N\rangle\right) \tag{12.21}$$

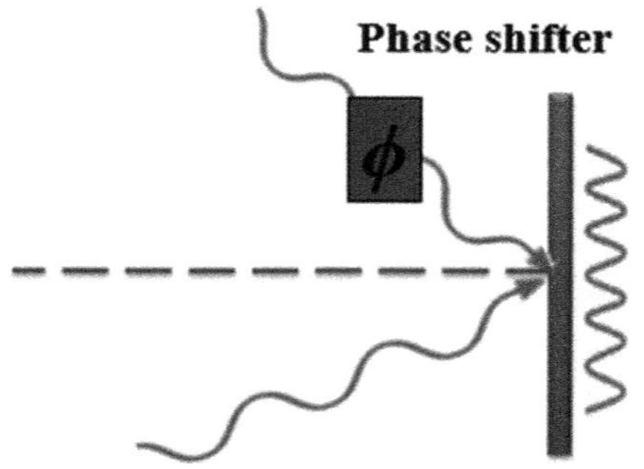

Figure 12.10: Path-entangled NOON state.

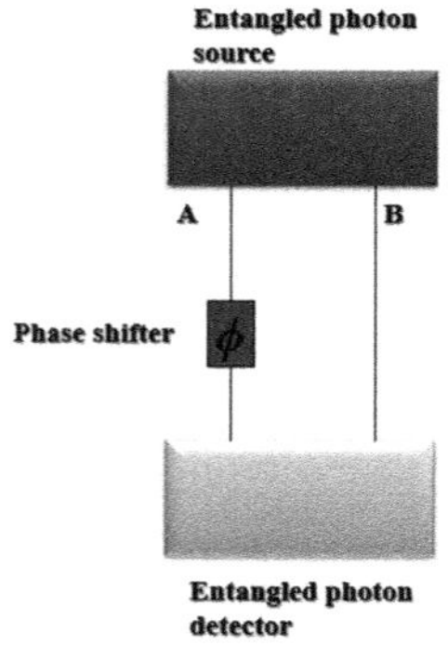

Figure 12.9: Quantum interferometry in holography with dual modes.

here, $e^{i\phi_1}$ and $e^{i\phi_2}$ represent the phase factors associated with the lengths traveled by the two halves of the entangled state through the interferometer. The interference between these two paths introduces a phase difference $\Delta\phi = \phi_1 - \phi_2$ that contributes to the overall quantum interference pattern (Figure 12.10). This expression captures the unique feature of entanglement in hologram interferometry. The entangled photons, having traversed different paths, exhibit interference effects that depend on the relative phases acquired during their journey. The entangled state $|\Psi\rangle$ evolves dynamically as it encounters the beam splitter, and the resulting interference pattern reflects the quantum correlations between the entangled photons. The ability of entangled states to manifest interference effects across different spatial paths is a key aspect of quantum information processing and quantum optics [20–25].

Let us assume the bosonic creation operator $a_\alpha^\dagger$ by:

$$a_\alpha^\dagger \left| n_1, ..., n_{\alpha-1}, n_\alpha, n_{\alpha+1}, ... \right\rangle = \sqrt{n_{\alpha+1}} \left| n_1, ..., n_{\alpha-1}, n_\alpha, n_{\alpha+1}, ... \right\rangle, \tag{12.22}$$

So the corresponding annihilation operator a_α can be given by:

$$a_\alpha \left| n_1, ..., n_{\alpha-1}, n_\alpha, n_{\alpha+1}, ... \right\rangle = \sqrt{n_\alpha} \left| n_1, ..., n_{\alpha-1}, n_\alpha, n_{\alpha+1}, ... \right\rangle, \tag{12.23}$$

Both equations 12.22 and 12.23 can be used to define the number operator $N_\alpha = a_\alpha^\dagger a_\alpha$ such that

$$N_\alpha \left| n_1, n_2, ..., n_\alpha, ... \right\rangle = n_\alpha \left| n_1, n_2, ..., n_\alpha, ... \right\rangle \tag{12.24}$$

One consequence of these commutation relations is that any multi-particle basis state can be formulated as [23]:

$$N_\alpha \left| n_1, n_2, ..., n_\alpha, ... \right\rangle = \left(a_1^\dagger\right)^{n_1} \left(a_2^\dagger\right)^{n_2} ....\left(a_\alpha^\dagger\right)^{n_\alpha} ...\left|0\right\rangle, \tag{12.25}$$

In this view, equation 12.21 can be expressed based on annihilation operators as:

$$\left|\psi_{NOON}^{(\phi)}\right\rangle = \frac{1}{\sqrt{2N!}}\left(\left(\hat{a}_1^\dagger\right)^N + e^{iN\phi}\left(\hat{a}_2^\dagger\right)^N \left|0\right\rangle_1 \left|0\right\rangle_2\right) \tag{12.26}$$

Let us assume the sensor observation D_s is given by

$$\hat{D}_s = \frac{1}{N!}\left(\left(\hat{a}_1^\dagger\right)^N \left|0\right\rangle\left\langle 0\right|\left(\hat{a}_2^\dagger\right)^N + \left(\hat{a}_2^\dagger\right)^N \left|0\right\rangle\left\langle 0\right|\left(\hat{a}_1^\dagger\right)^N\right) \tag{12.27}$$

The sensor observation measurements are important to estimate the phase shift based on annihilation operators. In this regard, the fluctuation in the sensor observations $\Delta\hat{D}_s$ can lead to error in the phase shift as follows:

$$\partial\phi = \frac{\Delta D_s}{\left|-N \sin N\varphi\right|} = N^{-1} \tag{12.28}$$

Equation 12.28 reveals that the Heisenberg constraint as the interferometric phase is considered as an extremely entangled state. In other words, equation 12.28 can be expressed as [25]:

$$\lim_{\alpha_1 \to 1} \lim_{\alpha_2 \to 1} \partial\phi = \frac{\Delta D_s}{\left|-N \sin N\varphi\right|} = N^{-1} \tag{12.29}$$

In the circumstance of the eye of a hurricane, the attenuation of the quantum interferometry must be occurring. In this understanding, let us consider to coherence set C and decoherence set d_c which represent tropical cyclonic and surrounding environment, respectively. In this view, the photon signals pass across the two sets and have different backscatter indices β_1 and β_2 with different attenuation coefficients μ_C and μ_{d_c}, respectively. In this sense, The expression for the propagation of an attenuated NOON state is formulated as:

$$\begin{aligned} \left|\psi_{\text{NOON}}\right\rangle &= \frac{1}{\sqrt{2N!}} e^{-i\omega\beta_1/c-\mu_c(0.5\omega)(NC)} \left(\hat{a}_1^{\dagger}\right)^N \left|0\right\rangle_1 \left|0\right\rangle_2 \\ &+ \frac{1}{\sqrt{2N!}} e^{-i\omega\beta_2/c-\mu_{d_c}(0.5\omega)(Nd_c)} \left(\hat{a}_2^{\dagger}\right)^N \left|0\right\rangle_1 \left|0\right\rangle_2 + \left|\Phi\right\rangle \end{aligned} \tag{12.30}$$

here $|\Phi\rangle$ is a state that represents those states that have been scattered outside the NOON basis due to the impact of chaotic tropical cyclonic turbulence. Moreover, c is the speed of light. Therefore, the dispersion of the phase shift is then estimated as:

$$\phi_0 \equiv \frac{\omega}{c}\left(\beta_2 d_c - \beta_1 C\right) \tag{12.31}$$

On the contrary, NOON state without attenuation can be given by:

$$\left|\psi_a\right\rangle = \frac{1}{\sqrt{2^N}}\left(\left|10\right\rangle + \left|01\right\rangle\right)^{\otimes N} \tag{12.32}$$

Equation 12.32 proves that separable states cannot be implemented to pulsate the standard quantum constraint even in the absence of coherence of tropical cyclonic turbulence flow. In other words, equation 12.32 can be expressed in terms of phase change as:

$$\lim_{\alpha_1 \to 1} \lim_{\alpha_2 \to 1} \partial\phi = \frac{1}{\sqrt{N}} \tag{12.33}$$

Equations (12.29) and (12.33) highlight the concept of the Heisenberg limit, often referred to as “super-sensitivity” when surpassing the shot-noise limit. In quantum mechanics, achieving super-sensitivity implies pushing measurement precision beyond classical limits. Or in a system with a state of minimum uncertainty, optimal performance is expected when there is maximal uncertainty in the photon number. The uncertainty is effectively distributed between the dual-photon routes, denoted as A and B. In the context of NOON states, uncertainty exists when all N photons are exclusively in mode A (with none in B) or when all N photons are in mode B (with none in A). This uncertainty persists even if A and B are illuminated at different time intervals. The NOON state introduces a unique form of uncertainty that goes beyond the classical understanding. In other words, the information suggests that in quantum interferometry, particularly with NOON states, achieving super-sensitivity involves managing uncertainty in the distribution of photons between different modes. The concept of “super-sensitivity” arises from optimizing the uncertainty in photon number, showcasing the unique and powerful properties of quantum states in surpassing classical limits [22–25].

12.7 Marghany 4D Quantized Hologram Interferometry Algorithm

According to the above perspective, the mathematical description of the two complex SAR hologram interferometry images $H(x,y,x,z,t)$ is formulated as:

$$H(x,y,x,z,t) = \varepsilon_0 c \left[S_{01}^2 \cos^2(kz - \omega_1 t) + \frac{S_{02}^2}{B_r^2} \cos^2(kr - \omega_2 t) + \frac{2S_{01}S_{02}}{B_r} \cos^2(kz - \omega_1 t)\cos(kr - \omega_2 t)\cos(\vartheta_{12}) \right] \quad (12.34)$$

Equation 12.34 outlines the utilization of two complex radar images, S_1 and S_2, to formulate a 3-D hologram radar image. This holographic representation incorporates two signal parameters, namely the wavenumber k, the baseline B_r representing the distance between the acquired Synthetic Aperture Radar (SAR) images during the acquisition time t, and the radial frequency. Moreover, within equation 12.34, there is a component exclusively utilizing the radar complex signals and another term related to interferometry, incorporating the cosine of the angle θ_{12} between the two vector amplitudes, which may be a function of position (x, y, z). In this context, the process of estimating phase maps for diverse holographic radar images can be achieved as follows:

$$\phi_{12}(x,y,z) = arc\tan\left(\frac{\sin(\phi_2(x,y,z)) \otimes \cos(\phi_1(x,y,z)) - \cos(\phi_2(x,y,z)) \otimes \sin(\phi_1(x,y,z))}{\cos(\phi_2(x,y,z)) \otimes \cos(\phi_1(x,y,z)) \oplus \sin(\phi_2(x,y,z)) \otimes \sin(\phi_1(x,y,z))} \right), \quad (12.35)$$

$$\phi_{32}(x,y,z) = arc\tan\left(\frac{\sin(\phi_3(x,y,z)) \otimes \cos(\phi_2(x,y,z)) - \cos(\phi_3(x,y,z)) \otimes \sin(\phi_2(x,y,z))}{\cos(\phi_3(x,y,z)) \otimes \cos(\phi_2(x,y,z)) \oplus \sin(\phi_3(x,y,z)) \otimes \sin(\phi_2(x,y,z))} \right), \quad (12.36)$$

$$\phi_{123}(x,y,z) = arc\tan\left(\frac{\sin(\phi_{12}(x,y,z)) \otimes \cos(\phi_{23}(x,y,z)) - \cos(\phi_{12}(x,y,z)) \otimes \sin(\phi_{23}(x,y,z))}{\cos(\phi_{12}(x,y,z)) \otimes \cos(\phi_{23}(x,y,z)) \oplus \sin(\phi_{12}(x,y,z)) \otimes \sin(\phi_{23}(x,y,z))} \right). \quad (12.37)$$

In this perspective, formulas 12.34 to 12.37 indicate that $\phi_{123}(x,y,z)$ varies from 0 to 2π without 2π cutoffs, allowing the creation of the image of the wrapped phase. Consequently, the reconstruction of the 3-D surface with precision using $\phi_{123}(x,y,z)$ alone may not yield accurate results, as $\phi_{123}(x,y,z)$ does not reveal 2π discontinuities. This is particularly true in SAR long-wavelength scenarios, where low quantity precision is common due to the nature of its coherence signal. To elaborate, the phase $\phi_1(x,y)$ has a wavelength ratio within three complex SAR images, causing discontinuities with an identical layout. In this scenario, within each $\frac{S_{12}(\lambda)}{S_1(\lambda)}$, there are discontinuities on $S_{12}(\lambda)$. In simpler terms, the conjugate of three complex SAR images forms discontinuities. In summary, the mathematical expression for the hologram interferometry image can be articulated as:

$$H_k(x,y,z) = \alpha(x,y,z) + \beta(x,y,z)\sin(\phi(x,y,z) + 0.5(\pi k)) + G(x,y,z). \quad (12.38)$$

Equation 12.38 illustrates the presence of white Gaussian noise $G(x,y,z)$, as equations 12.34 to 12.37 exhibit degradation in the computation of the absolute phase [24, 26]. The primary question that emerges is: how can one achieve 4-D phase unwrapping from 2-D phase unwrapping to construct a 4-D visualization of tropical cyclonic turbulence flow in hologram interferometry SAR data?

12.8 Marghany 4-D Quantized Phase Unwrapping Algorithm

In hologram interferometry, the quantized phase unwrapping process is a crucial step in reconstructing a comprehensive and accurate representation of a real object in four dimensions. According to Marghany [25], in holography, information about the object is encoded in the phase of the recorded interference pattern. The phase contains details about the object's surface shape, deformations, or any changes that occurred during the holographic recording. Therefore, The phase information is often wrapped or modulo 2π due to the limited dynamic range of the recording medium. This wrapping introduces ambiguity in interpreting the phase values. To this end, the 4-D phase unwrapping can be expressed as:

$$\Psi_3(x,y,z,t) = \phi(x,y,z,t) + 2\pi\left[int\frac{\phi_{12}(x,y,z,t)}{2\pi} \otimes \left(\frac{S_1(\lambda)}{S_2(\lambda)}\right) \otimes \left(\frac{S_2(\lambda)}{S_3(\lambda)}\right) + int\frac{\phi_{23}(x,y,z,t)}{2\pi} \otimes \left(\frac{S_2(\lambda)}{S_3(\lambda)}\right) \otimes \left(\frac{S_2(\lambda)}{S_3(\lambda)}\right) + G(\sigma)\right] \quad (12.39)$$

The innovation presented in equation 12.39 is grounded on the interchange of absolute phase estimates derived from three complex Synthetic Aperture Radar (SAR) images, each acquired at a distinct time. This innovative equation acknowledges the influence of white Gaussian noise, characterized by the standard deviation σ, a consideration crucially incorporated into the 4-D phase unwrapping approach [25]. Beyond the primary objective of 4-D phase unwrapping, equation 12.39 also endeavors to address and rectify discontinuities. To express the wrapped phase in this context, it is formulated as the Laplacian of the real phase, derived as follows:

$$\nabla^2\phi = \cos\Psi_w \nabla^2(\sin\Psi_w) - \sin\Psi_w \nabla^2(\cos\Psi_w) \quad (12.40)$$

here, the Laplace operator ∇^2 is applied, and by employing an inverse of the real phase ϕ, it can be approximated in 4-D through the compilation of quantum Fourier-domain forward and backward transformations, as ∇^2 outlined by Marghany [27]:

$$\phi'(i,j,k,t)_Q = QFT^{-1}\frac{\left[QFT[\cos\phi_w(QFT^{-1}[|p^2+q^2+r^2+s^2\rangle QFT(\sin\phi_w)\right]}{|p^2+q^2+r^2+s^2\rangle} - QFT^{-1}\frac{\left[QFT[\sin\phi_w(QFT^{-1}[|p^2+q^2+r^2+s^2\rangle QFT(\cos\phi_w)\right]}{|p^2+q^2+r^2+s^2\rangle} \quad (12.41)$$

Equation 12.41 illustrates that *QFT* and QFT^{-1} represent the quantum Fourier inverse and forward cosine or sine transforms, respectively, in 4-D of time-domain coordinates (*i, j, k*) and the Quantum Fourier-domain coordinates. Equation 12.41 encompasses three types of operations: quantum forward and inverse cosine transforms, trigonometric operations, and the masking expression. In contrast to the conventional discrete Fourier transform, which requires $O(n^2)$ gates, the discrete Fourier transform on 2^n amplitudes can be implemented as a quantum circuit using only $O(n^2)$ Hadamard gates and controlled phase shift gates, where n is the number of qubits. Furthermore, the exact *QFT* algorithms only need *O(nlogn)* gates to perform a calculation [23, 25, 27].

The specific equation for 4-D phase unwrapping is based on:

$$\begin{pmatrix} \partial\Phi_1 \\ \partial\Phi_2 \\ \partial\Phi_3 \\ \partial\Phi_4 \end{pmatrix} = \frac{2\pi}{\lambda} \begin{pmatrix} (H_s,\partial\zeta,U_r)_{1i} & (H_s,\partial\zeta,U_r)_{1j} & (H_s,\partial\zeta,U_r)_{1k} & (H_s,\partial\zeta,U_r)_{1t} \\ (H_s,\partial\zeta,U_r)_{2i} & (H_s,\partial\zeta,U_r)_{2j} & (H_s,\partial\zeta,U_r)_{2k} & (H_s,\partial\zeta,U_r)_{2t} \\ (H_s,\partial\zeta,U_r)_{3i} & (H_s,\partial\zeta,U_r)_{3j} & (H_s,\partial\zeta,U_r)_{3k} & (H_s,\partial\zeta,U_r)_{3t} \\ (H_s,\partial\zeta,U_r)_{4i} & (H_s,\partial\zeta,U_r)_{4j} & (H_s,\partial\zeta,U_r)_{4k} & (H_s,\partial\zeta,U_r)_{4t} \end{pmatrix} \otimes \begin{pmatrix} W^x_{i,j,k,t} \\ W^y_{i,j,k,t} \\ W^z_{i,j,k,t} \\ W^t_{i,j,k,t} \end{pmatrix} \quad (12.42)$$

In this context, W stands for the user-defined weights, Cartesian coordinates I, j, k, and t for the 4-D vector representation in the hyperspace holographic interferometry. Consistent with (Wnuk et al., 2019), the spatial variance of shear-wave turbulence flows significantly over time in this example of a 3D volume of a hologram image that is to be stored in 4D. Equation 12.42 shows how the entanglement of the wave function can be used to determine the areas with the highest and lowest entanglement levels based on the wave function of non-unwrapped areas. In other words, the wave function exhibits two unwrapping phase states, lowest-entanglement |0> and highest-entanglement |1>, and is a superb estimator of the quality of phase unwrapping images generated from InSAR data [19, 24]. This novel formula, called Marghany 4-D Quantized Hologram Interferometry, was developed by Marghany [19] and exploited in recent publications [25, 27].

In four-dimensional space, each tesseract possesses 80 neighbors, providing a robust criterion derived from the sum of 40 measurements. The construction of a 4-D hyperspace or hypercube from physical sea surface parameters, obtained through 4-D phase unwrapping, involves the implementation of a Hamming graph (Figure 12.12) formula. This formula is executed as a quantum walk, represented by the expressions:

$$G_{H_c} \oplus H_m \otimes \left|\phi_{G_{H_c}} \otimes \phi_{H_m}(t)\right\rangle = \left|\phi_{G_{H_c}}(t) \otimes \right|\phi_{H_m}(t)> \quad (12.43)$$

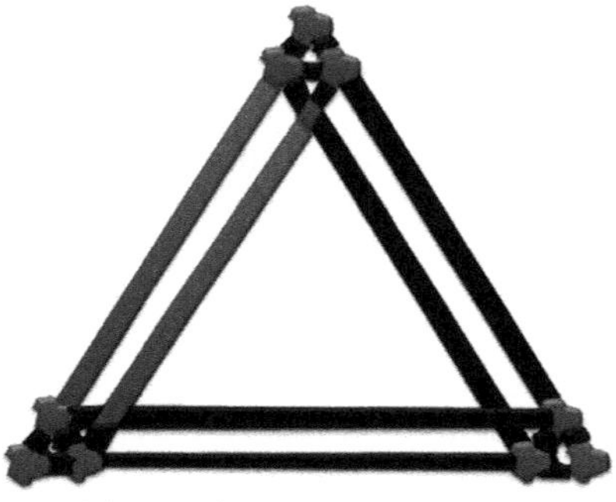

Figure 12.11: Hamming graph with H_m (3,3).

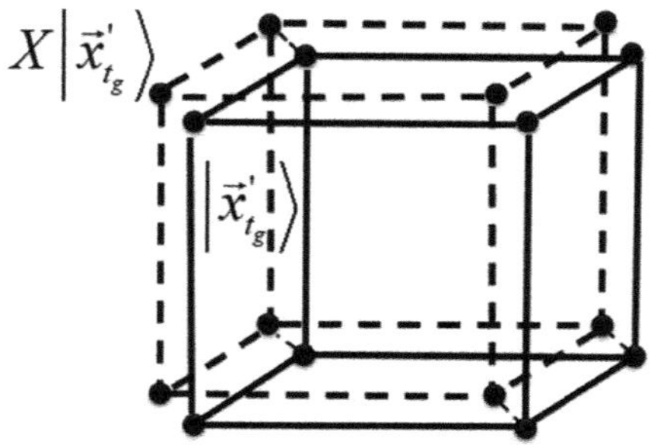

Figure 12.12: Encoding 3D data into a 4D hypercube.

In the realm of quantum phase unwrapping walks, denoted as $\phi_{G_{H_c}}$ and ϕ_{H_m}, initiated from vertices v_{m_1} and v_{m_2}, respectively. Therefore, equation 12.43 emerges as a pivotal component in the creation of hypercube hologram interferometry utilizing Hamming graphs. Broadly speaking, the Hamming graph serves as a potent tool for characterizing quantum uniform mixing. In the realm of quantum mechanics, the natural method to amalgamate dual systems is through the tensor product ⊗. In this context, the Cartesian graph artifact assumes a parallel role in facilitating quantum walks during the phase unwrapping process [25]. Therefore, the ultimate expression for the quantum walk, aiming for optimal exploration with a radius of 2r, can be elegantly presented through the pulsating sequence of the unitary operators $U''^{(+)}_m$:

$$\left|\phi^{(e)}{}_{G_{HC}\oplus H_m}(2r)\right\rangle = (U^{(+)}_m U''^{(+)}_m)^r \left|\phi^e{}_{G_{HC}}(t)\right\rangle \otimes \left|\phi_{H_m}(t)\right\rangle \quad (12.44)$$

where

$$U_m''^{(+)} = S^{(+)}C''^{(+)} \tag{12.44.1}$$

Here $C''^{(+)}$ is the coin operator, which acts on the total Hilbert space $H^C \otimes H^V$ and the S is a propagator of a quantum walk across the hypercube vertices V as:

$$S^{(+)} = \sum_{d,\vec{x}} \left\| d, \vec{x} \oplus \vec{e}_d \right\rangle \left| d, \vec{x} \right\rangle \right| \tag{12.44.2}$$

here d is the direction of propagation and $\vec{e}$ is the edge of the vertices and $\oplus$ denotes the bitwise addition modulo 2 operator. Moreover, $\vec{x}$ is the Hamming weight of an integer is the number of 1's in its binary string. The coin operator $C''^{(+)}$ is then determined from:

$$C''^{(+)} = C_0 \otimes M_A + (C_1 - C_0) \otimes \sum_{j=1}^{m} \left| \vec{x}_{tg}^{(j)} \right\rangle \left\langle \vec{x}_{tg}^{(j)} \right| \tag{12.45}$$

C_0 presents the n-dimensional Grover operator, which is also known as the Grover diffusion operator" and C_1 is chosen to be −1. However, due to the symmetry of the hypercube graph (Figure 12.13), the vertices can always be relabeled in such a way that the marked vertex becomes $\vec{x}_{tg}^{(j)} = 0$. To formalize the task of finding multiple marked vertices, let us denote the number of elements marked by the oracle by m, and their labels by $\vec{x}_{tg}^{(j)}$ and $j = 1, \ldots, m$ [28].

According to Marghany [23] Figure 12.13 explains how the 4-D is encoded into a 3-D hypercube. The effect of operator X is switching between phase unwrapping and vertices. In other words, the processors in the cubes of dimensions 1, 2, and 3 are categorized with integers, which are represented as binary numbers. Therefore, those dual processors are neighbors in dimension $\vec{x}_{tg}^{(j)}$ in a hypercube of dimension, a message can be routed between any pair of processors in at most X hops.

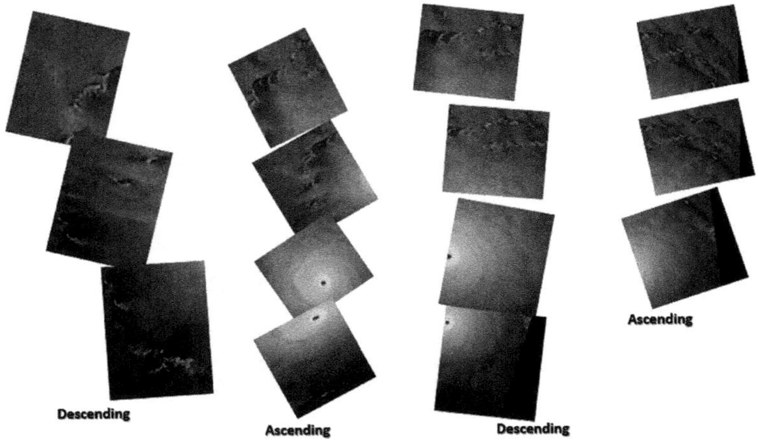

Figure 12.13: Ascending and descending scenes captured Typhoon Goni.

12.9 4D Hologram Interferometry of Tropical Cyclone

The RADARSAT-2 SAR data captured in the ScanSAR-wide mode of Typhoon Goni over the Pacific Ocean on October 30, 2020, as discussed in Chapter 11, Section 11.11, is employed in this context to reconstruct 4-D hologram interferometry. This image is a mosaic of 10 ScanSAR wide images, encompassing both 7 ascending and 7 descending scenes captured over various periods. This comprehensive compilation aims to capture the complete scenario of Typhoon Goni over the Pacific Ocean.

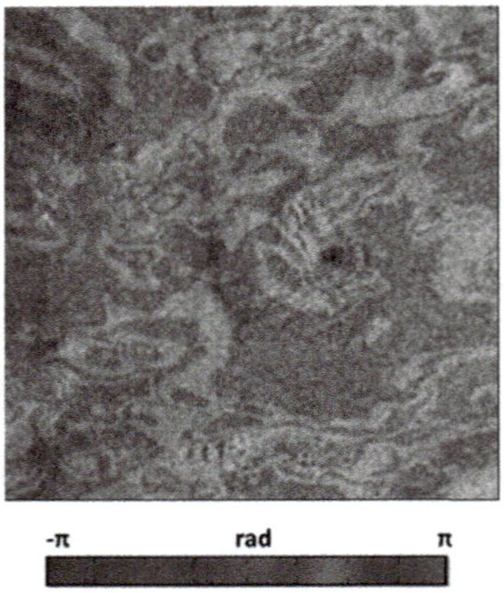

Figure 12.14: Two-dimensional hologram phase unwrapping.

Quantum hologram interferometry hinges on the crucial process of phase unwrapping. The application of the Hamming weight graph facilitates a comprehensive 4-D phase unwrapping, showcasing an exceptional ability to discern the unwrapping phase cycle within 2-D (see Figure 12.15), 3-D (see Figure 12.16), and 4-D hologram phase unwrappings (see Figure 12.17). The clarity and precision of hologram phase interferometry exhibit notable improvement with the transition from 2D to 3D and 4D.

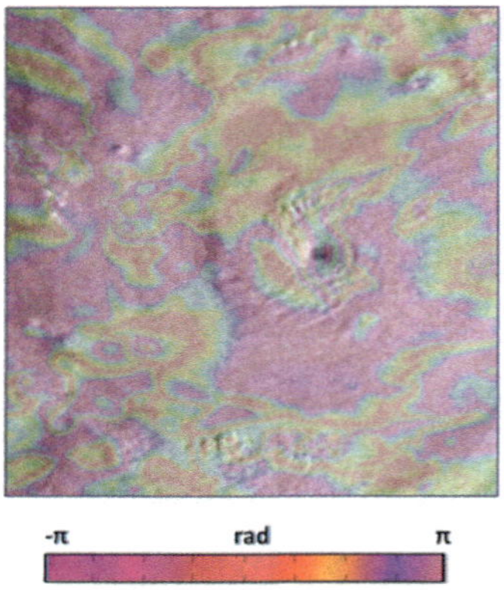

Figure 12.15: Three-dimensional hologram phase unwrapping.

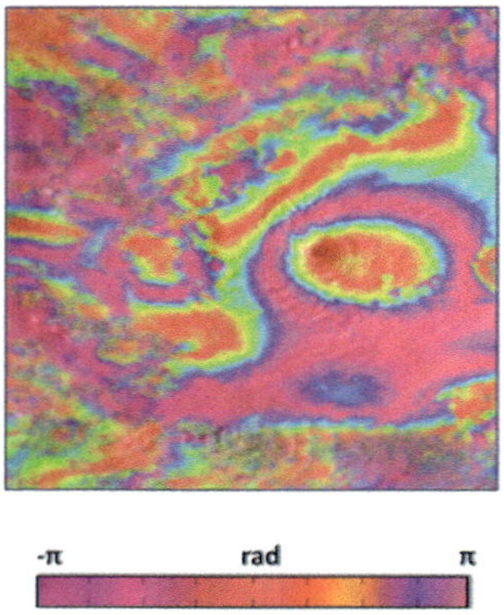

Figure 12.16: Four-dimensional hologram phase unwrapping.

Figure 12.17: 3-D visualization of typhoon Goni in RADARSAT-2 SAR image using 3-D quantum hologram phase unwrapping.

In the WideScanSAR mode, the achieved hologram phase interferometry cycles exhibit sharper characteristics in 3D and 4D compared to 2D. The most resilient and chaotic Typhoon Goni is identified and traced more effectively in 3D and 4D phase unwrapping than in 2D. This advancement underscores the enhanced capabilities and insights gained through the application of higher-dimensional phase unwrapping techniques in quantum hologram interferometry.

Certainly, the application of 4-D hologram phase unwrapping, integrating the Hamming weight and a quantum walk algorithm, outshines the excellence observed in both 2-D and 3-D hologram phase unwrapping. The paramount advantage of 4-D phase unwrapping lies in its ability to preserve a flawless phase cycle—sustaining coherence across the entire cycle of the phase along-track interferometry patterns. Furthermore, the utilization of Hamming weight algorithms has effectively minimized errors in the interferogram cycle, especially in regions with low coherence along the cyclone eye, where high decorrelation is prevalent, notably in WideScanSAR mode. This innovative approach not only advances the precision of hologram phase unwrapping but also addresses challenges associated with low coherence, presenting a breakthrough in achieving more accurate and reliable results, particularly in complex scenarios such as those encountered in ScanSARwide mode during cyclone monitoring.

In the comparative analysis between 3-D (Figure 12.18) and 4-D visualizations (Figure 12.19) in both ScanSAR-wide modes, it becomes evident that the identification of chaotic cyclone turbulence flow is significantly enhanced in the 4-D hologram representation. The 4-D hologram interferometry offers more comprehensive insights into the characteristics of turbulence formed by cyclones, showcasing features such as the generation of eddies and vorticities. The detailed temporal evolution and spatial distribution captured in the 4-D visualization contribute to a more nuanced understanding of the dynamic behavior of cyclone turbulence.

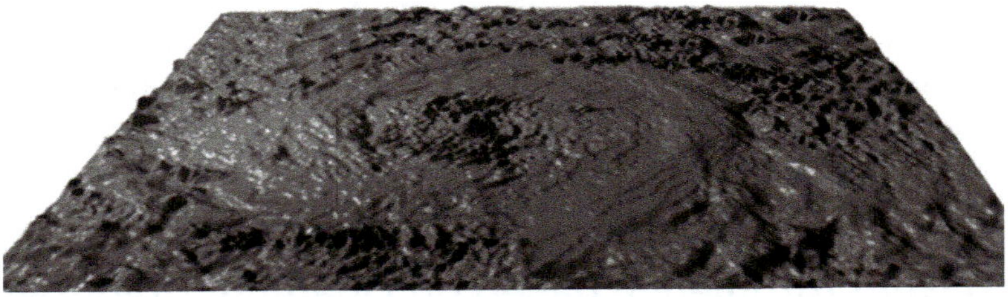

Figure 12.18: 3-D visualization of typhoon Goni in RADARSAT-2 SAR image using 3-D quantum hologram phase unwrapping.

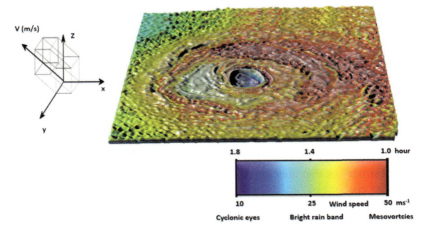

Figure 12.19: 4-D visualization of typhoon Goni in RADARSAT-2 SAR image using 4-D quantum hologram phase unwrapping.

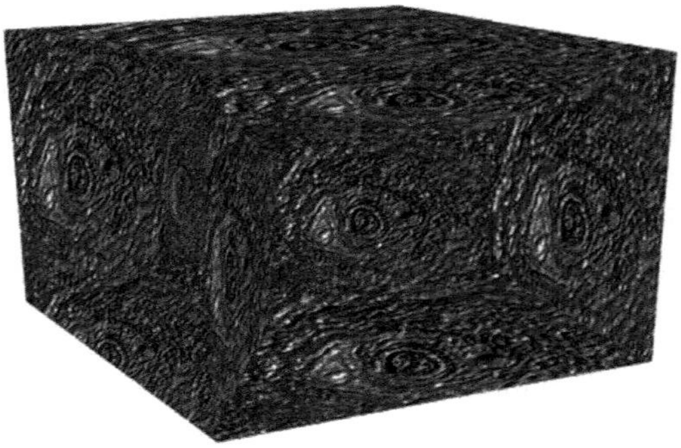

Figure 12.20: 4-D quantum hologram interferometry of typhoon Goni topology.

The 4-D simulation of the turbulence dynamic behavior of Typhoon Goni using quantum hologram interferometry provides a comprehensive and detailed representation of the typhoon's topology. In this simulation, the fourth dimension is depicted by wind speed velocity, showcasing how the wind speed changes as one moves away from the cyclone's eye over a distance of 324 km within 1.8 hours. The simulation, therefore, reveals a distinct pattern in the wind speed distribution. Mesovortices, identified as smaller-scale vortices within the typhoon, exhibit the highest wind velocities, reaching a maximum speed of 50 m/s. These regions of intense wind activity contribute to the dynamic and chaotic nature of the typhoon's turbulence.

Conversely, the cyclone eyes, characterized by a calmer and more stable atmospheric condition, show the lowest wind speeds, registering at 10 m/s. This observation aligns with the outcomes derived from the Quantum Non-dominated Sorting Genetic Algorithm II (QNSGA-II) in Chapter 11, corroborating the reliability and accuracy of the quantum hologram interferometry results. Overall, the 4-D simulation not only captures the spatial distribution of wind speeds within Typhoon Goni but also provides a temporal aspect, allowing for a dynamic portrayal of the turbulence behavior over both distance and time. This nuanced understanding contributes to a more comprehensive analysis of the typhoon's atmospheric dynamics.

12.10 4-D Goni Turbulence based on Quantum Hologram Hypercube Interferometry

In the application of quantum hologram interferometry to 14 RADARSAT-2 SAR scenes, a novel, and insightful perspective emerges when examining the chaotic turbulence of Typhoon Goni through the lens of the hypercube (Figure 12.21). The hypercube, serving as a higher-dimensional analog of a conventional cube, provides a distinctive vantage point to dissect and comprehend the intricate features exhibited by the turbulent flows within the typhoon. This unique perspective proves especially enlightening when contemplating the inherent duality present in these turbulent dynamics.

The hypercube's higher-dimensional representation allows for a more nuanced and comprehensive analysis of the complex interactions and behaviors within the turbulent flows. It serves as a powerful tool to capture and explore the multidimensional nature of the atmospheric phenomena associated with Typhoon Goni. This perspective is valuable in revealing patterns, correlations, and structures that may be challenging to discern in lower-dimensional analyses.

Moreover, the concept of duality within the turbulent flows is emphasized. Duality, in this context, refers to the coexistence and interaction of contrasting elements or characteristics within the turbulent dynamics of the typhoon. The hypercube's higher-dimensional framework facilitates a more in-depth exploration of these dual aspects, shedding light on the intricate interplay between different components of the turbulent flows.

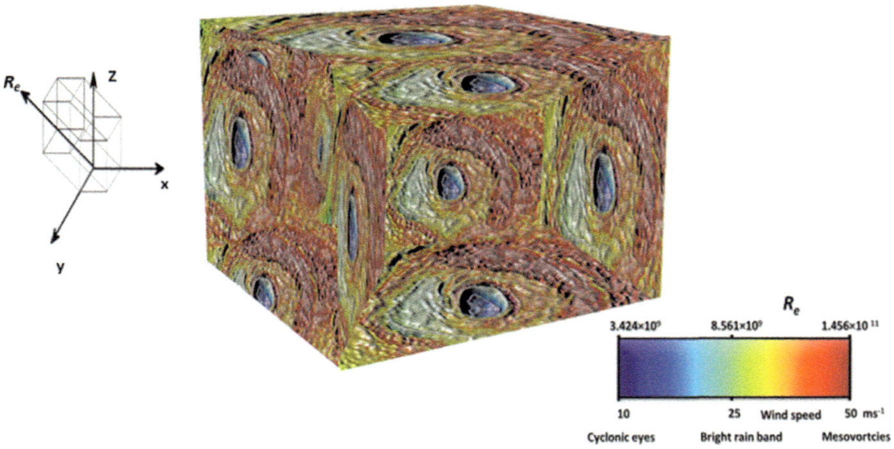

Figure 12.21: Hypercube of typhoon Goni Mobility in RADARSAT-2 SAR data.

In essence, the application of quantum hologram interferometry, coupled with the insightful viewpoint provided by the hypercube, enhances our understanding of the chaotic turbulence within Typhoon Goni. It allows us to delve into the complexities of atmospheric phenomena, uncovering hidden patterns and providing a more holistic comprehension of the dual nature inherent in turbulent flows.

Transitioning to the hypercube's viewpoint, the visualization of edge waves becomes a key focal point, as depicted in Figure 12.22 and the chaotic typhoon Goni generates various air turbulence flows such as vorticity, and eddies as well demonstrated in ScanSAR wide mode. In the exploration of Reynold's number within the dynamics of chaotic typhoon Goni turbulence, a notable convergence into the eye zone becomes apparent when comparing mesovortices within the cyclone eye. The distinction is particularly pronounced in the highest Reynold number, which attains a value of 1.456×10^{11} in the context of mesovortices. This numerical discrepancy signifies a heightened intensity of chaotic vorticity turbulence within the domain of typhoon Goni as opposed to the conditions observed in the narrow cyclone eye. The numerical values of Reynold numbers serve as quantitative indicators of the dynamic complexity inherent in chaotic cyclone turbulence, shedding light on the distinct characteristics and intensities present across cyclone topology structures.

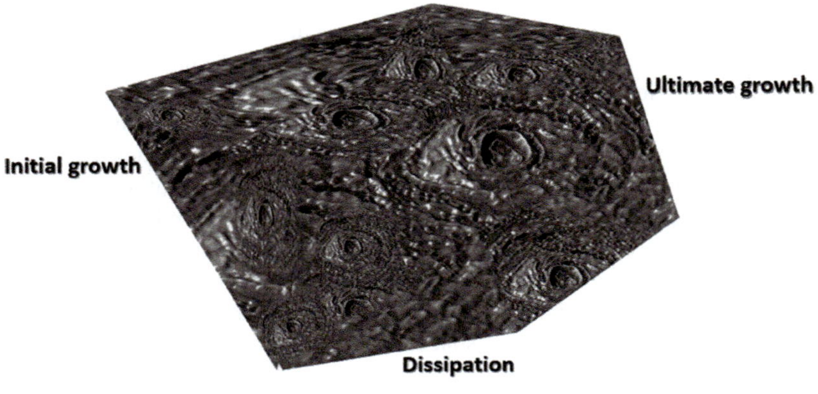

Figure 12.22: Development of 4-D chaotic typhoon Goni turbulence in RADARSAT-2SAR imagery.

Essentially, this 4-D exploration provides a profound insight into the capabilities of quantum hologram interferometry in unraveling the complexities of the chaotic turbulence exhibited by Typhoon Goni. The establishment of tesseracts within the global space signifies a harmonious interplay of dimensions, encapsulating the multidimensional nature of cyclonic turbulence with precision and depth.

The fundamental question that arises is whether a 4-D quantum hologram can reconcile the developmental stages of the cyclone over periods and geographical locations with substantial temporal gaps. In this perspective, the 4-D quantum, when compared to its 3D counterpart, demonstrates its ability to accommodate distinct periods of Typhoon Goni's development stages, as outlined in Figure 12.23. Indeed, the three distinct scenarios of cyclonic chaos seamlessly integrate into the 4-D quantum hologram.

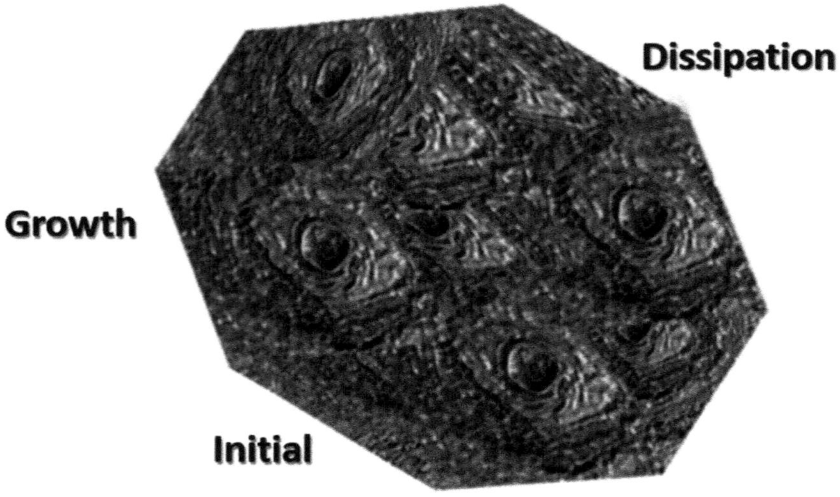

Figure 12.23: Unveiling cyclonic chaos across different stages and different spaces-Insights from 4-D quantum hologram interferometry.

The initial stage illustrates the origin of typhoon generation, initiated by small vorticities and arc cloud formation. The subsequent growth stage showcases the complete topology of arc clouds, mesovortices, eyewall, and eye. Dissipation occurs after the eyewall and eye are formed. In this understanding, the initial RADARSAT-2 SAR overpasses construct the first phase in time, while subsequent scenarios at different times and geographical locations are constructed in diverse temporal coordinates. This 4-D quantum hologram interferometry representation not only captures the dynamic evolution of Typhoon Goni but also allows for a comprehensive examination of its various stages, offering valuable insights into the temporal and spatial intricacies of cyclonic turbulence. The integration of distinct scenarios into a unified quantum hologram provides a holistic view of the typhoon's lifecycle, showcasing its developmental phases with clarity and precision.

This phenomenon can be elucidated through the lens of the relativity hypothesis. Einstein's theory of relativity posits that time is not an absolute entity but is intricately linked with space in a four-dimensional continuum known as spacetime. In this framework, events are defined not only by their spatial coordinates but also by their temporal placement. Thus, the 4-D quantum hologram, encompassing both time and space dimensions, aligns with the relativistic concept of spacetime, enabling the compilation and representation of diverse scenarios across different temporal and spatial coordinates [19, 25, 31]. The consistent coordination of unique cyclonic turbulent events by the 4-D quantum visualization is a testament to its adherence to the principles of the relativity hypothesis.

The depiction of initial stages of cyclonic occurrences in distinct temporal and geographical contexts, along with their anticipated developments, within the realm of 4-D quantum hologram interferometry is indeed significant. This peculiarity serves as a compelling demonstration and, in a way, validates the concept of quantum entanglement. While quantum entanglement is typically associated with the study of electron behavior, applying this concept to the dynamics of atmospheric turbulence systems across different regions and periods is a remarkable extension. The entanglement principle suggests that particles are connected in a way that the state of one particle immediately influences the state of another, regardless of the distance between them.

In the context of cyclonic dynamic systems, the 4-D quantum hologram's ability to encapsulate and display the interconnected evolution of cyclonic behaviors in different spatial and temporal domains aligns with the essence of quantum entanglement [30–31]. This broadens our understanding of quantum peculiarities beyond the microscopic realm, showcasing its relevance and applicability to naturally observable systems like atmospheric phenomena. The 4-D quantum hologram, acting as a visual representation of entangled cyclonic events, highlights the interconnected nature of these occurrences across diverse locations and periods, demonstrating the broader implications of quantum principles in deciphering complex natural phenomena.

The chapter establishes that these 4D images unveil the Reynolds number derived from 4D quantum hologram interferometry. The cyclonic turbulence spectra, recognized as evidence of quantum entanglement theory, serve as a key indicator of the complex and dynamic nature of cyclonic chaotic flows. The outcomes demonstrate the impact of 4D cyclonic chaotic flows in generating phenomena such as eddies, vorticity, and cyclonic turbulence boundary layer. This suggests that the quantum chaotic nonlinear system in cyclonic dynamics contributes to the formation of a quantized cyclonic turbulence boundary layer.

In essence, the integration of quantum principles with the 4D phase unwrapping of RADARSAT-2 SAR satellite data is articulated as Marghany's 4D quantized phase unwrapping algorithm. This algorithm stands as a promising and high-quality technique for the 4D reconstruction of cyclonic chaotic turbulence flows and other features within complex SAR data, encompassing various cyclonic development stages in such RADARSAT-2 SAR satellite data.

References

[1] Pavlov, D.G. (2004). Four-dimensional time. *Space-Time Structure. Algebra and Geometry*, 45.

[2] Heller, M. (1984). Temporal parts of four dimensional objects. *Philosophical Studies: An International Journal for Philosophy in the Analytic Tradition*, 46(3), 323–334.

[3] Nishimura, J. and Sugino, F. (2002). Dynamical generation of four-dimensional space-time in the IIB matrix model. *Journal of High Energy Physics*, 2002(05), 001.

[4] Weiskopf, D. (2001). Visualization of four-dimensional spacetimes. Doctoral dissertation, Tübingen, Univ., Diss., 2001.

[5] Kiritsis, E. and Kounnas, C. (1995). Curved four dimensional space time as infrared regulator in superstring theories. *Nuclear Physics B-Proceedings Supplements*, 41(1–3), 331–340.

[6] Roseman, D. (1998). Reidemeister-type moves for surfaces in four-dimensional space. *Banach Center Publications*, 42(1), 347–380.

[7] Davis, M.W. (2012). *The Geometry and Topology of Coxeter Groups*. (LMS-32). Princeton University Press.

[8] Björner, A. and Brenti, F. (2005). *Combinatorics of Coxeter Groups* (Vol. 231, pp. xiv+363). New York: Springer.

[9] Haglund, F. and Wise, D.T. (2010). Coxeter groups are virtually special. *Advances in Mathematics*, 224(5), 1890–1903.

[10] Björner, A. and Wachs, M.L. (1988). Generalized quotients in Coxeter groups. *Transactions of the American Mathematical Society*, 308(1), 1–37.

[11] Youla, D. and Gnavi, G. (1979). Notes on n-dimensional system theory. *IEEE Transactions on Circuits and Systems*, 26(2), 105–111.
[12] Hagedorn, T.R. (1999). On the existence of magic n-dimensional rectangles. *Discrete Mathematics*, 207(1–3), 53–63.
[13] Yagdjian, K. (2006). Global existence for the n-dimensional semilinear Tricomi-type equations. *Communications in Partial Differential Equations*, 31(6), 907–944.
[14] White, B. (1983). Existence of least-area mappings of n-dimensional domains. *Annals of Mathematics*, 179–185.
[15] Kreis, T. (2006). *Handbook of Holographic Interferometry: Optical and Digital Methods*. John Wiley & Sons.
[16] Toal, V. (2022). *Introduction to Holography*. CRC Press.
[17] Hariharan, P. (2002). *Basics of Holography*. Cambridge University Press.
[18] Kostuk, R.K. (2019). *Holography: Principles and Applications*. CRC Press.
[19] Marghany, M. (2018). *Advanced Remote Sensing Technology for Tsunami Modelling and Forecasting*. CRC Press.
[20] Zhang, S. (Ed.). (2013). *Handbook of 3D Machine Vision: Optical Metrology and Imaging*. CRC Press.
[21] Lanzagorta, M. (2011). *Quantum Radar Synthesis Lectures on Quantum Computing*. Morgan & Claypool Publishers.
[22] Luong, D., Damini, A., Balaji, B., Chang, C.S., Vadiraj, A.M. and Wilson, C. (2019). A quantum-enhanced radar prototype. *In:* 2019 IEEE Radar Conference (RadarConf) (pp. 1–6). IEEE.
[23] Chrapkiewicz, R., Jachura, M., Banaszek, K. and Wasilewski, W. (2016). Hologram of a single photon. *Nature Photonics*, 10(9), 576–579.
[24] Marghany, M. (2021). *Nonlinear Ocean Dynamics: Synthetic Aperture Radar*. Elsevier.
[25] Marghany, M. (2019). *Synthetic Aperture Radar Imaging Mechanism for Oil Spills*. Gulf Professional Publishing.
[26] Ou, Z.Y.J. (2007). *Multi-Photon Quantum Interference* (Vol. 43). New York: Springer.
[27] Marghany, M. (2022). *Remote Sensing and Image Processing in Mineralogy*. CRC Press.
[28] Best, A., Kliegl, M., Mead-Gluchacki, S. and Tamon, C. (2008). Mixing of quantum walks on generalized hypercubes. *International Journal of Quantum Information*, 6(06), 1135–1148.
[29] Potoček, V., Gábris, A., Kiss, T. and Jex, I. (2009). Optimized quantum random-walk search algorithms on the hypercube. *Physical Review A*, 79(1), 012325.
[30] Edery, A. (2003). Casimir energy of a relativistic perfect fluid confined to a D-dimensional hypercube. *Journal of Mathematical Physics*, 44(2), 599–610.
[31] Miller, M.A. (1995). Regge calculus as a fourth-order method in numerical relativity. *Classical and Quantum Gravity*, 12(12), 3037.

Index

A

Absorption, 49, 117, 121, 125, 129, 132, 198-200, 202-205, 207, 208, 212, 213, 283
Accelerated charge, 6, 61, 139
Ageostrophic-torque, 61
Air -Turbulence, 15-19, 22, 48, 55, 91, 118, 128, 280, 333
Alfvén waves, 179
Algorithm, 40, 146-148, 153, 169-171, 197, 202, 209, 213, 230, 231, 235-239, 241
Altocumulus, 15, 79, 183-186, 190, 212
Altostratus, 183-185, 211
Ampere's Law, 94, 97-99
Amplitude, 16, 48, 62, 68, 86, 87, 106, 114, 115, 117, 121, 204
Angstroms Å, 131
Angular momentum, 8, 47-51, 61, 77, 81
Angular velocity, 60, 70, 284, 315
Annihilation operators, 346
Aqua, 167-169, 212
Argon, 43, 44
Atmospheric tide, 281, 282-286
Atmospheric turbulence, 2, 40, 44, 125, 186, 241, 261, 263-265, 281, 307, 333, 357
Atom, 43-48, 50-54, 57, 91, 94, 106, 112, 113, 117-119, 123, 124, 125, 129-131, 134, 179, 199, 203, 204, 207, 209, 311
Attenuation, 135, 285, 318, 347
Auroral activity, 133
Avogadro constant, 54
Azimuthal velocity, 77

B

Backscatter, 125, 258, 261, 264, 265, 272-278, 292, 294, 298, 301, 317, 318, 328, 347
Barotropic mode, 65, 66
Barotropizes, 64
Beam, 104, 110, 119, 134, 138-140, 148, 149, 166, 168, 174, 176, 328, 344
Beam-width, 276
Bell-state, 206
Biot-Savart law, 95-97
Black body, 110-112, 125
Bloch sphere, 320, 321
Bohr's model, 119, 131
Boltzmann's constant, 107, 146, 272
Bose-Einstein, 76, 81, 108, 226
Bosons, 226
Bragg-scattering, 276, 277, 280, 298
Brunt-Väisälä frequency, 243, 244

C

Carbon dioxide, 43-46, 153, 161
Carrier frequency, 271
Cartesian components, 17, 33, 48, 338, 350
Cavity radiation, 109
Cavum clouds, 188, 190, 191
C-band, 264
Charge field, 92
Circuit, 147, 148, 170, 233, 254, 255, 349
Cirrus, 183-185, 187, 190, 198, 200, 213, 313, 329
Clear air, 14, 18, 19
Coherence, 167, 205, 303, 321, 322, 327, 347, 348, 353
Coordinates, 48, 106, 141, 270, 284, 286, 317, 336, 340, 341-343, 349, 350, 356
Coriolis effect, 59-61, 222, 286, 307, 309, 311, 314, 315
Cosine, 108, 348, 349
Coulomb-potential, 49, 94, 147, 208
Coxeter-Dynkin diagrams, 339
Cross product, 70, 122, 315
Cross section, 70, 272, 273, 275, 276, 277, 292, 293, 294, 297, 318, 328
Cumulonimbus, 15, 183, 185, 87, 91, 92, 94, 211, 213
Cumulus, 15, 28, 79, 183, 184, 186, 187, 190, 193, 194, 212, 213, 217-221, 224, 242
Cumulus cloud, 28, 73, 183, 184, 187, 190, 194, 213, 217-221, 224-226, 233, 235, 242
Curl, 69, 70-72, 94, 99, 183, 198, 342
Current, 2, 5, 8, 14, 15, 68, 74, 94-98, 102, 103, 114, 137, 139, 153, 163, 165, 175, 186, 190, 199, 217, 218, 238, 260, 284, 298, 313, 321, 336, 340, 341
Curvature, 36, 58-62, 67, 80, 82, 85, 86, 342

Cyclone, 8, 69, 72, 158, 160, 174, 286, 307, 309-311, 313, 336, 351, 353-356

D

de Broglie theory, 118, 119
Density, 3, 5, 18, 19, 26, 44, 45, 48, 50, 94, 100-103, 107, 108, 129, 134, 135, 137-139, 146, 170, 175, 198, 199, 207, 223, 228, 272, 283, 289, 296, 303, 314, 315, 344
Density Functional Theory (DFT), 207
Diffusion, 5-7, 25, 26, 28, 38-40, 57, 58, 71, 76, 80, 138, 223, 269, 285, 286, 351
Dimension, 3, 6, 8, 17, 25, 31, 36, 48, 62-66, 71, 72, 99, 102, 105, 106, 124, 169, 183, 222, 244, 246, 254, 260, 266, 269, 271, 275, 278, 284, 317, 333, 336-343
Dirac quantization, 105
Dirac spinor, 104, 105, 318
Direction, 4, 5, 12, 25, 36, 38-40, 48, 53, 83, 86, 94 -101, 110, 122, 124, 136, 141, 165, 166, 175, 188, 190, 201, 202, 217, 220, 242, 258, 266, 269, 272-277, 287, 307, 310, 313, 315, 325, 330, 340, 351
Discrete, 15, 23, 55, 56, 68, 82, 88, 112, 123, 167, 206, 270, 349
D-layer, 129, 130
Doppler shifts, 269
Dual, 14, 21, 22, 23, 39, 43, 51, 102, 106, 121-125, 141, 166, 170, 192, 204, 206, 213, 226, 241, 264, 274, 289, 295, 311, 313, 319, 322, 344, 350, 354, 355
Duality, 83, 85, 88, 106, 122, 124, 166, 295, 354

E

E- region, 132, 136, 144
Eddy diffusion, 39
Eddy diffusivities, 63
Edge detection, 217, 230, 231, 235-239, 259
Eigenvalue, 48, 101, 142, 146, 147, 170, 204, 207, 290
Einstein's light quantum, 117
Electricity, 19, 86, 91-94, 133, 139, 141, 161, 165, 176, 179, 203, 263, 273, 294, 318
Electrodes, 100, 101
Electromagnetic wave, 91, 92, 99, 100, 103, 106, 113, 139, 164, 165, 249, 250, 273, 274, 278, 280
Electron, 46-54, 66, 86, 91, 97, 103, 105, 109, 113-119, 121, 138-144, 147, 165, 175, 176, 204, 205, 208, 357
Energy, 2, 6, 19, 27, 38, 46, 91, 100, 30, 31, 40, 60, 61, 83, 85 ,91, 93, 95, 97, 211, 212, 222, 226, 230, 244, 245, 263, 264, 282, 283, 291, 295, 307, 309, 310, 315, 316, 327, 331
Enhanced Wall Treatment (EWT), 34
Enstrophy cascade rate, 30
Entanglement, 79, 88, 128, 141, 142, 144, 146, 147, 166, 169, 170, 179, 203, 205, 249, 250, 251, 294, 295, 345, 346, 350, 357
Entropy, 72, 143, 144, 146
Euclidean metric, 252, 338, 342
Euler formulas, 31
Exponentials, 57, 102, 103, 169, 251, 274, 283, 185, 290

F

Faraday's Law, 94, 99, 104
Fermi energy, 146
Fermions, 146
Feynman, 103-105, 142
Field, 3, 14, 25, 27, 43, 52-55, 58, 61, 62, 91-106, 136, 138, 161, 169, 179, 180, 193, 202, 204, 218, 219, 239, 250, 263, 265 274, 287, 290, 294, 295, 302, 309, 316, 342, 343
Flammagenitus clouds, 193, 194
Fluctus clouds, 189, 190
Flumen clouds, 192
Four-dimensional, 336-340, 350, 356
F-Region, 129, 137, 144
Frequency, 60, 67, 77, 81, 82, 106 -113, 128, 131, 135, 146, 148, 158, 166, 175, 179, 199, 230, 242, 243, 247, 263, 264, 268-271, 284-288, 290, 317, 318, 328, 343, 348
Froude number (Fr), 243, 245, 246

G

Gamma wave, 112, 120, 121, 122, 137, 232
Gates, 147, 170, 209, 231, 349
Gauss' Law, 93, 94
Gaussian, 27, 39, 167, 279, 292, 293, 348, 349
Geometrical optics, 134, 340
Geometry, 38, 40, 58, 60, 107, 139, 202, 264, 266, 267, 277, 318, 336, 339, 340, 342
Geostrophic velocity, 61
Gradient, 9, 12, 13, 17, 26, 32, 58, 59, 83, 88, 159, 218, 220, 221, 223, 235, 238, 243, 246, 288, 291, 315, 327
Gravitational force, 59, 165, 283, 288, 289
Gravity, 19, 23, 33, 55, 59, 67, 83, 92, 165, 223, 228, 246, 263, 276, 281, 282, 285, 286, 315, 319
Gross-Pitaevskii Equation (GPE), 77
Ground-range, 269
Grover operator, 351
Grover search, 322 351
Gyroharmonic, 149

H

HAARP, 125, 128, 136, 137-147, 153, 164, 170, 175, 177 -179, 180
Hadamard gate, 147, 349

Hamiltonian, 47, 48, 55, 57, 66, 104, 105, 141, 146-148, 170, 204, 208, 254, 294, 295
Harmonic oscillators, 295
Hartree-Fock, 147, 148
Heat index, 160, 161
Heatwave, 128, 157-162, 164, 166, 169-178, 180
Heisenberg's uncertainty principle, 124
Hermite polynomial, 170, 209, 286, 324
Hermitian operator, 208, 320
Hertz, 113, 114, 117, 264
HH-band, 300
Hilbert space, 169, 319, 341, 351
Homogenitus clouds, 194, 195
Homomutatus clouds, 195
Hurricane, 74, 75, 134, 137, 161, 263, 264, 307-309, 314, 317, 329

I

IKONOS satellite, 236
Illumination, 224, 266, 269, 278, 347
Imaging, 91, 106, 134, 167, 168, 197, 200, 201, 202, 205, 224, 255, 264, 265, 292, 295, 298, 301, 317, 328
Incident angle, 202, 277
Inertia-gravity waves, 67-69
Information, 43, 45, 74, 21, 25, 32, 33, 42, 43, 70-72, 75, 90, 97, 99, 202, 203, 205, 210, 224, 230, 243, 247, 265, 276, 295, 303, 319, 320, 333, 342, 344, 346, 347, 349
Infrared spectrum, 197, 200, 203, 224
Intensity, 15, 17, 22, 23, 26, 100, 108, 112-117, 123, 144, 145, 179, 180, 191, 192, 198, 209, 210, 230, 274, 275, 277, 278, 302, 310, 317, 327, 328, 343, 344, 355
Interference, 122, 23, 37, 203, 269, 76, 78, 79, 301, 19, 20, 32, 43, 44, 46, 49
Interferometry, 336, 343-357
Ionization, 129, 132 –136, 138, 140 -143, 145, 147, 153, 179, 180
Ionosphere, 45, 128, 129, 131, 132, 135-149, 151, 152, 170, 179, 180

J

James Clerk Maxwell, 91
Jets, 9, 19, 32
Joule loss, 49, 73, 107, 132, 163, 229

K

Ka-band, 264
Kármán vortex street, 218, 219, 222, 241, 224-244, 248, 250, 251-257, 261
Kelvin, 73, 81-83, 88, 107, 112, 129, 198, 229, 286
Kelvin waves, 81-83, 286
Kinetic energy, 8, 9, 27, 32, 36, 38, 47, 56, 64, 65, 73, 82, 14, 15, 22, 43, 44, 208, 222, 226, 244, 245, 283, 307
Kirchhoff, 110, 292
Kolmogorov speculation, 28
Kronecker product, 32
Ku-band, 264
k–ε turbulence model, 36, 40
k–ω models, 36, 40

L

Ladder operator, 46
Lagrangian model, 38, 39
Laminar, 3-6, 10, 11, 31, 33, 34, 36, 40, 63, 70
Langmuir dispersion, 146
Laplace operators, 32
Laplacian, 65, 349
Lapse rate, 15, 19, 221, 227, 228, 246
Laser, 141
L-band, 121, 264, 299
Light, 17, 19, 86, 91, 92, 101, 103, 130-134, 161, 166, 183, 189, 224, 229, 256, 263, 290, 314, 318, 322, 327
Loop current, 96
Lorentz, 94, 100, 101, 104, 105
Lyman series, 129, 130

M

Magnetic field, 52, 53, 86, 91, 92, 94, 98, 103, 105, 136, 144, 148, 158, 161, 180, 203, 204, 250
Magnetosphere, 133, 144, 179
Magnitude, 26, 34, 49, 54, 60, 94, 98, 100, 141, 242, 271, 331, 332
Marghany Operator $\mathfrak{M}^{\dagger}_{\vartheta_b} > 1$, 66
Marghany quantization of earth rotation, 62, 76, 80, 88
Marghany quantum energy wind-driven theory, 62, 76, 80, 88
Marghany Quantum Spectral Algorithms (MQSA), 125
Marghany' 4-D phase unwrapping algorithm, 349, 357
Marghany's Hologram Interferometry Theory, 36, 42-46, 48-53
Max Planck, 106, 109, 112
Maximum, 4, 10, 15, 22, 104, 106, 117, 119, 120, 140, 152, 158, 175, 209, 228, 229, 232, 243, 252, 272, 297, 303, 310, 311, 318, 326, 329, 331, 354
Maxwell's equations, 91, 99-106, 111, 113
Mermin-Kohn-Sham (MKS), 207
Mesovortices, 328, 354-356
Meteosat 8, 238
Microbursts, 21
Microwave, 104, 106, 120, 121, 203, 263 -266, 274,

278, 281, 318, 327
Millibar, 55, 83
Minimum, 35, 117, 124, 133, 140, 162, 208, 229, 245, 297, 339, 347
MODIS, 167- 173, 175, 210, 211, 212, 213, 224, 225, 232, 237, 257, 259
Modulation, 76, 298, 303, 317, 344
Momentum, 2, 6- 8, 12, 24, 25, 31, 47-51, 55, 69, 77, 81, 88, 100, 103, 104, 118, 142, 218, 222, 223, 286, 315
Murus clouds, 191, 192

N

Nanometer, 121, 122, 130, 165
NASA, 168, 169, 224, 264
Navier-Stokes equations, 31, 33, 34, 36, 222, 316
N-dimensional, 63, 340, 341, 351
Newtonian fluid, 10, 17, 31, 33
Newton's theory, 61
Nimbus, 15, 183, 185, 187, 191, 192, 211, 213
Non-dissipative, 9, 81
Nonlinear, 9, 31, 32, 38, 40, 65, 67, 77, 81, 82, 294, 295, 302, 303, 319, 333, 357
Non-superconducting, 74
Normalization radar cross-section, 272, 273, 275, 277, 292 -294, 297, 318
NSGA-II, 325, 329-333, 336, 354

O

Odd qubits, 147, 169, 170, 171, 172, 205, 209
Operator, 32, 47, 48, 52, 55, 65, 66, 71, 105, 147, 204, 206, 231, 249, 294, 314, 324, 343
Optical regime, 106, 169, 197
Optimal solution, 231, 304, 325-327, 332
Optimization, 231, 232, 254, 260, 261, 295, 303, 304, 325, 327, 331
Oracle, 351
Orthorhombic, 61, 169
Oscillator, 56, 109, 228, 295
Oxides, 44, 45, 129, 134
Ozone, 44, 45, 283

P

Pareto optimization, 331, 332
Particle Swarm Optimization (PSO), 281, 295, 297, 301-304
Particles, 5, 9, 24, 39, 51, 52, 54, 70, 74, 82, 85, 91, 99, 101, 103, 104, 117, 118, 129, 136, 141, 142, 161, 165, 197, 200, 201, 212, 226, 263, 274, 303, 304, 342, 357
Paschen-delta, 131
Pauli exclusion principle, 51
Periodic table, 62, 64
Photoelectric effect, 106, 113, 115, 117, 118, 122, 203
Photon, 49, 91, 101-106, 117, 118, 165, 199, 203, 208, 274, 344-347
Photovoltaic effect, 117, 118
Physical properties, 138, 151, 208, 210, 241, 249, 261
Planetary boundary layer (PBL), 218, 220
Plane-wave, 47, 99, 100, 146
Plank-quanta, 107
Plasma, 45, 128, 131, 133, 164, 165,1 72, 175, 179, 180
Polarimetry, 278, 332
Polarized wave, 278
Potential energy, 47, 48, 62, 65, 73, 76, 101, 118, 245, 294
Poynting theory, 100, 101
Prandtl number (Pr), 62
Pressure, 4, 5, 18, 21, 53, 72, 111, 138, 159, 166, 187, 191, 219, 223, 241, 281, 288, 291, 307, 309, 312-316
Propagation, 19, 83, 87, 91, 99, 100, 103, 132-135, 137, 141, 146, 152, 166, 177, 243, 247, 280, 286, 298, 294, 318, 345, 347
Pulse Repetition Interval (PRI), 269, 270

Q

QME, 321, 322
QNSGA-II, 325, 329, 330-333, 336, 354
Quanta, 56, 64, 106, 107, 109, 112 , 117, 121
Quantity, 6, 30, 45, 53-55, 63, 72, 116, 129, 135, 147, 165, 206, 209, 271, 272, 286, 295, 322, 337, 344, 348
Quantization, 57, 58, 74, 101, 105, 106, 109, 125, 128, 199
Quantization of Marghany wind generation mechanism, 57
Quantized Marghany-Heatwaves, 164
Quantized Marghany quasi-geostrophic turbulence, 66-69, 88
Quantum Approximate Optimization Algorithm (QAOA), 231, 232, 254, 255, 260, 261
Quantum electrodynamic (QED), 92
Quantum Vorticity States in Remote Sensing (QVSRS), 248
Quasi-Biennial Oscillation (QBO), 86, 87
Quasigeostrophic flow, 65
Qubits, 147, 169, 170, 171, 172, 232, 233, 252, 254, 260, 298, 303, 319-324, 333, 349

R

Radar, 106, 121, 138 ,145, 257-261, 263
Radar cross-section (RCS), 272, 292
RADARSAT-1, 297, 298
Radarsat-2, 327-331, 333, 336, 351, 352, 354-357
Radiation, 13, 46, 92, 106-113, 129, 153, 158, 162,

166, 197-200, 221, 225, 229, 272, 275, 318, 327
Radio wave, 106, 109, 120, 121, 122, 129, 132, 133, 135, 145, 152, 239, 245, 252, 275, 263
Range-resolution, 267-270
Rayleigh scattering, 81, 86, 106, 109, 112, 289
Receiver, 138, 139, 145, 205, 237, 264, 272-274, 277
Receiver Operating Characteristic (ROC) curve, 237
Reflection, 125, 129, 132, 133, 135, 148, 202, 207, 212, 273, 274, 276-279, 329
Relative Humidity (RH), 227, 228
Relativity theory, 54, 88
Resolution, 27, 31, 120, 167, 224, 236, 238, 250, 255, 256, 265, 279, 280, 317, 319, 327, 328
Resonant scattering, 276
Reynolds number, 3-7, 62, 67, 329, 331, 357
Reynolds-averaged Navier-Stokes (RANS), 31, 222
Richardson's Four-Thirds Power Law, 26
Rossby wave, 86, 158, 159, 281, 286
Rotation, 8, 9, 52, 77, 80, 121, 165, 199, 222, 243, 247, 248, 284, 287, 307, 309, 311, 314, 315, 320, 324, 325, 331
Rydberg R_H, 130

S

Schrödinger equation, 47, 48, 51, 56, 77, 81, 105, 118, 141, 249, 303
SeaWiFS, 255, 256
Sensors, 91, 125, 162, 163, 197, 202, 203, 205, 206, 224, 225, 241, 250, 258, 261, 317, 320, 327
Sentinel-1A, 257-259, 302
Slant range, 268
Solenoid field, 97, 98
Solution, 9, 26, 27, 31, 34, 73, 81, 99, 100, 106, 112, 120, 167, 224, 231, 236, 250, 255, 256, 295, 303, 304, 317, 319, 321, 325
Space-time, 55, 58, 59, 67, 85, 340, 341
Spatial resolution, 255, 267, 269, 279, 328
Speckles, 257, 278, 279, 298
Spectral, 23, 62, 66, 82, 111, 125, 167, 183, 197-203, 264 ,277, 329
Spectral signature, 125, 206, 211-213, 224, 226, 235, 249-251, 258, 259, 261, 329
Spectrometer, 121, 167
Spectrum, 120, 202, 203
Specular scattering, 273
Spin, 47, 52
Sporadic E, 132, 133, 134
Stefan-Boltzmann law, 110, 111
Stratosphere, 18, 19, 20, 44, 45, 68, 80, 81, 86, 282 286
Stratus, 183-185, 187, 190, 211, 212, 235
Streamline, 3, 5, 22, 36, 70, 81, 243, 244, 245
Stress-tensor, 82
Superposition, 68, 211, 241, 251, 259, 260
SWAP gate, 147
Swath, 92, 169, 266, 267, 269
Synthetic aperture radar (SAR), 263, 264, 266, 273, 277, 278, 281, 293, 294, 317, 317, 327, 328, 331, 333, 348, 349

T

TACSat4 satellite, 145
Taylor microscale, 29
Taylor series, 29
Teleported state, 207
Thermal, 15, 177, 179, 203
Thermal Infrared (TIR), 203, 211, 213, 256
Threshold, 4, 50, 51, 73, 115-117, 138, 231, 289
Thunderstorm, 2, 14, 18-21, 73, 83, 84, 174, 185, 187, 197, 263, 307, 309, 311, 313, 314
Time delay, 269, 271
Time evolution, 6, 27
Time-dependent, 31, 47, 57, 141, 208
Transmission line, 24, 32, 38, 100, 134, 138, 197, 266, 269, 270, 273, 278
Trapezoidal rule, 245
Tropopause, 18, 19, 45, 84, 86
Turbulence, 1-5, 43, 44, 50, 51, 91, 92, 128, 143, 186, 335
Turbulence kinetic energy (TKE), 29
Turbulent, 3-11, 15, 17, 56, 63, 82, 91, 143, 176, 179, 186, 222-224, 307, 317, 318, 322 -327, 331-333, 354-356
Typhoon, 261, 307, 309, 327, 328, 329, 333, 351-356

U

UHF antenna, 145
Ultraviolet, 112, 120, 121, 129, 130, 134, 161, 162
Uncertainty theory, 124, 125, 133, 149, 269, 295, 332
Unitary operators, 350
Unwrapping, 349-353, 357
UV/VIS, 113, 112, 122, 129, 132, 134

V

Vapour 168, 175
Variation, 5, 14, 33, 35, 40, 58, 75, 118, 133, 162, 200, 217, 223, 235, 238, 260, 264, 270, 277, 284, 301, 316, 319, 328, 329
Vector, 144, 170, 222, 231, 233, 248, 251, 278, 314, 320, 341, 348, 350
VH polarization, 278
Vibrational state, 199
VIIRS data, 256, 257
Viscosity, 3, 5-7, 9-11, 17, 25, 28-30, 32, 58, 62, 67, 81, 222
Viscous flow, 3, 33

Visible and near-infrared (VNIR), 213, 263
Visible light, 109, 112, 119, 120, 121, 198, 200, 224, 327, 328
Volume, 33, 38, 46, 74, 101, 106, 125, 143, 207, 273, 314, 318, 338, 350
Vortex Dissimilarity Distance (VDD), 252
Vorticity, 8, 34, 35, 38, 64, 65, 67, 80, 81, 89, 222, 241-244, 248, 311, 320, 329, 355, 357
VV-polarization, 278, 303

W

Wake, 9, 21, 22, 36, 69, 70, 222, 242, 243, 247, 263
Wave, 8, 9, 15, 16, 19, 47, 48, 50, 56, 62-69, 76, 91, 92, 99-113, 115, 117, 128- 139, 141, 149, 152, 158, 159, 187-190, 197, 224, 225, 229, 249, 250, 263, 264, 266, 285-295, 311, 317-319, 321, 327, 343, 344, 348, 350, 355
Wave function, 48, 50, 51, 56, 57, 66, 75-77, 81, 101, 102, 105, 146-148, 178, 204, 208, 209, 210, 226, 293, 294, 295, 321, 350
Wave propagation, 91, 99, 100, 132, 135, 137, 146, 152, 166, 177, 288, 289, 290, 294
Wave turbulence (WT), 16, 81-83, 88, 128, 147, 350
Wavefunction, 47, 56, 77, 124, 142, 171, 343, 344
Wavelength, 67, 88, 104, 106-113, 129-132, 134, 197, 198, 200-202, 209, 210, 224, 263, 264, 271, 272, 274, 289, 293, 298, 304, 328, 348
Wavenumber, 64-67, 285, 290-292, 348
Weight, 29, 55, 250, 252, 254, 260, 342, 350-353
Weighting matrix, 251
Weyl groups, 340
Wideband, 268
Wien's Displacement Laws, 110
Wien's frequency, 110-112
Wind speed, 17, 19, 22, 55, 57, 58, 61, 67, 80, 174, 187, 188, 218, 221, 224, 246, 248, 290-293, 298, 303, 304, 307, 310, 315, 327-329, 331, 354
Window kernel, 68, 170
Wrapping, 349

X

X-band, 169, 264
X-ray, 120-122, 129, 30, 32, 34

Y

y-axis, 172
Young's experiment, 122, 123
Young scattering, 123, 198
Young's slits, 123
y-plane, 166
y-polarized, 82

Z

z-direction, 97, 287
Z_c of frequency, 244, 245
Zero spin, 52
Zero-baseline, 348
Zeroth order, 140
Zone, 4, 6, 12-14, 18, 44, 45, 55, 57, 62, 129, 133, 135, 136, 152, 166, 167, 219, 222, 223, 230, 233, 237, 257, 281, 283, 355

9781032344584